Operator Theory
Advances and Applications
Vol. 115

Editor
I. Gohberg

Operator Theory
Advances and Applications
Vol. 113

Editor:
I. Gohberg

Complex Analysis, Operators, and Related Topics

The S. A. Vinogradov Memorial Volume

Victor P. Havin
Nikolai K. Nikolski
Editors

Springer Basel AG

Editors:
Victor P. Havin
Department of Mathematics
St. Petersburg University
Bibliotechnaia pl. 2
198904 Stary Peterhof, St. Petersburg
Russia

Nikolai K. Nikolski
Laboratoire de Mathématiques Pures
UFR de Mathématiques et Informatique
Université de Bordeaux I
351, cours de la Libération
33405 Talence Cedex
France

1991 Mathematics Subject Classification 47-06; 32-06

A CIP catalogue record for this book is available from the
Library of Congress, Washington D.C., USA

Deutsche Bibliothek Cataloging-in-Publication Data

Complex analysis, operators, and related topics : the S. A. Vinogradov memorial volume / Victor P. Havin ; Nikolai K. Nikolski ed. – Basel ; Boston ; Berlin : Birkhäuser, 2000
(Operator theory ; Vol. 113)
ISBN 978-3-7643-6214-0 ISBN 978-3-0348-8378-8 (eBook)
DOI 10.1007/978-3-0348-8378-8

Originally published by Birkhäuser Verlag in 2000

Printed on acid-free paper produced from chlorine-free pulp. TCF ∞
Cover design: Heinz Hiltbrunner, Basel

9 8 7 6 5 4 3 2 1

A Word from the Editors

This volume contains 28 articles dedicated to the memory of Stanislav A. Vinogradov. We also decided to include the English translations of two of Vinogradov's articles, true jewels of the volume, we believe. Each was written for a separate purpose: the first was for Vinogradov's PhD ("candidate") defense (1968, Leningrad State University) and the second for his "doctoral" defense (1983, Steklov Mathematical Institute, Leningrad). Both articles are thesis abstracts ("Avtoreferat" in Russian), in accordance with Russian rules of thesis defenses. Fewer than 80 copies of each were printed and the Russian originals are hardly accessible now. They have been translated by V.Havin for this volume. Our joint introductory article on Vinogradov's life and mathematics contains some comments and additions.

We would have been unable to prepare this memorial volume without generous assistance, both technical and mathematical, of our and S.A.Vinogradov's friends and colleagues

K. Abramenko	A. Aleksandrov
A. Baranov	D. Belyaev
V. Chebotareva	J. Choksi
E. Dyn'kin	K. Dyakonov
M. Gamal	V. Kapustin
P. Koosis	S. Kupin
A. Lodkin	A. Nersessian
V. Peller	A. Petrov
A. Plotkin	D. Sherman
A. Sudakov	I. Verbitskii
I. Videnskii	A. Volberg

We are very sorry to conclude this note with the sad news that Seva (Evsei) Dyn'kin unexpectedly died during the preparation of this volume. He was one of the most brilliant members of St.Petersburg Analysis Seminar and Vinogradov's circle, a very original and strong analyst and an unforgettable personality.

Victor Havin
Nikolai Nikolski

Contents

Stanislav A. Vinogradov
May 1, 1941–November 14, 1997

Operator Theory:
Advances and Applications, Vol. 113

Stanislav Aleksandrovich Vinogradov, His Life and Mathematics

V. P. Havin and N. K. Nikolski

This volume is dedicated to the memory of Slava Vinogradov, our dear friend. His untimely death has shaken many who knew and loved him.

Life

S.A.Vinogradov was born in Leningrad on May 1, 1941, on the eve of the German invasion. The siege began when he was a baby of 5 months, his father Aleksandr Antonovich fell at the front, and his mother Olga Ivanovna had to overcome indescribable hardships to save the child from starvation amidst artillery and aircraft bombardments. Postwar years were also hard; her income was scarce (she was a janitor). But she had luck with her son. People who knew him in the forties remember his early addiction to books (mainly classical Russian literature). He was successful at school, especially in mathematics. He joined the mathematical circle of the Palace of Young Pioneers where many Leningrad schoolchildren developed their talents under the guidance of university teachers and students. In his last high school year Stanislav was uncertain about which of the numerous Leningrad institutes to choose for his further education. His hesitation was over after his visit to the mathematics department of the Leningrad State University where G.M.Fichtenholz, a famous professor and brilliant lecturer, gave a talk for high school students (an elementary survey of Calculus). Vinogradov stated his decision as follows: "This is the right place for me to choose, since such people teach here".

He passed his entrance exams quite successfully and impressed his examiners. Slava quickly became one of the brightest students. The first author of this preface had him in his class and remembers Vinogradov's vivid and keen interest in the subject (power series in the spirit of Landau's "Darstellung und Begründung", the Hardy classes, and other spaces of analytic functions). His reaction was always sharp and swift. G. P. Akilov who was giving the course on Functional Analysis was impressed by the following episode. In one of his introductory lectures he proved that the intersection of any decreasing sequence of closed balls in a complete metric space is non-empty if the radii tend to zero; he had then mentioned that (maybe) the last condition is redundant. Vinogradov protested immediately, saying that this is so in a Banach space, but not in general.

The following reminiscence illustrates Slava's addiction to mathematics in his early student years. In the late 50's and the early 60's, every summer and fall, the university students were used as an unpaid agricultural labour force, picking potatos, cabbages, etc. Now, please try to imagine the surrealistic picture to which the second author, being Slava's closest university classmate, was a witness: a dark, cold and wet fall of 1959 in a kolkhoz lost amidst the impassability of the northern Leningrad region; a big room in a barrack where 25 guys sit on plank beds playing cards, shouting, smoking, drinking wine ... In the center of the room, under the only lamp, a sophomore, S.Vinogradov, is sitting cracking problems from the Polya and Szegö Problem Book, and, when tired of Polya and Szegö, studying a French memoir by R. Baire "Sur les fonctions de variables réelles", Ann. Mat. Pura ed Appl. (3), 3(1899), 1–122 ...

His talent was clear to us since his student years (the first author was his adviser, the second was his fellow student and closest friend). He was a born analyst,the one "with a square skull" as the saying goes. He liked abstract "soft" approaches, but his most natural inclination was an elegant "hard" estimate.

Vinogradov graduated from the university in 1963, being already the author of nice results on the (fashionable at the time) free interpolation phenomena (see below); one of these results was mentioned by J.-P. Kahane who included it in his book on absolutely convergent Fourier series.

Vinogradov's living conditions during this period were still quite painful. The second author remembers a rather anecdotal but colourful episode from the First Summer Analysis School in Uzhgorod (Carpathian mountains, 1964): invited to an official conference reception held in a high-standing restaurant, Stanislav shocked the waiters by ordering ... five milk buckwheat puddings one after another, instead of the proposed delicacies. In fact, this was an ineffaceable relic of siege starvation, the well-known phenomenon among the siege survivors who subconsciously harbored forever a fear of permanent hunger. At about the same time, Stanislav, already a PhD student at Leningrad University, was forced to take a second job, spending hours and hours at a noisy electrical "Reinmetall" calculator and doing long computations for an applied project directed by Prof. L. M. Kachanov. (By the way, a result of this involvement was that he found a mistake in an engineering algorithm, and that had far-reaching consequences!). Even much later, when already a university professor, he did not have his own desk for a long time and often worked locked in the bathroom of his small apartment, the only place where he could isolate himself for a while.

A brief and formal description of his further career follows: in 1963 he became a graduate student at Leningrad university; in 1966 he was appointed an assistant professor; his first thesis was defended in 1968; he became an associate professor in 1969 and in 1983 defended his second thesis. In 1987 he got his full professorship and the Chair of Analysis at Leningrad University.

He was an enthusiastic and conscientious teacher. Teaching, and methods to present mathematics to beginners in an optimal way combining clarity and brevity were his constant preoccupations, the theme of innumerable conversations and disputes with colleagues with whom he generously shared his pedagogical findings. Being very inventive he often contrived to find clever tricks, simplifying and perfecting even the most common points in standard courses, not to mention more intricate topics.

As to PhD students, he had few of them formally speaking (chronologically ordered, A. Kotochigov, M. Bludze, T. Andrianova, S. Shimorin, P. Stoilov, A. Petrov). But instead he continually and informally worked with half of the huge Leningrad Analysis seminar. Being a very enthusiastic solver of tooth-breaking puzzles, S.V. was easily captivated by somebody else's problems. Once involved, he often inspired colleagues with an idea leading them out of a dead end. As a result, many of Vinogradov's important achievements are scattered throughout his friends' published papers. Without trying to be exhaustive we give a list of such results at the end of this Preface.

His duties were varied, numerous, and not always pleasant. The most pleasant one (besides teaching) was the supervision of budding analysts, i.e. of their Master's and PhD theses. But his duties included some purely administrative tasks, elements of routine, red tape, etc. He accepted all that as an inevitable burden, his duty to his comrades and the department. His conduct and the decisions he had to take were invariably marked by deep innate decency. To illustrate this we describe a typically Soviet moral dilemma he had to confront after he became the Chairman of the Analysis Department. For a leader of this rank to act efficiently, to be in a position to protect his colleagues from (not always friendly) bosses or to take a serious part in decision making at the departmental level it was an absolute necessity to be a Party member. Many friends (including – alas! – the first author) put strong psychological pressure on him to convince him to comply with this imperative and forget his scruples in favour of the common interest. He agreed that these considerations were serious, but was extremely reluctant to fulfil what everybody wanted from him. He was not a political dissident, a fervent adversary of the regime, not at all. Having benefited fully from the advantages of the Soviet educational system which ensured his access to a high level of the scientific hierarchy, and having been brought up in communist ideology, he respected and accepted the old noble slogans like internationalism, social justice, etc. But at the same time he hated the stupid and hideous sides of everyday Soviet life. He knew that a Party member couldn't escape taking part (be it passively) in some disgraceful actions (or, at least, in their formal approval). This is why, when attacked by somebody insisting once again that his Party membership was a must, he would nod, but not accelerate the process. He lingered thus for approximately five years, postponing and promising, until perestroika and subsequent events solved the problem by just removing it. But apart from similar nuisances he was lucky – with his friends, his career, his family. It is a pleasure to remember merry May day parties (his birthday) at a picturesque table laid by Irina, Slava's wife, with his son Aleksandr, in the company of his friends ...

Stanislav was a very balanced personality, kind of a practical philosopher. In professional life, he always ignored any superficial "pseudo mathematical" fuss:

elections, honours, titles. He disliked trying to push himself into "prestigious" mathematical journals, where, to please varying mathematical fads, or an "ideological" point of view, or simply to serve some group interests, various protagonists sometimes try to teach you how and what should be done in Analysis. (The majority of Vinogradov's beautiful papers were published in the purely domestic "Zapiski Nauchnyh Seminarov LOMI (Steklov Institut of Mathematics, Leningrad)"). Instead, S.V. was confident of the intrinsic value of hard analysis, no matter which fashion was imposing its whims outdoors.

His insidious illness progressed slowly, inexorably. In the summer of 1997, speaking to one of us about his still continuing mathematical plans, he would add repeatedly, as a refrain, "if I have enough time ... " (although, formally speaking, the diagnosis was kept from him, due to Russian medical practice). His last months were hard, he struggled against excruciating pain. Starting in the early fall of 1997 he was physically incapable of lecturing at the university, but still prepared his lectures and thoroughly instructed the younger colleagues who replaced him. He would joke: "I am a home-lecturer" ... The first author will never forget his last mathematical phone conversation with Stanislav (it was about the de Branges space $H(E)$'s being isometric to $H^2 \ominus IH^2$ for an appropriate inner function I; Slava promptly proved the conjecture by an explicit construction connecting E and I ...). He was an unforgettable example of spiritual strength and courage.

Many will always miss Stanislav, a good friend, ready to help, clever (even wise, which quality he demonstrated more than once when supporting and advising his friends in delicate situations). He was a bright personality with a vast field of cultural interests. An admirer of the magnificent Russian language of N.Gogol's prose and a passionate music lover, he did not miss any interesting concerts in the Leningrad Philharmonia Main Hall. In his younger years, he was there almost every week: gallery tickets cost less than a modest lunch. S.A. was a deep connoisseur of classical vocal music and, having himself a beautiful baritone, he often sang in friendly gatherings with wine ...

But mathematics was above all.

Mathematics

To locate Vinogradov's interests on the map of mathematics we start with a well-known theme of classical analysis discussed with a great force of expression in an old article [L] by Paul Lévy under a striking and somewhat emotional title. It is devoted to some "unsolved and surely unsolvable" problems of Fourier analysis. P. Lévy mentions, among many other things, the problem of finding a non-tautological and palpable description of the sequences $c = (c_n)_{n\in\mathbb{Z}}$ of complex numbers such that

$$c_n = \hat{f}(n), \qquad n \in \mathbb{Z} \tag{1}$$

for some function $f \in C(\mathbb{T})$, the space of all functions continuous on the unit circle $\mathbb{T}$; here $\hat{f}(n)$ is the n-th Fourier coefficient of f,

$$\hat{f}(n) = \int_{\mathbb{T}} f(z)z^{-n}dm(z), \tag{2}$$

and m is normalized Lebesgue measure on $\mathbb{T}$. The meaning of the adjectives "palpable" and "non-tautological" is illustrated by the Riesz-Fischer description of sequences c satisfying (1) with $f \in L^2(\mathbb{T})$ in place of $C(\mathbb{T})$. And any practicing classical analyst can add a lot of his own examples of such "sure non-solvability".

Problem (1) evoked by P. Lévy is a trigonometrical moment problem, representative of the large family of "moment problems" including the Hausdorff, Stieltjes, and Hamburger "power" moment problems

$$c_n = \int_{\mathbb{R}} x^n d\mu(x) \qquad (n \in \mathbb{Z}_+), \tag{3}$$

corresponding to various classes of real measures μ on $\mathbb{R}$.

This subject is important and rich in impressive connections, ideas, applications, and deep results. However, looking at them leaves (as a rule) an aftertaste of dissatisfaction. If, in order to decide whether a concrete sequence c satisfies (3) or (1), you are invited to test the positiveness of an infinite family of quadratic forms (or to use another similar procedure), you feel a bit disappointed. But experience shows that nothing more feasible can be proposed, and the feeling of "unsolvability" persists.

The same uneasiness is often caused by the interpolation problems of complex analysis. If you need a description of traces of a class of analytic functions on a given set, be ready to get something indigestible.

These phenomena of "sure non-solvability" often stem from prohibitions imposed by the Uncertainty Principle of harmonic analysis, which makes the simultaneous command and coordination of "time" and "spectral" properties so difficult. Another (but closely related) source lies in the uniqueness properties of analytic functions which entail rigid (and often mysterious) interconnections between the values of a function at different points.

But usually there is a way to modify the setting of a "surely unsolvable" problem so as to make the solution quite satisfactory and transparent, with all obscure constraints disappearing as if by miracle. Turn once again to Lévy's example (1). A necessary condition of solvability is obvious:

$$\sum |c_n|^2 < +\infty. \tag{4}$$

It is very far from sufficient, and nothing palpable can be done beyond it. But if we moderate our appetites and replace (1) by

$$c_n = \hat{f}(n), \qquad n \in E \tag{5}$$

with a *sparse* (infinite) $E \subset \mathbb{Z}$, then (4) becomes sufficient for the solvability of (5)! The "truncated" trigonometrical moment problem (5) is *freely* solvable.

Such phenomena of liberation, involving the free solvability of an initially overdetermined problem, were central for Vinogradov and stimulated his research in many other directions.

Another important idea illuminating many of his results is the duality between bases and interpolation problems, or more generally, between the geometry of reproducing kernels for a given holomorphic space and the "freedom" of its traces on sparse subsets. He paid a lot of attention to this duality (see below), and his impact on its development was considerable. The third feature of his mathematical style is his belief in the tools of real analysis, in a kind of positive operator inequalities underlying holomorphic structures.

Our article is followed by the translation of the prefaces to Vinogradov's theses ([Vin1], [Vin2]). They form a good survey of large parts of his work, with motivation and historical background. We complement [Vin1], [Vin2] turning to results not covered by these texts.

Free interpolation. The meaning of this term has been partly explained in the preceding remarks, see also [Vin1], [Vin2]; it often means that the restriction space $X|\Lambda$ on a given set Λ is a lattice, see [Vin1], [Vin2]. Vinogradov's contribution to this theme includes, besides numerous new results, a less spectacular but maybe not less important point, namely, a proof of Carleson's free interpolation theorem for H^∞ based on the so-called "Vinogradov-Senichkin test" (see the next section devoted to embedding theorems). This proof, devised by Vinogradov in 1974, is technically simpler and shorter than the original one. Moreover, it gives the Carleson interpolation condition a new form leading to a precise two-sided estimate of the interpolating function. The qualitative and quantitative advantages of Vinogradov's proof as well as a shorter and quite elementary proof by P. Jones are discussed in detail in Appendix 1 to the monograph [Koo], the proof can be also found in [Nik] and [13] (numbers in square brackets refer to Vinogradov's list of publications in this volume).

In [30] free interpolation with multiplicities in H^∞ is studied. New proofs and generalizations of the Nikolskii-Vasyunin results are given there. Vinogradov's approach works perfectly in this more difficult situation too (including the most interesting case of unbounded multiplicities), where the Jones explicit interpolation formulas are impossible (as is proved in [30]). In particular, the H^p free interpolation of locally defined germs of the spaces $H^p(L^\epsilon)$ on the level sets L^ϵ is considered, where $L^\epsilon = \bigcup_{\lambda\in\Lambda}\{|b_\lambda(z)|^{k_\lambda} < \epsilon\}$ and the Blaschke product $B(z) = \prod_{\lambda\in\Lambda} b_\lambda(z)^{k_\lambda}$ satisfies the generalized Carleson condition. It is proved that to be "free", the interpolation cannot be ensured for sets smaller than L^ϵ, $0 < \epsilon < 1$.

Some Jones-like formulas are constructed in [33]. They yield a lot of new information on interpolating functions. In particular, any bounded function defined on a Carleson set $E \subset \mathbb{D}$ (= open unit disk) can be interpolated by a function in

$H^\infty \cap K_{B^2}$, where $K_{B^2} := H^2 \ominus B^2H^2$, and B is the Blaschke product vanishing precisely on E. An interesting new feature of free interpolation phenomena was discovered in [15]. This paper, also written in the spirit of the bases-interpolation duality, is commented on in the preface to Vinogradov's second thesis [Vin2]. However, here we would like to stress that it contains the very first result showing that even for usual Banach spaces $X \subset Hol(\mathbb{D})$ the "freely interpolated" lattice-type restrictions $X|\Lambda$, $\Lambda \subset \mathbb{D}$ can be completely different for different Λ's, and not even isomorphic as Banach spaces. In [15] this is done for the spaces

$$X = l^p_A = \{f \in Hol(\mathbb{D}) : f(z) = \sum_{n\geq 0} \hat{f}(n)z^n, \sum_{n\geq 0} |\hat{f}(n)|^p < \infty\}, \ p \neq 2$$

and their half plane analogs $\mathcal{F}L^p(0,\infty)$. (By the way, being a concrete analyst S.V. was very far from the snobbish opinions to the effect that "true" analysis is limited to the Lebesgue norms of L^2, L^1, and L^∞. He seriously contributed to L^p spaces, and, as A. Markus recalled to one of us, polemically said in one of many friendly discussions: "Let them write '$p \neq 2$' on my tomb".)

Vinogradov made some ingenious observations on bases of exponentials and, more generally, on reproducing kernels for the model spaces K_Θ. Namely, he found the first example of a 'sine-type function' (in B.Ya. Levin's sense) with zeros not lying in a strip, and gave the first sufficient conditions for a family of exponentials $(e^{i\lambda t})_{\lambda\in\Lambda}$ to admit completion to a Riesz basis of eponentials of the entire space $L^2(0,a)$, $a > 0$. See the last Section of this Preface for references and more details.

A famous example of free interpolation is the Rudin-Carleson theorem: if $E \subset \mathbb{T}$ is closed and $m(E) = 0$ then any $f \in C(E)$ coincides on E with a function from the disc-algebra $C_A(\mathbb{D})$, the set of functions continuous on $\mathbb{D}\cup\mathbb{T}$ and analytic in $\mathbb{D}$. In [23] Vingradov and Hruščev turn to similar phenomena in U_A,

$$U_A := \left\{ f \in C_A : \sum_{n\geqslant 0} \frac{f^{(n)}(0)}{n!} z^n \text{ converges uniformly in } \mathbb{D} \right\}.$$

One of their main tools is the far-reaching Vinogradov generalization of the Kolmogorov inequality for the Hilbert transform (see 0.9 in [Vin2]). The starting point of [23] is the following stronger version of the Rudin-Carleson theorem due to Oberlin: $U_A|E = C(E)$ if $E \subset \mathbb{T}$ is closed and $m(E) = 0$. The authors consider, besides E, a set $\Lambda \subset \mathbb{Z}_+$ and the operator

$$f \mapsto \left(f|E, \hat{f}|\Lambda\right) \in C(E) \times l^2(\Lambda), \ f \in U_A, \ \hat{f} := \left(f^{(k)}(0)/k!\right)_{k\in\mathbb{Z}_+}.$$

They prove its surjectivity provided that $m(E) = 0$ and that Λ satisfies the $\Lambda(2)$-condition (i.e. $||p||_{L^1(\mathbb{T})} \asymp ||p||_{L^2(\mathbb{T})}$ for any trigonometrical polynomial p). Thus, freedom of interpolation is double: we are free to interpolate the values prescribed on E and, simultaneously, the Maclaurin coefficients with indices in Λ.

Another result of [23] is the existence of a linear interpolating operator in the Oberlin theorem. Functions in H^∞ with uniformly bounded partial sums of their Maclaurin series are also studied (together with the multipliers of this class).

This article contains an extremely useful result on the Hilbert transform $\tilde{\mu}$ of a complex Borel measure μ on $\mathbb{T}$:

$$\tilde{\mu}(z) := p.v. \int_{\mathbb{T}} \frac{d\mu(t)}{t-z}$$

Theorem.

$$\lim_{y \to +\infty} y\, m(\{z \in \mathbb{T} : |\tilde{\mu}(z)| > y\}) = \frac{var\, \mu_s}{\pi};$$

where μ_s denotes the m-singular part of μ.

This theorem was applied to the study of multiplicative properties of Cauchy-type integrals in [Gol], and to that of the Cauchy problem for the Laplace equation in [ABGHV].

Vinogradov's influence was important for research on the free interpolation of analytic functions smooth (in various senses) up to the boundary. One of the initial results in this direction is due to his student A. Kotochigov [Kot], who considered free interpolation on $\mathbb{T}$ and in $\mathbb{D}$ of functions analytic in $\mathbb{D}$ and satisfying a Lipschitz condition or belong to a Sobolev space. These results stimulated works by E. Dyn'kin [Dy1], [Dy2] who found a complete description of sets of free interpolation. In [25] free interpolation in some Besov classes of analytic functions is considered. Free interpolation in the spaces of Cauchy type integrals and of their multipliers is one of the themes of [14].

To conclude this section we mention [29], where free interpolation is used as a tool to improve known estimates of $\|P_+ f\|_\infty$ for a function f meromorphic in $\mathbb{D}$ having at most n poles and with $\|f\|_\infty \leq 1$; P_+ being the Riesz projection.

Carleson-type embedding theorems. To understand properties of a function space X it is often advisable to look at an "accompanying" space, like, say, the dual space X'. Another example is the algebra mult$X$ of all multipliers of X, one of Vinogradov's favourites (see below). One more example is the set $\mathrm{carl}_p X$ of all positive measures μ such that $\int_E |f|^p \, d\mu < +\infty$ for any $f \in X$ (E is the domain of functions in X). If X is a Banach space, then, as a rule, $\mu \in \mathrm{carl}_p X$ satisfies

$$\|f\|_{L^p(\mu)} \leqslant c(\mu)\|f\|_X, \quad f \in X,$$

so that X is continuously embedded in $L^p(\mu)$ whence the term "an embedding theorem". A classical sample is the Carleson embedding theorem yielding a very clear and efficient description of $\mathrm{carl}_p H^p$ (where $H^p = H^p(\mathbb{D})$ is the usual Hardy space). This result is crucial for free interpolation in H^p (including for $p = +\infty$). Its initial proof, based on some delicate "dyadic" techniques, wasn't easy. V. Senichkin simplified it essentially [Sen], and then Vinogradov turned it into something *quite* simple having reduced it to what is now known as the Vinogradov-Senichkin test ([Nik], lecture VIII). This is an easy estimate of the norm of an integral operator with a positive kernel acting in $L^2(\mu)$, μ being an abstract measure; its most sophisticated ingredients are the Cauchy-Schwarz inequality and Fubini's theorem. The test or its elements had appeared in the past here and there in implicit form (e.g., in a manuscript of Luzin published in [Luz], p. 287–319, or in theorems on the quasi- or almost everywhere convergence of Fourier series whose coefficients belong to a weighted l^2-space [Carl, ch.V]); it is a tool ready to be used on many occasions. An impressive application is the proof of a new version of the Carleson embedding theorem. The description of $\mathrm{carl}_2 H^2$ appears in a disguised, seemingly implicit (but equivalent) form. At first glance this form is not as clear geometrically as Carleson's, but it turns out to be much more practical (see the discussions in [Nik], Lecture VIII, and in [Koo], Appendix 1). The generality and power of this approach to "L^2-embeddings" are demonstrated by Vinogradov's axiomatic schemes and further results on $\mathrm{carl}_2 X$ for many Hilbert spaces X of analytic functions ([Vin2],[34],[39]). Some of his ideas and results related to this scheme are explained and discussed in [MSh], §3.11; their influence is acknowledged in [Ver1], [Ver2], [VerK], and [TV]. The latter paper also contains impressive applications of the test to matrix valued Hunt-Muckenhoupt-Wheeden theorems, as well as interesting modifications and improvements of the test itself giving, in particular, the simultaneous continuity of two operators with different kernels; see theorems 7.3 and 7.4 there. From a more general point of view, we can say that the Vinogradov-Senichkin test is one of the first and most explicit manifestations of what we can call the Reproducing Kernel Thesis (RKT). The RKT says, "an operator $T : H \longrightarrow H$ defined on a Hilbert space $H \subset Hol(\mathbb{D})$ with the reproducing kernel k_z, $z \in \mathbb{D}$, is bounded if it is bounded on reproducing kernels, that is, if $\sup_{z\in\mathbb{D}} \|Tk_z\|/\|k_z\| < \infty$." Of course, without any hypothesis on a linear operator T, the thesis fails. But RKT works for many important operators related to analysis on H ; why and under which assumptions, is still a mystery. The Vinogradov-Senichkin test proves RKT for embedding operators $H^2 \subset L^2(\mu)$, and in fact gives more. The RKT is true for Toeplitz operators $Tf = T_\varphi f = P_+\varphi f$ on H^2, and the Bonsall-Power theorem [BP] says that it is also true for Hankel operators $Tf = H_\varphi f = P_-\varphi f$ on H^2 (here P_+ means the Riesz orthogonal projection onto H^2, and $P_- = I - P_+$). This is equivalent to saying that the known Garsia norm is equivalent to the usual BMO norm. R. Coifman, D. Luecking, R. Rochberg, K. Zhu, and many others have proved the RKT for Toeplitz and Hankel operators on many other spaces with reproducing kernels, including some spaces of functions of several variables (Bergman, Dirichlet, etc); see [Zhu] for results until 1990. The Axler-Chang-Sarason-Volberg theorem (see [Nik], for example) says that the RKT is true for semi-commutators of Toeplitz operators. The famous Muckenhoupt (A_2) condition is nothing but the RKT for the Riesz projection P_+ on a weighted space $L^2(\mathbb{T}, w\,dm)$, and, as is shown in [TV], the only way to describe matrix-valued Muckenhoupt weights is to use the RKT. We end this list by referring to the theory of the Berezin transform $z \longmapsto (Tk_z, k_z)$ (see [AZ], [Zhu]) which is closely related to the RKT, and pointing out once again that Vinogradov's approach to embedding theorems is still one of the most useful examples of the RKT.

A generalization of the Carleson embedding theorem is obtained in [42]; it is applicable to spaces A_p^1 (to be defined at the end of the next section).

Multipliers of spaces of analytic functions. Let X be a class of functions analytic in $\mathbb{D}$. A multiplier of X is by definition a function f analytic in $\mathbb{D}$ such that $f \cdot x \in X$ for any $x \in X$; the set of all multipliers of X is denoted by $\operatorname{mult} X$. Vinogradov studied mainly $\operatorname{mult} \ell^p_A$ and $\operatorname{mult} \mathcal{K}_A$ where ℓ^p_A is the space of functions $\sum\limits_{n\geq 0} c_n z^n$ with $\sum\limits_{n\geq 0} |c_n|^p < +\infty$, and $\mathcal{K}_A$ is the space of Cauchy-type integrals $\int\limits_{\mathbb{T}} \frac{d\mu(\zeta)}{\zeta - z}$ (in $\mathbb{D}$), formed for complex Borel measures μ on $\mathbb{T}$.

Results for the former class are described in [Vin2]. Here we only point out the short problem notes [36], [41], and briefly discuss the elegant results of [43], the last item in Vinogradov's list of publications.

Let V be a Banach space contained in $L^\infty(E)$ where E is a set equipped with a measure. Suppose $\|f\|_\infty \leq \|f\|_V$ for any $f \in V$. A function Φ analytic in a disc $\{z : |z - c| < r\}$ is said to act on V if

$$\{\Phi \circ f : f \in V,\ \|f - c\|_V < r\} \subset V.$$

Any such Φ acts on V provided that V is a Banach algebra with respect to pointwise multiplication. The authors of [43] proclaim the following dichotomy: either V is a Banach algebra or Φ does not act on V unless it is affine (i.e., unless $\Phi'' \equiv 0$). This is not a formal assertion but rather a useful heuristic principle confirmed by some precise statements and many concrete examples. Among other things a question asked in [Nik2] is answered in [43]: if $1 < p < 2$, then $1/f$ may not belong to ℓ^p_A for an $f \in \ell^p_A$ separated from zero (i.e. $\inf\limits_{\mathbb{D}} |f| > 0$). The above principle proves to be applicable to $V = U_A$ and $V = \ell^p_A \cap C_A$ $(1 < p < 2)$. The last example entails the following corollary: $\ell^p_A \cap C_A$ is not an algebra if $1 < p < 2$, an unpublished result that J.Bourgain obtained by a completely different method. Many similar and much subtler results are proved in [43]. It contains delicate norm estimates in $\operatorname{mult} X$ for various spaces X based on random power series techniques.

The properties of $\operatorname{mult} \mathcal{K}_A$ turned out to be rather mysterious. It is a proper subset of H^∞; its elements possess some regularity properties (radial boundary values exist everywhere on $\mathbb{T}$, partial sums of the Maclaurin series are uniformly bounded), but none of these can characterize the functions in $\operatorname{mult} \mathcal{K}_A$. First results obtained in [H1], [H2], [8] led to the following question: which inner functions are in $\operatorname{mult} \mathcal{K}_A$? (A function $I \in H^\infty$ is called inner if $\lim\limits_{r\to 1} |I(r\zeta)| = 1$ for almost all $\zeta \in \mathbb{T}$.) It was answered in [24]. To state this beautiful solution we need the following definition: a sequence $(a_n)_{n=1}^\infty$ of points in $\mathbb{D}$ is called a Frostman sequence if

$$\sup \left\{ \sum_{n=1}^{\infty} \frac{1 - |a_n|^2}{|\zeta - a_n|} : \zeta \in \mathbb{T} \right\} < +\infty.$$

Theorem [24]. *Let I be an inner function. The following are equivalent: (1) $I \in \operatorname{mult} \mathcal{K}_A$; (2) I is a Blaschke product whose zeros form a Frostman sequence.*

The proof is based (in part) on the results of [8] and [14], but contains new ingredients including free interpolation techniques (the problem itself does not mention any interpolation). As a byproduct [24] offers a nice construction of a Blaschke product converging everywhere on $\mathbb{T}$ with a prescribed closed nowhere dense set of points of discontinuity on $\mathbb{T}$, its Maclaurin series converging everywhere on $\mathbb{T}$ and the n-th Maclaurin coefficient being $O(\frac{1}{n})$. Another result in [24] is a complete description of the families $\left(\frac{1}{1-\bar{\lambda}z}\right)_{\lambda\in\Lambda}$ forming an unconditional basis for their closed linear span in $\operatorname{mult}\mathcal{K}_A$. The following lemma of [24] sheds new light on the class $\operatorname{mult}\mathcal{K}_A$: *a function $\varphi \in H^\infty$ is in $\operatorname{mult}\mathcal{K}_A$ iff the Toeplitz operator $T_{\overline{\varphi}}$ acts continuously in the disc-algebra C_A ($T_{\overline{\varphi}}(f) := P_+(f\overline{\varphi})$, $f \in C_A$, where P_+ is the M. Riesz projector: $P_+(p) = \sum_{n\geq 0} c_n z^n$ for trigonometrical polynomials $p = \sum_{n\in\mathbb{Z}} c_n z^n$).*

More generally, multipliers play an important role in the theory of Toeplitz operators. In particular, Vinogradov's results on the spaces mult l^p_A, partly presented in [Vin2], were essentially used in the exhaustive monograph [BS], Ch. 6. We mean first of all the results from [12], [21] on the membership problem (sufficient conditions), and also the divisibility property of mult l^p_A described in the next Section.

Let us conclude our discussion of multipliers by turning to the space A^1_p of functions f analytic in $\mathbb{D}$ and satisfying

$$\int_{\mathbb{D}} |f'(z)|^p(1-|z|^2)^{p-1}dxdy < +\infty,\ z = x+iy,\ 0 < p \leq 2;$$

thus, A^1_1 is the "analytic" Sobolev class $W^1_1(\mathbb{D})$, A^1_2 is the Hardy class $H^2(\mathbb{D})$, and $\ell^p_A \subset A^1_p \subset H^p(\mathbb{D})$. A function $f \in A^1_p$ can also be looked on as the Cauchy transform of a function in a Besov class on $\mathbb{T}$. In [42] a nice and efficient description of mult A^1_p is obtained; another result is a description of some inner functions in mult A^1_p.

Note that the partial result that $\operatorname{mult}(A^1_1) = H^\infty \bigcap\{f : |f'|\,dxdy$ is a Carleson measure $\} = \{f \in Hol(\mathbb{D}) : \sup_{\lambda\in\mathbb{D}} \|f\circ b_\lambda\|_{A^1_1} < \infty\}$ was obtained by Vinogradov much earlier and was first published in V.Peller's paper [Pel] (with references to S.V.), where it is applied to estimates of functions of Hankel operators.

Inner-outer factorization (division problems). Any function f of the Smirnov class (in particular, of any Hardy class $H^p(\mathbb{D})$) can be decomposed into the product of two factors:

$$f = I_f \cdot O_f$$

where I_f is inner and O_f is outer (O_f does not vanish in $\mathbb{D}$, and $\log|O_f|$ is the Poisson integral of its boundary values). This inner-outer factorization is unique up to a constant factor and is often called canonical. Its role in complex and harmonic analysis and in various applications is outstanding (see, e.g., [Nik], [Koo]; some historical remarks are in [NikH]). In many cases O_f inherits properties of f. For example, $O_f \in H^p$ or $O_f \in C_A$ whenever $f \in H^p$ or $f \in C_A$ respectively. The following precise and more general question is natural and interesting: given a class X of functions of the Smirnov class and an inner function I, is it true that if $f = I \cdot O \in X$ and O is analytic in $\mathbb{D}$, then $O \in X$? If so, we say that division by I is possible in X. Vinogradov was one of the first who solved this division problem for interesting pairs X, I (for the present state of this circle of problems see [Sh]).

There is a simple way to prove divisibility by an inner factor I: if the Toeplitz operator $T_{\bar{I}}$ acts in $X \subset H^1$, then division by I is possible in X (since $T_{\bar{I}}(IQ) = Q$), and the boundedness of $T_{\bar{I}}$ can often be obtained for free by duality. This approach yields a lot of divisibility theorems, but it fails in some "extreme" cases due to the unboundedness of the Hilbert transform in a concrete space, and it is just these cases that turn out to be the most interesting. They usually need delicate estimates to show that a certain regularity of f is due to O_f alone, not to a cancellation of singularities of I_f and O_f (I_f is, as a rule, very irregular).

One such subtle situation was the theme of [9] where the divisibility by any inner function was proved for $X = \{f \in H^1 : f' \in H^p\}$, $p = 1$ or $+\infty$ (if $1 < p < +\infty$, then this can be proved by the "soft" Toeplitz-operator method). Actually, [9] gives more than mere divisibility: a precise two-sided estimate of $\|f'\|_{H^1}$ is given in terms of independent parameters defining O_f and I_f ($|f| \mid \mathbb{T}$, the zeros of f, and the singular positive measure figuring in I_f).

The second part of [42] contains positive and negative results on the division problem in A^1_p (see the preceding section for the definition). In particular, division by the elementary singular inner function $I = \exp[a\frac{z+1}{z-1}]$ $(a > 0)$ is impossible in spite of the inclusion $I \in \text{mult}A^1_p$.

In [14] Vinogradov proved that in $X = \mathcal{K}_A$, division by an arbitrary inner factor is possible. A different proof was found later by A.B.Aleksandrov [Al]; interesting refinements are in [Pol]. In [14] a general scheme is proposed to prove the divisibility by an inner factor in a space of analytic functions.

Uniqueness theorems, zero sets. This is the theme of items [1] (discussed in [Vin1]), [11], and [38].

A set $E \subset \mathbb{D} \cup \mathbb{T}$ is said to be a uniqueness set for a class X of functions analytic in $\mathbb{D}$ if any $f \in X$ vanishing on E vanishes identically; for $\zeta \in \mathbb{T}$ the value $f(\zeta)$ is understood as $\lim_{r \nearrow 1} f(r\zeta)$.

In [11] the uniqueness sets for the class $H^1_1 := \{f \in H^1 : f' \in H^1\}$ are described. The description looks very familiar to anybody who has ever dealt with similar problems: a closed set $E \subset \mathbb{D} \cup \mathbb{T}$ is a uniqueness set for H^1_1 iff

$$\int_{\mathbb{T}} \log \rho_E dm = -\infty \tag{6}$$

where $\rho_E(\zeta)$ is the distance from ζ to E. The necessity of (6) follows immediately from the Carleson theorem on uniqueness sets for analytic functions smooth (say, C^∞) up to the boundary. The sufficiency would be quite easy if, in place of H^1_1 we had a class X consisting of Lipschitz functions (of any order): in that case (6) implies the divergence of $\int_{\mathbb{T}} \log|f|dm$ for any $f \in X$ vanishing on E. However, for $f \in H_1^1$ the continuity modulus cannot be controlled (although $H_1^1 \subset C_A$), and to deduce uniqueness from (6) is a delicate matter. Actually, in [11] a nice upper estimate on the geometrical mean $G(|f|) = \exp\left(\int_{\mathbb{T}} \log|f|\,dm\right)$ for $f \in H_1^1$ over $\mathbb{T}$ is proved; $G(|f|)$ is majorized by $G(\rho_E|f'|)$.

The article [38] is devoted to the role played by the Blaschke condition

$$\sum_{k=1}^{\infty}(1-|a_k|) < +\infty, \qquad a_k \in \mathbb{D}, \tag{7}$$

in uniqueness theorems. This condition is necessary for the existence of a non-zero function of the Nevanlinna class (not to mention the Hardy classes) vanishing at a_k, $k = 1, 2, \ldots$. However, a set $\{a_k\}$ satisfying (7) may well be a uniqueness set for a class X of Hardy functions. For example, this is the case if $X \subset C_A$ (= the disc algebra), and the closure of $\{a_k\}$ contains $\mathbb{T}$, which is compatible with (7). In [38] much subtler situations are considered when X contains functions everywhere discontinuous on $\mathbb{T}$ or even not possessing radial boundary values. A general method of constructing such classes X and uniqueness sets for them satisfying (7) is proposed. The list of classes X considered includes the spaces mult$\mathcal{K}_A^p$, $p = 1, +\infty$, where $\mathcal{K}_A^p$ denotes the set of all Cauchy transforms of functions in $L^p(\mathbb{T})$. Another example is the class of Cauchy transforms of the functions

$$f \in C^{\omega}(\mathbb{T}) := \{f \in C(\mathbb{T}) : \omega_f \le c_f\omega\},$$

where ω is some given increasing positive function and ω_f denotes the continuity modulus of f. An interesting case occurs when

$$\int_0 \frac{\omega(t)}{t}dt = +\infty$$

(otherwise our class is in C_A). The approach of [38] works for some analytic Besov classes and some classes of power series characterized by the rate of decay of their coefficients. Results of [38] are closely related to the division problems discussed in the preceding section.

Singular integrals. Vinogradov's generalization of the Kolmogorov weak type estimate of the Hilbert transform is stated and discussed in [Vin2]. Here we only mention some repercussions of this result.

Bourgain applied a modification of Vinogradov's theorem to the study of the space $U(\mathbb{T})$ of uniformly convergent Fourier series (see [B]). In [K1] this result of Vinogradov was used to prove a quantitative refinement of the Men'shov correction theorem: *for any $\epsilon \in (0,1)$ and $f \in C(\mathbb{T})$ there is a $g \in U(\mathbb{T})$ such that*

$$m(\zeta \in \mathbb{T} : g(\zeta) \ne f(\zeta)) < \epsilon,$$

and $\|g\|_U \le const \cdot \log\frac{1}{\epsilon}\|f\|_\infty$; this inequality is sharp. In [K2] Vinogradov's result is again called on and leads, this time, to a solution of the long-standing Szidon problem for U_A: *given $(x_n)_{n\in\mathbb{Z}_+} \in l^2$, $x_n \ge 0$, there is an $f \in U_A$ such that*

$|f^{(n)}(0)|/n! \geq x_n$, $n = 0, 1, 2, \ldots$. In [K3] Vinogradov-like estimates again emerge in a new context.

As a whole, Vinogradov's generalization of the Kolmogorov estimate and its consequences show that the spaces $U(\mathbb{T})$ and U_A of uniformly convergent Fourier series are much closer to the familiar $C(\mathbb{T})$ and C_A than it can be predicted from preceding experiences of classical analysis (properties of the Dirichlet kernels, the Lebesgue constants, etc.)

We also mention estimates of Cauchy-type integrals on curves obtained in [18]. They may seem obsolete after the immense progress made in this area in the eighties. We believe, however, that the concrete norm inequalities of [18] and their proofs may be of interest even now. For example, let $\Gamma = \omega(\mathbb{T})$ where ω is a diffeomorphism of $\mathbb{T}$ such that the Fourier series of $s \mapsto \frac{d}{ds}\omega(e^{is})$ is absolutely convergent. It is proved in [18] that the singular integral operator S_Γ with the Cauchy kernel acts continuously from $L^p(\Gamma)$ to $L^p(\Gamma)$ $(1 < p < +\infty)$. This is, of course, a very particular case of the famous Calderon theorem, but the simple (but clever) norm estimate of S_Γ given in [18] involves the Wiener-Levy theorem on absolutely convergent Fourier series and can hardly be deduced from general results embracing larger classes of $\Gamma' s$.

"Donated" theorems. As mentioned above many of Vinogradov's results are scattered through papers of his friends and colleagues. Always ready to help, he often turned out to be a kind of "mathematical ambulance". Probably S.V. is a Guinness Record Book champion for theorems published in other people's papers. We try to prove this by listing *some* of these theorems.

1) The Vinogradov-Senichkin Test commented on above is first published in [Nik3].
2) There exist $\zeta_n \in \mathbb{T}$ such that $\{r_n\zeta_n : n \geq 1\}$ is a dominating subset of the disk $\mathbb{D}$ if and only if $\sum_n(1 - r_n) = \infty$. First published in [Nik2].
3) Given $a > 0$ and a Carleson sequence $(\lambda_n)_{n\geq 1}$ for the upper half plane $\mathbb{C}_+ = \{z : \operatorname{Im} z > 0\}$ such that $\lim_n \operatorname{Im}(\lambda_n) = \infty$, the exponentials $(e^{i\lambda_n t})$ can be complemented up to a Riesz basis $(e^{i\lambda_n t})\bigcup(e^{i\lambda'_k t})$ of the space $L^2(0, a)$. First published in [HNP] as Theorem III.3.1. (Later on, this result was considerably developed and applied to frame theory, [S]).
4) Given a Blaschke sequence of radii r_n, i.e., with $\sum_n(1 - r_n) < \infty$, there exist $\zeta_n \in \mathbb{T}$ such that $\{\lambda_n = r_n\zeta_n : n \geq 1\}$ is a strong Carleson sequence ($\lim_n |B_{\lambda_n}(\lambda_n)| = 1$ where $B_{\lambda_n} = B/b_{\lambda_n}$). (This strengthens a well-known result of Naftalevich). First published in [NikV].
5) There exists a Levin sine-type function F whose zero set Λ is a Carleson sequence in $\mathbb{C}_+$ such that $\sup\{\operatorname{Im}(\lambda) : \lambda \in \Lambda\} = \infty$. First published in [HNP] as Theorem III.3.3 (solving a long standing problem).
6) The description of multA_1^1 mentioned above is first published in [Pe1].
7) A construction of Banach algebras (like $\operatorname{mult}(\mathcal{K}_A) + H^\infty$) needed for resolvent estimates of Toeplitz operators is first published in [Pe2], Section 4, Examples 3 and 4.

8) The Taylor coefficients $\hat{B}(n)$, $n \geq 0$ of a Blaschke product whose zeros form a weak Newman sequence satisfy $\hat{B}(n) = O(1/n)$ for $n \longrightarrow \infty$. First published in [Ver3] where the converse is also proved.
9) One of the main ideas of the paper [TV] by S.Treil and A.Volberg was suggested by Vinogradov according to the authors' acknowledgement (private communication). The idea is that to prove the Muckenhoupt-type weighted estimate for the Hilbert transform one can replace the (previously omnipresent) maximal function techniques by Carleson-type embedding theorems. S.Treil and A.Volberg applied the idea in masterly fashion, finding and proving a matrix valued Hunt-Muckenhoupt-Wheeden theorem, [TV].
10) A non-uniqueness theorem for holomorphic Gevrey classes $G_\alpha = \{f \in Hol(\mathrm{Im}(z) > 0) : |f^{(n)}(z)| \leq C_f Q_f^n n! n^{n/\alpha}\}$: a closed subset $E \subset \mathbb{R}$ whose finite complementary intervals satisfy the following Vinogradov condition (V_α)

$$\sum_k l_k^{1-\alpha} (\log^+ \frac{1}{l_k})^{\alpha+\epsilon} < \infty$$

(for some $\epsilon > 0$) is not a uniqueness set for G_α. Published in [Hru] as Theorem 2, this result is still the best known sufficient condition in terms of lengths of the complementary intervals.

* * *

Two adjectives, *elegant* and *exquisite*, come to mind when we try to assess Vinogradov's work as a whole. Elegance is a permanent feature of what he did and how he did it. Exquisiteness marks his very choices of subjects and objects: the space of power series uniformly convergent on $\mathbb{D}$ preferred to the insipid disc algebra and the three parametric Besov scale preferred to Sobolev's, not to mention the dull L^p-scale; he is attracted by mysterious subtleties of multipliers; he tries to divide by an inner factor precisely in the spaces whose resistance to this procedure is most vigorous; not satisfied with the mere interpolation of *values* of a function he wants to simultaneously interpolate its *Fourier coefficients* as well, etc. Such inclinations are not fashionable nowadays, and some colleagues treat them as, at the best, idle manipulations with inequalities, if not corpora delicti. But, even if we forget the intrinsic beauty of Vinogradov's mathematics, such seemingly "exquisite" topics and "technical" tricks are indispensable here and there in Analysis, fashion or no fashion. Powder has to be dry and the weapons sharp and ready.

Stanislav Vinogradov was a faithful votary and skillful master of Analysis.

References

[ABGHV] A. Aleksandrov, J. Bourgain, M. Giesecke, V. Havin, Yu. Vymenets, *Uniqueness and free interpolation for logarithmic potentials and the Cauchy problem for the Laplace equation in R 2*, Geometric and Functional Analysis, **5**(1995), no. 3, 529–571.

[Al] A.B. Aleksandrov, *On the (A)-integrability of boundary values of harmonic functions,* Mat. Zametki, **30**(1981), no. 1, 59–72. (Russian)

[AZ] Sh.Axler and D.Zheng, *The Berezin transform on the Toeplitz algebra,* preprint, 1996.

[BP] F.F. Bonsall and S.C. Power, *A proof of Hartman's theorem on compact Hankel operators,* Math. Proc. Camb. Phil. Soc. **78**(1975), 447–450.

[B] J. Bourgain, *Quelques propriétés linéaires topologiques de l'espace des séries de Fourier uniformément convergentes*, Seminaire de l'Initiation a l'Analyse (G. Choquet, M. Rogalski, J.Saint Raymond, Eds.), 22e année, Expose no. 14, Univ. Paris-6, 1982–83.

[BS] A. Böttcher and B. Silbermann, *Analysis of Toeplitz Operators*, Berlin, Akademie-Verlag 1989.

[Carl] L.Carleson, *Selected Problems on Exceptional Sets*, Van Nostrand, 1967

[Dy1] E.M. Dyn'kin, *Free interpolation sets in Holder classes*, Matem. Sb. **109**(1979), no. 5, 107–128. (Russian)

[Dy2] E.M. Dyn'kin, *The pseudoanalytic extension*, Journ. d'Analyse Math. **60**(1993), 45–70.

[Gol] M.G. Goluzina, *On multiplication and division of the Cauchy type integrals*, Vestnik LGU, **19**(1981), 8–15. (Russian)

[H1] V.P. Havin, *On analytic functions representable by a Cauchy-Stieltjes integral*, Vestnik LGU, **1**(1958), no. 1, 66–79. (Russian)

[H2] V.P. Havin, *Relations between certain classes of functions regular in the unit disc*, Vestnik LGU, **17**(1962), no. 1, 102–110. (Russian)

[Hru] S.V. Hruscev, *Sets of uniqueness for the Gevrey classes*, Arkiv Mat., **15**(1977), no. 2, 253–304.

[HNP] S.V. Hruscev, N.K. Nikolski, and B.S. Pavlov, *Unconditional bases of exponentials and of reproducing kernels*, Lect. Notes Math. (Springer) **864**(1981), 214–335.

[K1] S.V. Kislyakov, *The quantitative aspect of correction theorems*, Zapiski Nauchn. Seminarov LOMI, **92**(1979), 182–101. (Russian)

[K2] S.V. Kislyakov, *Fourier coefficients of boundary values of functions analytic in a disc and in a bidisc*, Trudy MIAN, **155**(1981), 77–94. (Russian)

[K3] S.V. Kislyakov, *Martingale transforms and uniformly convergent orthogonal series*, Zapiski Nauchn. Seminarov LOMI, **141**(1985), 18–38.

[Koo] P. Koosis, *Introduction to H_p Spaces*, 2nd ed. Cambridge Tracts on Mathematics. Cambridge University Press, 1998.

[Kot] A.M. Kotochigov, *Interpolation by analytic functions smooth up to the boundary*, Zapiski Nauchn. Seminarov LOMI, **30**(1972), 154–157. (Russian)

[L] P. Lévy, *Sur quelques problèmes actuellement irresolus et sans doute insolubles dans les théories des séries et des intégrales de Fourier*, J. Ecole Polytechnique, **145**(1939), 179–194.

[Luz] N.N. Luzin, *Integral and Trigonometrical Series*, N.K. Bari and D.E. Menshov Eds., GITTL, Moscow, Leningrad, 1951. (Russian)

[MSh] V.G. Maz'ya, T.O. Shaposhnikova, *Theory of Multipliers in Spaces of Differentiable Functions*, Pitman, 1985.

[Nik] N.K. Nikolskii, *Treatise on the Shift Operator*, Springer-Verlag, 1986.

[Nik2] N.K. Nikolskii, *Five problems on invariant subspaces*, Zapiski Nauchn. Seminarov. LOMI, **23**(1971), 115–127. (Russian)

[Nik3] N.K. Nikolski, *Lectures on the shift operator*, II, Zapiski Nauchn. Seminarov LOMI, **47**(1974), 90–119 (Russian); English transl.: J. Soviet Math. **9**(1978).

[NikH] N.K. Nikolskii, V.P. Havin, *Results of V.I. Smirnov on complex analysis and their subsequent development*, In the book: V.I.Smirnov. Selected works. Complex Analysis. Mathematical Theory of Difraction. Leningrad, Leningrad Univ. Press, 1988, 111–145. (Russian)

[NikV] N.K. Nikolski and A.L.Volberg, *Tangential and approximate free interpolation*, In: Analysis and Partial Differential Equations (ed. C.Sadosky), Marcel Dekker, NY, 1990, 277–299.

[O] V.L. Oleinik, *Embedding theorems for weighted classes of harmonic and analytic functions*, Zapiski Nauchn. Seminarov LOMI, **47**(1974), 120–137 (Russian); English transl.: J. Soviet Math. **9**(1978), 228–243.

[Pe1] V.V. Peller, *Estimates of functions of power bounded operators on Hilbert spaces*, J. Operator Theory, **7**(1982), 341–372.

[Pe2] V.V. Peller, *Spectrum, similarity, and invariant subspaces of Toeplitz operators*, Izvestia Akad. Nauk SSSR, Seria Mat., **50**(1986), no. 4 (Russian); English transl.: Math. USSR Izvestya, vol. **29**(1987), no. 1, 133–144.

[Pol] A.G. Poltoratskii, *Boundary behaviour of pseudocontinuable functions*, Algebra i Analiz, **5**(1993), no. 2, 189–210. (Russian)

[S] K. Seip, *On the connection between exponential bases and certain related sequences in* $L^2(-\pi,\pi)$, J. Funct. Anal., **130**(1995), 131–160.

[Sen] V.N. Senichkin, *On the Carleson lemma on estimates of integrals of analytic functions*, Vestnik LGU, **19**(1971), 52–56. (Russian)

[SS] H.S. Shapiro and A.L. Shields, *On some interpolation problems for analytic functions*, Amer. J. Math., **83**(1961), 513–532.

[Sh] N.A. Shirokov, *Analytic functions smooth up to the boundary*, Lecture Notes in Math., **v. 1312**, Springer-Verlag, 1988.

[TV] S. Treil and A. Volberg, *Wavelets and the angle between past and future*, J. Funct. Anal., **143**(1997), 269–308.

[Ver1] I. Verbitsky, *Embedding theorems for harmonic extensions of functions in Sobolev spaces*, Soviet Math. Dokl. **32**(1985), 486–489.(Russian)

[Ver2] I. Verbitsky, *Multipliers in spaces with fractional norms and inner functions*, Siber. Math. J. **26**(1985), 51–76.

[Ver3] I. Verbitsky, *On Taylor coefficients and* L^p*-moduli of continuity of Blaschke products*, Zapiski Nauchn. Seminarov LOMI, **107**(1982), 27–35 (Russian); English transl.: J. Soviet Math. **36**(1987), 314–319.

[VerK] I. Verbitsky, N. Kalton, *Non-linear equations and weighted norm inequalities*, to appear in TAMS.

[Vin1] S. A. Vinogradov, *Interpolation problems for analytic functions continuous in the closed disk and for functions whose sequence of coefficients is in l^p*, this volume.

[Vin2] S. A. Vinogradov, *Free interpolation in spaces of analytic functions*, this volume.

[Zhu] K.H. Zhu, *Operator theory in function spaces*, Monographs and Textbooks in Pure and Appl. Math., 139, Marcel Dekker Inc, N.Y., 1990.

Victor P. Havin
Department of Mathematics
St. Petersburg University
Bibliotechnaia pl., 2
198904, Stary Peterhof
St.Petersburg, Russia
and
Department of Math. & Stat.
McGill University
805, Sherbrooke ave. W
Montreal H3A 2K6
Canada

Nikolai N. Nikolski
Laboratoire de Mathématiques
Pures, UFR Math. & Info.
Université Bordeaux 1
351, cours de la Libération
33405 Talence France
and
Steklov Institute of Mathematics
St. Petersburg
Fontanka, 27
191011 St. Petersburg
Russia

List of Publications of S.A. Vinogradov

[1] S. A. Vinogradov, *On the interpolation and zeros of power series with a sequence of coefficients in l^p*, Dokl. Akad. Nauk SSSR, **160** (1965), no. 2, 262–266 (Russian); English transl. in Soviet Math. Dokl, **6** (1965), no. 1, 57–61.

[2] S. A. Vinogradov, *On the interpolation of power series which converge absolutely on the unit circle*, Vestnik Leningrad. Univ. **2** (1965), no. 7, 30–44 (Russian).

[3] S. A. Vinogradov, *The interpolation of power series whose sequence of coefficients is from l^p*, Funkcional. Anal. i Priložen., **1** (1967), no. 3, 83–85 (Russian).

[4] S. A. Vinogradov, *Interpolation problems for analytic functions continuous in the closed disk and for functions whose sequence of coefficients is in l^p*, Avtoreferat dissertatsii na soiskanie uchenoy stepeni kandidata fiziko-matematicheskih nauk, pp. 1–9, Leningradskiy Gosudarstvenniy Universitet, 1968 (Russian); English transl. in this volume.

[5] S. A. Vinogradov, *Interpolation problems for analytic functions continuous in the closed disc and for functions with coefficient sequence in l^p*, PhD Thesis. Leningrad State University, 1968, p. 1–207 (Russian).

[6] S. A. Vinogradov, *Paley singularities and Rudin–Carleson interpolation theorems for certain classes of analytic functions*, Dokl. Akad. Nauk SSSR, **178** (1968), no. 3, 511–514 (Russian); English transl. in Soviet Math. Dokl, **9** (1968), no. 1, 111–115.

[7] S. A. Vinogradov, *Interpolation theorems of Banach–Rudin–Carleson and norms of embedding operators for certain classes of analytic functions*, Zap. Naučn. Sem. Leningrad. Otdel. Mat. Inst. Steklov. (LOMI), **19** (1970), 6–54 (Russian); English transl. in Journ. of Soviet Math., **19** (1972), 1–28.

[8] S. A. Vinogradov, M. G. Goluzina, and V. P. Havin, *Multipliers and divisors of Cauchy-Stieltjes type integrals*, Zap. Naučn. Sem. Leningrad. Otdel. Mat. Inst. Steklov. (LOMI), **19** (1970), 55–78 (Russian); English transl. in Journ. of Soviet Math., **19** (1972), 29–42.

[9] S. A. Vinogradov and N. A. Širokov, *The factorization of analytic functions with derivative in H^p*, Zap. Naučn. Sem. Leningrad. Otdel. Mat. Inst. Steklov. (LOMI), **22** (1971), 8–27 (Russian); English transl. in Journ. of Soviet Math., **2** (1973), 68–83.

[10] S. A. Vinogradov, *Interpolation by power series that converge uniformly in the unit disc*, Zap. Naučn. Sem. Leningrad. Otdel. Mat. Inst. Steklov. (LOMI), **30** (1972), 5–14 (Russian); English transl. in Journ. of Soviet Math., **4** (1975), 303–312.

[11] S. A. Vinogradov and N. A. Širokov, *Zeros of analytic functions with a derivative in H^1*, Zap. Naučn. Sem. Leningrad. Otdel. Mat. Inst. Steklov. (LOMI), **30** (1972), 154–157 (Russian); English transl. in Journ. of Soviet Math., **4** (1975), 434–440.

[12] S. A. Vinogradov, *Multipliers of power series with sequence of coefficients from l^p*, Zap. Naučn. Sem. Leningrad. Otdel. Mat. Inst. Steklov. (LOMI), **39** (1974), 30–39 (Russian); English transl. in Journ. of Soviet Math., **8** (1977), 20–27.

[13] S. A. Vinogradov and V. P. Havin, *Free interpolation in H^∞ and in certain other classes of functions. I*, Zap. Naučn. Sem. Leningrad. Otdel. Mat. Inst. Steklov. (LOMI), **47** (1974), 15–54, 184–185, 191 (Russian); English transl. in Journ. of Soviet Math., **9** (1978), 137–171.

[14] S. A. Vinogradov, *Properties of multipliers of integrals of Cauchy–Stieltjes type, and some problems of factorization of analytic functions*, In: Mathematical programming and related questions (Proc. Seventh Winter School, Drogobych, 1974), Theory of functions and functional analysis, 5–39. Central Ekonom.-Mat. Inst. Akad. Nauk SSSR, Moscow, 1976 (Russian); English transl. in Amer. Soc. Transl., (2) **115** (1980), 1–32.

[15] S. A. Vinogradov, *Bases of exponential functions and free interpolation in Banach spaces with L^p-norm*, Zap. Naučn. Sem. Leningrad. Otdel Mat. Inst. Steklov. (LOMI), **65** (1976), 17–68, 203 (Russian); English transl. in Journ. of Soviet Math., **16** (1981), 1035–1060.

[16] S. A. Vinogradov and V. P. Havin, *Free interpolation in H^∞ and in certain other classes of functions. II*, Zap. Naučn. Sem. Leningrad. Otdel. Mat. Inst. Steklov. (LOMI), **56** (1976), 12–58, 195 (Russian); English transl. in Journ. of Soviet Math., **14** (1980), 1027–1065.

[17] S. A. Vinogradov, *Convergence almost everywhere of the Fourier series of functions in L^2, and the behavior of the coefficients of uniformly convergent Fourier series*, Dokl. Akad. Nauk SSSR, **230** (1976), no. 3, 508–511 (Russian); English transl. in Soviet Math. Dokl, **17** (1976), no. 5, 1323–1327.

[18] S. A. Vinogradov, *Continuity of perturbed integral operators, Cauchy type integrals, maximal operators*, Zap. Nauchn. Sem. Leningrad. Otdel. Mat. Inst. Steklov. (LOMI), **73** (1977), 24–34, 229–230 (1978) (Russian); English transl. in Journ. of Soviet Math., **34** (1986), 2033–2039.

[19] S. A. Vinogradov, *Multiplication and division of power series with coefficient sequence in l^p*, Zap. Nauch. Semin. Leningrad. Otdel. Mat. Inst. Steklov. (LOMI), **81** (1978), 260–262 (Russian); English transl. in Journ. of Soviet Math., **26** (1984), no. 5.

[20] S. A. Vinogradov, V. P. Havin, *Letter to the editors*, Zap. Nauch. Semin. Leningrad. Otdel. Mat. Inst. Steklov. (LOMI), **92** (1979), 317 (Russian).

[21] S. A. Vinogradov, *Multiplicative properties of power series with a sequence of coefficients from l^p*, Dokl. Akad. Nauk SSSR, **254** (1980), no. 6, 1301–1306 (Russian); English transl. in Soviet Math. Dokl, **22** (1980), no. 2, 560–565.

[22] S. A. Vinogradov, *Simultaneous interpolation by analytic functions with small sets of singularities*, In: "All-Union symposium on approximation theory of functions in complex region". Summaries of reports. Ufa, 1980, p. 27 (Russian).

[23] S. V. Hruščev and S. A. Vinogradov, *Free interpolation in the space of uniformly convergent Taylor series*, In: Complex analysis and spectral theory (Leningrad, 1979/1980), Lecture Notes in Math., **864** (1981), 171–213.

[24] S. V. Hruščev and S. A. Vinogradov, *Inner functions and multipliers of Cauchy type integrals*, Ark. Mat. **19** (1981), no. 1, 23–42.

[25] S. A. Vinogradov and A. M. Kotočigov, *Some remarks on interpolation by analytic functions that belong to Besov spaces B_p^0*, Zap. Nauchn. Sem. Leningrad. Otdel. Mat.

Inst. Steklov. (LOMI), **113** (1981), 215–217, 269 (Russian); English transl. in Journ. of Soviet Math., **22** (1983), 1840–1841.

[26] S. A. Vinogradov, *On free interpolation in Bergman spaces*, Zap. Nauchn. Sem. Leningrad. Otdel. Mat. Inst. Steklov. (LOMI), **113** (1981), 208–211, 268 (Russian); English transl. in Journ. of Soviet Math., **22** (1983), 1835–1837.

[27] S. A. Vinogradov, E. A. Gorin, and S. V. Hruščev, *Free interpolation in H^∞ in the sense of P. Jones*, Zap. Nauchn. Sem. Leningrad. Otdel. Mat. Inst. Steklov. (LOMI), **113** (1981), 212–214, 268 (Russian); English transl. in Journ. of Soviet Math., **22** (1983), 1838–1839.

[28] S. A. Vinogradov, *A strengthening of Kolmogorov's theorem on the conjugate function and interpolational properties of uniformly converging power series*, Trudy Mat. Inst. Steklov, **155** (1981), 7–40, 183 (Russian); English transl. in Proc. Steklov Inst. Math. **155** (1981), 3–37.

[29] S. A. Vinogradov and P. Stoilov, *Estimates of the norm of holomorphic components of meromorphic functions in the unit disc*, C. R. Acad. Bulgare Sci., **34** (1981), no. 7, 931–934 (Russian).

[30] S. A. Vinogradov and S. E. Rukshin, *On the free interpolation of germs of analytic functions in Hardy spaces*, Zap. Nauchn. Sem. Leningrad. Otdel. Mat. Inst. Steklov. (LOMI), **107** (1982), 36–45, 229 (Russian); English transl. in Journ. of Soviet Math., **36** (1987), 319–325.

[31] S. A. Vinogradov, *Free interpolation in spaces of analytic functions*, Avtoreferat dissertatsii na soiskanie uchenoy stepeni doktora fiziko-matematicheskih nauk, Leningradskoe Otdelenie Matematicheskogo Instituta im. V.A.Steklova AN SSSR, 1983 (Russian); English transl. in this volume.

[32] S. A. Vinogradov, *Free interpolation in spaces of analytic functions*, Doctor Thesis, Steklov Inst. Math., Leningrad, 1983 (Russian).

[33] S. A. Vinogradov, *Some remarks on free interpolation by bounded and slowly growing analytic functions*, Zap. Nauchn. Sem. Leningrad. Otdel. Mat. Inst. Steklov. (LOMI), **126** (1983), 35–46 (Russian); English transl. in Journ. of Soviet Math. **27** (1984), 2450–2458.

[34] S. A. Vinogradov, *Spaces of analytic multipliers and free interpolation on thin sets*, ICM, Warszawa. 1982, 1983, vol. V (Russian).

[35] S. A. Vinogradov, *Analogues of L. Carleson's embedding theorem for some spaces of analytic functions*, Zap. Nauchn. Sem. Leningrad. Otdel. Mat. Inst. Steklov. (LOMI), **141** (1985), 144–148, 190 (Russian); English transl. in Journ. of Soviet Math., **37** (1987), 1347–1349.

[36] S. A. Vinogradov, *Multiplicative properties of l^p_A*, Lecture Notes in Math., **1043** (1987), 572–574.

[37] S. A. Vinogradov, D. A. Vladimirov, M. K. Gavurin, S. M. Ermakov, B. M. Makarov, G. I. Natanson, M. Z. Solomyak, D. K. Faddeev, V. P. Khavin, *Grigoriĭ Mikhaĭlovich Fikhtengolts (on the 100th anniversary of his birth)*, Vestnik Leningrad. Univ. Mat. Mekh. Astronom, **3**(1988), 3–6 (Russian).

[38] S. A. Vinogradov, *The Blaschke condition and sets of uniqueness for some subclasses of Hardy spaces*, Vestnik Leningrad. Univ. Mat. Mekh. Astronom. 1990, vyp. 2, 18–22, 120 (Russian); English transl. in Vestnik Leningrad Univ. Math. **23** (1990), no. 2, 18–22.

[39] S. A. Vinogradov, *Analogues of L. Carleson's embedding theorem for some Hilbert spaces of analytic functions*, Zap. Nauchn. Sem. S.-Petersburg. Otdel. Mat. Inst. Steklov. (POMI), **206** (1993), 40–54, 174 (Russian); English transl. in J. Math. Sci., **80** (1996), no. 4.

[40] S. A. Vinogradov, *Multiplicative properties of some L^p-spaces of analytic functions*, Summaries of reports of the All-Russian seminar "Function theory". Syktyvkar GU, 1993, pp. 10–11 (Russian).

[41] S. A. Vinogradov, *Multiplicative properties of l^p_A*, Lecture Notes in Math., **1574** (1994), Part II, 283–285.

[42] S. A. Vinogradov, *Multiplication and division in the space of analytic functions with an area-integrable derivative, and in some related spaces*, Zap. Nauchn. Sem. S.-Petersburg. Otdel. Mat. Inst. Steklov. (POMI), **222** (1995), 45–77, 308 (Russian); English transl. in J. Math. Sci., **87** (1997), no. 5.

[43] S. A. Vinogradov and A. N. Petrov, *The converse to a theorem on the action of analytic functions, and multiplicative properties of some subclasses of the Hardy space H^∞*, Zap. Nauchn. Sem. S.-Petersburg. Otdel. Mat. Inst. Steklov. (POMI), **232** (1996), 50–72, 214 (Russian); English transl. in J. Math. Sci.

Operator Theory:
Advances and Applications, Vol. 113

Interpolation Problems for Analytic Functions Continuous in the Closed Disk and for Functions whose Sequence of Coefficients is in l^p

S. A. Vinogradov

Problems of the following type are of great interest in Analysis. Let $X(K)$ be a class of functions defined on a set K, and let $E \subset K$; let Ψ be a function defined on E; the problem is to know whether Ψ agrees on E with an element of $X(K)$. For many concrete and interesting situations this question has been given a satisfactory answer. There are a lot of examples illustrating this general setting, but we are going to describe only two, directly related to our subject.

1°. Let K be the unit disc $\mathbb{D} = \{z : |z| < 1\}$ of the complex plane, and let $X(K)$ be the class H^∞ of all functions regular and bounded in $\mathbb{D}$. Then the above general problem is equivalent to the well-known Nevanlinna-Pick problem [Akh].

2°. Another interesting realization of this general problem arises when K is the set of integers, $X(K)$ is the class of all sequences $\{\hat{f}(n)\}_{n=-\infty}^{+\infty}$ where

$$\hat{f}(n) = \frac{1}{2\pi} \int_{-\pi}^{\pi} f(t) e^{-int} dt \quad (n = 0, \pm 1, \pm 2, \dots)$$

for a continuous 2π-periodic function.

In that case the above general problem on the extension of a function Ψ is closely related to the trigonometric moment problem. In spite of their completeness and depth, known results related to the examples we have mentioned yield no practical way to check whether a specific Ψ defined on E can be extended to an element of $X(K)$ (see [B, Ch. II, §11], and [Akh]). Moreover, these results suggest that it is hardly possible to characterize the restrictions to E of the functions in $X(K)$ in more or less palpable terms (at least if we are interested in arbitrary sets E).

Therefore, it is reasonable to modify the statement of the problem as follows: denote by S the operator taking any $f \in X(K)$ to its restriction to E, and consider a class $Y(E)$ of functions defined on E; what conditions imposed on E imply $S(X(K)) = Y(E)$?

Of course, this question is interesting only if $Y(E)$ is a sufficiently simple or well studied function class. We turn again to the examples 1° and 2° keeping in mind our new question.

1′. Let $Y(E)$ be the class $l^\infty(E)$ of all functions bounded on E. D. Newman and L. Carleson proved the following

Theorem ([N, C]). *The relation* $S(H^\infty) = l^\infty(E)$ *is true if and only if*

$$\inf_{\xi\in E} \prod_{\eta\in E_\xi} \left|\frac{\xi-\eta}{1-\bar\xi\eta}\right| > 0, \qquad (*)$$

where $E_\xi = \{\eta \in E : \eta \neq \xi\}$.

Note that at present there are a lot of papers related to the Newman-Carleson theorem in one way or another where $(*)$ plays an essential role. We mention here only [Na, Ka, ShaSh], the closest to this thesis. More information can be found in [H].

2′. Let E consist of integers, and let $Y(E) = l^2(E)$, the class of all square summable families $\{x_n\}_{n\in E}$ of complex numbers. The well-known Banach theorem asserts (see [Ba]) that

$$S(X(K)) = l^2(E)$$

where $X(K)$ is the same as in 2° provided $E = -E$, and

$$\inf_{\substack{m,k\in E\\ 0<k<m}} \frac{m}{k} > 1.$$

Now, we turn to a new example.

3′. Let $K = \overline{\mathbb{D}}$ be the closed unit disc, $X(K) = C_A$ the class of all functions regular in $\mathbb{D}$ and continuous in $\overline{\mathbb{D}}$. Suppose E is a closed subset of the unit circle $\partial\mathbb{D} = \{z : |z| = 1\}$. W. Rudin and Carleson [Ho] proved that $S(C_A) = C(E)$ (= the class of all functions continuous on E) if and only if mes $E = 0$ (mes denotes the Lebesgue measure on $\partial\mathbb{D}$).

Note that in this case any palpable description of functions representable as $S(f)$ $(f \in C_A)$ is problematic if mes $E > 0$.

The present thesis is devoted to interpolation problems related mainly to the three theorems stated above, namely, to the Newman-Carleson theorem (example 1′), the Banach theorem (example 2′) and the Rudin-Carleson theorem (example 3′).

To solve these problems, tools from Functional Analysis are used (first of all the well-known Banach theorems on solvability of linear equations in Banach spaces).

The thesis consists of the Introduction, two chapters, and appendices where the proofs of purely technical facts are collected.

We will use the following notation: $\mathbb{D}$ denotes the open unit disc of the complex plane centered at the origin, $\overline{\mathbb{D}}$ is its closure; $l^p_A(1 \le p \le +\infty)$ is the space of all functions analytic in $\mathbb{D}$ with $\{\hat f(n)\}_{n=0}^\infty \in l^p$ where

$$\hat f(n) = \frac{f^{(n)}(0)}{n!}, \quad n = 0, 1, 2, \ldots.$$

Let $E \subset \mathbb{D}$; then S will denote the operator taking any function analytic in $\mathbb{D}$ to its restriction to E; $C(E)$ is the set of all functions uniformly continuous on E.

We turn now to a survey of the thesis. For the sake of brevity and clarity, here we often give our theorems a less general form than in the thesis.

Chapter I
Interpolation of power series whose sequence of coefficients is in l^p

In this chapter interpolation problems are considered for functions in $l_A^p (1 \leq p \leq +\infty)$. These problems are of the same type as that in example 1′. If $p \neq 2$, then l_A^p differs from the classes whose interpolation was studied in [N, C, Na, Ka, ShaSh]. Namely, it is characterized by a norm defined in terms of the moduli of coefficients only and whose relationship with the function values (to be interpolated) is rather implicit. For $p = 2$ the corresponding interpolation problem was solved completely in [ShaSh].

In §2 of the first chapter we show that there is no infinite $E \subset \mathbb{D}$ such that

$$S(l_A^1) = C(E) \tag{1}$$

Note that the unit circle contains infinite and even perfect sets E satisfying (1); see [KS].

In §2 restrictions of functions $f \in l_A^1$ to some infinite subsets of $\mathbb{D}$ are studied.

Now we state the principal result of §3. Let bv be the space of all sequences $x = \{x_k\}_{k=0}^{\infty}$ of complex numbers with finite norm

$$\|x\|_{bv} = \lim_{n \to \infty} |x_n| + \sum_{n=0}^{\infty} |x_n - x_{n+1}|;$$

denote by V_A the space of all functions f regular in $\mathbb{D}$ and such that

$$\sup_{0<p<1} \int_{-\pi}^{\pi} |f'(pe^{it})| dt < +\infty.$$

Put $B_\lambda = \{z \in \mathbb{D} : |1 - z| \leq \lambda(1 - |z|)\}$, $\lambda \geq 1$. It is known that $V_A \subset l_A^1$ [Ho].

Theorem 1. *Let $\{\xi_k\}_{k=0}^{\infty}$ be a sequence in $\mathbb{D}$ satisfying the following conditions: $\xi_k \neq \xi_m$ if $k \neq m$, and $|1 - \xi_{k+1}| \leq |1 - \xi_k|$, $k = 0, 1, \ldots$. If $\{\xi_k\}_{k=0}^{\infty} \subset B_\lambda$, then the following are equivalent:*

1. *$S_1(l_A^1) = bv$;*
2. *$S_1(V_A) = bv$;*
3. *$\{\xi_k\}_{k=0}^{\infty}$ satisfies $(*)$;*

S_1 denotes the operator defined by $S_1(f) = \{f(\xi_k)\}_{k=0}^{\infty}$, $f \in l_A^1$.

In §4 several assertions are proved closely related to the Newman-Carleson theorem (example 1′) and the Rudin-Carleson theorem (example 3′). In particular, the following result is obtained.

Theorem 2. *If $E \subset \overline{\mathbb{D}} \setminus \{1\}$ satisfies*

$$\sup_{\substack{\eta, \xi \in E \\ \eta \neq \xi, |1-\eta| < |1-\xi|}} \frac{|1-\eta|}{|1-\xi|} < 1,$$

then any Ψ uniformly continuous on E can be extended to a function f defined on the extended complex plane $\widehat{\mathbb{C}}$, regular in $\widehat{\mathbb{C}} \setminus \{1\}$ and such that $f|\overline{\mathbb{D}}$ is continuous, and

$$\hat{f}(n) = o\Big(\frac{1}{n}\Big) \quad (n \to +\infty),$$

where $\hat{f}(n) = f^{(n)}(0)/n!$.

In §5 results of §3 are applied to interpolation in l^p_A $(1 < p \leq +\infty)$. Here we quote two principal results of §5.

Let $h^p(E)$, $1 < p \leq +\infty$, be the class of all functions Ψ defined on a set E and such that

$$\sum_{\eta \in E} |\Psi(\eta)|^p (1 - |\eta|)^{p-1} < +\infty \quad \text{if} \quad 1 < p < +\infty,$$

and

$$\sup_{\eta \in E} |\Psi(\eta)|(1 - |\eta|) < +\infty \quad \text{if} \quad p = +\infty.$$

Theorem 3. *Suppose $E \subset B_\lambda$ for some $\lambda \geq 1$ and $1 < p \leq +\infty$. Then the following are equivalent.*

1. *$S(l^p_A) = h^p(E)$;*
2. *E satisfies $(*)$.*

Theorem 4. *If $E \subset \mathbb{D}$ satisfies*

$$\sup_{\substack{\xi, \eta \in E \\ \xi \neq \eta, |\eta| < |\xi|}} \frac{1 - |\xi|}{1 - |\eta|} < 1,$$

then $S(l^p_A) = h^p(E)$ $(1 < p < +\infty)$.

In §6 a non-zero function $f \in \bigcap_{p>2} l^p_A$ is constructed such that $f(z_k) = 0$, $k = 0, 1, 2, \ldots$, $z_k \neq z_m$ for $k \neq m$, $|z_k| < 1$, and

$$\sum_{k=1}^{\infty} (1 - |z_k|) = +\infty.$$

Chapter II
Norms of some linear operators in $C_A(G)$ and the Banach and Rudin-Carleson interpolation theorems

Let E be a closed non-empty subset of $\partial\mathbb{D}$, $E \neq \partial\mathbb{D}$. Denote by $A_C(E)$ the set of all functions f defined on the whole extended complex plane $\widehat{\mathbb{C}}$, regular in $\widehat{\mathbb{C}} \setminus E$ and such that $f|\overline{\mathbb{D}}$ is continuous.

Chapter II is devoted to interpolation problems for $A_C(E)$ in the spirit of the Banach theorem (example 2′) and the Rudin-Carleson theorem (example 3′). Moreover, the behaviour of the coefficients of the functions in $A_C(\{1\})$ is studied. We need the following notation. For a simply connected domain $G \subset \widehat{\mathbb{C}}$, we denote by $C_A(G)$ the space of all functions continuous in $\overline{G}$ and analytic in G; $C_A(G)$ is endowed with the norm

$$\|f\|_{C_A(G)} = \max_{z\in\overline{G}} |f(z)|, \quad f \in C_A(G).$$

For a set Q of non-negative integers, we put $Q_n = \{k \in Q : k \leq n\}$, $n = 0, 1, \ldots$; $d = \{d_k\}_{k\in Q}$ will denote a family of positive numbers, and $l^p(Q_n, d)$ $(0 < p < 2)$ is the space of families $x = \{x_k\}_{k\in Q_n}$ of complex numbers with the norm $\|x\|_{(p,d)} = \left(\sum_{k\in Q_n} |x_k|^p d_k^2\right)^{1/p}$ (if $0 < p < 1$, then $\|x\|_{(p,d)}$ is not a norm in the usual sense). Suppose $G \supset \mathbb{D}$ and denote by $J^p_{n,G}$ the linear operator from $C_A(G)$ to $l^p(Q_n, d)$ defined by

$$J^p_{n,G} f = \left\{\frac{\hat{f}(k)}{d_k}\right\}_{k\in Q_n}, \quad f \in C_A(G), \quad n = 0, 1, \ldots,$$

where $\hat{f}(n) = \frac{f^{(n)}(0)}{n!}$. We also need the norm $\|J^p_{n,G}\|$ of $J^p_{n,G}$:

$$\|J^p_{n,G}\| = \sup_{\|f\|_{C_A(G)}\leq 1} \left(\sum_{k\in Q_n} |\hat{f}(k)|^p d_k^{2-p}\right)^{1/p} \quad (0 < p < 2).$$

The main result of §2 of Chapter II is

Theorem 5. *Let*

$$G = \widehat{\mathbb{C}} \setminus (\{z \in \mathbb{C} : 1 \leq |z| \leq 1 + \varepsilon\} \cap \{z \in \mathbb{C} : |\arg z| \leq \varepsilon\})$$

$0 < \varepsilon < \pi$, *and let* $p \in (0, 2\pi)$. *Then there is a constant* $M = M(\varepsilon, p) \in (0, +\infty)$ *such that*

$$M\left(\sum_{k\in Q_n} d_k^2\right)^{\frac{1}{p}-\frac{1}{2}} \leq \|J^p_{n,Q}\| \leq \left(\sum_{k\in Q_n} d_k^2\right)^{\frac{1}{p}-\frac{1}{2}}, \quad n = 0, 1, \ldots.$$

Theorem 5 is a generalization of a theorem proved by Paley [P]. Theorem 5 is used to study the Maclaurin coefficients of functions in $A_C(\{1\})$. As an example we state

Theorem 6. *Suppose a family* $\{d_k\}_{k\in Q}$ *of positive numbers satisfies* $\sum_{k\in Q} d_k^2 = +\infty$. *Then there is a function* $f \in A_C(\{1\})$ *such that*

$$\sum_{k\in Q} |\hat{f}(k)|^p d_k^{2-p} = +\infty$$

for all $p \in (0,2)$.

Note that §2 of Chapter II is closely related to the articles [P, S, H1]. We generalize all results of [Ma1, Ma2] on the space of functions regular off $[1,+\infty)$ and continuous on $\overline{\mathbb{D}}$ to $A_C(\{1\})$. Some results of §2, Ch. II are applicable to the space of all functions regular in $\widehat{\mathbb{C}}\setminus\{1\}$ whose Maclaurin series converges uniformly in $\overline{\mathbb{D}}$.

In §3 of Ch. II some generalizations of the interpolation theorems due to Banach and Rudin-Carleson are given (examples 2′, 3′). The main result is as follows.

Theorem 7. *Let* E *be a closed subset of* $\partial\mathbb{D}$ *of zero length,* Q *an Hadamard set of non-negative integers (i.e.* $\inf\{\frac{m}{n} : m,n \in Q, 0 < m < n\} > 1$*). Then for any function* Ψ *continuous on* E *and for any square summable family* $x \in \{x_k\}_{k\in Q}$ *of complex numbers there is a function* $f \in A_C(E)$ *such that* $f(z) = \psi(z)$, $z \in E$, *and* $\hat{f}(k) = x_k$, $k \in Q$.

The main results of this thesis were published in [1, 2, 3][1]

I am deeply gratefull to my adviser Viktor Petrovich Havin for posing the problems and for his permanent interest in my work.

References

[Akh] N. I. Akhiezer, *The classical moment problem*, Fizmatgiz, Moscow, **1961** (Russian); English transl., Oliver & Boyd, Edinburgh, **1965**.

[B] N. K. Bari, *Trigonometrical Series*, Fizmatgiz, Moscow, **1961** (Russian); English transl., *A treatise on trigonometrical series*, vols. I–II, Pergamon Press, Oxford, **1964**.

[Ba] S. Banach, *Über einige Eigenschaftern der lacunären trigonometrischen Reihen*, Studia Math., **2** (1930), 207–220.

[C] L. Carleson, *An interpolation problem for bounded analytic functions*, Amer. J. of Math., **80** (1958), no. 4, 921–930.

[H] V. P. Havin, *Spaces of analytic functions*, in the book "Mathematical Analysis. 1964" Moscow, VINITI, **1966**. (Russian)

[H1] V. P. Havin, *On the norms of some operations in the space of polynomials*, Vestnik Leningr. Univ. Mat. Mekh. Astronom. **4** (1959), no. 19, 47–59. (Russian)

[Ho] K. Hoffman, *Banach spaces of analytic functions*, Prentice-Hall, Englewood Cliffs, N.J., **1962**.

[Ka] V. Kabaila, *On the interpolation of functions in the class* H_δ, Uspekhi Mat. Nauk, **13** (1958), no. 1, 181–188. (Russian)

[1]These numbers refer to List of Publications of S.A.Vinogradov, this volume.

[KS] J.-P. Kahane and R. Salem, *Ensembles parfaits et séries trigonométriques*, Hermann, Paris, **1963**.

[Ma1] A. S. Makhmudov, *On the Fourier and Taylor coefficients of contiuous functions*, in the book: "Questions of functional analysis and its applications", Baku, **1965**, 103–128. (Russian)

[Ma2] A. S. Makhmudov, *On the Fourier and Taylor coefficients of continuous functions*, Izv. AN Azerb. SSR, **1964**, no. 2–4, 60–75. (Russian)

[N] D. Newman, *Interpolation in* H^∞, Trans. Amer. Math. Soc., **92** (1959), no. 3, 501–507.

[Na] A. G. Naftalevich, *On the interpolation of functions of bounded characteristics*, Uch. Zapiski Vilnius. Univ., Math., Phys. and Chem., **5** (1956), 5–27. (Russian)

[P] R. E. A. C. Paley, *A note on power series*, Journ. Lond. Math. Soc., **7** (1932), 122–130.

[S] S. B. Stechkin, *An extremum problem for polynomials*, Izv. AN SSSR, Ser. Mat., **20** (1956), 765–774. (Russian)

[ShaSh] H. S. Shapiro and A. L. Shields, *On some interpolation problems for analytic functions*, Amer. J. of Math., **83** (1961), no. 3, 513–532.

Translated by V. P. Havin

Russian original: Avtoreferat dissertatsii
na soiskanie uchenoy stepeni
kandidata fiziko-matematicheskih
nauk, pp. 1–9,
Leningradskiy Gosudarstvenniy
Universitet, 1968.

Operator Theory:
Advances and Applications, Vol. 113

Free Interpolation in Spaces of Analytic Functions

S. A. Vinogradov

§1. General survey

1.1. Values taken by an analytic function at different parts of its domain are rigidly interrelated by the classical uniqueness theorem. This is why a function defined on an infinite part Λ of a domain $G \subset \mathbb{C}$ and taken at random cannot be interpolated, as a rule, by a function analytic in G ($f \in A(G)$, for short). In other words, denoting by R_Λ the operator of restriction to Λ, we may assert that $R_\Lambda(A(G))$ is, as a rule, a very special subset of $s(\Lambda)$, the set of all complex valued functions defined on Λ; $R_\Lambda(A(G))$ resists any description in clear terms. Another way to express this feeling is to say that the equation $R_\Lambda(x) = f$ (with a given $f \in s(\Lambda)$ and "unknown" $x \in A(G)$) is strongly overdetermined, it is solvable only under rare circumstances.

The case of Λ with no limit points in G is an exception. It is well known that in this case $R_\Lambda(A(G)) = s(\Lambda)$; this situation is an illustration of a general scheme due to M. Eidelheit and B. M. Makarov applicable to a large class of "freely solvable" moment problems in linear topological spaces.

Historically, this is probably the first example of a phenomenon which is the theme of our thesis, namely, the phenomenon of *free interpolation* by analytic functions. The rigid interconnection of values of an analytic functions is destroyed on a "sparse" $\Lambda \subset G$; functions of class $A(G)$ conceal their analyticity when considered on Λ only, any function defined on Λ being the trace of an element of $A(G)$. In short, interpolation by analytic functions on Λ turns out to be *free*. We are not going to define *formally* the notion of "free interpolation" in any way. The words in quotation marks only express our intention to study the subspaces $\mathcal{E}$ of $A(G)$ and the infinite sets $\Lambda \subset G$ such that $R_\Lambda(\mathcal{E})$ can be described in simple terms not depending on the notion of analyticity.

1.2. What happens if we replace $R_\Lambda(A(G))$ by $R_\Lambda(\mathcal{E})$ where $\mathcal{E}$ is a part of $A(G)$ determined, say, by some growth or smoothness restrictions imposed near the boundary? If

$$G = \mathbb{D} \stackrel{\text{def}}{=} \{ z \in \mathbb{C} : |z| < 1 \} \quad \text{or} \quad G = \mathbb{C}_+ \stackrel{\text{def}}{=} \{ z \in \mathbb{C} : \operatorname{Im} z > 0 \},$$

then $\mathcal{E}$ can be singled out by the behaviour of Maclaurin coefficients, resp. by the Fourier transform. Such restrictions make $R_\lambda(\mathcal{E})$ "less free" and intensify its dependence on the complex structure of G.

Suppose, for example, $G = \mathbb{D}$, $\mathcal{E} = H^\infty \stackrel{\text{def}}{=} \{ f \in A(\mathbb{D}) : \sup_{|z|<1} |f(z)| < +\infty \}$. In this case, clearly, $R_\Lambda(H^\infty) \subset l^\infty(\Lambda) \stackrel{\text{def}}{=} \{ x \in s(\Lambda) : \sup_{\lambda\in\Lambda} |x(\lambda)| < +\infty \}$ for any $\Lambda \subset \mathbb{D}$. As a rule, $R_\Lambda(H^\infty)$ is essentially smaller than $l^\infty(\Lambda)$, even for discrete Λ's satisfying the Blaschke condition $\sum_{\lambda\in\Lambda}(1-|\lambda|) < +\infty$. The description of $R_\Lambda(H^\infty)$ supplied by the classical results on the Nevanlinna-Pick problem is very complicated and rather cumbersome (for Λ infinite).

At the end of the fifties an important discovery was made as a result of the work by W. Hayman, D. Newman, and, first of all, L. Carleson: the equality

$$R_\Lambda(H^\infty) = l^\infty(\Lambda) \tag{1.1}$$

was proved for a sufficiently "sparse" $\Lambda \subset \mathbb{D}$. On such Λ's the interpolation by bounded analytic functions becomes "free"; no obstacle for an arbitrary $x \in s(\Lambda)$ to be interpolated by elements of H^∞ exists anymore, provided x is bounded. The "sparseness" of Λ means that

$$\inf\Big\{ \prod_{\xi\in\Lambda\setminus\{\lambda\}} \Big|\frac{\lambda-\xi}{1-\lambda\bar{\xi}}\Big| : \lambda \in \Lambda \Big\} > 0 \tag{NC}$$

(the Newman-Carleson condition). The statement of Carleson's result obtained in 1958 follows.

The Carleson theorem. *For any* $\Lambda \subset \mathbb{D}$, *relation* (1.1) *is equivalent to* (NC).

The Newman-Carleson condition can be given a clear geometric form applicable to many concrete sets Λ.

This example of free interpolation (for a set $\mathcal{E} \subsetneqq A(G)$) is historically not the first. We have started with just this example for the following reasons.

First, the statement of the Carleson theorem related to H^∞ (one of the most important spaces) is both simple and complete.

Second (and this is our main point), important connections have been established of the Carleson theorem (which is the first conclusive result on free interpolation in a *Banach* space of analytic functions) with various problems of Function Theory and Operator Theory. The Carleson theorem and its proof have attracted many analysts and initiated a new direction of research.

Let us give more examples (neglecting the chronology) connected with the Carleson interpolation theorem for H^∞ and directly related to the subject of the present thesis. But first we want to emphasize once again that in the extremely vast area dealing with interpolation by analytic functions we are interested just in the free interpolation. Therefore we do not mention important classical topics exposed in the well-known monographs by V. L. Goncharov, J. Walsh, A. O. Gelfond, M. A. Evgrafov et al.

1.3. The study of interpolation in the Hardy classes H^p was strongly influenced by the results of H. Shapiro and A. Shields (1962). The (NC)-condition turned out to be decisive in this situation as well, although the very notion of "freedom of interpolation" is not as clear for finite values of p as it is for $p = \infty$. Not analyzing this notion in detail, we only mention that "freedom" means here the "ideal character" of the space $R_\Lambda(H^p)$ (in the sense of the theory of symmetric function spaces). In particular, this "ideal character"[1] (and even weaker assumptions) imply (NC) which, in its turn, yields an explicit description of $R_\Lambda(H^p)$.

1.4. Suppose elements of $\mathcal{E}(\subset A(G))$ can be extended in a natural way to the boundary ∂G of G. Then a free interpolation problem on the subsets of ∂G arises by itself. A classical example is the Rudin-Carleson theorem. To state it we denote by C_A the so-called disc-algebra, i.e., the algebra of all functions continuous in $\overline{\mathbb{D}}$ and analytic in $\mathbb{D}$.

The Rudin-Carleson theorem. *Let Λ be a closed subset of the unit circle $\mathbb{T}$. The following are equivalent:*

1°. $R_\Lambda(C_A) = C(\Lambda)$;

2°. $\operatorname{mes}\Lambda = 0$.

Thus, the elements of C_A cannot be told apart from the functions which are merely continuous on Λ if Λ is compact and of zero length (and only in that case).

1.5. Many researchers, starting with N. Wiener and R. Paley, S. Szidon, and S. Banach studied free interpolation in the spaces $\hat{L}^p(\mathbb{T})$ $(1 \le p \le +\infty)$ and $\hat{C}(\mathbb{T})$, where

$$\hat{f}(z) \stackrel{\text{def}}{=} \frac{1}{2\pi}\int_{-\pi}^{\pi} f(e^{it})e^{-izt}dt, \quad z \in \mathbb{C}, \quad f \in L^1(\mathbb{T}),$$

$$\hat{X} \stackrel{\text{def}}{=} \{\hat{f} : f \in X\}, \quad X \subset L^1(\mathbb{T}).$$

(For $p = 2$ an essential progress in this direction has been made recently by B. S. Pavlov, S. V. Hrushchev, and N. K. Nikol'skiĭ (1979)). Let us stress an important difference between the work related to $p = 2$ and that dealing with $p \neq 2$. For $p = 2$ an old tradition assumed the set Λ of "knots" to be a more of less arbitrary part of $\mathbb{C}$, whereas in almost all papers on free interpolation in $\hat{L}^p(\mathbb{T})$ for $p \neq 2$ the set Λ used to consist of *integers* (real numbers). The research methods were also different: the case $p = 2$ was often treated from a "complex analytic" point of view, and for $p \neq 2$ "real" methods of the theory of trigonometrical series were exploited. It is in this context that S. Banach discovered (in some interesting particular cases) the relationship of free interpolation with problems concerning bases of functions. This relationship can be illustrated by the following result.

The Banach theorem. *Let $\Lambda \subset \mathbb{Z}$. Then equality $R_\Lambda(\hat{C}(\mathbb{T})) = l^2(\Lambda)$ (where $l^2(\Lambda) \stackrel{\text{def}}{=} \{x \in s(\Lambda) : \|x\|_{l^2(\Lambda)} \stackrel{\text{def}}{=} (\sum_{\lambda\in\Lambda}|x(\lambda)|^2)^{1/2} < +\infty\}$) occurs if and only if there exists a number $c \in (0,1]$ such that*

$$c\|x\|_{l^2(\Lambda)} \le \frac{1}{2\pi}\int_0^{2\pi}\left|\sum_{k\in\Lambda} x(k)e^{ikt}\right|dt \le \|x\|_{l^2(\Lambda)}, \quad x \in l^2(\Lambda).$$

[1]This means that $(x \in R_\Lambda(H^p), y \in s(\Lambda), |y| \le |x|) \implies (y \in R_\Lambda(H^p))$.

The last assertion put into a modern form means that the family $(e^{i\lambda t})_{\lambda\in\Lambda}$ of exponentials is the basis of its closed linear span in $L^1(\mathbb{T})$ equivalent to the standard basis of $l^2(\Lambda)$. This theorem and the Carleman theorem on free interpolation by the derivatives of an infinitely differentiable function are among the first assertions illustrating the link indicated above, which has been studied since in many forms.

1.6. The results described in Sections 1.2–1.5 are the starting point of the systematic study (undertaken by the author) of *free interpolation* in a broader context. One of the main goals of this study was the following *question*:

What "sparseness" of the set Λ ($\Lambda \subset G$ or $\Lambda \subset \operatorname{clos} G$, depending on the situation) ensures the freedom of interpolation on Λ by functions of a given class $\mathcal{E} \subset A(G)$? The freedom can be very often interpreted as the ideal character of $R_\Lambda(\mathcal{E})$ or the relation $R_\Lambda(\mathcal{E}) = C(\Lambda)$ (if Λ is a compact set in $\mathbb{C}$).

The thesis answers this question for many concrete spaces $\mathcal{E}$ (some are described in §2 below). The norms defining our spaces $\mathcal{E}$ are related to the values of the elements of $\mathcal{E}$ in a very *indirect* way. This feature is the main difference between our work and the examples discussed in 1.2–1.4 on one hand, and the majority of studies on free interpolation of the last 20 years on the other.

Problems of free interpolation can be interpreted in many ways and are closely related to many other problems of Complex Analysis. Their solutions often result in new questions (and sometimes in the answers to these questions), which turn out to be even more important. We are going to describe the main problems considered in the thesis and related directly to the free interpolation by analytic functions. At the same time we characterize our principal results; precise statements are mostly postponed until §2.

1.7. Bases. Let X be a Banach space, K an X-valued function analytic in a domain $G \subset \mathbb{C}$. We consider the following general problem on basic families: to describe the countable sets $\Lambda \subset G$ for which the vector family $\vec{\varepsilon} = (K(\lambda))/\|K(\lambda)\|)_{\lambda\in\Lambda}$ is an unconditional (symmetric) basis in $\operatorname{span}_X \vec{\varepsilon}$, the closure in X of the linear span of $\vec{\varepsilon}$. It is not hard to see that this problem generates an *equivalent* interpolation problem in a space $\mathcal{E} \subset A(G)$ emerging naturally; moreover, this interpolation is free. And, reciprocally, for a given Banach space $\mathcal{E} \subset A(G)$ with an ideal (symmetric) $R_\lambda(\mathcal{E})$ there exists a space X and an X-valued function analytic in G such that $\vec{\varepsilon}$ is an unconditional (symmetric) basis of $\operatorname{span}_X \vec{\varepsilon}$ (the choice of X and K is, of course, not unique).

Thus, our initial problem on basic families has turned into a free interpolation problem.

This aspect of free interpolation is one of our main themes (Chapter 2). Chapter 2 contains, in particular, a complete description of symmetric basic families of rational fractions $(1-|\lambda|)^{1/p}/(1-\bar{\lambda}z)$ in l^p_A, $p \in [1,2)\cup(2,+\infty)$ ($l^p_A \overset{\text{def}}{=} \{ f \in A(\mathbb{D}) : (f^{(n)}(0)/n!)_{n\geq 0} \in l^p \}$); next, examples of non-symmetric basic families of exponentials in $L^p(0,a)$ ($a \in (0,\infty]$, $p \in (1,2) \cup (2,+\infty)$) are constructed (for the first time). Analogous results on families of exponentials in the spaces $L^p((0,a),t^\gamma dt)$ are obtained.

The relationship of free interpolation with unconditional bases can be exploited in two opposite directions: first, to deduce the property of a family to be an unconditional basis from interpolation theorems; second, to study the free interpolation itself.

1.8. The Carleson embedding theorem and its analogs. Free interpolation in H^∞ and the corona problem led Carleson to the following result (widely known as "the Carleson embedding theorem").

Theorem. *Let μ be a non-negative Borel measure in $\mathbb{D}$, and let $p \in (0,+\infty)$. Then the following are equivalent:*

1°. $f \in H^p \Rightarrow \int_{\mathbb{D}} |f|^p d\mu < +\infty$;

2°. $\sup_{\zeta\in\mathbb{T}, r\in(0,1]} \frac{1}{r}\mu\{\, z \in \mathbb{D} : |z-\zeta| < r\,\} < +\infty$.

This theorem plays an essential role in other parts of Function Theory as well. This is why a series of its analogs was found recently for other classes of analytic (harmonic) functions.

The thesis (Chapter 1) contains a new approach to the Carleson type embedding theorems. This approach essentially simplifies the proof of the Carleson theorem, making it more elementary. Moreover, it results in new embedding theorems; some of them are applied in Chapter 4 to the multipliers of analytic Besov classes and to free interpolation in some spaces of analytic functions. Our approach to the Carleson embedding theorem yields a new perception of the Hardy-Littlewood theorem on the maximal function in L^2 (including the multidimensional case), resulting in a purely "Hilbertian" proof that avoids any Besikovich-type covering theorem and the Marcinkiewicz interpolation theorem.

1.9. *Singular integral operators* are the ubiquitons weaponry of Function Theory; they appear in problems of free interpolation as well. For instance, some interpolation problems for C_A (in their dual form) require the integral operator with the Cauchy kernel in the space $M(\mathbb{T})$ of all complex Borel measures on $\mathbb{T}$, and its estimates due to Smirnov-Kolmogorov. These estimates do not suffice for free interpolation in the space U_A of all power series converging uniformly in $\mathbb{D}$. In Chapter 1 the following generalization of the Smirnov-Kolmogorov theorem is proved, which is of independent interest.

Theorem. *The function $K\vec{\mu}$ defined by*

$$(K\vec{\mu})(z) \stackrel{\text{def}}{=} \sum_{n\geq 0} z^n \int_{\mathbb{T}} \frac{d\mu_n(\zeta)}{\zeta - z}, \quad z \in \mathbb{D},$$

belongs to all Hardy classes H^p with $p \in (0,1)$ for an arbitrary sequence $\vec{\mu} = (\mu_n)_{n>0}$ in $M(\mathbb{T})$ satisfying $\sum_{n>0} \operatorname{var} \mu_n < +\infty$.

In fact, Chapter 1 contains a stronger result (a "weak type (1,1) estimate") generalizing the Kolmogorov inequality for the conjugate function. The proof of these results is based upon the deep Carleson theorem on the almost everywhere convergence of the Fourier series of an L^2-function.

Our estimate of K has already resulted in numerous applications; several authors have used it (during the last 5 years) to prove various results on U_A generalizing known properties of the disc algebra C_A.

1.10. Simultaneous free interpolation by analytic functions with a small set of singularities. Dealing with simultaneous interpolation, we are interested not only in the trace $R_{\Lambda_0} f$ of the interpolating function $f \in \mathcal{E}$ ($\mathcal{E} \subset C_A$) on a set $\Lambda_0 \subset \operatorname{clos} \mathbb{D}$, but also in the trace of its Fourier transform on a set $\Lambda_1 \subset \mathbb{Z}_+$. To be more precise (and general), we have in mind the following setting.

Let $\mathcal{E}$ be a (normed or linear topological) space of functions analytic in a domain $G_0 \subset \mathbb{C}$, and let J be a linear operator taking $\mathcal{E}$ to $A(G_1)$, where $G_1 \subset \mathbb{C}$ is another domain; we want to determine "the sparseness" of Λ_0 and Λ_1 ensuring a sufficiently simple description of the space

$$X(\Lambda_0, \Lambda_1) \stackrel{\text{def}}{=} \{ (R_{\Lambda_0} f, R_{\Lambda_1} Jf) : f \in \mathcal{E} \}$$

not depending on the complex structure of the domains G_0, G_1.

The following particular case is considered in detail in the thesis: $\mathcal{E} \subset U_A$,

$$(Jf)(z) = \frac{1}{2\pi} \int_0^{2\pi} f(e^{it}) e^{-izt} dt, \quad z \in \mathbb{C}, \quad \Lambda_0 \subset \mathbb{T}, \quad \Lambda_1 \subset \mathbb{Z}_+.$$

The methods of proofs used in the thesis suggest the following general principle: the unterpolation can always be achieved by a function analytic on the whole plane $\mathbb{C}$ off a "small" set of singularities. This principle is valid even for problems of simultaneous interpolation. The next statement is an illustration.

Theorem. *Let Λ_0 be a closed subset of $\mathbb{T}$, and let $\Lambda_1 \subset \mathbb{Z}_+$ satisfy the inequality*

$$\sup \left\{ \frac{1}{n+1} \sum_{k \in \Lambda_1, k \leq n} k : n \in \mathbb{Z}_+ \right\} < +\infty$$

(e.g., $\Lambda_1 = \{ 2^k : k \in \mathbb{Z}_+ \}$). Then for any $\varphi \in C(\Lambda_0)$ and any square summable numerical family x indexed by Λ_1 there is a function f such that

1°. $f \in A(\widehat{\mathbb{C}} \setminus \Lambda_0)$;
2°. $f|\mathbb{D} \in U_A$;
3°. $R_\Lambda f = \varphi$, $R_{\Lambda_1} \hat{f} = x$.

1.11. Spaces of analytic multipliers and free interpolation on "thin" sets. Let X be a linear space of functions defined on a set E (with pointwise algebraic operations). Suppose f is a function defined on E and enjoying the following property: $f \cdot g \in X$ whenever $g \in X$. Then f is called *a multiplier* of X; the set of all multipliers of X is denoted by mult X.

Any progress in the understanding of mult X entails a better understanding of X itself (cf. the usefulness of the dual space). We also remind the reader that the inclusion $f \in \operatorname{mult} X$ can sometimes be interpreted as the boundedness of an appropriate singular integral operator.

Free interpolation is a useful tool for the study of some interesting algebras of multipliers. The definition of $\operatorname{mult} X$ is rather implicit whence the properties of multipliers are usually hidden very deep; it is difficult even to understand the interrelations between a given function class Y and $\operatorname{mult} X$. (Is it true that $Y \subset \operatorname{mult} X$, or that $\operatorname{mult} X \subset Y$?) In such situations it is advisable to compare the trace spaces $R_\Lambda(Y)$ and $R_\Lambda(\operatorname{mult} X)$ on an appropriate $\Lambda \subset E$ on which free interpolation by elements of Y and $\operatorname{mult} X$ is possible.

A large part of Chapter 4 is devoted to $\operatorname{mult} l^p_A$ (recall that $l^p_A \stackrel{\text{def}}{=} \{f \in A(\mathbb{D}) : (f^{(n)}(0)/n!)_{n\geq 0} \in l^p\}$). This class occurs in many problems related to interpolation, bases, operator theory and Banach algebras. It is also of certain interest from the point of view of the general theory of L^p-multipliers playing an important role in modern Harmonic Analysis. A lot of papers are devoted to the algebras $\operatorname{mult} l^2_A$ and $\operatorname{mult} l^1_A$ coinciding, resp., with H^∞ and l^1_A. We will mainly be interested in $p \neq 1, 2, \infty$, i.e. in the cases not included into the classical theory. We note that there is a non-trivial relationship of $\operatorname{mult} l^p_A$ ($p \in (1,2)$) with the modern spectral operator theory in L^p-spaces; this link was discovered by V. V. Peller, G. Weiss, and R. Coifman.

Using the technique of free interpolation, we prove several theorems (Chapter 4) whose general meaning can be summarized as follows: as a rule, an inner non-rational function does not belong to $\operatorname{mult} l^p_A$ ($p \in (1,2)$). We also consider the well-known problem of division by an inner function in l^p_A. This problem turns out to be related to free interpolation in l^p_A and the description of the Blaschke products that do not belong to $\operatorname{mult} l^p_A$. As an application, we give an example of a Blaschke product B such that

$$\{f \in A(\mathbb{D}) : Bf \in l^p_A\} \setminus l^p_A \neq \varnothing \quad (p \in (1,2) \cup (2, +\infty]).$$

Similar problems are considered for the classes U^∞_A, $\operatorname{mult} U^\infty_A$ where

$$U^\infty_A \stackrel{\text{def}}{=} \Big\{ f \in H^\infty : \sup_{z \in \mathbb{D}, n \in \mathbb{Z}_+} \Big| \sum_{k \geq n} \hat{f}(k) z^k \Big| < +\infty \Big\}.$$

Chapter 4 also contains a description of some multipliers of analytic Besov spaces; this description is based on analogs of the Carleson embedding theorem.

In Chapter 4 certain free interpolation theorems are proved for all multiplier algebras we have just mentioned; some other subclasses $\mathcal{E}$ of H^∞ and C_A are considered under this angle; they are usually endowed with a norm essentially bigger than the uniform norm. For such spaces free interpolation is possible on sets $\Lambda \subset \operatorname{clos} \mathbb{D}$ whose "discrete" part $\Lambda \cap \mathbb{D}$ is "thinner" than is required by the (NC)-condition, and the "continuous" part $\Lambda \cap \mathbb{T}$ is "thinner" than in the Rudin-Carleson theorem (see 0.4 above); the corresponding trace space $R_\Lambda(\mathcal{E})$ is either $l^\infty(\Lambda)$ or $C(\Lambda)$ (the latter possibility occurs when $\Lambda \subset \operatorname{clos} \mathbb{D}$ is compact and $\mathcal{E} \subset C_A$).

§2. Statements and discussion of main results

We turn now to a short survey of the thesis respecting the order of the chapters (which was not always the case in §1 above), and give precise statements of the principal results.

2.1. Chapter 1 is devoted to singular integrals and analogs of the Carleson embedding theorem. In particular, we prove the following assertions.

Theorem 1. *Let $r \in (0,1)$, and let*

$$(K_r\vec{\mu})(\zeta) \stackrel{\text{def}}{=} \sum_{n\geq 0} r^n\zeta^n \int_{\mathbb{T}} \frac{d\mu_n(\xi)}{1-\bar{\xi}r\zeta}, \quad \zeta \in \mathbb{T},$$

where $\vec{\mu} = (\mu_n)_{n\geq 0}$ is a sequence of complex Borel measures on $\mathbb{T}$ such that $\sum\limits_{n\geq 0} \operatorname{var}\mu_n < +\infty$. Then there exists a constant $\lambda_0 \in (0,+\infty)$ such that

$$\operatorname{mes}\{\,\zeta \in \mathbb{T} : |(K_r\vec{\mu})(\zeta)| > t\,\} \leq \frac{\lambda_0}{t} \cdot \sum_{n\geq 0} \operatorname{var}\mu_n$$

for any $t \in (0,+\infty)$, $r \in (0,1)$.

Theorem 2. *Let μ be a positive Borel measure in the upper half-plane $\mathbb{C}_+$, let Ω be a positive concave non-decreasing function on $(0,+\infty)$. Then the following are equivalent:*

$$f \in L^2\left(\frac{dt}{\Omega(t)}\right) \Rightarrow \int_{\mathbb{C}_+} |\mathcal{F}f|^2 d\mu < +\infty \tag{1}$$

(*where* $(\mathcal{F}f)(z) \stackrel{\text{def}}{=} \int_0^{+\infty} f(t)e^{-tz}dt, \operatorname{Im} z > 0$).

$$\sup\Big\{\frac{1}{\varepsilon}\Omega\Big(\frac{1}{\varepsilon}\Big)\cdot \mu J_{x,\varepsilon} : x \in \mathbb{R},\ \varepsilon > 0\Big\} < +\infty \tag{2}$$

(*where* $J_{x,\varepsilon} \stackrel{\text{def}}{=} \{\,z \in \mathbb{C} : |z-x| < \varepsilon,\ \operatorname{Im} z > 0\,\}$).

Theorem 3. *Let μ be a Borel non-negative measure in $\mathbb{C}_+$, $\gamma \in [0,+\infty)$, $\alpha \in (0,+\infty)$, $\Delta_n \stackrel{\text{def}}{=} \{\,z \in \mathbb{C} : \operatorname{Im} z \in [2^n, 2^{n+1})\,\}$, $n \in \mathbb{Z}$. Suppose*

$$\sup\left\{\int_{\mathbb{C}_+} \frac{(\operatorname{Im}\lambda)^\alpha}{|\bar{\lambda}-z|^{\alpha+\gamma+1}} d\mu(z) : \lambda \in \mathbb{C}\cap \operatorname{supp}\mu\right\} < +\infty.$$

Then

$$\sum_{n\in\mathbb{Z}} \left(2^{n(\gamma+1)} \int_{\Delta_n} |\mathcal{F}f|^2 d\mu\right)^{p/2} \cdot 2^{-n(\gamma+1)} \leq c^p \int_0^\infty |f(t)|^p \frac{dt}{t^{\gamma(p-1)}}, \quad p \in [2,+\infty),$$

$$\sup_{n\in\mathbb{Z}} \left(2^{n(\gamma+1)} \int_{\Delta_n} |\mathcal{F}f|^2 d\mu\right) \leq \sup_{t>0} |f(t)| \cdot t^{-\gamma};$$

the constant $c \in (0,+\infty)$ does not depend on f.

Recall that Theorem 1 is a stronger version of the Kolmogorov theorem on the conjugate function (see 1.9 above). The proof of Theorem 1 uses the almost everywhere convergence of the Fourier series of an L^2-function established by Carleson (the Hunt estimate of the supremum of partial sums is not needed). We also give a simple method to deduce the Hunt estimate (for $p > 2$) from Carleson's L^2-result. Theorem 1 and an analogous result from §1 of Chapter 1 are used in Chapter 3.

Theorems 2 and 3 are analogs of the Carleson embedding theorem (see 1.8). Besides these theorems, Chapter 1 contains analogous results for some other classes of analytic functions. The most part of the results of §§3–4 of Chapter 1 are needed in Chapters 2 and 4.

2.2. The main results of Chapter 2 are the following two theorems.

Theorem 4. *Suppose* $p,\ q \in [1,2) \cup (2,+\infty]$, $\frac{1}{p}+\frac{1}{q}=1$, $\alpha \in (0,+\infty)$, $\Lambda \subset \mathbb{D}$, $\varepsilon_\lambda(z) \overset{\text{def}}{=} \frac{(1-|\lambda|^2)^\alpha}{1-\bar{\lambda}z}$ $(z \in \mathbb{D}, \lambda \in \Lambda)$,

$$\mathcal{A}^p_\beta \overset{\text{def}}{=} \left\{ f \in A(\mathbb{D}) : \|f\|_{\mathcal{A}^p_\beta} \overset{\text{def}}{=} \left\| \left(\frac{\hat{f}(n)}{(n+1)^{1/p-\beta}} \right)_{n\geq 0} \right\|_{l^p} < +\infty \right\}, \quad \beta \in \mathbb{R},$$

$$\hat{f}(n) \overset{\text{def}}{=} \frac{f^{(n)}(0)}{n!}, \quad (\mathcal{U}_\alpha f)(\lambda)f)(\lambda) \overset{\text{def}}{=} (1-|\lambda|^2)^\alpha f(\lambda) \quad (\lambda \in \mathbb{D}, f \in A(\mathbb{D})).$$

Then the following are equivalent:

(1) $\vec{\varepsilon} = (\varepsilon_\lambda)_{\lambda\in\Lambda}$ *is a symmetric basis of* $\operatorname{span}(\vec{\varepsilon}, \mathcal{A}^p_\alpha)$;

(2) *there exist constants* $c_1,\ c_2 \in (0,+\infty)$ *such that*

$$c_1\|x\|_{l^p(\Lambda)} \leq \left\| \sum_{\lambda\in\Lambda} x(\lambda)\cdot\varepsilon_\lambda \right\|_{\mathcal{A}^p_\alpha} \leq c_2\|x\|_{l^p(\Lambda)}, \quad x \in l^p(\Lambda);$$

(3) $\mathcal{U}_\alpha(\mathcal{A}^q_{1-\alpha})$ *is a symmetric Banach space with respect to the norm*

$$\|y\| \overset{\text{def}}{=} \inf\{ \|f\|_{\mathcal{A}^q_{1-\alpha}} : \mathcal{U}_\alpha f = y,\ f \in \mathcal{A}^q_{1-\alpha} \};$$

(4) $\mathcal{U}_\alpha(\mathcal{A}^q_{1-\alpha}) = l^q(\Lambda)$;

(5) Λ *satisfies*

$$\inf\left\{ \left| \frac{\lambda-\xi}{1-\bar{\lambda}\xi} \right| : \lambda,\xi \in \Lambda,\ \lambda \neq \xi \right\} > 0, \tag{$\mathfrak{R}$}$$

$$\sup\left\{ \sum_{\lambda\in\Lambda, |\xi|\leq|\lambda|} \frac{1-|\lambda|}{1-|\xi|} : \xi \in \Lambda \right\} < +\infty. \tag{wN}$$

These theorems yield, in particular, criteria for a family of exponentials to form a symmetric basis in an L^p-space with a power weight (see 1.7 above). Chapter 2 also contains a complete characterization of coefficients of basic families of exponentials and rational fractions in the spaces $L^p((0,r), t^{\alpha p-1}dt)$ and $\mathcal{A}^p_\alpha$ $(\alpha \in (0,+\infty), r \in (0,+\infty], p \in [1,+\infty])$. Simple geometric conditions (imposed on $\Lambda \subset \mathbb{C}_+$) are given that are sufficient for the family $\vec{\varepsilon} = ((\operatorname{Im}\lambda)^\alpha e^{i\lambda t})_{\lambda\in\Lambda}$ to form an unconditional, but not symmetric basis in $\operatorname{span}(\vec{\varepsilon}, L^p(t^{\alpha p-1}dt))$ $(\alpha > 0, p \in (1,2)\cup(2,+\infty))$.

2.3. Chapter 3 is devoted to free interpolation and coefficient properties in the space U_A of all power series convergent uniformly in $\mathbb{D}$, and in some other similar spaces. Many results of the chapter are based on theorems of §1, Chapter 1. Let us quote some typical results of Chapter 3.

Theorem 6. *Let Γ be a finite union of open arcs of $\mathbb{T}$, and let Λ_-, Λ_+ be closed subsets of Γ, $\operatorname{mes}\Lambda_- = \operatorname{mes}\Lambda_+ = 0$. Suppose a set E of integers satisfies*

$$\sup\left\{\frac{1}{r}\sum_{k\in E,|k|\le r}|k| : r\in[1,+\infty)\right\} < +\infty. \tag{2.1}$$

Then for any pair φ_+, φ_-, $\varphi_\pm \in C(\Lambda_\pm)$ and any square summable family $x = (x(k))_{k\in E}$ of complex numbers there exists a function $f\in A(\widehat{\mathbb{C}}\setminus\operatorname{clos}\Gamma)$ such that

(1) *$f|\mathbb{D}\in U_A, f\left(\frac{1}{z}\right)|\mathbb{D}\in U_A$;*
(2) *$f_+|\Lambda_+ = \varphi_+, f_-|\Lambda_- = \varphi_-$ where*

$$f_\pm(\zeta) = \lim_{r\to 1\mp 0} f(r\zeta), \zeta\in\mathbb{T};$$

(3) *$\hat f_+(n) = x(n)$ $(n\in E\cap[0,+\infty)), \hat f_-(n) = x(n)$ $(n\in E\cap(-\infty,0))$ ($\hat h(n)$ stands for the nth Fourier coefficient of h).*

Theorem 7. *Let a set E of integers satisfy (2.1) and suppose Λ is a closed subset of $\mathbb{T}$ of positive length. Then for any square summable family $x = (x(n))_{n\in E}$ of complex numbers there exists a function f analytic and bounded on $\widehat{\mathbb{C}}\setminus\Lambda$ such that*

(1) *the partial sums of the Fourier series of f_- and f_+ are umformly bounded on $\mathbb{T}$;*
(2) *$\hat f_+(n) = x(n)$ $(n\in E\cap[0,+\infty)$, $\hat f_-(n) = x(n)$ $(n\in E\cap(-\infty,0))$.*

Theorem 8. *For any sequence $d = d(n))_{n\ge 0}$ of non-negative numbers satisfying*

$$\sum_{n\ge 0}(d(n))^2 = +\infty$$

there exists a function $f\in\widehat{\mathbb{C}}\setminus\{1\}$ such that

$$f|\mathbb{D}\in U_A,\quad \sum_{n\ge 0}|\hat f_+(n)|d(n) = +\infty.$$

Theorem 8 answers a question posed by S. B. Stechkin in 1956.

The most part of the theorems proved in Chapter 3 are valid not only for C_A and U_A, but for many other classes of analytic functions satisfying a short list of simple axioms.

2.4. We state here typical results of Chapter 4.

Theorem 9. *Let $(\theta_n)_{n\ge 0}$ be a sequence of inner functions, and let $(J_n)_{n\ge 0}$ be a sequence of mutually disjoint arcs of $\mathbb{T}$. Suppose $\theta_n(0) > 0$ $(n\in\mathbb{Z}_+)$, $\sum_{n\ge 0}(1-\theta_n(0)) < +\infty, \varnothing \ne \mathbb{T}\cap\sigma(\theta_n)\subset J_n$ $(n\in\mathbb{N})$ where $\sigma(\theta_n)$ is the spectrum of* *the inner function θ_n (i.e. the smallest closed set in whose complement $1/\theta_n$ is analytic). Then there is a sequence $(\xi_n)_{n\geq 0}$ ($\xi_n \in \mathbb{T}, n \in \mathbb{Z}_+$) such that*

$$\prod_{n\geq 0} \theta_n(\xi_n, z) \notin \bigcup_{1\leq p<2} \operatorname{mult} l^p_A$$

(see 1.6 and 1.11 for the definitions of l^p_A and $\operatorname{mult} l^p_A$).

In particular, for any sequence $(r_n)_{n\geq 0}$ of points of $(0,1)$ satisfying $\sum_{n=1}^{\infty}(1 - r_n) < +\infty$ there exists a sequence $(\xi_n)_{n\geq 0}$ of points of $\mathbb{T}$ such that the Blaschke product B,

$$B(z) \stackrel{\text{def}}{=} \prod_{n\geq 0} \frac{r_n - \overline{\xi_n} z}{1 - r_n \overline{\xi_n} z}$$

does not belong to $\bigcup_{1\leq p<2} \operatorname{mult} l^p_A$.

Theorem 10. *There exists an infinite Blaschke product B by which it is impossible "to divide" in l^p_A ($p \neq 2, 1 < p < \infty$), that is*

$$\{g \in A(\mathbb{D}) : B \cdot g \in l^p_A\} \setminus l^p_A \neq \varnothing.$$

Similar assertions are also proved for the spaces l^1_A, U_A, U^∞_A and some others.

Note that Theorems 9 and 10 are not as remote from free interpolation as they may seem. The proofs given in Chapter 4 exploit some results on free interpolation in an essential way.

Theorem 11. *Let $p \in (1,2)$ and let Λ be a subset of $\operatorname{clos}\mathbb{D}$. Then*

(1) *if $\Lambda \in \mathbb{D}$ and Λ satisfies the (NC)-condition (see 1.2), and*

$$\sum_{n\geq 0}\Bigg(\sum_{\substack{\lambda\in\Lambda,\\ 2^{-(n+1)}<1-|\lambda|\leq 2^{-n}}} (1-|\lambda|)^{2(1-\frac{1}{p})}\Bigg)^{p/2} < +\infty,$$

then $R_\Lambda(H^\infty \cap l^p_A) = l^\infty(\Lambda)$;

(2) *if Λ is closed, $\Lambda \cap \mathbb{D}$ satisfies (NC), and*

$$\sup\left\{\frac{\operatorname{dist}(\lambda, \Lambda\cap\mathbb{T})}{1-|\lambda|} : \lambda \in \Lambda\cap\mathbb{D}\right\} < +\infty,$$

and

$$\operatorname{mes}(\Lambda\cap\mathbb{T}) = 0, \quad \sum_{n\geq 0} (\operatorname{mes} I_n)^{p-1} < +\infty$$

where the I_n are components of $\mathbb{T}\setminus\Lambda$, then

$$R_\Lambda(C_A \cap l^p_A) = C(\Lambda).$$

Some other interpolation theorems are proved in Chapter 4. They are analogous to Theorem 11 and are applicable to certain subclasses of H^∞ or C_A (for example, to $\operatorname{mult} l^p_A$, $\operatorname{mult} U^\infty_A$).

Principal results of the thesis are published in [16, 28, 15, 13, 12, 22, 10, 21, 19, 35, 38, 39] [2].

Translated by V. P. Havin

Russian original:	Avtoreferat dissertatsii na soiskanie uchenoy stepeni doktora fiziko-matematicheskih nauk, Leningradskoe Otdelenie Matematicheskogo Instituta im. V.A.Steklova AN SSSR, 1983.

[2]These numbers refer to List of Publications of S.A.Vinogradov, this volume.

Contributed Papers

Operator Theory:
Advances and Applications, Vol. 113

On Embedding Theorems for Coinvariant Subspaces of the Shift Operator. I

Alexei B. Aleksandrov

Dedicated to the memory of S. A. Vinogradov

0. Introduction

In this paper we deal with embedding theorems for invariant subspaces of the backward shift operator $S^* : H^p \to H^p$, $(S^*f)(z) \stackrel{\text{def}}{=} (f(z) - f(0))z^{-1}$, where H^p denotes the Hardy space in the open unit disk $\mathbb{D}$ with $0 < p \le +\infty$. In connection with the Hardy spaces, the backward shift operator, and related questions, we refer to [13] and [15]. Every function $f \in H^p$ has finite angular boundary values almost everywhere on the unit circle $\mathbb{T}$. Denote by $f(\zeta)$ the angular boundary value of f at the point $\zeta \in \mathbb{T}$ if it exists. Besides, we set by definition $f(\zeta) = \infty$ if f does not have a finite angular boundary value at $\zeta \in \mathbb{T}$. A function $\Theta \in H^\infty$ is said to be inner if $|\Theta| = 1$ almost everywhere on $\mathbb{T}$. We continue each inner function Θ to the extended complex plane $\overline{\mathbb{C}}$ setting $\Theta(z) \stackrel{\text{def}}{=} \overline{\Theta(\overline{z}^{-1})}^{\,-1}$ for $|z| > 1$. Thus, Θ is holomorphic in $\mathbb{D}$ and meromorphic in $\mathbb{D}_- = \{z \in \overline{\mathbb{C}} : |z| > 1\}$. Let $\rho(\Theta)$ denote the boundary spectrum of the inner function Θ, i.e. the set of all $\zeta \in \mathbb{T}$ such that $\liminf_{|z|<1, z\to\zeta} |\Theta(z)| = 0$. It is easy to see that $\rho(\Theta)$ is a closed subset of $\mathbb{T}$, and Θ is meromorphic in $\overline{\mathbb{C}} \setminus \rho(\Theta)$. Denote by $\Theta^*(H^p)$ the set of all functions $f \in H^p$ admitting analytic continuation across $\mathbb{T} \setminus \rho(\Theta)$ and such that $f(\zeta) = \Theta(\zeta)\overline{g(\zeta)\zeta}$ almost everywhere on $\mathbb{T}$ for some $g \in H^p$. Note that the condition of the existence of analytic continuation across $\mathbb{T} \setminus \rho(\Theta)$ is fulfilled automatically if $p \ge 1$. We regard an arbitrary function $f \in \Theta^*(H^p)$ as a function $f : \overline{\mathbb{C}} \to \overline{\mathbb{C}}$ by setting $f(z) \stackrel{\text{def}}{=} z^{-1}\Theta(z)\overline{g(\overline{z}^{-1})}$ for $|z| > 1$. It is easy to see that f is meromorphic in $\overline{\mathbb{C}} \setminus \rho(\Theta)$, and its exterior boundary values coincide almost everywhere with $f|_{\mathbb{T}}$ (i.e. with the interior boundary values). The space $\Theta^*(H^p)$ is an invariant subspace of the operator $S^* : H^p \to H^p$. It is well known that for $p \ge 1$ the operator $S^* : H^p \to H^p$ has no other proper invariant subspaces. The case of $1 < p < \infty$ was considered in [10]; the extreme cases of $p = 1$ and $p = +\infty$ can be found in [1] and [2]. As usual in the case where $p = +\infty$ one should consider

This investigation was supported by part by the Russian Foundation for Fundamental Studies (Grant no. 96-01-00693).

the weak topology $\sigma(L^\infty, L^1)$ or replace H^∞ with the disk-algebra C_A. Also note that $C_A \cap \Theta^*(H^p)$ is dense[1] in $\Theta^*(H^p)$; the case where $p \geq 1$ is contained implicitly in [2], the case where $p < 1$ is reduced to the case where $p = 2$ by means of results of [1], where a complete description of the invariant subspaces of the backward shift operator was obtained for $p < 1$. In particular this description implies that if E is a proper invariant subspace of the operator $S^* : H^p \to H^p$ and $E \cap C_A$ is dense in E, then $E = \Theta^*(H^p)$ for some inner function Θ. It is clear that if Θ is a constant inner function, then $\Theta^*(H^p) = \{0\}$. In what follows we will consider only nonconstant inner functions.

With every Borel set $K \subset \mathbb{C}$ we associate the space $M(K)$ of all complex Borel measures on $\mathbb{C}$ such that $|\mu|(\mathbb{C} \backslash K) = 0$, where $|\mu|$ denotes the total variation of μ. Denote by $M_+(K)$ the set of all positive measures in $M(K)$. Let $\mu \in M_+(\overline{\mathbb{D}})$. We are interested in the following question. When do we have the inclusion

$$\Theta^*(H^p) \subset L^p(\mu) \tag{0.1}$$

or, more precisely, when is it true that $\int_{\overline{\mathbb{D}}} |f|^p d\mu < +\infty$ for[2] all $f \in \Theta^*(H^p)$? This question was considered in a number of papers. We quote only a little part of them (see [8], [9], [17], [5], [6], [11], [12]).

It is not difficult to check (see [6]) that for[3] $p < +\infty$ this question is equivalent to the following question. When does there exist a constant $C = C(p, \mu, \Theta)$ such that

$$\int_{\overline{\mathbb{D}}} |f|^p d\mu \leq C \|f\|^p_{H^p} \tag{0.2}$$

for all $f \in \Theta^*(H^p)$? It follows from results of [6] that it is sufficient to check inequality (0.2) for all $f \in C_A \cap \Theta^*(H^p)$.

We denote by $\|\mu\|_{\mathcal{C}_p(\Theta)}$ the smallest constant C satisfying (0.2) for all $f \in C_A \cap \Theta^*(H^p)$. We set $\|\mu\|_{\mathcal{C}_p(\Theta)} = \infty$ if such a constant C does not exist. Put $\|\mu\|_{\mathcal{C}_p(\Theta)} \stackrel{\text{def}}{=} \||\mu|\|_{\mathcal{C}_p(\Theta)}$ for any (not necessarily positive) measure $\mu \in M(\overline{\mathbb{D}})$. It is easy to see that the set $\mathcal{C}_p(\Theta) \stackrel{\text{def}}{=} \{\mu \in M(\overline{\mathbb{D}}) : \|\mu\|_{\mathcal{C}_p(\Theta)} < +\infty\}$ is a Banach space for any positive p.

A measure $\mu \in M(\overline{\mathbb{D}})$ is called a Carleson measure if

$$|\mu|(\mathcal{D}(\zeta, r)) \leq Cr \tag{0.3}$$

for all $r > 0$ and for all $\zeta \in \mathbb{T}$, where $\mathcal{D}(\zeta, r)$ denotes the open disk with center at ζ and radius r. Denote by $\|\mu\|_{\mathcal{C}}$ the smallest constant C satisfying (0.3) for all $r > 0$ and for all $\zeta \in \mathbb{T}$. This makes the space $\mathcal{C}$ of all Carleson measures a Banach space. The Carleson embedding theorem implies that $\mathcal{C} \subset \mathcal{C}_p(\Theta)$ for all $p > 0$ and for all inner functions Θ.

Now we pass to the description of the results of the paper. In §1 we consider relations between spaces $\mathcal{C}_p(\Theta^n)$ for various p and n. For example, we prove that

---

[1]This may be obtained also with the help of Theorem 6 in [6].

[2]Recall that by definition $f(\zeta) = \infty$ if f does not have a finite boundary values at ζ.

[3]In what follows we will suppose that $p < +\infty$ unless otherwise is stated.

$\mathcal{C}_p(\Theta^n)$ does not depend on n for $n > p^{-1}$ (Theorem 1.3). From this we obtain as a consequence that $\mathcal{C}_p(\Theta^n)$ does not depend on n if the inner function Θ does not take at least one value $w \in \mathbb{D}$. Moreover, we prove that $\mathcal{C}_q(\Theta) \subset \mathcal{C}_p(\Theta)$ if $1 < q < p$ (Theorem 1.5). It can be proved that this fails for $q = 1$. In §1 we obtain also estimates of the norm of the identity embedding operator $\mathcal{C}_q(\Theta) \subset \mathcal{C}_p(\Theta)$. In §2 we present some estimates (sharp in a sense) of the $\mathcal{C}_p(\Theta)$-norm of the δ-measure δ_a at the point $a \in \mathbb{D}$ in terms of a and $\Theta(a)$. In §3 we obtain a necessary and sufficient condition for $\mathcal{C} = \mathcal{C}_p(\Theta)$. This equality is fulfilled if and only if $p < 1$ and $\rho(\Theta) = \mathbb{T}$. Moreover, we prove that for $p \neq 1$ the space $\mathcal{C}_p(\Theta)$ can be rotation invariant only in degenerate cases. In §4 we investigate the space $\mathcal{C}_1(\Theta)$ for inner functions of a special type. In particular, we construct an infinite Blaschke product Θ such that $\|\delta_a\|_{\mathcal{C}_1(\Theta)} = o((1-|a|)^{-1})$ as $|a| \to 1-$ and $\mathcal{C}_1(\Theta)$ is rotation invariant.

With each inner function Θ we associate a family $\{\sigma_\alpha\}_{\alpha\in\mathbb{T}}$ of singular positive measures on $\mathbb{T}$ uniquely determined by the following formula

$$\operatorname{Re}\frac{\alpha + \Theta(z)}{\alpha - \Theta(z)} = \int_{\mathbb{T}} \frac{1-|z|^2}{|z-\zeta|^2}\,d\sigma_\alpha(\zeta) \quad (z \in \mathbb{D}).$$

The measures σ_α play an important role in operator theory, in particular, for the investigation of rank one perturbations of a unitary operator, see [7]. The Clark theorem (see [7]) implies that $\|\sigma_\alpha\|_{\mathcal{C}_2(\Theta)} = 1$ for any $\alpha \in \mathbb{T}$ and any inner function Θ.

We conclude this section with the list of notation.

$\mathbb{N}$ is the set of all positive integers,

$\mathbb{D}$ is the open unit disk,

$\mathbb{T}$ is the unit circle,

m is normalized Lebesgue measure on $\mathbb{T}$,

δ_a is the δ-measure at a point a,

∂G is the boundary of a set G,

$L^p(\mu)$, where μ is a measure, and ℓ^p denote the usual L^p-spaces,

$M(K)$ (for a Borel set $K \subset \mathbb{C}$) denotes the set all complex Borel measures μ on $\mathbb{C}$ supported on K,

$M_+(K)$ is the set of all positive measures in $M(K)$.

For $\mu \in M(\mathbb{T})$ we set $\mu_+(z) \stackrel{\text{def}}{=} \int_{\mathbb{T}} \dfrac{d\mu(\zeta)}{1-\overline{\zeta}z}$.

For an inner function Θ we set $P_\Theta f \stackrel{\text{def}}{=} f_+ - \Theta(\overline{\Theta}f)_+$, where $f \in L^1(m)$.

1. Preliminary remarks

We say that inner functions Θ and I are *equivalent* if there exists a function $F \in H^\infty$ such that $F^{-1} \in H^\infty$ and

$$\frac{F}{\overline{F}} = \frac{\Theta}{I} \tag{1.1}$$

almost everywhere on $\mathbb{T}$. Note that (1.1) implies that $F \in (z\Theta)^*(H^\infty)$ and $F^{-1} \in (zI)^*(H^\infty)$. Consequently, F admits analytic continuation across $\mathbb{T} \setminus \rho(\Theta)$ and $\mathbb{T} \setminus \rho(I)$. Hence, from (1.1) it follows that $\rho(\Theta) = \rho(I)$.

Theorem 1.1. *Let Θ and I be equivalent inner functions. Then $\mathcal{C}_p(\Theta) = \mathcal{C}_p(I)$ for all $p > 0$.*

Proof. It is enough to note that (1.1) implies $\Theta^*(H^p) = FI^*(H^p)$ for all $p > 0$. □

Also, we note that if an inner function Θ is a divisor of an inner function I, then $\mathcal{C}_p(I) \subset \mathcal{C}_p(\Theta)$ for all $p > 0$ since $\Theta^*(H^p) \subset I^*(H^p)$.

Theorem 1.2. *Let Θ be an inner function. Then $\mathcal{C}_p(\Theta^n) \subset \mathcal{C}_{np}(\Theta)$ and $\| \ \|_{\mathcal{C}_{np}(\Theta)} \le \| \ \|_{\mathcal{C}_p(\Theta^n)}$ for all $p > 0$ and all $n \in \mathbb{N}$.*

Proof. It is enough to note that if $f \in \Theta^*(H^{np})$, then $f^n \in (\Theta^n)^*(H^p)$. □

Theorem 1.3. *Let Θ be an inner function. Let $0 < p < +\infty$. Then the set $\mathcal{C}_p(\Theta^n)$ does not depend on $n \in \mathbb{N}$ for[4] $n > \frac{1}{p}$. In other words, $\mathcal{C}_p(\Theta^n) = \mathcal{C}_p(\Theta^k)$ for all $k, n \in \mathbb{N}$ such that $k, n > \frac{1}{p}$.*

Before proving Theorem 1.3 we deduce some consequences from this theorem.

Corollary 1. *If Θ is a singular inner function, then $\mathcal{C}_p(\Theta^s)$ does not depend on $s \in (0, +\infty)$.*

Proof. From Theorem 1.3 it follows that $\mathcal{C}_p(\Theta^s)$ does not depend on s, where s runs through the positive rational numbers. It remains to note that the scale $\{\mathcal{C}_p(\Theta^s)\}_{s>0}$ decreases. □

Corollary 2. *If an inner function Θ does not take at least one value $w \in \mathbb{D}$, then $\mathcal{C}_p(\Theta) = \mathcal{C}_p(\Theta^n)$ for all $p > 0$ and for all $n \in \mathbb{N}$.*

Proof. Consider the singular inner function $S = \dfrac{\Theta - w}{1 - \overline{w}\Theta}$. It is clear that Θ^n is equivalent to S^n for any $n \in \mathbb{N}$. Consequently, by Theorem 1.1, $\mathcal{C}_p(\Theta^n) = \mathcal{C}_p(S^n)$ for all $n \in \mathbb{N}$. Now, the result follows from Corollary 1. □

Corollary 3. *Let Θ and I be inner functions each of which does not take at least one value in $\mathbb{D}$. Assume that Θ is a divisor of I^n and I is a divisor of Θ^n for some $n \in \mathbb{N}$. Then $\mathcal{C}_p(\Theta) = \mathcal{C}_p(I)$ for all $p \in (0, +\infty)$.*

Proof. Clearly, $\mathcal{C}_p(\Theta) \supset \mathcal{C}_p(I^n)$ and $\mathcal{C}_p(I) \supset \mathcal{C}_p(\Theta^n)$. It remains to apply Corollary 2. □

---

[4]It should be noted that the condition $np > 1$ in this context appeared for the first time in [11], Theorem 1.3.

Corollary 4. *If $np > 1$, then $C_p(\Theta^n) \subset C_{np}(\Theta^n)$ for any inner function Θ.*

Proof. By Theorem 1.3, we have $C_{np}(\Theta^n) = C_{np}(\Theta)$. It remains to refer to Theorem 1.2. □

To prove Theorem 1.2 we need the following

Lemma 1.4. *Let Θ be an inner function. Let $\frac{1}{n} < p < +\infty$ with $n \in \mathbb{N}$. Then for any $f \in H^p$ there exists $g \in (\Theta^n)^*(H^p)$ such that $f - g \in \Theta H^p$.*

Proof. We may represent f in the form $f = \prod_{j=1}^n f_j$, where $f_j \in H^{np}$. Set $g_j = P_\Theta f_j$. Then $g_j \in \Theta^*(H^{np})$ and $f_j - g_j \in \Theta H^{np}$. To complete the proof, put $g = \prod_{j=1}^n g_j$. □

Corollary. *Under the assumptions of Lemma 1.4, we have*

$$(\Theta^N)^*(H^p) = \sum_{k=0}^{N-n} \Theta^k (\Theta^n)^*(H^p) \tag{1.2}$$

for any $N \geq n$.

Proof. The inclusion $\supset$ is evident. To prove the opposite inclusion we use induction on N. The base of induction (i.e. the case where $N = n$) is trivial. We prove that if the inclusion $\subset$ is fulfilled for $N = l$, then it is fulfilled also for $N = l + 1$ ($l \geq n$). Let $f \in (\Theta^{l+1})^*(H^p)$. Then, by Lemma 1.4, f may be represented in the form $f = g + \Theta h$, where $g \in (\Theta^n)^*(H^p) \subset (\Theta^{l+1})^*(H^p)$, $h \in H^p$. Thus, $\Theta h \in (\Theta^{l+1})^*(H^p)$, whence $h \in (\Theta^l)^*(H^p)$. It remains to apply the induction hypothesis to h. □

Proof of Theorem 1.3. It suffices to use (1.2). □

Theorem 1.5. *If $1 < q < p < +\infty$, then $C_q(\Theta) \subset C_p(\Theta)$ for any inner function Θ.*

Proof. We take a positive integer n such that $nq > p$. Let $\mu \in C_q(\Theta)$. Then $\mu \in C_{nq}(\Theta)$ because by Theorem 1.3 and Theorem 1.2 we have $C_q(\Theta) = C_q(\Theta^n) \subset C_{nq}(\Theta)$. Thus, the operator P_Θ acts continuously from $L^q(m)$ into $L^q(\mu)$ and from $L^{nq}(m)$ into $L^{nq}(\mu)$. Consequently, by the Riesz-Torin theorem, it acts continuously from $L^p(m)$ into $L^p(\mu)$. □

Remark. *It can be proved that Theorem* 1.5 *fails if we replace the condition $q > 1$ with the condition $q \geq 1$. Moreover, the set $\{p > 0 : \mu \in C_p(\Theta)\}$ may consist of any finite or infinite number of the connected component.*

The closed graph theorem implies that under the assumptions of Theorem 1.5 there exists a constant $C(p, q, \Theta)$ such that $\|\ \|_{C_p(\Theta)} \leq C(p, q, \Theta) \|\ \|_{C_q(\Theta)}$. Our next aim is to estimate this constant $C(p, q, \Theta)$. First, we consider the simplest case where $q = 2$. In this case we obtain rather sharp estimates. In [4] it was proved that $\|\sigma_\alpha\|_{C_p(\Theta)} \leq Cp$ for all $p \geq 2$ and $\alpha \in \mathbb{T}$, i.e. $\|\sigma_\alpha\|_{C_p(\Theta)} \leq Cp \|\sigma_\alpha\|_{C_2(\Theta)}$, because $\|\sigma_\alpha\|_{C_2(\Theta)} = 1$ by the Clark theorem. The same proof yields the following

Theorem 1.6. *Let Θ be an inner function. Then*

$$\| \ \|_{C_p(\Theta)} \leq Cp\| \ \|_{C_2(\Theta)} \tag{1.3}$$

for all $p \geq 2$, where C is an absolute constant. Moreover, if p is an even integer, then inequality (1.3) *is true for $C = \frac{1}{2}$.*

Proof. First let $p = 2n$, where $n \in \mathbb{N}$. Then $\| \ \|_{C_{2n}(\Theta)} \leq \| \ \|_{C_2(\Theta^n)} \leq n\| \ \|_{C_2(\Theta)}$, where the first inequality is true by Theorem 1.2, and the second inequality readily stems from the orthogonal decomposition $(\Theta^n)^*(H^2) = \bigoplus_{j=0}^{n-1} \Theta^j\Theta^*(H^2)$. Now in the case where $2 \leq p \leq 4$ the required inequality follows from the Riesz-Torin theorem because the norm of the projection P_Θ in $L^p(m)$ is uniformly bounded for $p \in [2, 4]$. In the case where $p > 4$ we can take a positive integer n such that $2 \leq \frac{p}{n} \leq 4$. Then we have $\| \ \|_{C_p(\Theta)} \leq \| \ \|_{C_{\frac{p}{n}}(\Theta^n)} \leq C\frac{p}{n}\| \ \|_{C_2(\Theta^n)} \leq Cp\| \ \|_{C_2(\Theta)}$.

□

Remark. *Similar arguments allow us to obtain also the following inequalities:*

1) $\| \ \|_{C_p(\Theta)} \leq \frac{Cp}{q-1}\| \ \|_{C_q(\Theta)}$ *if* $1 < q \leq \min(2, p)$;
2) $\| \ \|_{C_p(\Theta)} \leq (1 + \frac{p}{q})^{Cq\log q}\| \ \|_{C_q(\Theta)}$ *if* $2 \leq q \leq p$.

2. Norm estimates of point evaluations

The function $\dfrac{1 - \overline{\Theta(a)}\Theta(z)}{1 - \bar{a}z}$ is the reproducing kernel of $\Theta^*(H^2)$. This implies the relation

$$\|\delta_a\|_{C_2(\Theta)} = \frac{1 - |\Theta(a)|^2}{1 - |a|^2}. \tag{2.1}$$

In particular, $|f(a)|^2 \leq \dfrac{1 - |\Theta(a)|^2}{1 - |a|^2}\|f\|^2_{H^2}$ for any $f \in \Theta^*(H^2)$ and $a \in \mathbb{D}$.

The following theorem gives an estimate (sharp[5] in a sense) for the $C_p(\Theta)$-norm of the δ-measure at a point $a \in \mathbb{D}$ in terms of a and $\Theta(a)$ for $p \geq 1$.

Theorem 2.1. *Let Θ be an inner function. Let $f \in \Theta^*(H^p)$, and let $a \in \mathbb{D}$. Then there exists an absolute constant C such that*

1) *if $p = 1$, then*

$$|f(a)| \leq \frac{C}{1 - |a|^2}\left(\log\frac{2}{1 - |\Theta(a)|^2}\right)^{-1}\|f\|_{H^1};$$

2) *if $1 < p \leq 2$, then*

$$|f(a)|^p \leq C\frac{(1 - |\Theta(a)|^2)^{p-1}}{1 - |a|^2}\|f\|^p_{H^p};$$

[5]See the remark after the proof of Theorem 2.1.

3) *if* $p \geq 2$, *then*

$$|f(a)|^p \leq Cp \min\left(\frac{(1-|\Theta(a)|^2}{1-|a|^2}, \left(\frac{(1-|\Theta(a)|^2}{1-|a|^2}\right)^{p-1}\right) \|f\|^p_{H^p}.$$

Proof (*the cases of* $p > 1$). If $1 < p \leq 2$, then the required inequality stems from the following inequalities:

$$|f(a)|^p \leq \left\|\frac{1-\overline{\Theta(a)}\Theta}{1-\bar{a}z}\right\|^p_{H^{p'}} \|f\|^p_{H^p},$$

$$\left\|\frac{1-\overline{\Theta(a)}\Theta}{1-\bar{a}z}\right\|^{p'}_{H^{p'}} \leq \left\|\frac{1-\overline{\Theta(a)}\Theta}{1-\bar{a}z}\right\|^{p'-2}_{H^\infty} \left\|\frac{1-\overline{\Theta(a)}\Theta}{1-\bar{a}z}\right\|^2_{H^2}$$
$$\leq \frac{2^{p'-2}}{(1-|a|^2)^{p'-2}} \cdot \frac{1-|\Theta(a)|^2}{1-|a|^2},$$

where p' is the conjugate exponent. Now, let $p \geq 2$. Then Theorem 1.6 and (2.1) imply

$$|f(a)|^p \leq Cp \frac{1-|\Theta(a)|^2}{1-|a|^2} \|f\|^p_{H^p}.$$

Now, we prove that

$$|f(a)|^p \leq Cp \left(\frac{1-|\Theta(a)|^2}{1-|a|^2}\right)^{p-1} \|f\|^p_{H^p}.$$

Clearly, it suffices to proof this inequality for all f in a dense subset of $\Theta^*(H^p)$. By the Clark theorem, we have

$$f(a) = \int_{\mathbb{T}} f(\zeta) \frac{1-\Theta(a)\overline{\Theta(\zeta)}}{1-a\bar{\zeta}} dm(\zeta) = \int_{\mathbb{T}} f(\zeta) \frac{1-\Theta(a)\overline{\Theta(\zeta)}}{1-a\bar{\zeta}} d\sigma_\alpha(\zeta)$$

for all f in a dense subset of $\Theta^*(H^p)$. Hence, the Hölder inequality implies

$$|f(a)| \leq \left(\int_{\mathbb{T}} |f(\zeta)|^p d\sigma_\alpha(\zeta)\right)^{\frac{1}{p}} \left(\int_{\mathbb{T}} \left|\frac{1-\Theta(a)\overline{\Theta(\zeta)}}{1-a\bar{\zeta}}\right|^{p'} d\sigma_\alpha(\zeta)\right)^{\frac{1}{p'}}.$$

Now, we estimate each factor in the right-hand side of the latter inequality. The first factor was estimated in [4]: $\int_{\mathbb{T}} |f(\zeta)|^p d\sigma_\alpha(\zeta) \leq Cp\|f\|^p_{H^p}$ (see also Theorem 1.6). We estimate the second factor using the Hölder inequality:

$$\int_{\mathbb{T}} \left|\frac{1-\Theta(a)\overline{\Theta(\zeta)}}{1-a\bar{\zeta}}\right|^{p'} d\sigma_\alpha(\zeta) \leq \left(\int_{\mathbb{T}} \left|\frac{1-\Theta(a)\overline{\Theta(\zeta)}}{1-a\bar{\zeta}}\right|^2 d\sigma_\alpha(\zeta)\right)^{\frac{p'}{2}} \sigma_\alpha(\mathbb{T})^{1-\frac{p'}{2}}$$
$$= \left(\frac{1-|\Theta(a)|^2}{1-|a|^2}\right)^{\frac{p'}{2}} \left(\frac{1-|\Theta(0)|^2}{|\alpha-\Theta(0)|^2}\right)^{1-\frac{p'}{2}}.$$

Substituting $\alpha = -\frac{\Theta(0)}{|\Theta(0)|}$ we obtain

$$\int_{\mathbb{T}} \left| \frac{1-\Theta(a)\overline{\Theta(\zeta)}}{1-a\bar{\zeta}} \right|^{p'} d\sigma_\alpha(\zeta) \le \left(\frac{1-|\Theta(a)|^2}{1-|a|^2} \right)^{\frac{p'}{2}} \left(\frac{1-|\Theta(0)|}{1+|\Theta(0)|} \right)^{1-\frac{p'}{2}} \le \frac{1-|\Theta(a)|^2}{1-|a|^2},$$

because $\frac{1-|\Theta(0)|}{1+|\Theta(0)|} \le \frac{1-|\Theta(a)|^2}{1-|a|^2}$ for every holomorphic function $\Theta : \mathbb{D} \to \mathbb{D}$. Now it is clear that

$$|f(a)|^p \le C p \left(\frac{1-|\Theta(a)|^2}{1-|a|^2} \right)^{\frac{p}{p'}} \|f\|^p_{H^p} = C p \left(\frac{1-|\Theta(a)|^2}{1-|a|^2} \right)^{p-1} \|f\|^p_{H^p}.$$

□

Remark 1. *Letting $p \to \infty$, we obtain*

$$|f(a)| \le \min\left(1, \frac{1-|\Theta(a)|^2}{1-|a|^2}\right) \|f\|_{H^\infty}$$

for any $f \in \Theta^*(H^\infty)$. This is clear also from the proof of the theorem.

Remark 2. *If $\Theta(0) = 0$, then assertion 3) of Theorem 2.1 may be rewritten as follows:*

$$|f(a)|^p \le C p \frac{1-|\Theta(a)|^2}{1-|a|^2} \|f\|^p_{H^p},$$

provided $p \ge 2$. Moreover, the proof of Theorem 2.1 yields also the following inequality

$$|f(a)|^p \le C p \left(\frac{1-|\Theta(0)|}{1+|\Theta(0)|} \right)^{\frac{p}{2}-1} \left(\frac{1-|\Theta(a)|^2}{1-|a|^2} \right)^{\frac{p}{2}} \|f\|^p_{H^p},$$

provided $p \ge 2$.

To prove Theorem 2.1 for $p = 1$ we need the following

Lemma 2.2. *Let $a \in \mathbb{D}$ and $f \in \Theta^*(H^1)$, where Θ is an inner function. Then*

$$f(a) \int_{\mathbb{T}} \frac{dm(z)}{|1-\Theta(a)z|^q} = \int_{\mathbb{T}} \frac{1-\Theta(a)\overline{\Theta(\zeta)}}{|1-\Theta(a)\overline{\Theta(\zeta)}|^q} \frac{f(\zeta)dm(\zeta)}{1-a\bar{\zeta}}$$

for any $q \in \mathbb{C}$.

Proof. We may suppose that $f \in \Theta^*(H^2)$. Then using the Clark theorem [7] and Proposition 1 in [3] we have

$$\begin{aligned} &\int_{\mathbb{T}} \frac{1-\Theta(a)\overline{\Theta(\zeta)}}{|1-\Theta(a)\overline{\Theta(\zeta)}|^q} \frac{f(\zeta)dm(\zeta)}{1-a\bar{\zeta}} \\ &\quad = \int_{\mathbb{T}} \left(\int_{\mathbb{T}} f(\zeta) \frac{1-\Theta(a)\overline{\Theta(\zeta)}}{1-a\bar{\zeta}} d\sigma_\alpha(\zeta) \right) \frac{dm(\alpha)}{|1-\Theta(a)\bar{\alpha}|^q} \\ &\quad = f(a) \int_{\mathbb{T}} \frac{dm(z)}{|1-\Theta(a)z|^q}. \end{aligned}$$

□

Proof of Theorem 2.1 (*the case where* $p = 1$). It is enough to apply Lemma 2.2 for $q = 1$ and to observe that $\left\|\frac{1}{1-\Theta(a)z}\right\|_{H^1} \geq \text{const} \log \frac{2}{1-|\Theta(a)|^2}$. □

Remark. *For each fixed* $p \geq 1$ *Theorem* 2.1 *cannot be refined. More precisely, the following assertion holds. For any* $p \geq 1$ *and any* $a, \lambda \in \mathbb{D}$ *there exists an inner function* Θ *with* $\Theta(a) = \lambda$ *and a non-zero function* $f \in \Theta^*(H^p)$ *such that*

1) $|f(a)| \geq \frac{C}{1-|a|^2}\left(\log \frac{2}{1-|\lambda|^2}\right)^{-1} \|f\|_{H^1}$ *for* $p = 1$;

2) $|f(a)|^p \geq C(p)\frac{(1-|\lambda|^2)^{p-1}}{1-|a|^2}\|f\|_{H^p}^p$ *for* $1 < p \leq 2$;

3) $|f(a)|^p \geq C(p)\min\left(\frac{1-|\lambda|^2}{1-|a|^2}, \left(\frac{1-|\lambda|^2}{1-|a|^2}\right)^{p-1}\right)\|f\|_{H^p}^p$ *for* $p \geq 2$.

Proof. First, we consider two special cases where $a = 0$ and where $\lambda = 0$. In these cases Θ is an arbitrary inner function with $\Theta(a) = \lambda$. If $a = 0$, we set $f = 1 - \overline{\Theta(0)}\Theta$. It is easy to check that f satisfies statements 1), 2) and 3) of the remark. If $\lambda = 0$, we set $f = \frac{1}{1-\bar{a}z}$. Clearly, f satisfies statements 2) and 3) of the remark. In the general case we can construct the required function f only under a special choice of an inner function $\Theta = \Theta_{a,\lambda,p}$. If $p > 1$, we may choose a required inner function Θ among the Blaschke product of order 1 setting $f = \frac{1}{1-\bar{b}z}$ and $\Theta = \zeta_0\frac{z-b}{1-\bar{b}z}$, where $b \in \mathbb{D}$ and ζ_0 are taken in such a way that $\Theta(a) = \lambda$, $b/a \in \mathbb{R}$ and $a\zeta_0/\lambda \in \mathbb{R}$, provided $a\lambda \neq 0$. It is easy to see that there exist precisely two such inner functions. One of them is such that $|b| = \frac{|a|+|\lambda|}{1+|a\lambda|}$; the other is such that $|b| = \frac{||a|-|\lambda||}{1-|a\lambda|}$. The former is suitable for $1 < p \leq 2$, the latter is suitable for $2 \leq p < +\infty$. However, for $p = 1$ the Blaschke products of order 1 allows us to construct a non-zero function $f \in \Theta^*(H^1)$ satisfying only the inequality

$$|f(a)| \geq \frac{C}{1-|a|^2}\left(\log \frac{2}{(1-|a|^2)(1-|\lambda|^2)}\right)^{-1} \|f\|_{H^1}.$$

To construct the required function for $p = 1$ we take as Θ a Blaschke product of order 2. If $a = 0$, then we may set $\Theta_0 = \frac{\lambda - z^2}{1-\bar{\lambda}z^2}$ and $f_0 = \frac{1}{1-\bar{\lambda}z^2}$. In the general case set $\Theta(z) = \Theta_0(\varphi(z))$ and $f(z) = f_0(\varphi(z))\varphi'(z)$ where $\varphi(z) = \frac{a-z}{1-\bar{a}z}$. Then $\Theta(a) = \lambda$, $f(a) = (|a|^2-1)^{-1}$, $\|f\|_{H^1} = \|f_0\|_{H^1}$, and the required inequality is again obvious. □

Lemma 2.2 implies also the following

Theorem 2.3. *Let Θ be an inner function. Then $\|\delta_0\|_{C_p(\Theta)} = \|(1-\overline{\Theta(0)}z)^{-1}\|_{H^p}^{-p}$ for all $p \geq 1$.*

Proof. Applying Lemma 2.2 to $a = 0$ and $q = 2 - p$ we have

$$f(0)\int_{\mathbb{T}} \frac{dm(z)}{|1-\Theta(0)z|^{2-p}} = \int_{\mathbb{T}} \frac{1-\Theta(0)\overline{\Theta(\zeta)}}{|1-\Theta(0)\overline{\Theta(\zeta)}|^{2-p}} f(\zeta)\, dm(\zeta). \tag{2.2}$$

Consequently,

$$|f(0)|\int_{\mathbb{T}} \frac{dm(z)}{|1-\Theta(0)z|^{2-p}} \leq \left(\int_{\mathbb{T}} |1-\Theta(0)\overline{\Theta(\zeta)}|^{p}\, dm(\zeta)\right)^{\frac{1}{p'}} \|f\|_{H^p} \tag{2.3}$$

by the Hölder inequality, where p' is the conjugate exponent. Note that the integrand on the right-hand side of (2.2) is positive and

$$\left|\frac{1-\Theta(0)\overline{\Theta(\zeta)}}{|1-\Theta(0)\overline{\Theta(\zeta)}|^{2-p}}\right| = |f(\zeta)|^{p-1}$$

for $f = 1-\overline{\Theta(0)}\Theta(\zeta)$. Consequently, equality occurs in (2.3) for $f = 1-\overline{\Theta(0)}\Theta(\zeta)$. Thus,

$$\begin{aligned}\|\delta_0\|_{C_p(\Theta)} &= (1-|\Theta(0)|^2)^p\left(\int_{\mathbb{T}} |(1-\overline{\Theta(0)}\Theta(\zeta)|^p dm(\zeta)\right)^{-1}\\ &= \left(\int_{\mathbb{T}} |1-\overline{\Theta(0)}z|^p dm(z)\right)^{-1}\end{aligned}$$

because the transformation $\dfrac{\Theta(0)-\Theta}{1-\overline{\Theta(0)}\Theta}$ preserves the Lebesgue measure m. □

The author does not know whether Theorem 2.3 is true for $p < 1$.

3. The spaces $C_p(\Theta)$ and rotations

We note that the equality $C = C_p(\Theta)$ can be fulfilled only for $p < 1$. Indeed, by the open mapping theorem the equality $C = C_p(\Theta)$ implies the equivalence of the norms $\|\ \|_C$ and $\|\ \|_{C_p(\Theta)}$ but $\lim_{|\Theta(a)|\to 1} \dfrac{\|\delta_a\|_{C_p(\Theta)}}{\|\delta_a\|_C} = 0$ for $p \geq 1$ by Theorem 2.1.

The following theorem yields a necessary and sufficient condition for the equality $C = C_p(\Theta)$.

Theorem 3.1. *Let Θ be an inner function. Let $0 < p < +\infty$. Then the following statements are equivalent*

1) $C_p(\Theta) = C$,
2) *there exists a positive number C such that*

$$\sup\{|f(a)|^p : f \in \Theta^*(H^p),\ \|f\|_{H^p} \leq 1\} \geq \frac{C}{1-|a|^2}$$

for all $a \in \mathbb{D}$,

3) $\rho(\Theta) = \mathbb{T}$ *and* $0 < p < 1$.

Proof. We prove that 1) $\Longrightarrow$ 2). The statement 2) means that $\|\delta_a\|_{\mathcal{C}_p(\Theta)} \geq C\|\delta_a\|_{\mathcal{C}}$ for all $a \in \mathbb{D}$. To prove this inequality it suffices to apply the open mapping theorem. Now, we prove that 2) $\Longrightarrow$ 3). As noted above, the inequality $p < 1$ follows from Theorem 2.1. Suppose that $\rho(\Theta) \neq \mathbb{T}$. Then for $\zeta \in \mathbb{T} \setminus \rho(\Theta)$ we have

$$\sup\{|f(r\zeta)|^p : f \in \Theta^*(H^p),\ \|f\|_{H^p} \leq 1,\ 0 \leq r < 1\} < +\infty,$$

which contradicts statement 2). It remains to prove that 3) $\Longrightarrow$ 1). It suffices to check that $\int_{\overline{\mathbb{D}}} |f|^p d\mu \leq C(p,\mu) \int_{\mathbb{T}} |f|^p dm$ for any positive measure $\mu \in \mathcal{C}_p(\Theta)$ and any polynomial f provided $0 < p < 1$ and $\rho(\Theta) = \mathbb{T}$. Take a positive integer n greater than the degree of f. Then $\dfrac{f}{1-\zeta z^n} \in \Theta^*(H^p)$ for all $\zeta \in \mathbb{T}$. Consequently,

$$\int_{\overline{\mathbb{D}}} \frac{|f(z)|^p}{|1-\zeta z^n|^p} d\mu(z) \leq \|\mu\|_{\mathcal{C}_p(\Theta)} \int_{\mathbb{T}} \frac{|f(z)|^p}{|1-\zeta z^n|^p} dm(z)$$

for all $\zeta \in \mathbb{T}$. Averaging this inequality over $\zeta \in \mathbb{T}$ we obtain

$$\int_{\overline{\mathbb{D}}} |f|^p d\mu \leq C(p,\mu) \int_{\mathbb{T}} |f|^p dm.$$ □

Thus, if the equivalent statements of Theorem 3.1 are fulfilled, then $\mathcal{C}_p(\Theta)$ is rotation invariant. Moreover, clearly, $\mathcal{C}_p(\Theta)$ is rotation invariant in the case where Θ is a finite Blaschke product. The following theorem asserts that if $p \neq 1$, then $\mathcal{C}_p(\Theta)$ is rotation invariant only in these two special cases.

Theorem 3.2. *Let Θ be an inner function. Let $0 < p < +\infty$, $p \neq 1$. Then the following statements are equivalent*

1) *$\mathcal{C}_p(\Theta)$ is rotation invariant,*
2) *Θ is a finite Blaschke product or $\mathcal{C}_p(\Theta) = \mathcal{C}$.*

Proof. Clearly, 2) implies 1). We prove that 1) implies 2). First, we suppose that $p < 1$. Then $\delta_\zeta \in \mathcal{C}_p(\Theta)$ for $\zeta \in \mathbb{T}$ if and only if $\zeta \notin \rho(\Theta)$. Consequently, 1) implies that $\rho(\Theta)$ is rotation invariant, i.e. $\rho(\Theta) = \varnothing$ or $\rho(\Theta) = \mathbb{T}$. In the first case Θ is a finite Blaschke product. In the second case $\mathcal{C}_p(\Theta) = \mathcal{C}$ by Theorem 3.1. Now let $p > 1$. Condition 1) implies the inequality $C^{-1}\|\delta_a\|_{\mathcal{C}_p(\Theta)} \leq \|\delta_{a\zeta}\|_{\mathcal{C}_p(\Theta)} \leq C\|\delta_a\|_{\mathcal{C}_p(\Theta)}$ for all $a \in \mathbb{D}$ and $\zeta \in \mathbb{T}$, where $C = C(\Theta, p)$ is a positive constant. This inequality and Theorem 2.1 imply that

$$\|\delta_a\|_{\mathcal{C}_p(\Theta)} = o((1-|a|)^{-1}) \quad \text{as} \quad |a| \to 1. \tag{3.1}$$

Clearly,

$$\|\delta_a\|_{\mathcal{C}_p(\Theta)} \geq \frac{1-|\Theta(a)|^2}{1-|a|^2} \left\| \frac{1-\overline{\Theta(a)}\Theta}{1-\bar{a}z} \right\|_{H^p}^{-p}.$$

Consequently,
$\|\delta_a\|_{\mathcal{C}_p(\Theta)} \geq C(p)(1-|a|)^{-1}$ provided $|\Theta(a)| < \frac{1}{2}$. Now (3.1) implies that $|\Theta(a)| \geq \frac{1}{2}$ if a is close enough to $\mathbb{T}$, i.e. Θ is a finite Blaschke product. □

Remark 1. *There exists an infinite Blaschke product B such that $C_1(B)$ is rotation invariant (see §4). Thus, the condition $p \neq 1$ is essential in Theorem 3.2.*

Set $\Lambda_p(\Theta) \overset{\text{def}}{=} \{\zeta \in \mathbb{T} : \mathcal{C}_p(\Theta) = \mathcal{C}_p(\Theta_\zeta)\}$, where $\Theta_\zeta(z) \overset{\text{def}}{=} \Theta(\zeta z)$. Clearly, $\Lambda_p(\Theta)$ is a (not necessarily closed) subgroup of $\mathbb{T}$. Note that $\Lambda_p(\Theta) = \mathbb{T}$ if and only if $\mathcal{C}_p(\Theta)$ is rotation invariant. Thus, by Theorem 3.2 the equality $\Lambda(\mathcal{C}_p(\Theta)) = \mathbb{T}$ for $p \neq 1$ can be fulfilled only in exceptional cases. In particular, if $\Lambda(\mathcal{C}_p(\Theta)) = \mathbb{T}$ and $p > 1$, then Θ is a finite Blaschke product.

Remark 2. *There exists a singular inner function Θ such that $\bigcap_{p>0} \Lambda_p(\Theta)$ is a subgroup of $\mathbb{T}$ of the power of the continuum.*

Proof. Let $\{n_k\}_{k\geq 1}$ be a sequence in $\mathbb{N}$ such that n_k is a divisor of n_{k+1} $(k \geq 1)$ and $\sum_{k\geq 1} \frac{n_k}{n_{k+1}} < +\infty$. Consider the Riesz product $\mu = \prod_{k\geq 1}(1 + \frac{1}{2}\operatorname{Re} z^{n_k})$. It is well known that μ is a positive singular measure on $\mathbb{T}$. Set

$$\Theta(z) = S_\mu(z) \overset{\text{def}}{=} \exp\left(\int_{\mathbb{T}} \frac{z+\zeta}{z-\zeta}\, d\mu(\zeta)\right).$$

It is clear that $\Theta_\zeta = S_{\mu_\zeta}$, where $\mu_\zeta = \prod_{k\geq 1}(1 + \frac{1}{2}\operatorname{Re}(\zeta z)^{n_k})$. Denote by X the set of all $\zeta \in \mathbb{T}$ satisfying $\sum_{k\geq 1} |\zeta^{n_k} - 1| < +\infty$. We prove that $X \subset \Lambda_p(\Theta)$ for all $p > 0$ and X is of the power of the continuum. It is easy to see that

$$(1 + |\zeta^{n_k} - 1|)^{-1} \leq \frac{1 + \frac{1}{2}\operatorname{Re}(\zeta z)^{n_k}}{1 + \frac{1}{2}\operatorname{Re} z^{n_k}} \leq 1 + |\zeta^{n_k} - 1|.$$

Consequently, for any $\zeta \in X$ there exists a function h_ζ in $L^\infty(\mu)$ such that $\mu_\zeta = h_\zeta \mu$ and $h_\zeta^{-1} \in L^\infty(\mu)$. Thus, $X \subset \Lambda_p(\Theta)$ by Corollary 3 to Theorem 1.3. To prove that X is of the power of the continuum it suffices to note that $\exp(2\pi i \sum_{k\in N} n_k^{-1}) \in X$ for any subset N of $\mathbb{N}$. □

Using the Frostman theorem we may construct also an infinite Blaschke product B such that $\bigcap_{p>0} \Lambda_p(B)$ is a subgroup of $\mathbb{T}$ of the power of the continuum.

Remark 3. *For any countable subset Λ of $\mathbb{T}$ there exists an inner function Θ such that $\Lambda \subset \Lambda_p(\Theta)$ for all $p > 0$.*

Proof. It is enough to note that we can choose a sequence $\{n_k\}_{k\geq 1}$ in the proof of Remark 2 in such a way that $\Lambda \subset X$. □

4. BMO-interpolating Blaschke products and embedding theorems

A set $\Lambda \subset \mathbb{D}$ is said to be *BMO-interpolating* if for any function $\varphi : \Lambda \to \mathbb{C}$ with $\varphi(z) = O\left(\log \frac{1}{1-|z|}\right)$ there exists a function[6] $f \in BMO_A$ such that $f|_\Lambda = \varphi$. Clearly, a BMO-interpolating set cannot be a uniqueness set for H^2. Consequently,

[6] In connection with the space BMO_A we refer to [13].

$\sum_{\lambda\in\Lambda}(1-|\lambda|) < +\infty$ for any BMO-interpolating set Λ. A Blaschke product B with simple zeros is said to be *BMO-interpolating* if $B^{-1}(0)$ is BMO-interpolating.

Remark. *We will see that any BMO-interpolating set is an interpolating set (see Corollary 1 to Theorem 4.2). Sundberg [16] described the set $\{\{f(\lambda)\}_{\lambda\in\Lambda} : f \in BMO_A\}$ for any interpolating set $\Lambda \subset \mathbb{D}$. But it is not clear whether his result allows one to obtain a description of BMO_A-interpolating sets. Kotochigov [14] investigated BMO-interpolating sets contained in a Stolz angle. Other examples of BMO-interpolating sets were considered by Dyakonov [12].*

It is easy to see that a Blaschke product $B(z) = \prod_{n\geq 1} \frac{\overline{a}_n}{|a_n|}\frac{a_n - z}{1-\overline{a}_n z}$ is BMO-interpolating if and only if

$$\sum_{n\geq 1} |\alpha_n| \log \tfrac{2}{1-|a_n|} \leq C \Big\| \sum_{n\geq 1} \frac{\alpha_n}{1-\overline{a}_n z}\Big\|_{H^1}$$

for any sequence $\{\alpha_n\}_{n\geq 1}$ with $\alpha_n = 0$ for n large enough. In other words, B is BMO-interpolating if and only if $\left\{\frac{1}{(1-\overline{a}_n z)\log \frac{2}{1-|a_n|}}\right\}_{n\geq 1}$ is an l^1-basis in $B^*(H^1)$. From this we deduce

Theorem 4.1. *Let $B(z) = \prod_{n\geq 1} \frac{\overline{a}_n}{|a_n|}\frac{a_n - z}{1-\overline{a}_n z}$ be a BMO-interpolating Blaschke product. Then $\mu \in M_+(\overline{\mathbb{D}})$ belongs to $\mathcal{C}_1(B)$ if and only if $\int_{\overline{\mathbb{D}}} \frac{d\mu(z)}{|1-\overline{a}_n z|} \leq C \log\frac{2}{1-|a_n|}$ for all $n \geq 1$. Moreover,*

$$C_1\|\mu\|_{\mathcal{C}_1(B)} \leq \sup_{n\geq 1}\left(\log\tfrac{2}{1-|a_n|}\right)^{-1} \int_{\overline{\mathbb{D}}} \frac{d\mu(z)}{|1-\overline{a}_n z|} \leq C_2\|\mu\|_{\mathcal{C}_1(B)}. \qquad \square$$

Corollary 1. *Let B and the zeros $\{a_n\}_{n\geq 1}$ satisfy the assumptions of Theorem 4.1. Suppose that the sequence $\{|a_n|\}_{n\geq 1}$ increases monotonically. Then*

$$\|\delta_a\|_{\mathcal{C}_1(B)} \leq C \max\left(\frac{1}{(1-|a_n|)\log\frac{2}{1-|a_n|}}, \frac{1}{(1-|a|)\log\frac{2}{1-|a_{n+1}|}}\right)$$

for all $a \in \mathbb{D}$ with $|a_n| \leq |a| \leq |a_{n+1}|$.

Proof. If $k \leq n$, then

$$\frac{1}{|1-a\overline{a}_k|\log\frac{2}{1-|a_k|}} \leq \frac{1}{(1-|a_k|)\log\frac{2}{1-|a_k|}} \leq C\frac{1}{(1-|a_n|)\log\frac{2}{1-|a_n|}}.$$

If $k > n$, then

$$\frac{1}{|1-a\overline{a}_k|\log\frac{2}{1-|a_k|}} \leq \frac{1}{(1-|a|)\log\frac{2}{1-|a_k|}} \leq \frac{1}{(1-|a|)\log\frac{2}{1-|a_{n+1}|}}.$$

It remains to apply Theorem 4.1. $\square$

Corollary 2. *Under the assumptions of Theorem* 4.1

$$C_1\|\delta_{a_n}\|_{C_1(B)} \le \frac{1}{(1-|a_n|)\log\frac{2}{1-|a_n|}} \le C_2\|\delta_{a_n}\|_{C_1(B)}$$

for all $n \in \mathbb{N}$. □

Corollary 3. *Under the assumptions of Theorem* 4.1

$$\|\delta_a\|_{C_1(B)} \le \frac{C}{(1-|a|)\log\frac{2}{1-|a|}}$$

for all $a \in \mathbb{D}$. □

Theorem 4.2. *Let* Θ *be an inner function which is not a finite Blaschke product. Assume that* $\|\delta_a\|_{C_1(\Theta)} = o\Big(\frac{1}{1-|a|}\Big)$ *as* $|a| \to 1$. *Then* Θ *is a Blaschke product with* $\lim_{n\to\infty}(1-|a_n|^2)|\Theta'(a_n)| = 1$ *where* $\{a_n\}_{n\ge1}$ *is the sequence of zeros of* Θ.

First, we deduce some consequences from this theorem.

Corollary 1. *Let* B *be a BMO-interpolating Blaschke product. Then* B *is an interpolating Blaschke product.*

Proof. It suffices to apply Corollary 3 to Theorem 4.1, and then Theorem 4.2. □

Corollary 2. *Let* Θ *be an inner function. Assume that* $C_1(\Theta)$ *is rotation invariant. Then there exists a finite Blaschke product* b *such that* Θb^{-1} *is an interpolating Blaschke product.*

Proof. The proof of Theorem 3.2 shows that $\|\delta_a\|_{C_1(\Theta)} = o\Big(\frac{1}{1-|a|}\Big)$ as $|a| \to 1$. It remains to apply Theorem 4.2. □

Before proving Theorem 4.2, we introduce some notation and prove auxiliary assertions. With every inner function Θ we associate a function $d_\Theta : \mathbb{D} \to [0,1)$ defined by the formula $d_\Theta(z) = \inf\{\max(|\varphi(z)|, |\psi(z)|) : \Theta = \varphi\psi\}$, where the infimum is taken over all representations of Θ in the form of the product of two inner functions $\Theta = \varphi\psi$. It is easily seen that this infimum is attained.

Lemma 4.3. *Let* Θ *be an inner function. Then*

$$\|\delta_a\|_{C_1(\Theta)} \ge \frac{(1-d_{z\Theta}^2(a))^2}{1-|a|^2}$$

for all $a \in \mathbb{D}$.

Proof. We may represent the function $z\Theta$ in the form $z\Theta = \varphi\psi$ in such a way that $\max(|\varphi(a)|, |\psi(a)|) \le d_{z\Theta}(a)$. Set

$$f(z) = \frac{(1-\overline{\varphi(a)}\varphi(z))(1-\overline{\psi(a)}\psi(z))}{(1-\bar{a}z)^2}.$$

It is easy to see that

$$f \in \Theta^*(H^1), \quad \|f\|_{H^1} \le \frac{4}{1-|a|^2}, \quad \text{and} \quad |f(a)| \ge \frac{(1-d^2_{z\Theta}(a))^2}{(1-|a|^2)^2}.$$

Consequently,

$$\|\delta_a\|_{C_1(\Theta)} \ge \frac{|f(a)|}{\|f\|_{H^1}} \ge \frac{(1-d^2_{z\Theta}(a))^2}{1-|a|^2}.$$

□

Lemma 4.4. *Let B be a Blaschke product. Suppose that $|B'(a)|(1-|a|^2) \ge \tau$ for any root a of B, where τ is positive constant. Then $d_B \ge \dfrac{1-\sqrt{1-\tau^2}}{\tau}$ everywhere in $\mathbb{D}$*

Proof. Let a be a root of B. Set $\varphi_a(z) = \frac{a-z}{1-\bar{a}z}$, $B_a = \varphi_a^{-1}B$. Denote by K_a the disk $\{z \in \mathbb{D} : |\varphi(z)| \le \varepsilon\}$, where $\varepsilon = \dfrac{1-\sqrt{1-\tau^2}}{\tau}$. Clearly, $|B_a(a)| = |B'(a)|(1-|a|^2) \ge \tau$. Consequently, $|B_a| \ge \frac{\tau-\varepsilon}{1-\tau\varepsilon} = \varepsilon$ everywhere in K_a. Thus, $|B| = |B_a||\varphi_a| \ge \varepsilon^2$ everywhere on the boundary ∂K_a of K_a. By the maximum principle we have $|B| \ge \varepsilon^2$ in $\mathbb{D} \setminus \bigcup_{B(a)=0} K_a$. Consequently, $d_B \ge \varepsilon$ everywhere in $\mathbb{D} \setminus \bigcup_{B(a)=0} K_a$. It remains to prove that $d_B \ge \varepsilon$ everywhere in K_a. Let $B = \varphi\psi$, where φ and ψ are inner functions. Clearly, $\varphi(a) \ne 0$ or $\psi(a) \ne 0$. To be definite assume that $\varphi(a) \ne 0$. Then $|\varphi| \ge |B_a| \ge \varepsilon$ everywhere in K_a. Consequently, $d_B \ge \varepsilon$ in K_a.

□

Corollary. *Let Θ be an inner function that is not a finite Blaschke product. Then the following two assertions are equivalent*

1) $\lim_{|z|\to 1-} d_\Theta(z) = 1$,
2) *Θ is a Blaschke product with $\lim_{n\to\infty}(1-|a_n|^2)|\Theta'(a_n)| = 1$ where $\{a_n\}_{n\ge 1}$ is the sequence of zeros of Θ.*

Proof. We prove that 1) $\Longrightarrow$ 2). We may represent Θ in the form $\Theta = SB$ where S is a singular inner function and B is a Blaschke product. Assume that $S \ne$ const. Then $\liminf_{|z|\to 1-} d_\Theta(z) \le \liminf_{|z|\to 1-} d_S(z) = 1$, and we get a contradiction. Consequently, Θ be a Blaschke product. To prove the required equality it remains to note that $d_\Theta(a_n) = (1-|a_n|^2)|\Theta'(a_n)|$. Now we prove that 2) $\Longrightarrow$ 1). Denote by B_N the Blaschke product with the sequence of zeros $\{a_n\}_{n\ge N}$. Fix $\tau \in (0,1)$. Clearly, there exists $N \in \mathbb{N}$ such that $(1-|a_n|^2)|B_N'(a_n)| \le \tau$ for all $n \ge N$. By Lemma 4.4 we have $d_{B_N} \ge \dfrac{1-\sqrt{1-\tau^2}}{\tau}$. Consequently, $\liminf_{|z|\to 1} d_\Theta = \liminf_{|z|\to 1} d_{B_N} \ge \dfrac{1-\sqrt{1-\tau^2}}{\tau}$. It remains to note that τ is an arbitrary number in $(0,1)$.

□

Remark. *The example of the Blaschke product $B(z) = z\dfrac{\tau - z}{1-\tau z}$ shows that the constant $\dfrac{1-\sqrt{1-\tau^2}}{\tau}$ in Lemma 4.4 is sharp.*

Proof of Theorem 4.2. By Lemma 4.3 we have $\lim_{|a|\to 1} d_{zB}(a) = 1$, whence $\lim_{|a|\to 1} d_B(a) = 1$. It remains to apply the corollary to Lemma 4.4. □

Denote by $\widetilde{\mathcal{C}}$ the set of all measures in $M(\overline{\mathbb{D}})$ such that

$$\int_{\overline{\mathbb{D}}} \frac{d|\mu|(\zeta)}{|1-\bar{a}\zeta|} \le C \log \frac{2}{1-|a|} \tag{4.1}$$

for all $a \in \mathbb{D}$. Clearly, $\mathcal{C} \subset \widetilde{\mathcal{C}}$.

Let Ω be a subset of $\mathbb{D}$. Then Ω is said to be *exhaustive* if for any $a \in \mathbb{D}$ with $|a| \ge \frac{1}{2}$ there exists $b \in \Omega$ such that $|a-b| \le C(1-|a|)$ and $1-|b| \ge (1-|a|)^{1+\delta}$ where $C = C(\Omega)$ and $\delta = \delta(\Omega)$ are positive constants.

It is to easy to check that if Ω is an exhaustive set, then $\sum_{\lambda\in\Omega}(1-|\lambda|)^{\varepsilon} = +\infty$ for some positive ε. In fact we may set $\varepsilon = (1+\delta)^{-1}$.

Lemma 4.5. *Let $\Omega \subset \mathbb{D}$. Then Ω is exhaustive if and only if for any $a \in \mathbb{D}$ there exists a $z \in \Omega$ satisfying*

$$|1-\bar{a}z| \log \tfrac{2}{1-|z|} \le C(1-|a|) \log \tfrac{2}{1-|a|}. \tag{4.2}$$

Proof. It is enough to prove that (4.2) implies that $|z-a| \le C(1-|a|)$ and $1-|z| \ge (1-|a|)^{1+\delta}$ provided $|a| \ge \frac{1}{2}$. Clearly, $|1-\bar{a}z| \ge 1-|a|$. Consequently, (4.2) implies that $\log \frac{2}{1-|z|} \le C \log \frac{2}{1-|a|}$. Thus, $1-|z| \ge (1-|a|)^{1+\delta}$ provided $|a| \ge \frac{1}{2}$. Moreover, (4.2) implies that $(1-|z|) \log \frac{2}{1-|z|} \le C(1-|a|) \log \frac{2}{1-|a|}$. Consequently, $1-|z| \le C_1(1-|a|)$, whence $\log \frac{2}{1-|a|} \le C_2 \log \frac{2}{1-|z|}$. Thus, $|1-\bar{a}z| \log \frac{2}{1-|z|} \le C(1-|a|) \log \frac{2}{1-|a|} \le C_3(1-|a|) \log \frac{2}{1-|z|}$, whence $|1-\bar{a}z| \le C(1-|a|)$. The last inequality clearly implies that $|z-a| \le C(1-|a|)$. □

Theorem 4.6. *Let Θ be an inner function. Assume that the set $\{|\Theta| < \varepsilon\}$ is exhaustive for some $\varepsilon \in (0,1)$. Then $\mathcal{C}_1(\Theta) \subset \widetilde{\mathcal{C}}$.*

Proof. Let μ be a positive measure in $\mathcal{C}_1(\Theta)$. We need to prove that μ satisfies (4.1). Take a point $b \in \mathbb{D}$ such that $|a-b| \le C(1-|a|)$, $1-|b| \ge (1-|a|)^{1+\delta}$ and $|\Theta(b)| < \varepsilon$. Then $|1-\bar{a}z|^{-1} \le C\left|\dfrac{1-\overline{\Theta(b)}\Theta}{1-\bar{b}z}\right|$ everywhere in $\overline{\mathbb{D}}$. Consequently,

$$\begin{aligned}\int_{\overline{\mathbb{D}}} \frac{d\mu(\zeta)}{|1-\bar{a}\zeta|} &\le C \int_{\overline{\mathbb{D}}} \left|\frac{1-\overline{\Theta(b)}\Theta}{1-\bar{b}z}\right| d\mu(\zeta) \le C \left\|\frac{1-\overline{\Theta(b)}\Theta}{1-\bar{b}z}\right\|_{H^1} \\ &\le C \log \frac{2}{1-|b|} \le C \log \frac{2}{1-|a|}.\end{aligned}$$ □

Theorem 4.7. *Let B be an infinite BMO-interpolating Blaschke product. Then the following statements are equivalent.*

1) *$\mathcal{C}_1(B)$ is rotation invariant.*
2) *$\mathcal{C}_1(B) = \widetilde{\mathcal{C}}$.*
3) *$B^{-1}(0)$ is exhaustive.*
4) $\|\delta_a\|_{\mathcal{C}_1(B)} \ge \dfrac{C}{(1-|a|)\log \frac{2}{1-|a|}}$ *for all $a \in \mathbb{D}$.*

Proof. Let $\{a_n\}_{n\geq 1}$ be the sequence of zeros of B. We prove that 4) $\Longrightarrow$ 3). By Theorem 4.1, statement 4) implies the inequality

$$\frac{1}{|1-\bar{a}_n a|\log\frac{2}{1-|a_n|}} \geq \frac{C}{(1-|a|)\log\frac{2}{1-|a|}}.$$

Thus, $B^{-1}(0)$ is exhaustive by Lemma 4.5. The implication 3) $\Longrightarrow$ 2) is an evident consequence of Theorems 4.1 and 4.6. The implication 2) $\Longrightarrow$ 1) is trivial. It remains to prove that 1) $\Longrightarrow$ 4). Set $r_n = |a_n|$. We may assume that the sequence $\{r_n\}_{n\geq 1}$ is monotone increasing. First, we prove that $\log\frac{2}{1-r_n} \geq C\log\frac{2}{1-r_{n+1}}$ for all $n \in \mathbb{N}$. It is clear that $\|\delta_a\|_{C_1(B)} \geq C\|\delta_{a_n}\|_{C_1(B)} \geq C(1-r_n)^{-1}\log\frac{2}{1-r_n}$ for any $a \in \mathbb{D}$ with $|a| = r_n$. Thus, by Theorem 4.1, for any $a \in \mathbb{D}$ with $|a| = r_n$ there exists $k = k(a) \in \mathbb{N}$ satisfying

$$\frac{1}{|1-\bar{a}_k a|\log\frac{2}{1-r_k}} \geq \frac{C}{(1-r_n)\log\frac{2}{1-r_n}}. \tag{4.3}$$

Inequality (4.3) implies that $|a-a_k| \leq C_1(1-r_n)$ (see the proof of Lemma 4.5). Clearly, $m\{\zeta \in \mathbb{T} : |\zeta r_n - a_k| \leq C_1(1-r_n)\} \leq C_2(1-r_n)$ if n is large enough. Thus, the relation $1-r_n = o(\frac{1}{n})$ as $n \to +\infty$ implies that $k(a) > n$ for some $a \in \mathbb{D}$ with $|a| = r_n$. Inequality (4.3) with $k > n$ clearly implies that

$$\log\frac{2}{1-r_n} \geq C\log\frac{2}{1-r_k} \geq C\log\frac{2}{1-r_{n+1}}$$

if n is large enough. Clearly,

$$\|\delta_a\|_{C_1(B)} \geq \frac{C}{|1-\bar{a}_j a|\log\frac{2}{1-r_j}}$$

for any $a \in \mathbb{D}$ and any $j \geq 1$. Using 1), we obtain

$$\|\delta_a\|_{C_1(B)} \geq \frac{C}{(1-r_j|a|)\log\frac{2}{1-r_j}}$$

for any $a \in \mathbb{D}$ and any $j \geq 1$. Thus, if $r_n \leq |a| \leq r_{n+1}$, then

$$\|\delta_a\|_{C_1(B)} \geq \frac{C}{(1-r_n|a|)\log\frac{2}{1-r_{n+1}}} \geq \frac{C}{(1-|a|)\log\frac{2}{1-|a|}}.$$ □

Let $\Lambda \subset \mathbb{D}$ be representable in the form $\Lambda \stackrel{\text{def}}{=} \bigcup_{k\in\mathbb{N}} A_k$, where A_k is the set of all vertices of a regular n_k-gon inscribed in the circle $r_k\mathbb{T}$ with $0 < r_k < 1$. Suppose that the sequence $\{r_k\}_{k\geq 1}$ tends monotonically to 1.

Theorem 4.8. *Suppose that the sequences $\{n_k\}_{k\geq 1}$ and $\{r_k\}_{k\geq 1}$ satisfy the following conditions*

1) $n_{k+1} \leq C(1-r_k)^{-\gamma}$,
2) $n_{k+2} \geq C^{-1}(1-r_k)^{-1}$,
3) $n_{k+2} \leq C\dfrac{1-r_k}{(1-r_{k+2})^{\gamma}}$,

where $C > 0$ and $0 < \gamma < \frac{1}{2}$. Then Λ is BMO-interpolating and exhaustive.

We need some auxiliary assertions.

Lemma 4.9. *Let* $0 < \delta < 1$. *Set* $\Delta_a = \Delta_a(\delta) \stackrel{\text{def}}{=} \{\zeta \in \mathbb{T} : |\zeta - \frac{a}{|a|}| < (1-|a|)^\delta\}$, *where* $a \in \mathbb{D}$. *Then*

$$\int_{\Delta_a} \frac{dm(\zeta)}{|1-\bar{a}\zeta|} \sim (1-\delta) \int_{\mathbb{T}} \frac{dm(\zeta)}{|1-\bar{a}\zeta|} \quad (|a| \to 1-).$$

We omit the standard proof of this lemma.

Corollary. *For any* $\delta \in (0, \frac{1}{2})$ *there exists* $\tau > 0$ *such that*

$$\int_{\Delta_a} \frac{dm(\zeta)}{|1-\bar{a}\zeta|} \geq (\tfrac{1}{2}+\tau) \int_{\mathbb{T}} \frac{dm(\zeta)}{|1-\bar{a}\zeta|}$$

for all $a \in \mathbb{D}$ *close enough to* $\mathbb{T}$. □

Lemma 4.10. *Let* $\Lambda \subset \mathbb{D}$. *Suppose that there exists a positive* τ *and a family* $\{\Sigma_a\}_{a\in\Lambda}$ *of pairwise disjoint subsets of* $\mathbb{T}$ *such that*

$$\int_{\Sigma_a} \frac{dm(\zeta)}{|1-\bar{a}\zeta|} \geq (\tfrac{1}{2}+\tau) \int_{\mathbb{T}} \frac{dm(\zeta)}{|1-\bar{a}\zeta|}$$

for all $a \in \Lambda$. *Then* Λ *is BMO-interpolating.*

The proof of this lemma can be found in [12].

Proof of Theorem 4.8. Fix number $\delta \in (\gamma, \frac{1}{2})$. First, we prove that Λ is exhaustive. Let $a \in \mathbb{D}$ with $r_k \leq |a| \leq r_{k+1}$. Denote by b a point in A_{k+3} that is nearest to $\frac{ar_{k+3}}{|a|}$. Clearly, $\left|b - \frac{ar_{k+3}}{|a|}\right| < \frac{\pi r_{k+3}}{n_{k+3}}$. Consequently, $|a-b| \leq \left|a - \frac{ar_{k+3}}{|a|}\right| + \left|\frac{ar_{k+3}}{|a|} - b\right| < 1-|a| + \frac{\pi r_{k+3}}{n_{k+3}} \leq 1-|a| + C(1-r_{k+1}) \leq (C+1)(1-|a|)$. Note that conditions 1) and 2) of Theorem 4.8 imply the inequality $(1-r_{k+1})^\gamma \leq C^2(1-r_k)$, whence $(1-r_{k+1})^\delta \leq 1-r_k$ for k large enough. Consequently, $1-|b| = 1-r_{k+3} \geq (1-r_k)^{\delta^{-3}} \geq (1-|a|)^{\delta^{-3}}$ for k large enough. Now using Lemma 4.10 we prove that Λ is BMO-interpolating. Clearly, it suffices to prove that $\Lambda_l \stackrel{\text{def}}{=} \bigcup_{k>l} A_k$ is BMO-interpolating for l large enough. Let Δ_a denote the same as in Lemma 4.9. Note that if $\Delta_a \cap \Delta_b \neq \varnothing$ with $|a| = |b| = r_k$, then $\left|\frac{a}{|a|} - \frac{b}{|b|}\right| < 2(1-r_k)^\delta \leq 2(1-r_{k-1})^\delta$. Consequently, by condition 1) of Theorem 4.8, the sets Δ_a with $a \in A_k$ are pairwise disjoint for k large enough. For $a \in A_k$, we set $\Sigma_a = \Delta_a \setminus \bigcup_{b\in\Lambda_k} \Delta_b$. Clearly, the sets Σ_a with $a \in \Lambda_l$ are pairwise disjoint for l large enough. We prove that $m(\Delta_a \setminus \Sigma_a) \leq C(1-r_k)$ for $a \in A_k$. We have $\Delta_a \setminus \Sigma_a \subset \bigcup_{b\in A_{k+1}} (\Delta_a \cap \Delta_b) \cup \bigcup_{b\in\Lambda_{k+1}} \Delta_b$. Clearly, $m\big(\bigcup_{b\in\Lambda_{k+1}} \Delta_b\big) \leq \sum_{b\in\Lambda_{k+1}} m(\Delta_b) \leq C\sum_{j\geq k+2} n_j(1-r_j)^\delta \leq C\sum_{j\geq k+2}(1-r_{j-2}) \leq C(1-r_k)$ because condition 3) of Theorem 4.8 implies that $\lim_{j\to+\infty} \frac{1-r_{j+2}}{1-r_j} = 0$. To estimate the Lebesgue measure of $\bigcup_{b\in A_{k+1}} (\Delta_a \cap \Delta_b)$ we note that for k large enough there exists at most one $b \in A_{k+1}$ such that $\Delta_a \cap \Delta_b \neq \varnothing$. Indeed, if $\Delta_a \cap \Delta_b \neq \varnothing$, then clearly $\left|\frac{a}{|a|} - \frac{b}{|b|}\right| < (1-r_k)^\delta + (1-r_{k+1})^\delta \leq 2(1-r_k)^\delta$. Consequently, if

$\Delta_a \cap \Delta_{b_1} \neq \varnothing$ and $\Delta_a \cap \Delta_{b_2} \neq \varnothing$ with $b_1 \neq b_2$, then $4C^{-1}(1-r_k)^\gamma \leq \dfrac{4}{n_{k+1}} \leq |b_1 - b_2| \leq 4(1-r_k)^\delta$, and we get a contradiction for k large enough. Thus, $m\big(\bigcup_{b\in A_{k+1}}(\Delta_a \cap \Delta_b)\big) \leq C(1-r_{k+1})^\delta \leq \dfrac{C}{n_{k+2}} \leq C(1-r_k)$ for k large enough. Now using the corollary to Lemma 4.9 we have

$$\begin{aligned}\int_{\Sigma_a} \frac{dm(\zeta)}{|1-\overline{a}\zeta|} &\geq \int_{\Delta_a} \frac{dm(\zeta)}{|1-\overline{a}\zeta|} - \int_{\Delta_a\setminus\Sigma_a} \frac{dm(\zeta)}{|1-\overline{a}\zeta|}\\ &\geq (\tfrac{1}{2}+\tau)\int_{\mathbb{T}} \frac{dm(\zeta)}{|1-\overline{a}\zeta|} - \frac{m(\Delta_a\setminus\Sigma_a)}{1-r_k}\\ &\geq (\tfrac{1}{2}+\tau)\int_{\mathbb{T}} \frac{dm(\zeta)}{|1-\overline{a}\zeta|} - C \geq \frac{1+\tau}{2}\int_{\mathbb{T}} \frac{dm(\zeta)}{|1-\overline{a}\zeta|}\end{aligned}$$

for $a \in A_k$ and k large enough. It remains to apply Lemma 4.10. □

Example. *Set* $B(z) \stackrel{\text{def}}{=} \prod_{k\geq 1} \dfrac{r_k^{n_k} - z^{n_k}}{1 - r_k^{n_k} z^{n_k}}$, *where* $r_k = 1 - 2^{-3^{k+1}}$ *and* $n_k = 2^{3^{k-1}}$. *Then* $\mathcal{C}_1(B) = \widetilde{\mathcal{C}}$. *In particular,* $\mathcal{C}_1(B)$ *is rotation invariant.*

Proof. It is easy to check that the sequences $\{r_k\}_{k\geq 1}$ and $\{n_k\}_{k\geq 1}$ satisfy condition 1), 2) and 3) of Theorem 4.8. Consequently, B is a BMO-interpolating Blaschke product and $B^{-1}(0)$ is exhaustive. It remains to apply Theorem 4.7. □

References

[1] A. B. Aleksandrov, *Invariant subspaces of the backward shift operator in the space* H^p $(p \in (0,1))$, Zap. Nauchn. Sem. Leningrad. Otdel. Mat. Inst. Steklov. (LOMI), **92** (1979), 7–29. (Russian)

[2] A. B. Aleksandrov, *Invariant subspaces of the shift operators. An axiomatic approach,* Zap. Nauchn. Sem. Leningrad. Otdel. Mat. Inst. Steklov. (LOMI), **113** (1981), 7–26. (Russian)

[3] A. B. Aleksandrov, *Multiplicity of boundary values of inner functions,* Izvestia AN Arm. SSR, Matem., **22 (5)** (1987), 490–503. (Russian)

[4] A. B. Aleksandrov, *Inner functions and related spaces of pseudocontinuable functions,* Zap. Nauchn. Sem. Leningrad. Otdel. Mat. Inst. Steklov. (LOMI), **170** (1989), 7–33. (Russian)

[5] A. B. Aleksandrov, *A simple proof of the Volberg-Treil theorem on the embeddings of co-invariant subspaces of the shift operator,* Zap. Nauchn. Sem. S. Petersburg. Otdel. Mat. Inst. Steklov. (POMI), **217** (1994), 26–35. (Russian)

[6] A. B. Aleksandrov, *On existence of angular boundary values of pseudocontinuable functions,* Zap. Nauchn. Sem. S. Petersburg. Otdel. Mat. Inst. Steklov. (POMI), **222** (1995), 5–17. (Russian)

[7] D. Clark, *One dimensional perturbation of restricted shifts,* J. Anal. Math., **25** (1972), 169–191.

[8] W. S. Cohn, *Carleson measures for functions orthogonal to invariant subspaces,* Pacific J. Math., **103** (**2**) (1982), 347–364.

[9] W. S. Cohn, *Radial embedding theorems for invariant subspaces,* Complex Variables Theory Appl., **17** (**1-2**) (1991), 33–42.

[10] R. G. Douglas, H. S. Shapiro and A. L. Shields, *Cyclic vectors and invariant subspaces for the backward shift operator,* Ann. Inst. Fourier, **20** (**1**) (1970), 37–76.

[11] K. M. Dyakonov, *Smooth functions and co-invariant subspaces of the shift operator,* Algebra i Analiz, **4** (**5**) (1992), 117–147. (Russian)

[12] K. M. Dyakonov, *Moment problems for bounded functions,* Communications in Analysis and Geometry, **2** (**3**) (1994), 533–562.

[13] J. B. Garnett, *Bounded analytic functions,* Academic Press, New York, 1981.

[14] A. M. Kotochigov, *The free interpolation by analytic functions,* Thesis, Leningrad 1981. (Russian)

[15] N. K. Nikol'skiĭ, *Treatise on shift operator. Spectral function theory,* Grundlehren Math. Wiss., **273** (1986), Springer-Verlag Berlin–New York

[16] C. Sundberg, *Values of BMOA functions on interpolating sequences,* Michigan Math. J. **31** (1984), 21–30.

[17] A. L. Volberg and S. R. Treil, *Embedding theorems for invariant subspaces of the backward shift operator,* Zap. Nauchn. Sem. Leningrad. Otdel. Mat. Inst. Steklov. (LOMI), **149** (1986), 38–51. (Russian)

St. Petersburg Branch
Steklov Mathematical Institute,
Russian Academy of Sciences,
Fontanka 27
191011 St. Petersburg, Russia
E-mail address: alex@pdmi.ras.ru

Operator Theory:
Advances and Applications, Vol. 113

Continuous and Compact Embeddings Between Star-invariant Subspaces

Konstantin M. Dyakonov

Dedicated to the memory of S. A. Vinogradov

Abstract. Given an inner function θ on the upper half-plane $\mathbb{C}_+$, let $K_\theta^p \stackrel{\text{def}}{=} H^p \cap \theta\,\overline{H^p}$ be the corresponding star-invariant subspace of the Hardy space $H^p = H^p(\mathbb{C}_+)$. It has been previously shown by the author that the (continuous) embedding relation $K_\theta^p \subset K_\theta^q$ holds, for $1 < p < q < \infty$, if and only if $\theta' \in L^\infty(\mathbb{R})$. In this paper, we first give an extended version of the above result and then establish a compactness criterion for the embedding operator involved. Namely, we prove that the inclusion map $id : K_\theta^p \hookrightarrow K_\theta^q$ is compact if and only if $\theta' \in C_0(\mathbb{R})$. Similar results are obtained on the embeddings $K_\theta^p \subset C_0(\mathbb{R})$ and $K_\theta^p \subset \text{BMO}$.

1. Introduction and results

Let $\mathbb{C}_+$ denote the half-plane $\{ z \in \mathbb{C} : \operatorname{Im} z > 0 \}$ and $H^p = H^p(\mathbb{C}_+)$ the corresponding Hardy space of analytic functions (see [G], Chap. II). We also treat H^p as a subspace of $L^p = L^p(\mathbb{R})$ and endow it with the natural L^p-norm $\|\cdot\|_p$. Further, suppose θ is an *inner function* on $\mathbb{C}_+$; that is, $\theta \in H^\infty$ and the boundary values $\theta(x) = \lim_{y\to 0+} \theta(x+iy)$ have modulus 1 for almost all $x \in \mathbb{R}$.

Of concern are the *star-invariant subspaces*

$$K_\theta^p \stackrel{\text{def}}{=} H^p \cap \theta\,\overline{H^p}, \quad p \geq 1, \tag{1.1}$$

and the way in which their properties depend on those of θ. By "star-invariant" we mean invariant under the semigroup $\{ \mathcal{L}_a : a > 0 \}$ of left (or backward) shifts; here $\mathcal{L}_a$ is given in Fourier coordinates by

$$(\mathcal{L}_a f)^\wedge(\xi) = \hat{f}(\xi + a) \cdot \chi_{(0,\infty)}(\xi) \quad (f \in H^p,\ \xi \in \mathbb{R}).$$

(For $p > 2$, the Fourier transform is defined via distributions.) The prefix "star-" is due to the fact that $\mathcal{L}_a = \mathcal{R}_a^*$, where $\mathcal{R}_a$ is the right (or forward) shift given by $(\mathcal{R}_a f)^\wedge(\xi) = \hat{f}(\xi - a)$. It now follows from Beurling's theorem on $\mathcal{R}_a$-invariant subspaces (cf. [G], Chap. II) that (1.1), with θ an inner function, provides in fact the general form of a closed star-invariant subspace in H^p.

Now consider, for a given $a > 0$, the (singular) inner function $\mathcal{S}(z) = \mathcal{S}_a(z) \stackrel{\text{def}}{=} e^{iaz}$. The Paley–Wiener theorem [HJ, p. 174] shows that $K^p_{\mathcal{S}} = H^p \cap \mathcal{E}^p_a$, where $\mathcal{E}^p_a$ stands for the class of entire functions f having exponential type $\le a$ and satisfying $f\big|_{\mathbb{R}} \in L^p$. Further, it is well known (see [N], Chap. 3) that $\mathcal{E}^p_a \subset \mathcal{E}^q_a$ if $1 \le p < q \le \infty$; moreover, one has then

$$\|f\|_q \le 2a^{1/p-1/q}\,\|f\|_p, \quad f \in \mathcal{E}^p_a. \tag{1.2}$$

In particular, we see that $K^p_{\mathcal{S}} \subset L^q$ for $p < q$, the inclusion map being continuous.

Motivated by this result, we now turn to generic inner functions θ and ask two questions:

(I) When does the (continuous) embedding $K^p_\theta \subset L^q$ hold for some/any exponents $p < q$? (Equivalently, when do we have $K^p_\theta \subset K^q_\theta$ for $p < q$?)

(II) When does it happen, in addition, that the embedding operator is compact?

Actually, part (I) was settled by the author in [D1]; see also [D2]. Besides, we pointed out in [D1] some partial results related to the compactness problem.

In this paper we first state, as Theorem 1, an extended version of our earlier result that answers question (I). Then we provide, in Theorem 2, a complete solution to (II). Also involved in the two theorems are the embeddings $K^p_\theta \subset C_0$ and $K^p_\theta \subset \mathrm{BMO}$. Here $C_0 = C_0(\mathbb{R})$ is the class of continuous functions tending to 0 at infinity, and $\mathrm{BMO} = \mathrm{BMO}(\mathbb{R})$ is the space of functions with bounded mean oscillation (see [G], Chap. VI). The BMO-norm will be denoted by $\|\cdot\|_*$, as usual.

Theorem 1. *Let θ be an inner function on $\mathbb{C}_+$. Given $1 < p < q < \infty$, the following are equivalent:*

(i.1) $\theta' \in L^\infty$.

(ii.1) $K^p_\theta \subset C_0$.

(iii.1) $K^p_\theta \subset \mathrm{BMO}$.

(iv.1) $K^p_\theta \subset L^q$.

In what follows, we write "$X \subset Y$ compactly" to mean that $X \subset Y$ and the arising embedding operator is compact.

Theorem 2. *Let θ be an inner function on $\mathbb{C}_+$. Given $1 < p < q < \infty$, the following are equivalent:*

(i.2) $\theta' \in C_0$.

(ii.2) $K^p_\theta \subset C_0$ *compactly.*

(iii.2) $K^p_\theta \subset \mathrm{BMO}$ *compactly.*

(iv.2) $K^p_\theta \subset L^q$ *compactly.*

A few remarks are in order.

(a) By the Closed Graph Theorem, each of the conditions (ii.1)–(iv.1) entails the corresponding norm inequality (i.e., the inclusion maps are automatically continuous).

(b) In the case $1 = p < q < \infty$, the implications (i.j) $\Longrightarrow$ (ii.j) $\Longrightarrow$ (iii.j) $\Longrightarrow$ (iv.j) are still true (and have the same proofs) for $j = 1, 2$. Under additional hypotheses on the zeros of θ, we are able to prove that (iv.j), with $p = 1$, implies (i.j).

(c) Precisely speaking, condition (i.1) should be understood as follows: the boundary function $x \mapsto \lim_{y \to 0+} \theta(x + iy)$, defined initially for almost all $x \in \mathbb{R}$, is absolutely continuous and has bounded derivative. Some useful restatements of (i.1) are given in Lemma 1 below. In particular, (i.1) implies that θ, and hence also θ', is analytic on $\mathbb{R}$. Thus, when passing from (i.1) to (i.2), the only additional assumption we impose is that $\theta'(x) \to 0$ as $x \to \pm\infty$.

(d) Consider, once again, the inner function $\mathcal{S}_a(z) = e^{iaz}$, $a > 0$. Clearly, $\mathcal{S}_a' \in L^\infty \setminus C_0$. Applying Theorem 1 with $\theta = \mathcal{S}_a$, we essentially recover (1.2) for $1 < p < q \le \infty$, while Theorem 2 tells us that the embedding operators involved fail to be compact. As another example of an inner function θ with $\theta' \in L^\infty \setminus C_0$, one can take an infinite Blaschke product whose zeros $\{z_j\}$ lie in some strip $\{a < \operatorname{Im} z < b\}$, with $0 < a < b < \infty$, and satisfy $\inf_{j \ne k} |z_j - z_k| > 0$.

(e) Now let $\{z_j = x_j + iy_j\}$ be a sequence of points in $\mathbb{C}_+$ such that

$$|x_j| \le \text{const} \cdot y_j, \quad \frac{y_j}{j} \uparrow \infty, \quad \sum_j \frac{1}{y_j} < \infty, \tag{1.3}$$

and let $B = B_{\{z_j\}}$ stand for the corresponding Blaschke product. This time, the formula

$$|B'(t)| = 2 \sum_j \frac{y_j}{|t - z_j|^2}, \quad t \in \mathbb{R},$$

in conjunction with (1.3), enables us to infer that $B' \in C_0$.

(f) Let $\mathbb{D}$ denote the unit disk. It is easily shown that, for the (suitably defined) star-invariant subspaces $K_\theta^p = K_\theta^p(\mathbb{D})$, the inclusion $K_\theta^p \subset L^q(\partial\mathbb{D})$ with $1 \le p < q \le \infty$ is only possible if θ is a finite Blaschke product on $\mathbb{D}$. In this case K_θ^p $(= K_\theta^q)$ is finite-dimensional, so the inclusion map is of course compact. On the other hand, for p and q as above and for any θ, we have the (trivial) reverse inclusions $K_\theta^q \subset L^p(\partial\mathbb{D})$. A recent result of A. B. Aleksandrov [A] says the arising embedding operators are always compact. In connection with embedding theorems of the form $K_\theta^p(\mathbb{D}) \subset L^q(\mathbb{D}, \mu)$, where μ is a measure on $\mathbb{D}$, see also [D3] and the references therein.

The rest of the paper is organized as follows. In Section 2, we collect a number of preparatory lemmas to be employed later. In Sections 3 and 4, we prove Theorems 1 and 2, respectively.

The author thanks S. V. Kislyakov for a helpful suggestion concerning the proof of Lemma 6.

2. Preliminaries

The first of our lemmas is known (cf. [G], Exercise 8 in Chap. IX). A short proof is included, however, for the reader's convenience.

Lemma 1. *Let θ be an inner function on $\mathbb{C}_+$. The following are equivalent.*

(i) $\theta' \in L^\infty$.

(ii) $\theta' \in H^\infty$.

(iii) *For some* $\delta > 0$,

$$\inf\{\,|\theta(z)| : 0 < \operatorname{Im} z < \delta\,\} > 0.$$

Proof. (i) $\Longrightarrow$ (ii). If (i) holds with $\operatorname{ess\,sup}_{\mathbb{R}} |\theta'| = M$, then

$$|\theta(t+h) - \theta(t)| \le M|h| \quad \text{for} \quad t, h \in \mathbb{R}.$$

Since the function $z \mapsto \theta(z+h) - \theta(z)$ is the Poisson integral of its boundary values, we get

$$|\theta(z+h) - \theta(z)| \le M|h| \quad \text{for} \quad z \in \mathbb{C}_+, \quad h \in \mathbb{R}.$$

This in turn implies $|\theta'(z)| \le M$ for all $z \in \mathbb{C}_+$, so that $\theta' \in H^\infty$ and, moreover, $\|\theta'\|_{H^\infty} = M (= \|\theta'\|_\infty)$.

(ii) $\Longrightarrow$ (iii). Under the current assumption, θ must be analytic across $\mathbb{R}$. Now, given $z = x + iy$ and $\delta > 0$ with $x \in \mathbb{R}$ and $0 < y < \delta < 1/\|\theta'\|_\infty$, we have

$$1 - |\theta(z)| = |\theta(x)| - |\theta(z)| \le |\theta(x) - \theta(z)| \le y\|\theta'\|_\infty \le \delta\|\theta'\|_\infty.$$

Therefore,

$$|\theta(z)| \ge 1 - \delta\|\theta'\|_\infty > 0 \quad \text{for} \quad 0 < \operatorname{Im} z < \delta.$$

(iii) $\Longrightarrow$ (i). Let σ be the value of the infimum involved. It follows again that θ is analytic on $\mathbb{R}$. Moreover, by the symmetry principle, θ has an analytic extension to $\{\operatorname{Im} z > -\delta\}$ satisfying $|\theta(z)| \le 1/\sigma$. Now fix $t \in \mathbb{R}$ and consider the analytic function

$$\varphi_t(z) \stackrel{\text{def}}{=} \frac{\theta(z) - \theta(t)}{z - t}$$

on the disk $\{\,z : |z - t| < \delta\,\}$. The maximum principle yields

$$|\theta'(t)| = |\varphi_t(t)| \le \sup_{|z-t|=\delta} |\varphi_t(z)|$$

$$\le \sup_{|z-t|=\delta} \frac{|\theta(z)| + |\theta(t)|}{|z-t|} \le \frac{1}{\delta}\left(\frac{1}{\sigma} + 1\right).$$

Since $t \in \mathbb{R}$ has been arbitrary, we conclude that $\theta' \in L^\infty$. □

In what follows, we make repeated use of the *reproducing kernels*

$$k_z(t) \stackrel{\text{def}}{=} \frac{1-\overline{\theta(z)}\,\theta(t)}{t-\overline{z}} \quad (t \in \mathbb{R},\ z \in \mathbb{C}_+).$$

One easily verifies that, for any $z \in \mathbb{C}_+$, k_z lies in all K^p_θ spaces with $1 < p \leq \infty$ and has indeed a certain reproducing property (see (2.2) below). Now if θ is analytic in a neighborhood of a boundary point $x \in \mathbb{R}$, then we similarly define

$$k_x(t) \stackrel{\text{def}}{=} \frac{1-\overline{\theta(x)}\,\theta(t)}{t-x}, \quad t \in \mathbb{R}.$$

Again, we have $k_x \in \bigcap_{1<p\leq\infty} K^p_\theta$. Furthermore, k_x is nicely behaved (analytic) at x, once we agree that

$$k_x(x) \stackrel{\text{def}}{=} \lim_{t\to x} k_x(t) = -\overline{\theta(x)}\,\theta'(x).$$

Lemma 2. *If θ is an inner function with $\theta' \in H^\infty$ and if $f \in K^p_\theta$ $(1 \leq p < \infty)$, then $f \in C(\mathbb{R})$ and one has, for every $x \in \mathbb{R}$,*

$$f(x) = \frac{1}{2\pi i}\int f(t)\,\overline{k_x(t)}\,dt. \tag{2.1}$$

Proof. Fix $x \in \mathbb{R}$, and let $z = x + iy$ with $y > 0$. We have then

$$f(z) = \frac{1}{2\pi i}\int f(t)\,\overline{k_z(t)}\,dt. \tag{2.2}$$

This (well-known) representation formula is due to Cauchy's theorem and the inclusion $f\,\overline{\theta} \in \overline{H^p}$.

Since $\theta' \in H^\infty$, it follows that θ is analytic across $\mathbb{R}$, and so is f. (See [DSS], p. 58, for the fact that K^p_θ-functions are analytic wherever θ is.) Therefore, $f \in C(\mathbb{R})$ and

$$\lim_{y\to 0} f(x+iy) = f(x). \tag{2.3}$$

Next, we note that

$$|k_z(t)| = \left|\frac{\theta(z)-\theta(t)}{z-t}\right| \leq \|\theta'\|_\infty \tag{2.4}$$

and

$$|k_z(t)| \leq \frac{2}{|t-x|} \tag{2.5}$$

for all $t \in \mathbb{R}$. Consequently, the function

$$g_x(t) \stackrel{\text{def}}{=} \min\left(\|\theta'\|_\infty, \frac{2}{|t-x|}\right), \quad t \in \mathbb{R},$$

is a majorant of class $\bigcap_{1<q\leq\infty} L^q$ for the family $\{\,|k_{x+iy}| : y > 0\,\}$. Hence

$$|f(t)\,k_z(t)| \leq |f(t)|g_x(t), \quad t \in \mathbb{R},$$

the right-hand side being integrable and independent of y. The dominated convergence theorem now shows, together with (2.3), that (2.1) is readily derived from (2.2) upon letting $y \to 0$. □

Lemma 3. *If* $1 < p \le \infty$ *and* θ *is an inner function with* $\theta' \in H^\infty$, *then*

$$\sup_{x \in \mathbb{R}} \|k_x\|_p \le N_{p,\theta} < \infty,$$

where $N_{\infty,\theta} \stackrel{\text{def}}{=} \|\theta'\|_\infty$ *and*

$$N_{p,\theta} \stackrel{\text{def}}{=} \left(2\|\theta'\|_\infty^p + \frac{2^{p+1}}{p-1}\right)^{1/p} \quad \textit{for} \quad 1 < p < \infty.$$

Proof. Fix $x \in \mathbb{R}$ and observe that (2.4) and (2.5) hold for $z = x$, so that

$$|k_x(t)| \le \|\theta'\|_\infty, \quad t \in \mathbb{R}, \tag{2.6}$$

and

$$|k_x(t)| \le \frac{2}{|t-x|}, \quad t \in \mathbb{R}. \tag{2.7}$$

In particular, (2.6) yields the $p = \infty$ case of the lemma.

Further, for $1 < p < \infty$ one has

$$\|k_x\|_p^p = \int |k_x(t)|^p \, dt = \int_{|t-x|<1} + \int_{|t-x|\ge 1}.$$

Using (2.6) and (2.7) to estimate the two integrals, one gets

$$\begin{aligned} \|k_x\|_p^p &\le 2\|\theta'\|_\infty^p + 2^p \int_{|t-x|\ge 1} \frac{dt}{|t-x|^p} \\ &= 2\|\theta'\|_\infty^p + \frac{2^{p+1}}{p-1} = N_{p,\theta}^p. \quad \square \end{aligned}$$

Lemma 4. *Let* $1 < q < \infty$ *and* $\delta > 0$, *and let* θ *be an inner function with* $\theta' \in H^\infty$. *Then*

$$\sup\left\{ \|k_{x+h} - k_x\|_q : x \in \mathbb{R},\ h \in (-\delta, \delta) \right\} \le c_q \|\theta'\|_\infty \delta^{1/q},$$

where $c_q > 0$ *is a constant depending only on* q.

Proof. Fix $x \in \mathbb{R}$ and $h \in \mathbb{R}$. Recalling (2.6), we get

$$|k_{x+h}(t) - k_x(t)| \le 2\|\theta'\|_\infty, \quad t \in \mathbb{R}. \tag{2.8}$$

Besides, a bit of computation gives

$$k_{x+h}(t) - k_x(t) = h\frac{k_{x+h}(t)}{t-x} - \theta(t)\frac{\overline{\theta(x+h)} - \overline{\theta(x)}}{t-x}.$$

Each of the two terms on the right is bounded in modulus by $\|\theta'\|_\infty |h|/|t-x|$, and so

$$|k_{x+h}(t) - k_x(t)| \le \frac{2\|\theta'\|_\infty |h|}{|t-x|}, \quad t \in \mathbb{R}. \tag{2.9}$$

Using (2.8) and (2.9), we now obtain

$$\int |k_{x+h}(t) - k_x(t)|^q \, dt = \int_{|t-x|<\delta} + \int_{|t-x|\geq\delta}$$

$$\leq (2\|\theta'\|_\infty)^q \cdot 2\delta + (2\|\theta'\|_\infty |h|)^q \cdot \int_{|t-x|\geq\delta} \frac{dt}{|t-x|^q} \tag{2.10}$$

$$= 2^{q+1}\|\theta'\|_\infty^q \left(\delta + \frac{|h|^q}{(q-1)\delta^{q-1}}\right).$$

Finally, for $|h| < \delta$, (2.10) yields

$$\|k_{x+h} - k_x\|_q \leq 2^{1+1/q}\left(\frac{q}{q-1}\right)^{1/q} \|\theta'\|_\infty \delta^{1/q},$$

which proves the lemma. □

Lemma 5. *Let θ be an inner function with $\theta' \in C_0$. For $A > 0$, set*

$$\beta(A) \stackrel{\text{def}}{=} \sup\{\, |k_x(t)| : t,\ x \in \mathbb{R},\ |x| \geq A\,\}.$$

Then $\beta(A) \to 0$ as $A \to +\infty$.

Proof. Given $A > 0$, let us pick a pair of points t, $x \in \mathbb{R}$ with $|x| \geq A$ and distinguish two cases.

Case 1: $|t - x| \geq A/2$. Then

$$|k_x(t)| \leq \frac{2}{|t-x|} \leq \frac{4}{A}. \tag{2.11}$$

Case 2: $|t - x| < A/2$. Then

$$|k_x(t)| = \left|\frac{\theta(t) - \theta(x)}{t - x}\right| \leq \sup\{\, |\theta'(\xi)| : \xi \in I_{t,x}\,\}, \tag{2.12}$$

where $I_{t,x}$ stands for the segment with endpoints t and x. Since $|x| \geq A$ and $|t-x| < A/2$, it follows that $I_{t,x} \subset \mathbb{R} \setminus [-A/2, A/2]$. Consequently, the supremum in (2.12) is dominated by the quantity

$$\eta(A) \stackrel{\text{def}}{=} \sup\{\, |\theta'(\xi)| : \xi \in \mathbb{R},\ |\xi| \geq A/2\,\},$$

and so

$$|k_x(t)| \leq \eta(A). \tag{2.13}$$

Now that the two cases have been studied, a combination of (2.11) and (2.13) yields

$$\beta(A) \leq \frac{4}{A} + \eta(A).$$

Hence, sending A to ∞ and recalling that $\theta' \in C_0$, we arrive at the desired conclusion. □

Lemma 6. *Given $1 < p < q < \infty$, there is a constant $c_{p,q} > 0$ such that every $f \in L^p \cap \mathrm{BMO}$ satisfies*

$$\|f\|_q \le c_{p,q} \|f\|_p^{p/q} \|f\|_*^{1-p/q}. \tag{2.14}$$

Proof. It is known that L^q coincides with the intermediate space $(\mathrm{BMO}, L^p)_{p/q,q}$ in the real method of interpolation. (This is essentially contained in [RS].) Now (2.14) follows by general properties of interpolation spaces (see [T], Chap. 1, Sect. 1.3.3). □

Remark. *Using the relation* $(\mathrm{BMO}, \mathcal{M})_{1/q,q} = L^q$ *(see [RS]), where $\mathcal{M}$ is the space of finite Borel measures on $\mathbb{R}$, one similarly verifies (2.14) for $p = 1$ and $1 < q < \infty$.*

3. Proof of Theorem 1

(i.1) $\Longrightarrow$ (ii.1). Given $f \in K_\theta^p$ and $x \in \mathbb{R}$, Lemmas 2 and 3 yield

$$|f(x)| = \left| \frac{1}{2\pi i} \int f(t)\, \overline{k_x(t)}\, dt \right| \le \|f\|_p \, \|k_x\|_{p'} \le N_{p',\theta} \, \|f\|_p$$

(here, as usual, $p' \stackrel{\mathrm{def}}{=} p/(p-1)$). Thus,

$$\|f\|_\infty \le N_{p',\theta} \, \|f\|_p, \quad f \in K_\theta^p, \tag{3.1}$$

and we arrive at the continuous embedding relation $K_\theta^p \subset C(\mathbb{R}) \cap L^\infty$. To verify the sharper inclusion $K_\theta^p \subset C_0$, observe that K_θ^p coincides with the L^p-closed linear hull of the family $\{\, k_x : x \in \mathbb{R} \,\}$. (Indeed, the dual of K_θ^p can be identified [C] with $K_\theta^{p'}$, and the only function in $K_\theta^{p'}$ that annihilates every k_x is zero.) In view of (3.1), the linear combinations $\{\sum_{j=1}^N \lambda_j \, k_{x_j}\}$ are also dense in K_θ^p with respect to the sup-norm. Since each k_x is in C_0, while C_0 is a closed subspace of L^∞, we conclude that $K_\theta^p \subset C_0$, as desired.

(ii.1) $\Longrightarrow$ (iii.1). This is obvious, since $C_0 \subset \mathrm{BMO}$.

(iii.1) $\Longrightarrow$ (iv.1). Use Lemma 6 and the given estimate

$$\|f\|_* \le \mathrm{const} \|f\|_p, \quad f \in K_\theta^p,$$

to infer that

$$\|f\|_q \le c_{p,q} \|f\|_p^{p/q} \, \|f\|_*^{1-p/q} \le \mathrm{const} \|f\|_p.$$

(iv.1) $\Longrightarrow$ (i.1). Assuming, finally, that $K_\theta^p \subset L^q$ and applying the corresponding norm inequality to the reproducing kernels

$$k_z(t) = \frac{1 - \overline{\theta(z)}\, \theta(t)}{t - \overline{z}} \quad (t \in \mathbb{R},\ z \in \mathbb{C}_+),$$

we get

$$\|k_z\|_q \le \mathrm{const} \|k_z\|_p. \tag{3.2}$$

Since

$$1 - |\theta(z)| \le \left|1 - \overline{\theta(z)}\,\theta(t)\right| \le 2,$$

while the quantity $(\int |t-z|^{-p}\,dt)^{1/p}$ is comparable to $y^{-1/p'}$ (here and below $y = \operatorname{Im} z$), (3.2) implies

$$\frac{1 - |\theta(z)|}{y^{1-1/q}} \le \frac{\tilde{c}}{y^{1-1/p}}, \tag{3.3}$$

where $\tilde{c} > 0$ depends on p and q but not on z.

Rewriting (3.3) as

$$|\theta(z)| \ge 1 - \tilde{c}\, y^{1/p-1/q},$$

we see that

$$\inf\{\, |\theta(z)| : 0 < y < \delta \,\} > 0, \tag{3.4}$$

whenever $\delta > 0$ satisfies $\tilde{c}\,\delta^{1/p-1/q} < 1$. By Lemma 1, (3.4) is equivalent to (i.1), and the proof is complete.

4. Proof of Theorem 2

(i.2) $\Longrightarrow$ (ii.2). We have to show that the unit ball

$$B^p_\theta \overset{\text{def}}{=} \{\, f \in K^p_\theta : \|f\|_p \le 1 \,\}$$

is a relatively compact subset of C_0. Since (i.2) implies (i.1), Theorem 1 tells us that B^p_θ is bounded in the sup-norm. Thus, in view of the well-known relative compactness criterion for C_0, it remains to check that

$$\sup\{\, |f(x+h) - f(x)| : f \in B^p_\theta,\ x \in \mathbb{R},\ |h| < \delta \,\} \to 0 \ \text{ as } \ \delta \to 0 \tag{4.1}$$

(equicontinuity of B^p_θ-functions on $\mathbb{R}$) and

$$\sup\{\, |f(x)| : f \in B^p_\theta,\ |x| \ge A \,\} \to 0 \ \text{ as } \ A \to \infty \tag{4.2}$$

("equicontinuity at ∞").

To this end, let us first fix $f \in B^p_\theta$ and $\delta > 0$. Observe that, for any $x, h \in \mathbb{R}$ with $|h| < \delta$, one has

$$|f(x+h) - f(x)| = \left| \frac{1}{2\pi i} \int f(t)\, \left[\overline{k}_{x+h}(t) - \overline{k}_x(t)\right] dt \right|$$

$$\le \|f\|_p \left\|k_{x+h} - k_x\right\|_{p'} \le \left\|k_{x+h} - k_x\right\|_{p'} \le c_p \|\theta'\|_\infty\, \delta^{1/p'}.$$

(We have successively used Lemma 2, Hölder's inequality, and Lemma 4.) This proves (4.1).

Further, for $f \in B^p_\theta$ and $x \in \mathbb{R}$, we have

$$|f(x)| = \left| \frac{1}{2\pi i} \int f(t)\, \overline{k}_x(t)\, dt \right| \le \|f\|_p\, \|k_x\|_{p'}$$

$$\leq \|k_x\|_{p'} = \left(\int |k_x(t)|^{(p'+1)/2} |k_x(t)|^{(p'-1)/2} \, dt \right)^{1/p'} \tag{4.3}$$

$$\leq \left(\int |k_x(t)|^{(p'+1)/2} \, dt \right)^{1/p'} \cdot \sup_{t \in \mathbb{R}} |k_x(t)|^{(p'-1)/(2p')}$$

$$\leq c_{p,\theta} \cdot \sup_{t \in \mathbb{R}} |k_x(t)|^{1/(2p)},$$

where the last step relies on Lemma 3. We now look at the resulting inequality in (4.3) and pass to the supremum over $|x| \geq A$. This gives

$$\sup_{|x| \geq A} |f(x)| \leq c_{p,\theta} \cdot \sup \left\{ |k_x(t)|^{1/(2p)} : t \in \mathbb{R}, \ |x| \geq A \right\}$$

$$= c_{p,\theta} \cdot \beta(A)^{1/(2p)}, \tag{4.4}$$

where $\beta(A)$ is the same as in Lemma 5. Since $\theta' \in C_0$, the lemma says $\lim_{A \to \infty} \beta(A) = 0$, and so (4.4) yields (4.2).

(ii.2) $\Longrightarrow$ (iii.2). This is obvious, since we have $K^p_\theta \subset C_0 \subset \mathrm{BMO}$, where the first embedding is compact and the second continuous.

(iii.2) $\Longrightarrow$ (iv.2). Let $\{f_n\}$ be a sequence of functions in B^p_θ. Since B^p_θ is relatively compact in BMO, we can pick a subsequence $\{g_j\} = \{f_{n_j}\}$ with $\|g_j - g_k\|_* \to 0$ as $j, k \to \infty$. We now invoke Lemma 6 to write

$$\left\| g_j - g_k \right\|_q \leq c_{p,q} \left\| g_j - g_k \right\|_p^{p/q} \left\| g_j - g_k \right\|_*^{1-p/q}$$

$$\leq 2^{p/q} c_{p,q} \left\| g_j - g_k \right\|_*^{1-p/q}$$

and conclude that $\{g_j\}$ is also a Cauchy sequence in L^q. It follows that B^p_θ is relatively compact in L^q, and (iv.2) is established.

(iv.2) $\Longrightarrow$ (i.2). Clearly, condition (iv.2) implies (iv.1) and hence also (i.1). The implication (i.1) $\Longrightarrow$ (ii.1), with p replaced by q, convinces us now that $K^q_\theta \subset C_0$. Denoting by $c_{q,\theta}$ the norm of the arising embedding operator, we find that

$$\left| \theta'(x) \right| = \left| k_x(x) \right| \leq \left\| k_x \right\|_\infty \leq c_{q,\theta} \left\| k_x \right\|_q$$

$$= c_{q,\theta} \left(\int_{|t|<A} \left| k_x(t) \right|^q dt + \int_{|t| \geq A} \left| k_x(t) \right|^q dt \right)^{1/q}, \tag{4.5}$$

whenever $x \in \mathbb{R}$ and $A > 0$.

We proceed by estimating the two integrals in (4.5) under the additional assumption that $|x| \geq 2A$. First, if $|t| < A$ then

$$|x - t| \geq |x| - |t| \geq 2A - A = A,$$

whence $|k_x(t)| \leq 2/A$, and so

$$\int_{|t|<A} \left| k_x(t) \right|^q dt \leq \left(\frac{2}{A} \right)^q \cdot 2A = \frac{2^{q+1}}{A^{q-1}}. \tag{4.6}$$

Next, our current hypothesis (iv.2) tells us that B_θ^p is a relatively compact subset of L^q. Recalling the classical relative compactness criterion for L^q (see [Y]), we deduce, in particular, that B_θ^p-functions are L^q-equicontinuous at ∞. In other words, the quantity

$$\delta(A) \stackrel{\text{def}}{=} \sup\left\{ \int_{|t|\geq A} |f(t)|^q \, dt : f \in B_\theta^p \right\}$$

satisfies

$$\lim_{A\to\infty} \delta(A) = 0. \tag{4.7}$$

Consequently, since $k_x/N_{p,\theta} \in B_\theta^p$ (here $N_{p,\theta}$ is the constant appearing in Lemma 3), we obtain

$$\int_{|t|\geq A} |k_x(t)|^q \, dt \leq N_{p,\theta}^q \cdot \delta(A). \tag{4.8}$$

Finally, substituting (4.6) and (4.8) into (4.5) gives

$$\sup_{|x|\geq 2A} |\theta'(x)| \leq c_{q,\theta}\left(\frac{2^{q+1}}{A^{q-1}} + N_{p,\theta}^q \cdot \delta(A)\right)^{1/q}.$$

Sending A to ∞ and taking (4.7) into account, we eventually conclude that $\lim_{|x|\to\infty} \theta'(x) = 0$, as desired.

References

[A] A. B. Aleksandrov, *Lacunary series and pseudocontinuations*, Zap. Nauchn. Semin. POMI, **232** (1996), 16–32.

[C] W. S. Cohn, *Radial limits and star-invariant subspaces of bounded mean oscillation*, Amer. J. Math., **108** (1986), 719–749.

[DSS] R. G. Douglas, H. S. Shapiro, and A. L. Shields, *Cyclic vectors and invariant subspaces for the backward shift operator*, Ann. Inst. Fourier, **20** (1970), 37–76.

[D1] K. M. Dyakonov, *Moduli and arguments of analytic functions from subspaces in H^p that are invariant for the backward shift operator*, Sibirsk. Mat. Zh., **31** (1990), No. 6, 64–79; English transl., Siberian Math. J., **31** (1990), 926–939.

[D2] K. M. Dyakonov, *Entire functions of exponential type and model subspaces in H^p*, Zap. Nauchn. Semin. LOMI, **190** (1991), 81–100; English transl., J. Math. Sci., **71** (1994), 2222–2233.

[D3] K. M. Dyakonov, *Embedding theorems for star-invariant subspaces generated by smooth inner functions*, J. Funct. Anal., **157** (1998), 588–598.

[G] J. B. Garnett, *Bounded Analytic Functions*, Academic Press, New York, **1981**.

[HJ] V. Havin and B. Jöricke, *The Uncertainty Principle in Harmonic Analysis*, Springer-Verlag, Berlin and New York, **1994**.

[N] S. M. Nikolskii, *Approximation of Functions of Several Variables and Imbedding Theorems*, Springer-Verlag, Berlin, **1975**.

[RS] N. M. Rivière and Y. Sagher, *Interpolation between L^∞ and H^1, the real method*, J. Funct. Anal., **14** (1973), 401–409.

[T] H. Triebel, *Interpolation Theory, Function Spaces, Differential Operators*, VEB Deutscher Verlag der Wissenschaften, Berlin, **1977**.

[Y] K. Yosida, *Functional Analysis*, Springer-Verlag, Berlin, **1965**.

Steklov Institute of Mathematics,
St. Petersburg Branch (POMI),
Fontanka 27,
St. Petersburg, 191011,
Russia
E-mail address: dyakonov@pdmi.ras.ru

Operator Theory:
Advances and Applications, Vol. 113

Rational Functions in Bergman Spaces

Evsey Dyn'kin

Dedicated to the memory of Prof. S. A. Vinogradov

Abstract. An inversion of the classical Hardy-Littlewood theorem on embedding of Hardy spaces into Bergman spaces is proved for the subset of rational functions of a given degree. Many other embeddings for such functions follow from this result. For the Hardy-Littlewood theorem itself, a refinement in terms of rational approximation is proved. As an application, a sharp rational approximation theorem in the Bergman norm is obtained.

1. Introduction

The weighted Bergman space $A_p^\alpha(\mathbb{D})$ is the collection of all functions f analytic in the unit disc $\mathbb{D}$ and such that

$$\|f\|_{A_p^\alpha} = \left(\int_{\mathbb{D}} |f(z)|^p (1-|z|)^\alpha \, dxdy\right)^{\frac{1}{p}} < \infty.$$

Here $1 < p < \infty$ and $\alpha > -1$, so any function analytic in the closed disc belongs to all A_p^α. If $\alpha = 0$, one obtains the standard Bergman space of analytic functions with finite area L^p norm.

Rational functions r of any degree with poles outside $\mathbb{D}$ belong to A_p^α. (The degree of a rational function is the total number of its poles on the Riemann sphere counted with multiplicities.) However, not always the location of the poles affects the A_p^α norm essentially. Indeed, for the Cauchy kernel

$$r(z) = \frac{1}{z-(1+\varepsilon)}$$

the norm

$$\|r\|_{A_p^\alpha} \asymp \frac{1}{\varepsilon^{1-\frac{2+\alpha}{p}}}, \qquad \varepsilon \to +0,$$

only if $\alpha+2 < p$. Otherwise, if $\alpha+2 > p$, the norm does not depend on ε essentially and remains bounded as $\varepsilon \to 0$.

1991 *Mathematics Subject Classification.* Primary 26C15, 30D15, 41A17, 41A20.
Key words and phrases. Finite Blaschke products, rational functions, Hardy-Littlewood embedding theorem, rational approximation.
This research was supported by the fund for the promotion of sponsored research at the Technion.

Here, we consider the first case only, that is, throughout the paper we assume

$$1 < 2 + \alpha < p. \tag{1}$$

In particular, for $\alpha = 0$ this means $p > 2$.

We do not discuss the limit case $\alpha + 2 = p$.

This paper is a continuation of [5]. In [5], the following phenomenon was discovered: sharp embedding theorems are invertible on the subset of rational functions of a given degree. This subset, being infinite-dimensional, is by no means linear or compact, so this invertibility of the embedding theorems is unexpected.

Here, we investigate the most fundamental embedding theorem for Bergman spaces – the Hardy-Littlewood embedding theorem [3].

Theorem 1.1 (Hardy-Littlewood embedding theorem). *For any $p > 1$ and $\alpha > -1$*

$$H^\sigma(\mathbb{D}) \subset A_p^\alpha(\mathbb{D})$$

provided

$$\sigma = \frac{p}{2+\alpha}. \tag{2}$$

Here H^σ is the well-known Hardy space in the disc [3, 6].

It turns out that the Hardy-Littlewood theorem also is invertible on rational functions of a given degree n. Theorem 4.1 of this paper asserts that under condition (1) for such functions we have

$$\|r\|_{H^\sigma} \le C n^{\frac{1+\alpha}{p}} \|r\|_{A_p^\alpha}.$$

The exponent $\frac{1+\alpha}{p}$ is sharp.

A lot of other "inverse" estimates follows from this. Thus, we obtain a sharp Bernstein-type inequality for rational functions with poles outside $\mathbb{D}$:

$$\|r\|_{B_q^s(\mathbb{T})} \le C n^{\frac{1}{q} - \frac{1}{p}} \|r\|_{A_p^\alpha(\mathbb{D})},$$

where the $B_q^s(\mathbb{T})$ stands for the Besov space on the unit circle [4, 10] and

$$s = \frac{1}{q} - \frac{2+\alpha}{p} = \frac{1}{q} - \frac{1}{\sigma} > 0. \tag{3}$$

Another corollary is the following inversion of the well-known Sobolev embedding theorem:

$$\|r\|_{W_q^1(\mathbb{D})} \le C n^{\frac{1}{2}} \|r\|_{L^p(\mathbb{D})}$$

for $p > 2$ and

$$\frac{1}{q} = \frac{1}{p} + \frac{1}{2}.$$

All the exponents are sharp.

The proof of Theorem 4.1 involves unexpectedly complicated tools. We make use of the famous Carleson corona construction [6] and hyperbolic area estimate for level sets of finite Blaschke products from [5]. For the reader's convenience, we repeat here its proof. Together with the Carleson construction, applied to finite Blaschke products too, this leads to the proof of Theorem 4.1.

Our next objective is a quantitative refinement of the Hardy-Littlewood theorem. For any function $f \in A_p^\alpha$ one can define its best approximations $R_n(f)_{A_p^\alpha}$ in the A_p^α norm by rational functions of degree n. Theorem 5.1 of this paper asserts that if $f \in H^\sigma$, then

$$R_n(f)_{A_p^\alpha} \le C \frac{\|f\|_{H^\sigma}}{n^{\frac{1+\alpha}{p}}}.$$

Therefore, the Hardy-Littlewood embedding has some additional regularity, which cannot be expressed in terms of classical width estimates, because the approximation in our case is not linear.

The proof of Theorem 5.1 follows Pekarskii's method of rational approximation [9] and refers to his result on approximation in the H^σ norm.

As an application of Theorems 4.1 and 5.1 we obtain a new sharp rational approximation theorem in the Bergman norm. Theorem 6.1 of this paper asserts that for a function $f \in A_p^\alpha$ the inclusion $f \in B_q^s(\mathbb{T})$ is true if and only if

$$\sum_1^\infty \frac{1}{n^{\frac{q}{p}}} R_n(f)_{A_p^\alpha}^q < \infty.$$

Here $q > 0$ and s is defined by (3) again.

Section 2 of the paper is devoted to preliminary information.

In Section 3 estimates for finite Blaschke products are exposed.

In Section 4 we prove Theorem 4.1 (an inversion of the Hardy-Littlewood embedding theorem), and in Section 5 we prove Theorem 5.1 (its refinement).

Theorem 6.1 (concerning rational approximation in the A_p^α norm) is proved in the last Section 6.

The results presented here have a long history. During all my work, I enjoyed help and advice of my friend and teacher Prof. S. A. Vinogradov. This paper is dedicated to his memory.

2. Preliminaries

2.1. Notation

$z = x + iy$ and $\zeta = \xi + i\eta$ are complex variables.

$\mathbb{D} = \{z : |z| < 1\}$ is the unit disc, $\mathbb{T} = \{z : |z| = 1\}$ is the unit circle.

$\frac{\partial}{\partial \bar{z}} = \frac{1}{2}\left(\frac{\partial}{\partial x} + i\frac{\partial}{\partial y}\right)$.

C and c are various constants, not necessarily the same within a formula.

$\rho(z) = |1 - |z||$ is the distance from z to $\mathbb{T}$.

A Whitney disc is a disc of the form

$$\Delta = \Delta(z) = \{\zeta : |\zeta - z| < \frac{1}{2}\rho(z)\},$$

where $z \notin \mathbb{T}$.

The norm in $L^\sigma(\mathbb{T})$ or in the Hardy space $H^\sigma(\mathbb{D})$ is always calculated relative to normalized Lebesgue measure:

$$\|f\|_{L^\sigma} = \left(\frac{1}{2\pi}\int_{\mathbb{T}} |f(e^{i\theta})|^\sigma \, d\theta\right)^{\frac{1}{\sigma}}.$$

For a function $f \in L^\sigma(\mathbb{T})$ $(\sigma > 1)$, f_+ stands for its Cauchy integral, or its Riesz projection to H^σ [3, 6]. In particular,

$$\|f_+\|_{H^\sigma} \le C\|f\|_{L^\sigma}. \tag{4}$$

As usual,

$$\sigma' = \frac{\sigma}{\sigma - 1}$$

is the conjugate exponent.

2.2. A partition of the annulus

Choose small $\varepsilon > 0$, fix an integer L such that $2^{-L+1} < \varepsilon$, and for each $n = 2, 3, \ldots$, define

$$Q^n_{jk} = \{re^{i\theta} : 2^{-n-L}j \le 1 - r \le 2^{-n-L}(j+1), \quad 2^{-n-L}k \le \frac{\theta}{2\pi} \le 2^{-n-L}(k+1)\},$$

where

$$1 \le j \le 2^{L+1} - 1, \quad 0 \le k \le (2^L - 1)\cdot 2^n.$$

We can enumerate all "dyadic squares" Q^n_{jk} into a sequence Q_k, $k = 1, 2, \ldots$. The Q_k form a partition of the annulus

$$K = \{z : \frac{1}{2} < |z| < 1\} = \bigcup_{k=1}^{\infty} Q_k \tag{5}$$

and for each of them

$$\operatorname{diam}(Q_k) \le \varepsilon\rho_k,$$

where

$$\rho_k = \operatorname{dist}(Q_k, \mathbb{T}).$$

2.3. Rational approximation and Besov spaces

For any f in a Hardy space $H^\sigma(\mathbb{D})$ or in a Bergman space $A_p^\alpha(\mathbb{D})$, one can define its best rational H^σ approximations

$$R_n(f)_{H^\sigma} = \inf \|f - r\|_{H^\sigma}, \qquad n = 1, 2, \ldots,$$

or its best rational A_p^α approximations

$$R_n(f)_{A_p^\alpha} = \inf \|f - r\|_{A_p^\alpha}, \qquad n = 1, 2, \ldots,$$

respectively. Here the infimum is taken over all rational functions r of degree n with poles outside $\mathbb{D}$.

The definition of the classical Besov spaces $B_q^s(\mathbb{T})$, $q > 0$, $s > 0$, can be found in [10, 4].

In particular, the well-known embedding theorem says that

$$B_q^s(\mathbb{T}) \subset L^\sigma(\mathbb{T}) \tag{6}$$

for

$$\frac{1}{\sigma} = \frac{1}{q} - s. \tag{7}$$

The following rational approximation theorem is due to Pekarskii [8, 9]:

Theorem 2.1 (Pekarskii Theorem). *Let s, q and σ satisfy* (7) *and let $f \in H^\sigma$. Then*

$$f \in B_q^s(\mathbb{T})$$

if and only if

$$\sum_{n=1}^{\infty} \frac{1}{n} [n^s R_n(f)_{H^\sigma}]^q < \infty.$$

2.4. Muckenhoupt condition

Under our assumption (1), the weight $\rho(z)^\alpha$ satisfies the well-known Muckenhoupt condition (A_p) in the plane [6, 7]. In particular, this means that the Calderón-Zygmund singular integral operator

$$\varphi \mapsto \int \frac{\varphi(\zeta)}{(\zeta - z)^2}\, d\xi d\eta \tag{8}$$

is bounded in the weighted L^p norm

$$\|\varphi\|_{p,\alpha} = \left(\int_{\mathbb{R}^2} |\varphi(z)|^p \rho(z)^\alpha \, dx dy \right)^{\frac{1}{p}}.$$

Another example is the Marcinkiewicz integral operator

$$\psi \mapsto \int \frac{\rho(\zeta)}{(|\zeta - z| + \rho(\zeta))^3} \psi(\zeta)\, d\xi d\eta, \tag{9}$$

which is also bounded in this norm [2].

2.5. Carleson measures and the corona construction

A positive measure μ in $\mathbb{D}$ is called a Carleson measure [6] if

$$\mu(Q) \leq \varkappa\, \mathrm{diam}(Q)$$

for each "square" of the form

$$Q = \{re^{i\theta} : 0 < 1 - r < a, \quad |\theta - \theta_0| < a\}, \qquad 0 < a < 1, \quad 0 < \theta_0 < 2\pi.$$

The least possible value of the constant $\varkappa$ is called the Carleson norm of μ.

The well-known Carleson embedding theorem [6] asserts that

$$\|f\|_{L^\sigma(\mu)} \leq C\varkappa \|f\|_{H^\sigma} \tag{10}$$

for any $\sigma > 0$ and any $f \in H^\sigma(\mathbb{D})$, with some absolute constant C.

The famous Carleson corona construction may be summarized in the following statement ([6], Ch. 8, Section 5, Theorem 5.1).

Theorem 2.2 (Carleson Theorem). *Let $\delta > 0$. If a function f is analytic in $\mathbb{D}$ and $|f| \leq 1$, then there exists a function $\psi \in C^\infty(\mathbb{D})$ such that*

(i) $$0 \leq \psi \leq 1,$$

(ii) $$\psi(z) = 1 \quad \text{if} \quad |f(z)| \geq \delta,$$

(iii) $$\psi(z) = 0 \quad \text{if} \quad |f(z)| < \varepsilon(\delta),$$

(iv) $$|\nabla\psi(z)| \leq \frac{c}{1 - |z|}, \qquad z \in \mathbb{D},$$

and the measure

$$\left|\frac{\partial\psi}{\partial\bar{z}}\right| dxdy$$

is a Carleson measure with Carleson norm $A(\delta)$.

Here the constants $\varepsilon(\delta)$ and $A(\delta)$ depend on δ only.

2.6. Maximal estimates

Denote by $\mathcal{M}f(e^{i\theta})$ the Hardy-Littlewood maximal function of $f \in L^\sigma(\mathbb{T})$. It is well known that

$$\|\mathcal{M}f\|_{L^\sigma} \leq C\|f\|_{L^\sigma}, \qquad \sigma > 1. \tag{11}$$

Another standard fact is the so-called nontangential estimate: if $f \in H^\sigma(\mathbb{D})$, then

$$|f(z)| \leq C\,\mathcal{M}f(e^{i\theta}) \tag{12}$$

for any $z \in \mathbb{D}$ in the sector

$$\Gamma(e^{i\theta}) = \{z : |z - e^{i\theta}| < 2\rho(z)\}. \tag{13}$$

2.7. An elementary identity

Differentiating the identity

$$\frac{1}{z-\zeta}=\sum_{l=1}^{m}\frac{(\zeta-\zeta_0)^{l-1}}{(z-\zeta_0)^l}+\frac{(\zeta-\zeta_0)^m}{(z-\zeta_0)^m(z-\zeta)}$$

in ζ, one obtains

$$\frac{1}{(\zeta-z)^2}=\sum_{l=2}^{m}(-1)^l\frac{(l-1)(\zeta-\zeta_0)^{(l-2)}}{(\zeta_0-z)^l}$$

$$+(-1)^{m+1}\frac{m(\zeta-\zeta_0)^{m-1}}{(\zeta_0-z)^m(\zeta-z)}+(-1)^m\frac{(\zeta-\zeta_0)^m}{(\zeta_0-z)^m(\zeta-z)^2}.$$

If $|\zeta-\zeta_0|<|\zeta-z|$, this means that

$$\left|\frac{1}{(\zeta-z)^2}-\sum_{l=2}^{m}(-1)^l\frac{(l-1)(\zeta-\zeta_0)^{(l-2)}}{(\zeta_0-z)^l}\right|$$

$$\leq\frac{(m+1)|\zeta-\zeta_0|^{m-1}}{|\zeta_0-z|^m|\zeta-z|}. \tag{14}$$

3. Finite Blaschke products

3.1. Hyperbolic area of level sets

Let

$$B(z)=\prod_{k=1}^{n}\frac{z_k-z}{1-\bar{z}_kz}$$

be a Blaschke product of degree n (for the unit disc).

Define the level set

$$E=\{z\in\mathbb{D}:|B(z)|<\frac{1}{2}\}. \tag{15}$$

Remark 3.1. *If $z\in E$, then $|B|\leq 4/5$ on the whole Whitney disc $\Delta(z)$* (cf. [6], Ch. 1, Section 1.1).

The following equality is well known:

$$\int_{\mathbb{D}}|B'(z)|^2\,dxdy=\pi n. \tag{16}$$

Indeed, by the Green formula

$$\int_{\mathbb{D}}|B'(z)|^2\,dxdy=\int_{\mathbb{D}}B'(z)\overline{B'(z)}\,dxdy=\frac{1}{2i}\int_{\mathbb{T}}B'(z)\overline{B(z)}\,dz$$

$$=\frac{1}{2i}\int_{\mathbb{T}}\frac{B'(z)}{B(z)}\,dz=\pi n.$$

We use this result in the proof of the following estimate [5].

Theorem 3.2.

$$\int_{\mathbb{D}} \left(\frac{1 - |B(z)|}{1 - |z|} \right)^2 dxdy \le 8\pi(n+1).$$

Proof. The contribution of the small disc $\{ |z| < 1/2 \}$ to the integral is obviously less than π. The rest of the integral in polar coordinates is

$$\int_{1/2}^{1} \frac{r}{(1-r)^2}\, dr \int_0^{2\pi} (1 - |B(re^{i\theta})|)^2 \, d\theta$$

$$\le \int_{1/2}^{1} \frac{r}{(1-r)^2}\, dr \int_0^{2\pi} |B(e^{i\theta}) - B(re^{i\theta})|^2 \, d\theta$$

$$\le \int_0^{2\pi} d\theta \int_{1/2}^{1} \frac{r}{(1-r)^2}\, dr \left(\int_r^1 |B'(se^{i\theta})| \, ds \right)^2$$

By the well-known Hardy inequality ([10], Appendix A.4) this is less than

$$4 \int_0^{2\pi} d\theta \int_{1/2}^{1} |B'(se^{i\theta})|^2 ds \le 8 \int_{\mathbb{D}} |B'(z)|^2 \, dxdy = 8\pi n.$$

The last step is due to (16). □

Some important properties of level sets of B follow from Theorem 3.2.

Corollary 3.3.

$$\int_E \frac{dxdy}{(1-|z|)^2} \le 32\pi(n+1).$$

This estimate of the hyperbolic area of E may be written in a different form.

Corollary 3.4. *The level set E may be covered by at most Cn Whitney discs, with some absolute constant C.*

Proof. By the well-known Whitney construction ([10], Ch. 6, Section 1) one can cover the whole of $\mathbb{D}$ by Whitney discs with bounded multiplicity of intersections. Each Whitney disc intersecting E contributes a positive constant to the integral of Theorem 3.2 (see the remark above). □

3.2. An application of the corona construction

We apply the Carleson theorem of Subsection 2.5 to our finite Blaschke product B and $\delta = \frac{1}{2}$.

In the notation of the theorem, define

$$B^*(z) = \frac{\psi(z)}{B(z)}, \qquad z \in \mathbb{D}. \tag{17}$$

The definition is consistent because $\psi = 0$ near the zeroes of B.

The following result is an immediate consequence of the Carleson theorem.

Theorem 3.5. *Let B be a finite Blaschke product, and let*

$$E = \{z \in \mathbb{D} : |B(z)| < \frac{1}{2}\}$$

be its level set. There exists a function $B^* \in C^\infty(\bar{\mathbb{D}})$ such that

(i)
$$B^*(z) = \frac{1}{B(z)}$$

outside E, in particular, in a neighborhood of $\mathbb{T}$.

(ii)
$$|B^*| \le M.$$

(iii) *The Cauchy-Riemann derivative*

$$\frac{\partial B^*}{\partial \bar{z}}$$

vanishes outside E, and on E we have

$$\left|\frac{\partial B^*}{\partial \bar{z}}\right| \le \frac{C}{1-|z|}.$$

(iv) *The measure*

$$\left|\frac{\partial B^*}{\partial \bar{z}}\right| dxdy$$

is a Carleson measure with Carleson norm A.

Here M, A and C are positive absolute constants.

In particular, the support of $\frac{\partial B^*}{\partial \bar{z}}$ can be covered by at most Cn Whitney discs, and all three constants M, A and C do not depend on the degree n of our Blaschke product.

4. An inversion of the Hardy-Littlewood embedding theorem

4.1. The main result

Consider the Bergman space $A_p^\alpha(\mathbb{D})$ with $1 < 2+\alpha < p$. Recall that the norm in A_p^α is defined as

$$\|f\|_{A_p^\alpha} = \left(\int_{\mathbb{D}} |f(z)|^p (1-|z|)^\alpha \, dxdy\right)^{\frac{1}{p}}, \tag{18}$$

and the corresponding exponent σ is

$$\sigma = \frac{p}{2+\alpha}, \tag{19}$$

so that $1 < \sigma < p$ and $\sigma = \frac{p}{2}$ in the main case $\alpha = 0$.

The Hardy-Littlewood embedding theorem asserts that under these conditions

$$H^\sigma(\mathbb{D}) \subset A_p^\alpha(\mathbb{D}).$$

The following result is an inversion of the Hardy-Littlewood theorem for rational functions of a given degree.

Theorem 4.1. *Let r be a rational function of degree n with all its poles outside $\mathbb{D}$. Then*

$$\|r\|_{H^\sigma} \le Cn^{\frac{1+\alpha}{p}} \|r\|_{A_p^\alpha}.$$

Proof. Let B be the finite Blaschke product of degree $n+2$ with all poles of r and two extra poles at ∞. The quotient

$$f = \frac{r}{B}$$

is an analytic function in $\mathbb{C} \setminus \mathbb{D}$ vanishing at ∞.

Let ψ be a function of the dual space $L^{\sigma'}(\mathbb{T})$. Consider the Riesz projection

$$\varphi = (\psi B)_+ \in H^{\sigma'}.$$

Now

$$\int_{\mathbb{T}} \psi r \, dz = \int_{\mathbb{T}} \psi B f \, dz = \int_{\mathbb{T}} \varphi f \, dz,$$

because f is analytic outside $\mathbb{D}$ and vanishes at ∞.

We may apply Theorem 3.5 of Section 3 to our Blaschke product B and get a new function B^* in $\mathbb{D}$ coinciding with $1/B$ in a neighborhood of $\mathbb{T}$. Thus,

$$\int_{\mathbb{T}} \varphi f \, dz = \int_{\mathbb{T}} \varphi r / B \, dz = \int_{\mathbb{T}} \varphi r B^* \, dz.$$

By the Green formula, the latter integral equals

$$2i \int_{\mathbb{D}} \varphi r \frac{\partial B^*}{\partial \bar z} \, dx dy.$$

But

$$1 = \frac{1}{\sigma} + \frac{1}{\sigma'} = \frac{1}{\sigma'} + \frac{1}{p} + \frac{1+\alpha}{p},$$

and one can apply the Hölder inequality to obtain

$$\begin{aligned} \left| \int_{\mathbb{T}} \psi r \, dz \right| &\le 2 \int_{\mathbb{D}} |\varphi(z)| \left| \frac{\partial B^*}{\partial \bar z} \right|^{\frac{1}{\sigma'}} \cdot |r(z)| (1-|z|)^{\frac{\alpha}{p}} \cdot \left| \frac{\partial B^*}{\partial \bar z} \right|^{\frac{1}{\sigma}} (1-|z|)^{-\frac{\alpha}{p}} \, dx dy \\ &\le 2 \left(\int_{\mathbb{D}} |r(z)|^p (1-|z|)^\alpha \right)^{\frac{1}{p}} \left(\int_{\mathbb{D}} |\varphi|^{\sigma'} \left| \frac{\partial B^*}{\partial \bar z} \right| \right)^{\frac{1}{\sigma'}} \\ &\quad \times \left(\int_{\mathbb{D}} \left| \frac{\partial B^*}{\partial \bar z} \right|^{\frac{2+\alpha}{1+\alpha}} (1-|z|)^{-\frac{\alpha}{1+\alpha}} \right)^{\frac{1+\alpha}{p}} . \end{aligned} \tag{20}$$

The first factor is just $\|r\|_{A_p^\alpha}$. The second does not exceed $\|\varphi\|_{H^{\sigma'}}$, because the measure involved is a Carleson measure of a bounded norm by Theorem 3.5 of Section 3. As for the third, we know that the domain of integration here may be covered by at most $N = Cn$ Whitney discs with radii ρ_k, $k = 1, 2, \ldots, N$. The contribution of each disc is dominated by a constant multiple of

$$\rho_k^{-\frac{2+\alpha}{1+\alpha}} \cdot \rho_k^{-\frac{\alpha}{1+\alpha}} \cdot \rho_k^2 = 1.$$

Therefore the last factor in (20) does not exceed

$$C \cdot n^{\frac{1+\alpha}{p}},$$

and taking (4) into account, one obtains

$$\left| \int_{\mathbb{T}} \psi r \, dz \right| \leq C n^{\frac{1+\alpha}{p}} \|r\|_{A_p^\alpha} \|\psi\|_{L^{\sigma'}}.$$

By duality, this estimate is precisely the desired one. □

Remark 4.2. *The exponent* $\frac{1+\alpha}{p}$ *in Theorem* 4.1 *is sharp. Indeed, take* $r(z) = z^n$ *as a test function. Its* H^σ *norm is* 1, *whereas the* A_p^α *norm of* r *is*

$$\left(\int_0^1 r^{np+1} (1-r)^\alpha \, dr \right)^{\frac{1}{p}} \asymp n^{-\frac{1+\alpha}{p}}.$$

4.2. Corollaries

4.2.1. An inner-outer estimate for rational functions. Once again, consider the function $f = r/B$ occurring in the proof of Theorem 4.1. This is a rational function with all poles inside $\mathbb{D}$ and vanishing at ∞. On $\mathbb{T}$ the Blaschke product is unimodular, and so $|f| = |r|$. Therefore,

$$\|f\|_{H^\sigma} \leq C n^{\frac{1+\alpha}{p}} \|r\|_{A_p^\alpha}.$$

In view of the Hardy-Littlewood embedding theorem (for $\mathbb{C} \setminus \mathbb{D}$, of course), this gives

Corollary 4.3. *Let* r *be a rational function of degree* n *with poles outside* $\mathbb{D}$. *Let* B *be the Blaschke product of degree* $n+2$ *with the same poles as* r *and two additional poles at infinity. Let* $f = r/B$.

Then for any p *and* α, $0 < 2 + \alpha < p$,

$$\left(\int_{\mathbb{C}\setminus\mathbb{D}} |f(z)|^p (|z|-1)^\alpha \, dxdy \right)^{\frac{1}{p}} \leq C n^{\frac{1+\alpha}{p}} \left(\int_{\mathbb{D}} |r(z)|^p (1-|z|)^\alpha \, dxdy \right)^{\frac{1}{p}}.$$

The constant C *depends on* p *and* α. *The exponent* $\frac{1+\alpha}{p}$ *is sharp.*

Remark 4.4. *The statement of Corollary* 4.3 *is a sharp version of Theorem* 4 *from* [5]. *In the latter* $\alpha = 0$, *but the exponent is equal to* 1/2 (*instead of* $(1+\alpha)p^{-1}$, *as presented above*). *The above proof is due to F. Nazarov* (*private communication*). *The initial proof repeated the proof of Theorem* 4.1 *itself.*

4.2.2. Embedding in Besov spaces. Let $q > 0$ and

$$s = \frac{1}{q} - \frac{2+\alpha}{p} = \frac{1}{q} - \frac{1}{\sigma} > 0.$$

Consider the Besov space $B_q^s(\mathbb{T})$ (see Subsection 2.3).

The well-known Bernstein type inequality for rational functions [8, 5] asserts that in our notation

$$\|r\|_{B^s_q(\mathbb{T})} \le C n^s \|r\|_{H^\sigma}.$$

Hence Theorem 4.1 immediately gives

Corollary 4.5. *Let r be a rational function of degree n with poles outside $\mathbb{D}$. Then*

$$\|r\|_{B^s_q(\mathbb{T})} \le C n^{\frac{1}{q}-\frac{1}{p}} \|r\|_{A^\alpha_p(\mathbb{D})}.$$

4.2.3. An inversion of the Sobolev embedding theorem. The Sobolev embedding theorem [10] says that

$$W^1_q(\mathbb{D}) \subset L^p(\mathbb{D}), \qquad 2 < p < \infty, \qquad \frac{1}{q} = \frac{1}{p} + \frac{1}{2}.$$

In order to obtain its inversion for a rational function r of degree n with poles outside $\mathbb{D}$, take $\alpha = 0$ in the previous corollary. Now

$$s = \frac{1}{q} - \frac{2}{p} = 1 - \frac{1}{q}.$$

It is well known [10] that for functions harmonic in $\mathbb{D}$ the $B_q^{1-\frac{1}{q}}(\mathbb{T})$ norm is equivalent to the norm in the Sobolev space $W^1_q(\mathbb{D})$. Therefore, Corollary 4.5 tells us that

$$\|r\|_{W^1_q(\mathbb{D})} \le C n^{\frac{1}{q}-\frac{1}{p}} \|r\|_{L^p(\mathbb{D})}.$$

But in our case

$$\frac{1}{q} - \frac{1}{p} = \frac{1}{2}$$

and we obtain

Corollary 4.6. *Let $p > 2$ and*

$$\frac{1}{q} = \frac{1}{p} + \frac{1}{2}.$$

For any rational function r of degree n with poles outside $\mathbb{D}$ we have

$$\|r\|_{W^1_q(\mathbb{D})} \le C n^{\frac{1}{2}} \|r\|_{L^p(\mathbb{D})}.$$

In a weaker form, with the exponent 1 instead of 1/2, Corollary 4.6 was proved in [5]. The exponent 1/2 is sharp, as one can check taking $r(z) = z^n$.

5. A refinement of the Hardy-Littlewood embedding theorem

5.1. Statement of the results

According to the Hardy-Littlewood embedding theorem,

$$H^\sigma(\mathbb{D}) \subset A^\alpha_p(\mathbb{D}),$$

where, as before,

$$\sigma = \frac{p}{2+\alpha}$$

and

$$1 < 2 + \alpha < p.$$

In particular, for any $f \in H^\sigma$ one can consider its best rational approximations $R_n(f)_{A_p^\alpha}$ in the A_p^α norm (see Subsection 2.3).

Theorem 5.1. *If $f \in H^\sigma(\mathbb{D})$, then*

$$R_n(f)_{A_p^\alpha} \le C \frac{\|f\|_{H^\sigma}}{n^{\frac{1+\alpha}{p}}}, \qquad n = 1, 2, \dots,$$

where the constant C depends on p and α only.

The proof of Theorem 5.1 occupies the next three subsections.

5.2. A maximal estimate

In the proof of Theorem 5.1 we will follow the approximation method due to Pekarskii [9].

Consider the partition (5)

$$K = \bigcup Q_k$$

of the annulus $K = \{z : 1/2 < |z| < 1\}$ into disjoint "dyadic squares" with small $\varepsilon > 0$ to be chosen later.

Let $f \in H^\sigma(\mathbb{D})$; in the sequel we suppose $\|f\|_{H^\sigma} = 1$.

Renumber the squares Q_k in such a way that the quantities $f_k \rho_k^{\frac{1}{\sigma}}$ (where $f_k = \sup_{Q_k} |f|$) decrease.

Lemma 5.2.

$$f_k \rho_k^{\frac{1}{\sigma}} \le \frac{C}{k^{\frac{1}{\sigma}}}. \tag{21}$$

Proof. Fix $\lambda > 0$ and set

$$E = \{re^{i\theta} : (1-r)^{\frac{1}{\sigma}} \mathcal{M}f(e^{i\theta}) > \lambda\}$$

($\mathcal{M}f$ is the Hardy-Littlewood maximal function for f on $\mathbb{T}$). E is a subset of $\mathbb{D}$ and if $z = re^{i\theta} \in E$, then

$$1 - r \ge \left(\frac{\lambda}{\mathcal{M}F(e^{i\theta})}\right)^\sigma.$$

The maximal theorem yields the following estimate for the hyperbolic area of E:

$$\begin{aligned}\int_E \frac{1}{(1-|z|)^2}\, dxdy &\le \text{const} + \int d\theta \int_{\left(\frac{\lambda}{\mathcal{M}f(e^{i\theta})}\right)^\sigma}^{\frac{1}{2}} \frac{d\rho}{\rho^2} \\ &\le \frac{1}{\lambda^\sigma} \int \mathcal{M}f(e^{i\theta})^\sigma\, d\theta \le \frac{C_1}{\lambda^\sigma}.\end{aligned}$$

Consider now a square Q_k and $z = re^{i\theta} \in Q_k$, so that $1 - r > \rho_k$. If the parameter ε in the definition of the squares Q_k is small enough, then for each

$z = re^{i\theta} \in Q_k$ the whole of Q_k is contained in the sector $\Gamma(e^{i\theta})$ (cf. Subsection 2.2). The nontangential estimate (12) gives

$$|f(\zeta)| \leq C_2 \mathcal{M} f(e^{i\theta})$$

for all points $\zeta \in Q_k$. In particular, this estimate is fulfilled for f_k. Therefore, if

$$f_k \rho_k^{\frac{1}{\sigma}} > C_2 \lambda,$$

then

$$(1-r)^{\frac{1}{\sigma}} \mathcal{M} f(e^{i\theta}) > \lambda,$$

that is, $Q_k \subset E$.

However, each Q_k contributes a positive constant to the hyperbolic area of E, and so, in our enumeration rule, this means that

$$k \leq C_3 \frac{1}{\lambda^\sigma}.$$

Choosing a good λ, we obtain the desired estimate. □

5.3. Method of approximation

Let F be a function analytic in $\mathbb{D}$ such that $F(0) = 0$ and $F' = f$. One can extend F onto $\mathbb{C} \setminus \mathbb{D}$ by the formula

$$F(z) = F(z^*), \qquad |z| > 1,$$

where

$$z^* = \frac{1}{\bar{z}}$$

is the standard symmetry with respect to the unit circle. For this extension,

$$\left| \frac{\partial F}{\partial \bar{z}} \right| \asymp |f(z^*)|$$

if $1 < |z| < 2$.

Now, the Cauchy-Green formula for F gives

$$f(z) = \int_{1<|\zeta|<2} \frac{\varphi(\zeta)}{(\zeta - z)^2} \, d\xi d\eta + f_1(z) = g(z) + f_1(z), \qquad z \in \mathbb{D},$$

where

$$|\varphi(\zeta)| \asymp |f(\zeta^*)|$$

and f_1 is a function analytic and bounded by an absolute constant in the disc $2\mathbb{D}$. (Recall that $\|f\|_{H^\sigma} = 1$.)

The f_1 term, being analytic in a larger disc, is approximable by polynomials of degree n exponentially. Therefore, we must only estimate best A_p^α approximations of g.

However,

$$g(z) = \int_{1<|\zeta|<2} \frac{\varphi(\zeta)}{(\zeta - z)^2} \, d\xi d\eta = \sum_k \int_{Q_k^*} \frac{\varphi(\zeta)}{(\zeta - z)^2} \, d\xi d\eta = \sum_k g_k(z),$$

where Q_k^* stands for the image of Q_k under the symmetry $z \mapsto z^*$.

Fix n and set

$$G_1(z) = \sum_{k=1}^{n} g_k(z)$$

and

$$G_2(z) = \sum_{k=n+1}^{\infty} g_k(z) = \int_{\bigcup_{k>n} Q_k^*} \frac{\varphi(\zeta)}{(\zeta - z)^2}\, d\xi d\eta.$$

We intend to approximate G_1 constructively and to estimate G_2.

5.4. Estimate of the term G_2

By (8),

$$\|G_2\|_{A_p^\alpha} \le C \left(\int_{\bigcup_{k>n} Q_k^*} |\varphi(\zeta)|^p (|\zeta| - 1)^\alpha \, d\xi d\eta \right)^{\frac{1}{p}}.$$

However, $|\varphi| \le C f_k$ and $|\zeta| - 1 \asymp \rho_k$ on Q_k^*, whereas the area of Q_k^* is of order ρ_k^2. Therefore

$$\int_{\bigcup_{k>n} Q_k^*} |\varphi(\zeta)|^p (|\zeta| - 1)^\alpha \, d\xi d\eta \le C \sum_n^\infty f_k^p \rho_k^{2+\alpha} = C \sum_n^\infty f_k^p \rho_k^{\frac{p}{\sigma}}$$

$$\le C \sum_n^\infty \frac{1}{k^{\frac{p}{\sigma}}} = C \sum_n^\infty \frac{1}{k^{2+\alpha}} \le C \frac{1}{n^{1+\alpha}}.$$

Here we have used the estimate (21).

Thus,

$$\|G_2\|_{A_p^\alpha} \le C \frac{1}{n^{\frac{1+\alpha}{p}}}. \tag{22}$$

5.5. Approximation of the term G_1

Set

$$m_k = \left[\sqrt{\frac{n}{k}} \right] + 1, \qquad k = 1, 2, \dots, n.$$

We construct an approximation of each g_k by a rational function r_k of degree m_k, so that G_1 is approximated by the sum

$$r = \sum_1^n r_k,$$

that is, by a rational function of degree

$$N = \sum_1^n m_k \le n + \sum_1^n \sqrt{\frac{n}{k}} \le 3n.$$

To this end, recall the elementary identity from Subsection 2.7. Fix a point $\zeta_k \in Q_k^*$ for each k, and define

$$r_k(z) = \int_{Q_k^*} \left(\sum_{l=2}^{m_k} (-1)^l \frac{(l-1)(\zeta - \zeta_k)^{(l-2)}}{(\zeta_k - z)^l} \right) \varphi(\zeta)\, d\xi d\eta.$$

This is a rational function with one pole ζ_k of order m_k. By (14), for $z \in \mathbb{D}$ we have

$$|g_k(z) - r_k(z)| \le \int_{Q_k^*} \frac{(m_k+1)|\zeta - \zeta_k|^{m_k-1}}{|\zeta_k - z|^{m_k}|\zeta - z|} |\varphi(\zeta)| \, d\xi d\eta.$$

However, by definition $\operatorname{dist}(Q_k^*, \mathbb{T}) \ge \rho_k$ and $\operatorname{diam}(Q_k^*) \le 4\varepsilon\rho_k$. If we choose $\varepsilon < 1/4$ so small that

$$(m+1)\varepsilon^{m-1} \le e^{-m}$$

for all $m \ge 2$, then

$$|g_k(z) - r_k(z)| \le C\, e^{-m_k} \int_{Q_k^*} \frac{|\zeta| - 1}{|\zeta - z|^3} |\varphi(\zeta)| \, d\xi d\eta.$$

Thus

$$|G_1(z) - r(z)| \le C \int_{\bigcup_1^n Q_k^*} \frac{|\zeta| - 1}{|\zeta - z|^3} \psi(\zeta) \, d\xi d\eta,$$

where

$$\psi(\zeta) = e^{-m_k} |\varphi(\zeta)| \le C\, e^{-m_k} f_k \quad \text{on } Q_k^*.$$

But for $z \in \mathbb{D}$ and $\zeta \notin \mathbb{D}$

$$|\zeta - z| \ge |\zeta| - 1,$$

and our operator

$$\psi \mapsto \int_{\mathbb{C} \setminus \mathbb{D}} \frac{|\zeta| - 1}{|\zeta - z|^3} \psi(\zeta) \, d\xi d\eta, \qquad z \in \mathbb{D},$$

is majorized by the Marcinkiewicz integral operator (9). Thus,

$$\|G_1 - r\|_{A_p^\alpha}^p \le C \int_{\bigcup_1^n Q_k^*} \psi(\zeta)^p (|\zeta| - 1)^\alpha \, d\xi d\eta \le C \sum_1^n e^{-pm_k} f_k^p \rho_k^{2+\alpha},$$

which, in view of (21), is less than

$$C \sum_1^n \frac{1}{k^{2+\alpha}} e^{-p\sqrt{\frac{n}{k}}} \le C \int_1^n \frac{1}{k^{2+\alpha}} e^{-p\sqrt{\frac{n}{k}}} \, dk.$$

The change of variable $k = \frac{n}{t}$ turns this integral into

$$\frac{1}{n^{1+\alpha}} \int_1^n t^\alpha e^{-pt^{\frac{1}{2}}} \, dt = O\left(\frac{1}{n^{1+\alpha}}\right),$$

because

$$\int_1^n t^\alpha e^{-pt^{\frac{1}{2}}} \, dt \le \int_1^\infty t^\alpha e^{-pt^{\frac{1}{2}}} \, dt < \infty.$$

Collecting the estimates, we obtain

$$R_{3n}(f)_{A_p^\alpha} \le \frac{C}{n^{\frac{1+\alpha}{p}}},$$

which is equivalent to the desired result.

Theorem 5.1 is proved. □

6. Rational approximation in the A_p^α norm

The previous results allows us to obtain a new sharp rational approximation theorem for the approximation in the Bergman norm. As usual, we discuss the Bergman spaces $A_p^\alpha(\mathbb{D})$ with $1 < 2+\alpha < p$.

Theorem 6.1. *Let $q > 0$, and let*

$$s = \frac{1}{q} - \frac{2+\alpha}{p} > 0.$$

If $f \in A_p^\alpha(\mathbb{D})$, then $f \in B_q^s(\mathbb{T})$ if and only if

$$\sum_{n=1}^{\infty} \frac{1}{n^{\frac{q}{p}}} R_n(f)_{A_p^\alpha}^q < \infty. \tag{23}$$

Proof.

The "only if" part. Let $f \in B_q^s$. Then $f \in H^\sigma(\mathbb{D})$ (see Subsection 2.3). Due to the Pekarskii theorem in the same section,

$$\sum_1^{\infty} \frac{1}{n} [n^s R_n(f)_{H^\sigma}]^q < \infty$$

precisely for our usual σ:

$$\frac{1}{\sigma} = \frac{1}{q} - s = \frac{2+\alpha}{p}.$$

Let the rational function r of degree n be the best H^σ approximation to f. Theorem 5.1 says that

$$R_{2n}(f)_{A_p^\alpha} \le R_n(f-r)_{A_p^\alpha} \le C \frac{R_n(f)_{H^\sigma}}{n^{\frac{1+\alpha}{p}}},$$

and so

$$\sum_{n=1}^{\infty} \frac{1}{n^{\frac{q}{p}}} R_n(f)_{A_p^\alpha}^q < \infty.$$

The "if" part. Let the series (23) converge.

Chose two different exponents q_1 and q_2, $0 < q_1 < q < q_2 < \frac{p}{2+\alpha}$. Then by Corollary 4.5

$$\|r\|_{B_{q_i}^{s_i}} \le C_i n^{\beta_i} \|r\|_{A_p^\alpha(\mathbb{D})}, \qquad i = 1, 2,$$

where

$$\beta_i = \frac{1}{q_i} - \frac{1}{p}, \qquad s_i = \frac{1}{q_i} - \frac{2+\alpha}{p}, \qquad i = 1.2,$$

and r is an arbitrary rational function of degree n with poles outside $\mathbb{D}$.

Therefore Corollary 1 on page 129 in [1] asserts that for any t, $0 < t < 1$, if

$$\sum_{n=1}^{\infty} \frac{1}{n} [n^{\beta_t} R_n(f)]^q < \infty,$$

then

$$f \in [B_{q_1}^{s_1}, B_{q_2}^{s_2}]_{tq},$$

where

$$\beta_t = (1-t)\beta_1 + t\beta_2,$$

and $[X_1, X_2]_{tq}$ denotes the corresponding real interpolation space between (quasi)-Banach spaces X_1 and X_2.

Choosing t so that

$$\frac{1}{q} = \frac{1-t}{q_1} + \frac{t}{q_2},$$

and taking into account that

$$[B_{q_1}^{s_1}, B_{q_2}^{s_2}]_{tq} = B_q^s,$$

where

$$s = (1-t)s_1 + ts_2 = \frac{1}{q} - \frac{2+\alpha}{p},$$

we obtain the desired result. □

References

[1] Yu. A. Brudnyi, *Adaptive approximation of functions with singularities,* Trans. Moscow Math. Soc., **55** (1994), 123–186.

[2] A. P. Calderón, *On an integral of Marcinkiewicz,* Studia Math., **57** (1976), 279–284.

[3] P. L. Duren, *Theory of H^p spaces,* Academic Press, New York, **1970**.

[4] E.M. Dyn'kin, *On the classes B_p^s for* $0 < p < 1$, Soviet Math Dokl., **29** No. 2 (1984), 357–360.

[5] E. M. Dyn'kin, *Inequalities for rational functions,* J. Approx. Theory, **91** (1997), 349–367.

[6] J. B. Garnett, *Bounded analytic functions,* Academic Press, New York, **1981**.

[7] J. Garsia-Cuerva and J.-L. Rubio de Francia, *Weighted norm inequalities and related topics,* North Holland, **1985**.

[8] A. A. Pekarskii, *An inequality of Bernstein type for derivatives of rational functions and inverse theorems of rational approximation,* Math. USSR–Sb. **52** (1985), 557–574.

[9] A. A. Pekarskii, *Classes of analytic functions defined by best rational approximations in H^p,* Math. USSR-Sb., **55** (1986), 1–18.

[10] E. Stein, *Singular integrals and differentiability properties of functions,* Princeton Univ. Press, Princeton N. J., **1970**.

Operator Theory:
Advances and Applications, Vol. 113

S. A. Vinogradov, As I Remember Him

E. Gladkova (Shtern)

My memories of Stanislav Aleksandrovich Vinogradov take me back to the time when I was a student in the 9th grade. In May 1961 I heard on the radio that a new Mathematical School for Youth (MSY) under the supervision of the Department of Mathematics and Mechanics of Leningrad State University was about to open. I had known that the Department (we used to call it "Math-Mech") always organized a lot of activities for high school students, and was very supportive of them. But now it was to unite all of the different groups under one roof and to give the Mathematical School its best teachers and university students who were to dedicate their time and knowledge to the students of the School. One of these teachers was Slava Vinogradov.

Unlike the teaching of so-called "required" knowledge that was given to us in school, the MSY revealed how beautiful, graceful, and elegant is the art of mathematics. This impression developed as a response to the passionate teaching demonstrated by our student-professors: by stolid Boris Lurie; by tireless Slava Vinogradov; by Volodja Lifshic, who was almost our age but already a student at the university; by heavy smokers Lev Slucman, Misha Gordin, Rafa Valski, and Semen Belinski; and by many others. Certainly they were not yet grand masters like Professor Isidor Pavlovich Natanson, who addressed the students of MSY in their first lecture. But all of them were definitely professionals from the intellectual and creative community that we were to join through clubs of mathematics and mechanics for high school students.

MSY became for me not only a school for mathematics but also a vivid example of truly exciting work and friendship. I still think of my first math teachers with great appreciation and gratitude. Throughout the years they were always serious about working with high school students. Math-Mech became our second home. While reading the newsletter "This Week at Math-Mech" I always felt very proud that our teachers S. Vinogradov and B. Lurie were the best graduate students in the Department. Once, Slava brought us to the annual festival, the Math-Mech Day, and I remember the turmoil of the excited people in the Department gym, and lots of poems flying from a stage constructed from school desks piled together.

In 1964 we came again to Math-Mech as university students, and once again Slava, that is, Stanislav Aleksandrovich Vinogradov, was teaching our Group 11. At that time he was already a Ph.D. student at the university. He was teaching us and learning from us himself in his ever open-minded manner. He did not hide from us, but on the contrary demonstrated the stages of the teaching process he developed. Later, when I taught the course on Mathematical Analysis myself, I realized that I was doing my best to apply his methods.

Stanislav Aleksandrovich never erected any boundaries between himself as teacher and us as students. Our communication was always remarkably frank and simple. We perceived him as a good friend, and he definitely had a deep influence on the formation of my scientific interests, as I attended the same seminars as he did and the special courses he was teaching.

For a short period of time I had the opportunity to be his co-worker while working at math qualification improvement courses for engineers. And still his appeal lay in his sincere and friendly attitude. Those characteristics were symbols for me of Math-Mech for many years. The news of his premature death was for me a deep personal loss.

Nakhimova str. 3-1-100
Saint-Petersburg
199406, Russia
E-mail address: gld@inf.lcse.spb.se

Operator Theory:
Advances and Applications, Vol. 113

The Bellman Functions and Sharp Weighted Inequalities for Square Functions

S. Hukovic, S. Treil, A. Volberg

Dedicated to the memory of S. A. Vinogradov

Abstract. We give a sharp estimate of the norm of the S-functions on $L^2(w)$ in terms of the A_2 "norm" of w.

0. Introduction

Weighted norm inequalities for singular integral operators appear naturally in many areas of analysis, probability, operator theory, etc.

The one weight case is now pretty well understood, and the answers are given by the famous Helson–Szegö theorem and the Hunt–Muckenhoupt–Wheeden theorem. The fist one states that the Hilbert transform H is bounded in the weighted space $L^2(w)$ if and only if w can be represented as $w = \exp\{u + Hv\}$, where $u,\ v \in L^\infty$, $\|v\|_\infty < \pi/2$.

The Hunt–Muckenhoupt–Wheeden theorem states that the Hilbert transform H is bounded in $L^p(w)$ if and only if the weight w satisfies the so-called Muckenhoupt A_p condition

$$\sup_I \Big(\frac{1}{|I|}\int_I w\Big)\cdot\Big(\frac{1}{|I|}\int_I w^{-1/(p-1)}\Big)^{p-1} < \infty, \tag{A_p}$$

where the supremum is taken over all intervals I. This condition is also necessary and sufficient for the boundedness of a wide class of singular integral operators, as well as for the boundedness of the maximal operator M,

$$Mf(x) = \sup_{I\ni x}\frac{1}{|I|}\int_I |f|;$$

here the supremum is taken over all intervals I containing x.

It is worth mentioning that there in no direct proof of the equivalence of the Helson–Szegö condition and the Muckenhoupt condition A_2.

Notice that there is an important operator (called by different names: square function, Lusin's function, area integral, and shortly S-function), which plays a

---

1991 *Mathematics Subject Classification.* Primary 42B20, 42A50, 47B35.
Partially supported by the NSF grants DMS 9622936, ... , and MSRI program of the Fall 1997.

crucial role in the part of Harmonic Analysis known as the Littlewood-Paley theory (see e.g. [St1], [St2]).

In particular, the S-function is very helpful in proving that certain singular integral operators (for example, the Hilbert transform H) are bounded on, say, weighted $L^p(w)$. We restrict ourselves to the case $p = 2$. It is known that the S-function is bounded on $L^2(w)$ if and only if $w \in A_2$.

In this article we will be interested in *how the norms of H or S are bounded in terms of w.*

This question (especially for H) used to reappear from time to time since at least [St3]. There were several recent works which were primarily concerned with this question of *sharp weighted estimates of the Hilbert transform and S-function*: [Bu1], [Bu2], [Hu], [FP]. Let us also mention [NT], which is very pertinent to the subject. In [Bu2] the sharp weighted estimate for maximal operator was obtained. Also a certain estimate for singular integrals of Calderón-Zygmund type (say, for the Hilbert transform) appeared in that paper. These latter estimates are probably not sharp. In [Bu1] a certain weighted estimate for the S-function was obtained. Later, it was sharpened in [Hu]. The present paper is devoted to a sharp A_2 weighted estimate for the S-function. We hope that its methods (especially the use of Bellman functions) might help to find sharp A_2 weighted estimate for the Hilbert transform, which are still unknown (see [FP], p. 356). Let us mention that in [FP] sharp estimates of the Hilbert transform and the S-function were obtained in terms of the A_1 norm of the weight. They were used to give sharp growth estimates for the norms of these operators in *unweighted* L^p. As a corollary, [FP] contains a proof of Chang-Wilson-Wolff inequality (see [ChWW]): the exponential square integrability of functions with uniformly bounded S-function. It would probably be interesting to see the connections between Chang-Wilson-Wolff inequality and our sharp A_2 weighted estimate for the S-function, which we are going to present now.

One warning: the definition of the "norm" $\|w\|_{A_2}$ below differs from the definitions in [Bu1], [Bu2], [FP], [Hu].

0.1. Statement of results

Let us introduce the gauge for $w \in A_2$, which will be called the "norm" of w:

$$\|w\|_{A_2} := \sup_I \langle w\rangle_I^{1/2} \langle w^{-1}\rangle_I^{1/2}.$$

The term "norm" can be explained partially by the fact that the quantity $\langle w\rangle_I^{1/2} \langle w^{-1}\rangle_I^{1/2}$ is the usual operator norm of

$$f \mapsto \langle f\rangle_I \chi_I$$

as an operator on $L^2(w)$.

We will also need another "norm" of w. Let $P_z(t)$ denote the Poisson kernel, and let $w(z) := \int w(t)P_z(t)dt$. Notice that $w^{-1}(z) > w(z)^{-1}$ unless w is a constant function. Put

$$\|w\|_{A_{2,inv}} := w(z)^{1/2}w^{-1}(z)^{1/2}.$$

It is well known that $\|w\|_{A_2}$ is finite if and only if $\|w\|_{A_{2,inv}}$ is. Moreover, in [Hu] sharp estimates

$$c\|w\|_{A_2} \leq \|w\|_{A_{2,inv}} \leq C\|w\|^2_{A_2}.$$

were proved.

Let us introduce our S-functions. Let $\mathcal{D}$ is the set of dyadic intervals. We consider the dyadic S-function:

$$S_d(f)(t) := \Big(\sum_{I\in\mathcal{D},t\in I} |\langle f\rangle_{I_-} - \langle f\rangle_{I_+}|^2\Big)^{1/2},$$

where I_-, I_+ are the left and the right halves of I. Also, we need the harmonic S-function:

$$S(f)(t) := \Big(\int_{\Gamma(t)} |\nabla f(z)|^2\, dxdy\Big)^{1/2}.$$

We denote

$$\|S_d\|_w := \sup_{f\neq 0} \frac{\|S_d(f)\|_{L^2(w)}}{\|f\|_{L^2(w)}},$$

and

$$\|S\|_w := \sup_{f\neq 0} \frac{\|S(f)\|_{L^2(w)}}{\|f\|_{L^2(w)}}.$$

Let $\mathcal{L}^2(w)$ denote the space of measurable functions in the disc that are square integrable with respect to measure $w(z)\, ydxdy$. Obviously,

$$\|S(f)\|_{L^2(w)} \leq c\,\|\nabla f\|_{\mathcal{L}^2(w)}, \tag{0.1}$$

where c is an absolute constant. This is merely because $1/y\int_{x-y}^{x+y} w(t)dt \leq c\,w(x+iy)$.

We denote

$$\|\nabla\|_w := \sup_{f\neq 0} \frac{\|\nabla f\|_{\mathcal{L}^2(w)}}{\|f\|_{L^2(w)}}.$$

Theorem 0.1. *There exists $C < \infty$ such that*

$$\|S_d\|_w \leq C\|w\|^2_{A_2}.$$

Theorem 0.2. *There exists $c > 0$ such that for any $Q \geq 1$ there exists $w \in A_2$, $\|w\|^2_{A_2} = Q$, satisfying*

$$\|S_d\|_w \geq cQ.$$

Theorem 0.3. *There exists $C < \infty$ such that*

$$\|S\|_w \leq C\|w\|^2_{A_{2,inv}}. \quad \textit{Moreover,} \quad \|\nabla\|_w \leq C\|w\|^2_{A_{2,inv}}.$$

Theorem 0.4. *There exists* $c > 0$ *such that for any* $\mathcal{Q} \geq 1$ *there exists* $w \in A_2$, $\|w\|^2_{A_{2,inv}} = \mathcal{Q}$, *satisfying*

$$\|\nabla\|_w \geq c\mathcal{Q}. \quad \textit{Moreover,} \quad \|S\|_w \geq c\mathcal{Q}.$$

0.2. The plan of the paper

In Section 1 we reduce the claim of Theorem 0.1 to a weighted embedding theorem of Sawyer's type. In Section 2 we prove this embedding theorem with a right estimate of the embedding constant. This is done by constructing an appropriate Bellman function for the embedding theorem. In Section 3 we use Theorem 0.1 to construct another Bellman function. In the same section we apply this Bellman function to prove Theorem 0.3. In Section 4 we give an alternative proof of Theorem 0.3. In Section 5 we prove Theorem 0.2. Finally, in Section 6 we prove Theorem 0.4.

1. Reduction to a weighted embedding theorem

Let w be an A_2-weight. Then so is w^{-1}. Both weights have the same "norm". We denote $Q := \|w\|^2_{A_2} = \sup_I \langle w \rangle_I \langle w^{-1} \rangle_I$. In what follows c, C denote arbitrary finite positive constants not depending on Q.

Obviously, Theorem 0.1 is equivalent to the following estimate for every $J \in \mathcal{D}$ and every real-valued bounded measurable function f:

$$\frac{1}{|J|} \sum_{I \subset J} |\langle f \rangle_{I_-} - \langle f \rangle_{I_+}|^2 \langle w^{-1} \rangle_I |I| \leq CQ^2 \langle f^2 w^{-1} \rangle_J. \tag{1.1}$$

Now we are going to reshape the left-hand side on the basis of the idea of introducing the Haar functions $\{h_I\}_{I \in \mathcal{D}}$. We remind the reader that h_I vanishes outside of I, and on I we have

$$h_I = |I|^{-1/2}(\chi_{I_-} - \chi_{I_+}).$$

It is very well known that, after a suitable normalization, $\{h_I\}_{I \in \mathcal{D}}$ forms a Riesz basis in $L^2(w)$ if $w \in A_2$. We are going now to use this fact implicitly.

To make our transformation we need

1.1. Disbalanced Haar functions

Let u denote an arbitrary positive integrable function (weight). We represent h_I as follows:

$$x_I h_I = h_I^u + A_I \chi_I. \tag{1.2}$$

Here, for brevity, we have skipped the superscript u for the constants x_I, A_I, which are determined from requirement that the *disbalanced Haar functions* h_I^u form an orthonormal basis in $L^2(u)$. In other words we have two equations to determine two constants:

$$\int_I (h_I^u)^2 u dx = 1, \int_I h_I^u u dx = 0.$$

Now we can calculate

$$x_I = \frac{\langle u\rangle_I^{1/2}}{\langle u\rangle_{I_-}^{1/2}\langle u\rangle_{I_+}^{1/2}}, \tag{1.3}$$

$$A_I = x_I \frac{\langle u\rangle_{I_+} - \langle u\rangle_{I_-}}{\langle u\rangle_I}\frac{1}{|I|^{1/2}}. \tag{1.4}$$

1.2. Embedding theorem

Now to estimate the left-hand side of (1.1) we take into account that

$$|\langle f\rangle_{I_-} - \langle f\rangle_{I_+}|^2 |I| = |(f, h_I)|^2,$$

and we can use (1.2) with $u = w$ to write

$$\text{Left side of (1.1)} \le 2\sum_{I\subset J} |(f, h_I^w)|^2 \frac{\langle w^{-1}\rangle_I}{x_I^2} + 2\sum_{I\subset J} |(f,\chi_I)^2 \frac{A_I^2}{x_I^2}\langle w^{-1}\rangle_I. \tag{1.5}$$

By (1.3)

$$\Sigma_1 := \sum_{I\subset J} |(f, h_I^w)|^2 \frac{\langle w^{-1}\rangle_I}{x_I^2} \le C\sum_{I\subset J} |(fw^{-1}, h_I^w)_{L^2(w)}\langle w\rangle_I\langle w^{-1}\rangle_I.$$

Now we remind the reader that $\{h_I^w\}_{I\in\mathcal{D}}$ is an orthonormal system in $L^2(w)$. Thus we get

$$\Sigma_1 \le cQ\,\|fw^{-1}\|_{L^2(J,w)}^2 = cQ\,\langle f^2 w^{-1}\rangle_J. \tag{1.6}$$

By (1.4)

$$\Sigma_2 := \sum_{I\subset J} |(f,\chi_I)|^2 \frac{A_I^2}{x_I^2}\langle w^{-1}\rangle_I \le \sum_{I\subset J} |\langle f\rangle_I|^2 \left|\frac{\langle w\rangle_{I_+} - \langle w\rangle_{I_-}}{\langle w\rangle_I}\right|^2 \langle w^{-1}\rangle_I |I|. \tag{1.7}$$

We want to prove that

$$\Sigma_2 \le cQ^2\,|\langle f^2 w^{-1}\rangle_J. \tag{1.8}$$

Looking at (1.7) we can see that its left-hand side is the square of the norm of the embedding $f \mapsto \{f_I\}_{I\in\mathcal{D}}$, which looks almost like the Carleson embedding theorem (see e.g. [Ga]). The distinction is that f is in *the weighted space* $L^2(w^{-1})$. Such theorems were considered by Sawyer.

The common feature of the Sawyer embedding theorems is that embedding result of type (1.8) is true if and only if it is true uniformly for the test functions

$$f := w\chi_\ell,$$

where ℓ runs over all dyadic subintervals of J (some of Sawyer's theorems involve testing the dual operator as well).

Thus we have reduced Theorem 0.1 to an embedding theorem (1.8) of Sawyer type. In the next section we will prove (1.8) first for the test functions $f = w\chi_\ell$ and then for all f's. Then Theorem 0.1 will be finished.

2. A Bellman function for weighted embedding theorem and the proof of Theorem 0.1

The next lemma proves (1.8) for our test functions.

Lemma 2.1. *Let $w \in A_2$, and let $Q = \|w\|_{A_2}^2$. Then for any dyadic interval J*

$$\frac{1}{|J|}\sum_{I\subset J} |\langle w\rangle_{I_+} - \langle w\rangle_{I_-}|^2 \langle w^{-1}\rangle_I |I| \le cQ^2 \langle w\rangle_I. \tag{2.1}$$

Proof. If B is a function of several real variables then let d^2B denote the Hessian form of B. The proof is based on a construction of a Bellman function for the above inequality. Namely, we are going to construct a function $B = w\eta(wv)$ of two real variables w, v, which is defined in the domain $\{w, v : wv \ge 1\}$, and such that

$$1 \le wv \le Q \quad \text{implies} \quad 0 \le B \le cQ^2 w;$$

$$1 \le wv \le 2Q \quad \text{implies} \quad -d^2B \ge 0;$$

$$1 \le wv \le Q \quad \text{implies} \quad -d^2B \ge Cv\,(dw)^2.$$

Here is such a function

$$B(w,v) = w\Big(-\frac{4Q^2}{wv} - wv + 4Q^2 + 1\Big) = (4Q^2+1)w - \frac{4Q^2}{v} - w^2 v. \tag{2.2}$$

Then the first property of B is obvious. It is easy to compute

$$-d^2B = 2v(dw)^2 + 4w\,dw\,dv + \frac{8Q^2}{v^3}(dv)^2.$$

Obviously this form is positive definite on $\Omega_2 := \{w, v > 0 : 1 \le wv \le 2Q\}$ because the determinant and the diagonal elements of the corresponding matrix are nonnegative on Ω_2.

On $\Omega_1 := \{w, v > 0 : 1 \le wv \le Q\}$ one has a better estimate obviously:

$$-d^2B \ge \frac{3}{2}v(dw)^2.$$

All properties of B being checked, we pass to the proof of (2.1). First, we prove that for any three points $(w, v), (w_+, v_+), (w_-, v_-)$ in Ω_1 such that

$$w = \frac{w_+ + w_-}{2}, \quad v = \frac{v_+ + v_-}{2}$$

one has

$$B(w,v) - \frac{B(w_+,v_+) + B(w_-,v_-)}{2} \ge c(w_+ - w_-)^2 v. \tag{2.3}$$

To prove (2.3) consider $w(t) := \frac{(1-t)w_- + (1+t)w_+}{2}$, $v(t) := \frac{(1-t)v_- + (1+t)v_+}{2}$, and put $b(t) := B(w(t), v(t))$. The parameter t always runs over $[-1, 1]$. It is easy to see that $(w(t), v(t)) \in \Omega_2$. Thus, the negativity of d^2B on this domain shows that b is a concave function on $[-1,1]$. Notice also that either for all $t \in [0,1]$ or for all $t \in [-1,0]$

$$(w(t), v(t)) \in \Omega_1,$$

implying a qualified concavity of b:

$$-b''(t) \geq Cv\,(w_+ - w_-)^2 \tag{2.4}$$

for all $t \in [0,1]$ or for all $t \in [-1,0]$.

So, (2.3) follows from the concavity of b and from (2.4):

$$B(w,v) - \frac{B(w_+,v_+) + B(w_-,v_-)}{2} = b(0) - \frac{b(1)+b(-1)}{2}$$
$$= -\int_{-1}^{1} b''(t)(1-|t|)dt \geq Cv\,(w_+ - w_-)^2 \int_0^1 (1-|t|)dt = Cv\,(w_+ - w_-)^2.$$

Now we apply (2.3) for proving (2.1). First, we choose $w = \langle w\rangle_J$, $v = \langle w^{-1}\rangle_J$, $w_+ = \langle w\rangle_{J_+}$, $w_- = \langle w\rangle_{J_-}$, $v_+ = \langle w^{-1}\rangle_{J_+}$, $v_- = \langle w^{-1}\rangle_{J_-}$ to get

$$C(\langle w\rangle_{J_+} - \langle w\rangle_{J_-})^2\langle w^{-1}\rangle_J + \frac{1}{2}B(\langle w\rangle_{J_+}, \langle w^{-1}\rangle_{J_+}) + \frac{1}{2}B(\langle w\rangle_{J_-}, \langle w^{-1}\rangle_{J_-})$$
$$\leq B(\langle w\rangle_J, \langle w^{-1}\rangle_J).$$

Then we write this inequality replacing subsequently J by each dyadic subinterval of J: J_-, J_+, etc. Combining these inequalities we finally get

$$C\sum_{I\subset J} \frac{|I|}{|J|}(\langle w\rangle_{I_+} - \langle w\rangle_{I_-})^2\langle w^{-1}\rangle_I \leq B(\langle w\rangle_J, \langle w^{-1}\rangle_J). \tag{2.5}$$

Combining (2.5) with the inequality

$$B(\langle w\rangle_J, \langle w^{-1}\rangle_J) \leq 4Q^2\langle w\rangle_J,$$

we get (2.1). Lemma 2.1 is proved. □

The next result is probably very well known. This is simply a version of a Sawyer type theorem. The important thing is the behavior of constants.

Theorem 2.2. *Let v be an arbitrary weight, and let α_I be an arbitrary sequence of nonnegative numbers enumerated by dyadic intervals. If*

$$\forall J \in \mathcal{D}, \ \frac{1}{|J|}\sum_{I\subset J}\langle v\rangle_I^2\alpha_I \leq C\langle v\rangle_J, \tag{2.6}$$

then

$$\forall J \in \mathcal{D}, \ \frac{1}{|J|}\sum_{I\subset J}\langle f\rangle_I^2\alpha_I \leq 4C\langle f^2v^{-1}\rangle_J. \tag{2.7}$$

Proof. Consider the domain $D := \{(X, x, M, v) \geq 0 : x^2 \leq Xv,\ M \leq v\}$ and let B_0 be a function on D such that

$$0 \leq B_0 \leq X,\ -d^2 B_0 \geq 0,\ \frac{\partial B_0}{\partial M} \geq \frac{1}{4}\frac{x^2}{v^2}. \tag{2.8}$$

For example,

$$B_0(X, x, M, v) = X - \frac{x^2}{v + M}.$$

Lemma 2.3. *Let s and $s_\pm$ be quads in D. Let also*

$$s - \frac{s_+ + s_-}{2} = (0, 0, \alpha, 0),$$

with $\alpha \geq 0$. Then

$$B_0(s) - \frac{B_0(s_+) + B_0(s_-)}{2} \geq \frac{1}{4}\frac{x^2}{v^2}\alpha, \tag{2.9}$$

Proof. Let $M' = \frac{M_+ + M_-}{2}$. Then $M - M' = \alpha$. Put $s(t) := \frac{(1-t)s_- + (1+t)s_+}{2}$ and $b(t) := B_0(s(t))$. Then

$$\begin{aligned} B_0(s) &- \frac{B_0(s_+) + B_0(s_-)}{2} \\ &= B_0(X, x, M, v) - B_0(X, x, M', v) + B_0(X, x, M', v) \\ &\quad - \frac{B_0(X_+, x_+, M_+, v_+) + B_0(X_-, x_-, M_-, v_-)}{2} \\ &\geq \frac{1}{4}\frac{x^2}{v^2}\alpha + \left(b(0) - \frac{b(1) + b(-1)}{2}\right) \geq \frac{1}{4}\frac{x^2}{v^2}\alpha. \end{aligned}$$

The last inequality uses (2.8) and proves (2.9). □

We continue the proof of Theorem 2.2. We are going to prove (2.7) by using (2.9). First fix J and denote

$$X = \langle f^2 v^{-1} \rangle_J, \qquad x = \langle f \rangle_J, \qquad M = C^{-1}\frac{1}{|J|}\sum_{I \subset J} \langle v \rangle_I^2 \alpha_I,$$

$$X_\pm = \langle f^2 v^{-1} \rangle_{J_\pm}, \qquad x_\pm = \langle f \rangle_{J_\pm}, \qquad M_\pm = C^{-1}\frac{1}{|J_\pm|}\sum_{I \subset J_\pm} \langle v \rangle_I^2 \alpha_I.$$

Then apply (2.9) to these numbers. (This is possible because the above quads lie in D by (2.6).) Then replace J subsequently by J_-, J_+, then by their sons, et cetera. We keep applying (2.9). As a result we get

$$\frac{1}{4C}\sum_{I \subset J} \frac{\langle f \rangle_I^2}{\langle v \rangle_I^2} \frac{|I|}{|J|} \frac{1}{|I|} \langle v \rangle_I^2 \alpha_I \leq B_0(X, x, M, v).$$

Using the fact that

$$B_0(X, x, M, v) \leq X = \langle f^2 v^{-1} \rangle_J,$$

we get (2.7). Theorem 2.2 is completely proved. □

3. From dyadic to harmonic

Lemma 2.1 and Theorem 2.2 were proved by concocting concave functions with special properties (Bellman functions of the corresponding problems). However, Theorem 0.1 was proved in a different way. No Bellman function showed up (instead, for its proof we combined Lemma 2.1, Theorem 2.2, and a trick with disbalanced Haar functions). And for a good reason — it is not an easy task to invent a Bellman function for the problem (1.1) without a kind of analysis we have done in Sections 1, 2. In other words, it is hard to come up with the analytic expression for the concave function in the following theorem without the analysis which has been done. So the next theorem is an existence theorem *for a while.* In the next section we write down the analytic expression of one of such functions. It is obtained by careful analysis of what we were doing in Sections 1, 2 on the level of manipulating with formulas. After looking at the analytic expression of our function in Section 4, the reader will understand why we prefer to follow first a less formal approach presented below.

Theorem 3.1. *Let* $Q > 1$. *On the domain* $\mathcal{O} := \{s = (X, x, w, v) > 0 : x^2 < Xw, 1 < wv < Q\}$ *there exists a concave function* $\mathcal{B}$ *such that*

$$0 \le \mathcal{B} \le cQ^2\, X, \tag{3.1}$$

and such that for any points $s, s_+, s_- \in \mathcal{O}$, $s = \frac{s_+ + s_-}{2}$ *we have*

$$\mathcal{B}(s) - \frac{\mathcal{B}(s_+) + \mathcal{B}(s_-)}{2} \ge Cv\,(x_+ - x_-)^2. \tag{3.2}$$

The analytic expression for one of such $\mathcal{B}$ can be found in Section 4. Then, given this expression, all the rest is calculus. But we have already explained that it seems to be tricky to find such expressions — that is why we prefer first to give a proof of existence *without knowing the analytic expression.*

Proof. We fix $s = (X, x, w, v) \in \mathcal{O}$ and consider the class $\mathcal{S}$ of all positive bounded functions f, w supported on J such that

$$\langle f^2 w^{-1}\rangle_J = X,\ \langle f\rangle_J = x,\ \langle w\rangle_J = w,\ \langle w^{-1}\rangle_J = v,$$

$$\mathcal{B}(s) := \sup\left\{\frac{1}{|J|}\sum_{I\subset J} |\langle f\rangle_{I_+} - \langle f\rangle_{I_-}|^2 \langle w^{-1}\rangle_I |I| : (f, w) \in \mathcal{S}\right\}. \tag{3.3}$$

Now let $f^\pm, w^\pm$ be some pairs of functions in $\mathcal{S}_\pm$ almost providing the supremums for $\mathcal{B}(s_\pm)$. They are supported on J. We rescale them in order to place them on $J_\pm$. Take f^+, w^+ and rescale them to the interval J_+ preserving their averages. Call the new functions F^+, W^+ sitting on J_+. So, for example, $\langle F^+\rangle_{J_+} = \langle f^+\rangle_J$. The same is true for W^+, w^+. Do the same for f^-, w^- obtaining F^-, W^- sitting on J_-. Finally, we introduce two functions F, W by

$$F|J_- := F^-,\ F|J_+ := F^+, \qquad W|J_- := W^-,\ W|J_+ := W^+.$$

Obviously, the pair (F, W) belongs to $\mathcal{S}$ by construction and because $s = \frac{s_- + s_+}{2}$.

Since $(F, W) \in \mathcal{S}$, it follows that

$$\mathcal{B}(s) \geq \frac{1}{|J|} \sum_{I \subset J} |\langle F \rangle_{I_+} - \langle F \rangle_{I_-}|^2 \langle W^{-1} \rangle_I |I| = \frac{1}{2}\Big(\frac{1}{|J_-|} \sum_{I \subset J_-} \cdots + \frac{1}{|J_+|} \sum_{I \subset J_+} \cdots \Big). \tag{3.4}$$

Now the first sum is $\frac{1}{|J|} \sum_{I \subset J} |\langle f^- \rangle_{I_+} - \langle f^- \rangle_{I_-}|^2 \langle (w^-)^{-1} \rangle_I |I|$ by our rescaling law. By the choice of f^-, w^- this expression is as close to $\mathcal{B}(s_-)$ as we wish. By the same reason the second sum is as close to $\mathcal{B}(s_+)$ as we wish. Taking this into account and coming back to (3.4), we get

$$\mathcal{B}(s) \geq \frac{1}{2}(\mathcal{B}(s_-) + \mathcal{B}(s_+)) - \eta,$$

with arbitrary small η. Thus (3.2) is proved.

As to (3.1), this inequality is actually equivalent to Theorem 0.1, which has already been proved. □

Lemma 3.2. *Let K be a compact subset of $\mathcal{O}$. There exists an infinitely differentiable function $\mathcal{B}_K$ in a neighborhood of K such that*

$$0 \leq \mathcal{B}_K \leq c\mathcal{Q}^2 X, \tag{3.5}$$

and such that for all $s \in K$ the Hessian form $d^2\mathcal{B}_K$ satisfies

$$-d^2\mathcal{B}_K(s) \geq Cv\,(dx)^2. \tag{3.6}$$

Proof. Let $100\varepsilon < \operatorname{dist}(K, \partial\mathcal{O})$. Consider the convolution of B and $\frac{1}{\varepsilon^2}\phi(\frac{z}{\varepsilon})$, where ϕ is a bell shape infinitely differentiable function with support in the unit disc. It is now very easy to see that this convolution (we call it $\mathcal{B}_K$) satisfies (3.1), (3.2) — maybe with slightly different constants c, C. This is just because $\mathcal{B}$ and all its small shifts satisfy these inequalities in a small neighborhood of K. Obviously $\mathcal{B}_K$ is infinitely differentiable. On the other hand for any infinitely differentiable function (3.2) implies (3.6). To see this it suffices to consider the Taylor expansion of this function and to plug it into (3.2). □

Now we are ready to prove Theorem 0.3.

Proof. We like to think that everything is happening in the unit disc. Let $\mathcal{Q} > 1$. Consider w on the unit circle such that $\|w\|^2_{A_{2,inv}} = \mathcal{Q}$. We consider only smooth w. Also consider a smooth function f on the unit circle such that $f \neq w$. Let $\frac{f^2}{w}(z)$, $f(z)$, $w(z)$, $w^{-1}(z)$ be the harmonic extensions of the functions $\frac{f^2}{w}$, f, w, w^{-1} from the unit circle to the unit disc. Notice that if z runs over the set $|z| \leq r$, then $s = (\frac{f^2}{w}(z), f(z), w(z), w^{-1}(z))$ runs over a compact subset $K(r) \subset \mathcal{O}$. Put

$$b(z) := \mathcal{B}_{K(r)}\Big(\frac{f^2}{w}(z), f(z), w(z), w^{-1}(z)\Big).$$

For any $z, |z| \le r$, we have

$$-\Delta b(z) \ge C w^{-1}(z)\, |\nabla f(z)|^2. \tag{3.7}$$

This follows from (3.6) by direct calculation (see [NT]).

Now we apply Green's formula to b in the disc $|z| \le r$:

$$\begin{aligned} cQ^2 \int_{|z|=r} \frac{f^2}{w} dm &\ge \int_{|z|=r} b\, dm - b(0) \\ &= -\frac{1}{2\pi} \int_{|z|<r} \Delta b(z) \log \frac{r}{|z|} dxdy \\ &\ge C \int_{|z|<r} |\nabla f(z)|^2 w^{-1}(z) \log \frac{r}{|z|} dxdy. \end{aligned}$$

The left-hand side is bounded by $cQ^2 \int_{|z|=1} \frac{f^2}{w} dm$. Now, making r to go to 1, we prove Theorem 0.3 (just interchange w and w^{-1}). □

4. The analytic expression for a Bellman function. Alternative proof of Theorem 0.3

4.1. The analytic expression for a Bellman function

Theorem 4.1. *Let $Q > 1$. In the domain $\mathcal{O} := \{s = (X, x, w, v) > 0 : x^2 < Xw, 1 < wv < Q\}$ the function $\mathcal{B}$ given by the formula*

$$\mathcal{B} := (1 + \frac{1}{Q})X - \frac{x^2}{Qw} - \frac{Q^2 x^2}{Q^2 w + (4Q^2+1)w - w^2 v - 4Q^2/v} \tag{4.1}$$

satisfies

$$0 \le \mathcal{B} \le 2X, \tag{4.2}$$

and its Hessian satisfies

$$-d^2\mathcal{B} \ge C \frac{1}{Q^2} v (dx)^2. \tag{4.3}$$

Proof. This is certainly a direct computation, but we sketch it. First of all the right inequality in (4.2) is obvious. To see the left one, we introduce the notation

$$M(w, v) := (4Q^2 + 1)w - w^2 v - 4Q^2/v$$

and notice that in $\{(w, v) : 1 < wv < Q\}$ we have

$$M(w, v) \ge 0 \tag{4.4}$$

We put

$$\mathcal{B}_1 := X - \frac{x^2}{w}; \qquad \mathcal{B}_2 := X - \frac{x^2}{w + M(w,v)/Q^2}.$$

Clearly,

$$\mathcal{B} = \frac{1}{Q}\mathcal{B}_1 + \mathcal{B}_2.$$

It is also clear now (because of (4.4)) that $\mathcal{B}_1(X,x,w)$, $\mathcal{B}_2(X,x,w,v)$ are nonnegative as soon as $x^2 \le Xw$—in particular, they are nonnegative in $\mathcal{O}$. So (4.2) is proved.

To prove (4.3), we estimate $d^2\mathcal{B}_1$ and $d^2\mathcal{B}_2$ separately. Direct calculation shows that

$$-d^2\mathcal{B}_1 = 2\frac{x^2}{w}\Big(\frac{dx}{x} - \frac{dw}{w}\Big)^2. \tag{4.5}$$

On the other hand notice that $\mathcal{B}_2(X,x,w,v) = \mathcal{B}_1(X,x,w+M(w,v)/Q^2)$. If we denote $w' := w + M(w,v)/Q^2$ we can now see, using (4.5), that

$$-d^2\mathcal{B}_2 = 2\frac{x^2}{w'}\Big(\frac{dx}{x} - \frac{dw'}{w'}\Big)^2 + \frac{\partial\mathcal{B}_1}{\partial w'}(X,x,w')(-d^2M(w,v)/Q^2). \tag{4.6}$$

The first term is nonnegative and we omit it. Also

$$\frac{\partial\mathcal{B}_1}{\partial w'}(X,x,w') = \frac{x^2}{w'^2} \ge \frac{1}{36}\frac{x^2}{w^2} \tag{4.7}$$

just because $w' = w + M(w,v)/Q^2 \le 6w$ (see the definition of M).

The estimate for the Hessian of M was done already in the beginning of Section 2. This is

$$-d^2M(w,v) \ge \frac{3}{2}v(dw)^2. \tag{4.8}$$

Combining (4.6), (4.7), (4.8) we finally get

$$-d^2\mathcal{B}_2 \ge C\frac{1}{Q^2}vx^2\Big(\frac{dw}{w}\Big)^2. \tag{4.9}$$

Now

$$-d^2\mathcal{B} = -\frac{1}{Q}d^2\mathcal{B}_1 - d^2\mathcal{B}_2\,.$$

Taking into account that $\frac{1}{w} \ge \frac{1}{Q}v$ in our domain $\mathcal{O}$ and combining (4.5), (4.9) we obtain

$$-d^2\mathcal{B} \ge \frac{1}{Q^2}vx^2\Big(\frac{dx}{x} - \frac{dw}{w}\Big)^2 + \frac{1}{Q^2}vx^2\Big(\frac{dw}{w}\Big)^2.$$

Thus,

$$-d^2\mathcal{B} \ge \frac{1}{Q^2}v(dx)^2\,,$$

which is (4.3). Theorem 4.1 is completed. □

4.2. Alternative proof of Theorem 0.3

It is now clear how one can organize an alternative proof of Theorem 0.3. Given smooth strictly positive functions f, w on the circle $\mathbb{T}$, we denote $Q = \sup_{z\in\mathbb{D}} w(z)w^{-1}(z)$, and we wish to have a nonnegative function b on the disc continuous up to the boundary and such that $b|\mathbb{T} \leq 2f^2w^{-1}|\mathbb{T}$ and $-\Delta b(z) \geq C\frac{1}{Q^2}w^{-1}(z)(\nabla f(z))^2$.

Then we apply Green's formula to b:

$$2\int_{\mathbb{T}}\frac{f^2}{w}dm \geq \int_{\mathbb{T}} bdm - b(0) = \frac{1}{2\pi}\int_{\mathbb{D}}\Delta b(z)\log\frac{1}{|z|}dxdy$$
$$C\frac{1}{Q^2}\int_{\mathbb{D}}|\nabla f(z)|^2 w^{-1}(z)\log\frac{1}{|z|}dxdy\,,$$

to get the claim of Theorem 0.3.

Here is a required b:

$$b(z) := \Big(1+\frac{1}{Q}\Big)(f^2w^{-1})(z) - \frac{f(z)^2}{Qw(z)}$$
$$-\frac{Q^2f(z)^2}{Q^2w(z)+(4Q^2+1)w(z)-w(z)^2w^{-1}(z)-4Q^2/w^{-1}(z)}\,.$$

The required estimate on $b|\mathbb{T}$ is obvious. The estimate on the Laplacian of b follows immediately from Theorem 4.1.

5. The sharpness of Theorem 0.1

We remind the reader that we denote by $Q(w_0)$ the square of "norm" $\|w_0\|_{A_2}$ of the weight w_0. In this section we show that Theorem 0.1 is sharp, that is, we prove that there exists a weight w_0 and a function f_0 such that

$$\|S_d f_0\|_{L^2(w_0)} \geq cQ(w_0)\|f_0\|_{L^2(w_0)}. \tag{5.1}$$

In this section c denotes any constant that does not depend on the norm of the weight.

Let $f_0 = t^{-\alpha}\chi_{[0,1]}$; we put $w_0 = |t|^\alpha$ for $t \in [-1,1]$, $w_0 = 1$ otherwise. Then for $1/2 \leq \alpha < 1$ we have $Q(w_0)^2 = c\frac{1}{(1-\alpha)^2}$ and $\|f_0\|^2_{L^2(w_0)} = \int_0^1 t^{-\alpha} = \frac{1}{1-\alpha}$. Thus

$$Q(w_0)^2\|f_0\|^2_{L^2(w_0)} = \frac{1}{(1-\alpha)^3}. \tag{5.2}$$

To estimate the left-hand side of 5.1 from below we calculate:

$$\begin{aligned}\|S_d f_0\|^2_{L^2(w_0)} &= \sum_{I\in\mathcal{D}} (f_{I+} - f_{I-})^2 w_0(I)\\ &\ge \sum_{I\in\mathcal{D}, 0\in I} (f_{I+} - f_{I-})^2 w_0(I)\\ &= 4\sum_{n=1}^{\infty} (f_{[0,2^{-n}]} - f_{[0,2^{-n+1}]})^2 w_0([0,2^{-n+1}])\\ &\ge c\frac{1}{(1-\alpha)^2}\frac{1}{1+\alpha}\sum_{n=1}^{\infty}\Big(\frac{1}{2^{1-\alpha}}\Big)^n\\ &\ge c\frac{1}{(1-\alpha)^2}\frac{2^{1-\alpha}}{2^{1-\alpha}-1}.\end{aligned}$$

Now notice that, as $\alpha \to 1$, the quantities $\frac{1}{1-\alpha}$ and $\frac{1}{2^{1-\alpha}-1}$ are equivalent. Thus, as $\alpha \to 1$,

$$\|S_d f_0\|_{L^2(w_0)} \ge \frac{1}{(1-\alpha)^3}. \tag{5.3}$$

Thus, we have proved that for our choice of w_0 and f_0 (5.1) is true as $\alpha \to 1$.

6. The sharpness of Theorem 0.3

Consider the same pair $f_0 = t^{-\alpha}\chi_{[0,1]}$ and set $w_0 = |t|^\alpha$ for $t \in [-1,1]$, $w_0 = 1$ otherwise. This will be an example showing the sharpness of Theorem 0.3. Here we need the $A_{2,inv}$ "norm" of power weights, so we first state a lemma that shows that for all power weights in A_2 their classical and invariant norms are equivalent.

Lemma 6.1. *If $w(t) = t^\alpha$, where $\alpha \in (-1,1)$, then $\|w\|^2_{A_{2,inv}} = c\frac{1}{1-\alpha^2}$.*

Proof. Obviously we can consider only $\alpha \in [1/2, 1)$. Notice that $\|w\|^2_{A_2} = \frac{1}{1-\alpha^2}$ and that $\|w\|_{A_{2,inv}} \ge \frac{1}{4}\|w\|_{A_2}$. Thus, to prove the lemma, it suffices to show that there exists a constant K such that $\|w\|^2_{A_{2,inv}} \le K(1-\alpha^2)^{-1}$. Define $g(a) = \sup_c w(a+ic)w^{-1}(a+ic)$. First we show that $g(0) \le K(1-\alpha^2)^{-1}$, and then we show that $g(a) \le g(0)$ for all $a \in \mathbb{R}$. Let c be arbitrary, and let $I_k = [2^{k-1}c, 2^k c)$. We have

$$\begin{aligned}w(0,c) &= \int_{\mathbb{R}} \frac{c}{c^2+t^2}|t|^\alpha dt = 2\int_0^\infty \frac{c}{c^2+t^2} t^\alpha dt\\ &= 2\sum_{k=1}^{\infty}\int_{I_k}\frac{c}{c^2+t^2}t^\alpha dt \le \sum_{k=1}^{\infty}\frac{c}{c^2+2^{2k-2}c^2}\int_{I_k} t^\alpha dt\\ &\le 2c^\alpha\frac{2^{\alpha+1}-1}{\alpha+1}\sum_{k=1}^{\infty}(2^{\alpha-1})^{k-1} = 2c^\alpha\frac{2^{\alpha+1}-1}{\alpha+1}\frac{2^{1-\alpha}}{2^{1-\alpha}-1}.\end{aligned}$$

Similarly,

$$w^{-1}(0,c)=\int_{\mathbb{R}}\frac{c}{c^2+t^2}|t|^{-\alpha}dt\le 2c^{-\alpha}\frac{2^{-\alpha+1}-1}{-\alpha+1}\frac{2^{1+\alpha}}{2^{1+\alpha}-1}. \tag{6.1}$$

Thus, $g(0)\le\frac{16}{1-\alpha^2}$. To show that $g(a)\le g(0)$ for any a, we first make a change of variables and write

$$g(a)=g(-a)=\int_{\mathbb{R}}\frac{c}{c^2+u^2}|u+a|^{\alpha}du\int_{\mathbb{R}}\frac{c}{c^2+u^2}|u+a|^{-\alpha}du. \tag{6.2}$$

Thus,

$$\frac{g'(a)}{c^2\alpha}=\int_{\mathbb{R}}\frac{|u+a|^{\alpha-1}}{c^2+u^2}\int_R\frac{|u+a|^{-\alpha}}{c^2+u^2}-\int_{\mathbb{R}}\frac{|u+a|^{\alpha}}{c^2+u^2}\int_R\frac{|u+a|^{-\alpha-1}}{c^2+u^2}. \tag{6.3}$$

Recall that we consider only $\alpha\in[1/2,1)$. Set $p=\frac{2\alpha+1}{2\alpha}, p'=2\alpha+1, r=\frac{2\alpha^2}{2\alpha+1}$ and $s=\frac{-\alpha-1}{2\alpha+1}$, then $r+s=\alpha-1$, and

$$\begin{aligned}\int_{\mathbb{R}}\frac{|u+a|^{\alpha-1}}{c^2+u^2}du&\le\left(\int_R\frac{|u+a|^{rp}}{c^2+u^2}du\right)^{\frac{1}{p}}\left(\int_{\mathbb{R}}\frac{|u+a|^{sp'}}{c^2+u^2}du\right)^{\frac{1}{p'}}\\&=\left(\int_{\mathbb{R}}\frac{|u+a|^{\alpha}}{c^2+u^2}du\right)^{\frac{1}{p}}\left(\int_{\mathbb{R}}\frac{|u+a|^{-\alpha-1}}{c^2+u^2}du\right)^{\frac{1}{p'}}.\end{aligned}$$

Set $t=\frac{\alpha}{2\alpha+1}$ and $v=\frac{-2\alpha^2-2\alpha}{2\alpha+1}$, here $t+v=-\alpha$ and

$$\int_R\frac{|u+a|^{-\alpha}}{c^2+u^2}du\le\left(\int_{\mathbb{R}}\frac{|u+a|^{\alpha}}{c^2+u^2}du\right)^{\frac{1}{p'}}\left(\int_{\mathbb{R}}\frac{|u+a|^{-\alpha-1}}{c^2+u^2}du\right)^{\frac{1}{p}}. \tag{6.4}$$

The last two inequalities combined lead to the desired inequality $g'(a)\le 0$ for all a. □

We remind the reader that we have used the notation $\mathcal{Q}(w_0)$ for the square of invariant "norm" $\|w_0\|_{A_{2,inv}}$ of the weight w_0. We want to show that there exists $c>0$ such that

$$\|\nabla f_0(z)\|^2_{\mathcal{L}^2(w_0)}\ge\mathcal{Q}(w_0)^2\|f_0\|^2_{L^(w_0)}. \tag{6.5}$$

For $\alpha\in[1/2,1)$, the right-hand side of (6.5) is as follows:

$$\mathcal{Q}(w_0)^2\|f_0\|^2_{L^(w_0)}=c(1-\alpha)^{-3}.$$

Now we only need to estimate the left-side of (6.5):

$$\begin{aligned}\|\nabla f_0\|^2_{\mathcal{L}^2(w_0)} &= \int_{\mathbb{R}^2_+} |\nabla f(z)|^2 w_0(z) y\,dx\,dy \\ &\geq \int_{-\infty}^{0}\int_0^{\frac{-x}{4}} |\nabla f(z)|^2 w_0(z) y\,dx\,dy \\ &\geq \int_0^\infty \int_0^{\frac{x}{4}} \left(\int_0^1 \frac{(x+s)^2-y^2}{((x+s)^2+y^2)^2} s^{-\alpha} ds\right)^2 w_0(z) y\,dy\,dx \\ &\geq c\int_0^\infty \int_0^{\frac{x}{4}} \left(\int_0^1 \frac{1}{(x+s)^2} s^{-\alpha} ds\right)^2 w_0(z) y\,dy\,dx\end{aligned}$$

Now notice that for x fixed,

$$\begin{aligned}\int_0^1 \frac{1}{(x+s)^2} s^{-\alpha} ds &= x^{1-\alpha}\int_0^1 \left(\frac{x}{s}\right)^\alpha \frac{1}{(x+s)^2}\frac{ds}{x} \\ &= x^{-1-\alpha}\int_0^{\frac{1}{x}} u^{-\alpha}(1+u)^{-2} du \geq c x^{-1-\alpha}(1-\alpha)^{-1}.\end{aligned}$$

Hence,

$$\begin{aligned}\|\nabla f_0\|^2_{\mathcal{L}^2(w_0)} &\geq c(1-\alpha)^{-2}\int_0^\infty\int_0^{\frac{x}{4}} x^{-2-2\alpha} w_0(z) y\,dy\,dx \\ &\geq c(1-\alpha)^{-2}\int_0^1 t^\alpha \left(\int_0^\infty\int_0^{\frac{x}{4}} \frac{y^2}{y^2+(x+t)^2} x^{-2-2\alpha} dx\,dy\right) dt \\ &= c(1-\alpha)^{-2}\sum_{j=1}^\infty \int_{I_j} t^\alpha \left(\int_0^\infty\int_0^{\frac{x}{4}} \frac{y^2}{y^2+(x+t)^2} x^{-2-2\alpha} dx\,dy\right) dt \\ &\geq c(1-\alpha)^{-2}\sum_{j=1}^\infty \left(\frac{1}{2^j}\right)^{1+\alpha} 2^{2\alpha j} = c(1-\alpha)^{-2}\frac{1}{2^{1-\alpha}-1}.\end{aligned}$$

In the last inequality, we have used the following estimates for $t \in I_j$:

$$\begin{aligned}&\left(\int_0^\infty\int_0^{\frac{x}{4}} \frac{y^2}{y^2+(x+t)^2} x^{-2-2\alpha} dx\,dy\right) dt \\ &\qquad \geq \int_{2^{-j}}^1 r^{-1-2\alpha} dr \geq c 2^{2\alpha j}.\end{aligned}$$

To finish the proof of Theorem 0.3 we need to show that there exists $c > 0$ such that

$$\|S(f_0)\|^2_{L^2(w_0)} \geq \mathcal{Q}(w_0)^2 \|f_0\|^2_{L(w_0)}. \tag{6.6}$$

The calculation repeats the one just made. The only change is that now we will not have the Poisson extension $w_0(x+iy)$ of w_0. Instead, we have

$1/y \int_{x-y}^{x+y} w_0(t)\,dt$. This change amounts to replacing

$$\int_{I_j} t^\alpha \left(\int_0^\infty \int_0^{\frac{x}{4}} \frac{y^2}{y^2 + (x+t)^2} x^{-2-2\alpha} dx dy \right) dt$$

by

$$\int_{I_j} |t|^\alpha \left(\int_{|t|}^{4/3|t|} \int_{|x-t|}^{1/4|x|} x^{-2-2\alpha}\, dy dx \right) dt.$$

As in the previous calculation this integral is approximately $2^{-j(1-\alpha)}$. So, we are done with proving (6.6) and thus with proving Theorem 0.3.

References

[Bu1] St. Buckley, *Summation conditions on weights,* Mich. Math. J., **40** 1993, 153–170.

[Bu2] St. Buckley, *Estimates for operator norms on weighted spaces and reverse Jensen's inequality,* Tran. Amer. Math. Soc. **340** (1993), 253–272.

[ChWW] S.Y. A. Chang, J.M. Wilson, T.H. Wolff, *Some weighted inequalities concerning the Schrödinger operators*, Comment. Math. Helvetici **60** (1985), 217–246.

[FP] R. Fefferman, J. Pipher, *Multiparameter operators and sharp weighted inequalities* , Amer. J. of Mathematics, **119** (1997), 337–369.

[Ga] J.B. Garnett, "Bounded Analytic Functions". Acad. Press, NY, 1981.

[Hu] S. Hukovic, Thesis, Brown University, 1998.

[NT] F. Nazarov, S. Treil, *The weighted norm inequalities for Hilbert transform are now trivial,* C. R. Acad. Sci. Paris, Série I, **323**, (1996), 717–722.

[NTV] F. Nazarov, S. Treil, A. Volberg, *The Bellman functions and two weight inequality for Haar multipliers*, Preprint MSRI 1997, 1–30.

[St1] E. Stein, "Singular Integrals and Differentiability Properties of Functions", Princeton Univ. Press, Princeton, 1970.

[St2] E. Stein, "Harmonic Analysis: Real-Variable Methods, Orthogonality, and Oscillatory Integrals", with the assistance of Timothy S. Murphy, Princeton Math. Ser. **43**, Monographs in Harmonic analysis, iii, Princeton Univ. Press, Princeton, 1993.

[St3] E. Stein, Thesis, Univ. of Chicago, 1953.

Department of Mathematics,
Michigan State University
East Lansing, MI
48824, USA

Operator Theory:
Advances and Applications, Vol. 113

Multiplicative Chaos and Multimeasures

Jean-Pierre Kahane

Dedicated to the memory of S. A. Vinogradov

It is painful to think that we shall not see Stanislav Alexandrovitch again. It is sad to see the bright school to which he belonged dispersed all over the world. On the other hand, never was the Leningrad school of analysis better known than now, precisely because its members are everywhere, and this volume will contribute to draw the attention of a wide mathematical audience to the work and heritage of S. A. Vinogradov.

I had much in common with his work, but the present paper is not exactly along his lines. However, dealing with harmonic analysis and with operators, it would not be difficult to find a connection. The meaning of the dedication is that I hope that the matter would have pleased him, and could please others as well.

The paper is organized as follows. The first three sections deal with random multiplications, multifractal analysis and multimeasures, and their relations: they are purely expository. The last section is devoted to multiplicative chaos and multimeasures; part of it is expository, and it contains also new results.

1. Random multiplications

We are given (T, d), a metric space, that we assume to be locally compact and separable, and $(\Omega, \mathcal{A}, P)$ a Lebesgue probability space. We consider a sequence of independent weight functions, that we denote by $P_n(\omega, t)$, or simply $P_n(t)$; that means that each $P_n(\omega, t)$ is a positive $T \times \Omega$-measurable function, and that the $P_n(\omega, \cdot)$ are independent on Ω; we assume moreover $EP_n(t) = 1$ for each $n(= 1, 2, \dots)$ and t $(\in T)$. We use the term of weight functions because we let the $P_n(t)$ operate by multiplication on positive Radon measures on T, $\sigma(dt)$. We denote by $M^+(T)$ and $M_1^+(T)$ the sets of positive Radon measures and positive probabilities on T, respectively.

The partial products

$$Q_n(t) = P_1(t)P_2(t) \cdots P_n(t)$$

form a positive martingale, and the integrals

$$\int_B Q_n(t)\sigma(dt) \qquad (B \text{ borelian } \subset T) \tag{1}$$

as well, whenever $\sigma \in M^+(T)$ and $\sigma(B) < \infty$. These integrals converge to a limit, $S(B)$, almost surely. Thus S is a random measure and we write

$$S = * \lim Q_n \sigma = Q\sigma.$$

Q is an operator which carries ordinary measures into random measures.

There are two extreme cases. On one hand S may be the zero-measure; then we say that Q dies on σ, or that σ is Q-singular. On the other hand the integrals (1) can be convergent in $L^1(\Omega)$ whatever B, or, the same, $ES(B) = \sigma(B)$ for each B, or simply $ES = \sigma$; in that case we say that Q lives on σ, or that σ is Q-regular.

The starting point of the theory is that each $\sigma \in M^+(T)$ can be decomposed into two orthogonal measures:

$$\sigma = \sigma_s + \sigma_r$$

where σ_s is Q-singular and σ_r Q-regular. Of course $\sigma_r = ES$ [10].

The main problems are 1) to find concrete conditions for σ to be Q-regular (sufficient, necessary, both if possible); 2) σ being Q-regular, to study S (local behaviour, dimensional properties, multifractal analysis, distributional properties of the $S(B)$).

Here are three important examples.

A) Random coverings. Here we are given a sequence of independent Borel sets on T, $B_n(\omega)$. Denoting their indicator functions by $\chi_n(t, \omega)$, we define

$$P_n(\omega, t) = \frac{1 - \chi_n(t, \omega)}{1 - P(t \in B_n(\omega))}.$$

If σ is Q-regular, the support of σ cannot be covered a.s. by the $B_n(\omega)$. In several concrete cases, such as $T = \mathbb{T} = \mathbb{R}/\mathbb{Z}$ and $B_n(\omega) = (\omega_n, \omega_n + \ell_n)$, where the ω_n are independent random variables Lebesgue-distributed on $\mathbb{T}$, and ℓ_n a given sequence of positive numbers, most questions above have a very neat solution [11]. The starting point is a covering problem of Dvoretzky [4].

B) Multiplicative cascades. Here the starting point is a model proposed by B. Mandelbrot for turbulence phenomena [17, 18]. The space T is

$$T_b = \{1, 2, \cdots, b\}^{\mathbb{N}}$$

equipped with the ultrametric distance $d(t, s) = b^{-\nu}$ when t and s have in common the first ν coordinates. We define

$$P_n(t) = \sum_C W_C 1_C(t)$$

where C runs over all balls ("cells") or diameter b^{-n}, and the W_C are independent positive random variables (for all C and all n) having the same distribution as a positive random variable, W, such that $EW = 1$. The data are b and W. The questions above were considered when $\sigma = \lambda$, the Haar measure on T_b. The Q-regularity of λ, the dimensional properties of S and the existence of moments of order $p > 1$ for $S(T)$ are settled [15, 12]. The more refined questions on local behaviour, multifractal analysis and distributional properties of $S(B)$ are still under investigation [1].

C) Gaussian multiplicative chaos. Here

$$P_n(t) = \exp\left(X_n(t) - \frac{1}{2}E(X_n(t))^2\right),$$

where the $X_n(t)$ are independent centered Gaussian processes defined on T. Again, this is related to turbulence, and corresponds to a model sketched by Kolmogorov and established by B. Mandelbrot [16]. The theory is expounded in [9], and I shall come back to it.

2. Multifractal analysis and multimeasures

Given a bounded measure $\mu \in M^+(T)$ ($\mu(T) < \infty$), its local behaviour raised two kinds of dimensional analysis.

The first (rough) dimensional analysis is as follows. We consider the potential of order α

$$U_\alpha(t) = \int_T \frac{\mu(ds)}{(d(s,t))^\alpha} \qquad (\alpha > 0)$$

and the indicator function of the t-set where it is infinite:

$$\chi_\alpha(t) = \begin{cases} 1 & \text{if } U_\alpha(t) = \infty, \\ 0 & \text{if } U_\alpha(t) < \infty. \end{cases}$$

The α-singular part of μ is defined as

$$\mu_\alpha(dt) = \chi_\alpha(t)\mu(dt).$$

(The α-regular part of μ is $1-\mu_\alpha$; it is orthogonal to μ_α). The dimension spectrum of μ is the positive measure which is the derivative of $f(\alpha) = \mu_\alpha(T)$. The infimum and supremum of the support of the dimension spectrum are called lower dimension and upper dimension, low-dim μ and up-dim μ. Thus

$$\text{low-dim}\,\mu = \sup\{\alpha : \mu_\alpha(T) = 0\}$$
$$\text{up-dim}\,\mu = \inf\{\alpha : \mu_\alpha(T) = \mu(T)\}.$$

Under mild conditions it is not difficult to see that μ is carried by a Borel set whose Hausdorff dimension is up-dim μ and that all Borel sets whose Hausdorff dimension is strictly less than low-dim μ have μ-measure zero. Other equivalent definitions can be given [14, 5, 8]. The usual additional condition on (T, d) is the "doubling condition": each ball of radius $2r$ can be covered by N balls of radius r, N depending only on (T, d). Then (T, d) is what Coifman and Weiss called a "homogeneous space". From now on that will be our assumption.

When the lower and upper dimensions of μ are the same we say that μ is unidimensional and we speak of the dimension of μ.

Here begins a finer dimensional analysis. Let us assume that μ is unidimensional and $\dim\mu = D$. At each point $t \in T$ we can define the Hölder exponent of μ as

$$\alpha(t) = \liminf_{r\to 0}(\log\mu(B(t,r))/\log r),$$

where $B(t,r)$ is the ball with center at t and radius r. Let T_α be the t-set where $\alpha(t) = \alpha$, and $D_\alpha = \dim T_\alpha$. The so-called multifractal analysis of μ consists of studying D_α as a function of α. The assumption on μ implies

$$\lim_{r\to 0}(\log\mu(B(t,r))/\log r) = D \quad \mu\text{-a.e.},$$

therefore $\alpha(t) = D$ on a Borel set of dimension D.

There are many ways to perform the multifractal analysis of a unidimensional measure μ. Usually there is a direct estimate of D_α from above through a box counting argument. The lower estimate is obtained through an auxiliary measure, μ_α, such that $\mu_\alpha(T_\alpha) > 0$. The main difficulty is to construct the right one-parameter family of measures, so that μ can be scanned by using these measures. Such a family can be called a multimeasure. A more precise definition is given in the last lesson of my *Montreal* course [13]: given two measures σ and μ ($\in M^+(T)$), a multimeasure associated with σ and μ is a family $S^{(r)}$ of measures such that $S^{(0)} = \sigma$, $S^{(1)} = \mu$, and

$$\lim_{\rho\to 0}\frac{\log\mu(B_\rho(t))}{\log\sigma(B_\rho(t))} = \alpha_r \quad S^{(r)}\text{-a.e.} \tag{2}$$

for some α_r ($0 \le r \le 1$). When $T = \mathbb{R}^d$ and σ is the Lebesgue measure on $\mathbb{R}^d$, it means that $S^{(r)}$ is carried by $T_{d\alpha_r}$, therefore $D(d\alpha_r) \ge$ low-dim $S^{(r)}$.

From now on I shall be interested in the construction of multimeasures. They actually give an information on the set $E_\alpha(\subset T_\alpha)$, where

$$\lim_{r\to 0}(\log\mu(B(t,r)/\log r) = \alpha.$$

Sometimes "multifractal analysis" applies to the study of $\dim E_\alpha$ as a function of α. Multimeasures are well adapted to this study.

The literature on multifractal analysis is very large. In many cases the one-parameter family (μ_α) is given by the so-called Gibbs measure [3, 22, 21]. Fathi Ban Nasr showed that the Frostman measures can be used for this purpose [2]. Fan Ai-Hua investigated the case when the given measure is an infinite product [6, 7]. When moreover μ is a random measure it is natural to look for a random multimeasure; that is what we shall do now.

3. Random multiplications and multimeasures

Suppose that we are given (T,d) with the doubling condition, $(\Omega,\mathcal{A},P)$ and the $P_n(t)$ ($= P_n(\omega,t)$) as in section 1. We defined the operator Q, acting on $M^+(T)$, by the formulas

$$\begin{cases} \lim\limits_{n\to\infty} \displaystyle\int_B Q_n(t)\sigma(dt = S(B) \quad \text{a.s.} \\ S = Q\sigma, \end{cases}$$

where $Q_n(t) = (P_1P_2\cdots P_n)(t)$, $\sigma \in M^+(T)$, B borelian and $\sigma(B) < \infty$.

Let us write

$$P_{r,n}(t) = (P_n(t))^r / E(P_n(t))^r$$

for $r > 0$. When r is given, replacing $(P_n(t))$ by $(P_{r,n}(t))$ provides another operator, Q_r, defined by

$$\begin{cases} \lim\limits_{n\to\infty} \displaystyle\int_B Q_{r,n}(t)\sigma(dt) = S_r(B) \quad \text{a.s.} \\ S_r = Q_r\sigma. \end{cases}$$

When σ is given, the S_r $(0 \le r \le 1)$ provide a natural candidate for a multimeasure associated with σ and S. Moreover, in several interesting cases, a formula of the type (2) holds with $\mu = S$ whenever r is given between 0 and 1 ([13], lesson 4, [1]).

However, much work is still to be done. When σ is given, S_r is well defined for each given r, and also when r belongs to a countable set, for example, when r is rational. Is it possible, with probability 1, to define S_r as above for all r in $[0,1]$? When that is done, is it true that (2) (with $\mu = S$) holds for all r?

No general answer is known. For multiplicative cascades, Julien Barral succeeded in answering both questions in a positive way under mild conditions on b and W ([1] for the first question, oral communication for the second). It is a remarkable achievement. I tried to do the same for the Gaussian multiplicative chaos, and the results, expounded below, are far from being so satisfactory.

4. Multiplicative chaos and multimeasures

From now on, multiplicative chaos means Gaussian multiplicative chaos, as in [9]. The $X_n(t)$ $(= X_n(t,\omega))$ are independent centered Gaussian processes on T, with covariance

$$p_n(t,s) = E(X_n(t)X_n(s)).$$

For technical reasons we always assume $p_n(t,s) \ge 0$. The sums

$$Y_n(t) = (X_1 + X_2 + \cdots + X_n)(t)$$

are also centered Gaussian processes, and their covariances are

$$q_n(t,s) = (p_1 + p_2 + \cdots p_n)(t,s).$$

We define

$$q(t,s) = \sum_1^\infty p_n(t,s) \quad (\le \infty).$$

An interesting case is

$$q(t,s) = \frac{u}{2}\ell(\psi(t,s)), \tag{3}$$

where $u > 0$,

$$\ell(x) = \int_1^\infty e^{-yx}\frac{dy}{y}\left(= \log^+\frac{1}{x} + O(1)\right) \qquad (x > 0)$$

and $\psi(t,s)$ is a kernel of negative type on T, that is, a positive and continuous function defined on T^2, such that $\psi(t,t) = 0$ $(t \in T)$ and one of the following equivalent conditions is satisfied:

a) $e^{-y\psi(t,s)}$ is a covariance kernel (or kernel of positive type) for all $y > 0$,
b) there exists a continuous mapping θ from T into a Hilbert space such that $\psi(t,s) = \|\theta(t) - \theta(s)\|^2$ (square of the Hilbert norm),
c) whatever may be $c_1, \dots, c_k$ complex such that $c_1 + \cdots + c_k = 0$ and $t_1, \dots, t_k$, $s_1, \dots, s_k$ in T,

$$\sum c_i \bar{c}_j \psi(t_i, s_j) \leq 0$$

holds. In particular, when T is a Euclidean space and $\psi(t,s) = \|t-s\|^2$, we obtain

$$q(t,s) = u \log^+ \frac{1}{\|t-s\|} + O(1).$$

Here is a key lemma.

Lemma. *Let $p(t,s)$ and $p'(t,s)$ be two kernels of positive type (covariance functions), $X(t)$ and $X'(t)$ two Gaussian centered processes such that $E(X(t)X(s)) = p(t,s)$ and $E(X'(t)X'(s)) = p'(t,s)$, $P(t) = \exp(X(t) - \frac{1}{2}p(t,t))$ and $P'(t,s) = \exp(X'(t) - \frac{1}{2}p'(t,t)$. We assume $p'(t,s) \leq p(t,s)$ and we are given a convex function on $\mathbb{R}^+$, f, and $\sigma \in M^+(T)$. Then*

$$Ef\left(\int P'(t)\sigma(dt)\right) \leq Ef\left(\int P(t)\sigma(dt)\right) \qquad \text{([9], Lemma 1).}$$

It is possible to derive from this lemma the main theorem: the law of the operator Q depends only on $q(t,s)$ (and not on the way $q(t,s)$ is decomposed as a sum of kernels $p_n(t,s)$) ([9], Theorem 1). In the particular case when $T = \mathbb{R}^d$ and

$$q(t,s) = u \log^+ \frac{1}{\|t-s\|} + O(1), \tag{4}$$

the d-dimensional Lebesgue measure λ is Q-singular if $u \geq 2d$ and Q-regular if $u < 2d$; moreover, if $u < 2d$ and K is a Borel set such that $0 < \lambda(K) < \infty$,

$$E((Q\lambda)(K))^h < \infty \Leftrightarrow uh < 2d \qquad \text{([9], Theorem 4).}$$

Assuming (4) again, with $T = \mathbb{R}^d$, and $u < 2d$, given $\sigma \in M^+(T)$, the dimension spectrum of $Q\sigma$ is obtained a.s. in the following way: translate the dimension spectrum of σ to the left by $u/2$ and kill what is at the left of O,

$$\dim \cdot \text{spectrum}\, Q\sigma = 1_{\mathbb{R}^n}\left(\dim \cdot \text{spectrum}\, \sigma - \frac{u}{2}\right)$$

(the definition of the dimension spectrum is given in 2).

This last result is valid also for a general T when $q(t,s)$ is given by (3) ([9], Theorem 7).

A few results can be obtained for general kernels $q(t,s)$. Here is the easiest and most complete: the condition

$$\iint_{K^2} \exp q(t,s)\sigma(dt)\sigma(ds) < \infty \tag{5}$$

is necessary and sufficient in order to have

$$0 < E((Q\sigma)(K))^2 < \infty \qquad \text{([9], Theorem 5).}$$

A sufficient condition in order to have

$$0 < E((Q\sigma)(K))^{2m} < \infty \qquad (m \in \mathbb{N})$$

is

$$\iint_{K^2} \exp(mq(t,s))\sigma(dt)\sigma(ds) < \infty \qquad \text{([9], (117)).}$$

I asked the question, whether

$$\begin{cases} \sigma = \sum_1^\infty \sigma_j \\ \iint_{K^2} \exp\left(\frac{1}{2}q(t,s)\right)\sigma_j(dt)\sigma_j(ds) < \infty \qquad (j = 1, 2, \dots) \end{cases} \tag{6}$$

is either a sufficient or a necessary condition for σ to be Q-regular. Actually, it was proved by Sato and Tamashiro ([23], see also [20]) that it is neither necessary nor sufficient. It remains a challenging question: under which additional conditions (6) is necessary and sufficient in order that σ is Q-regular? For example, it is the case when $T = \mathbb{R}^d$, $\sigma = \lambda$ (Lebesgue measure) and $q(t,s)$ satisfies (4).

Let us consider now the multimeasure problem. We give $\sigma \in M^+(T)$ and a multiplicative chaos operator Q, defined as

$$Q\sigma(B) = S(B) = \lim_{n\to\infty} \int_B Q_n(t)\sigma(dt) \quad \text{a.s.}$$

For each r, $0 < r < 1$, we define

$$Q_r\sigma(B) = S_r(B) = \lim_{n\to\infty} \int_B Q_{r,n}(t)\sigma(\alpha t) \quad \text{a.s.} \tag{7}$$

where

$$Q_{r,n}(t) = (Q_n(t))^r / E(Q_n(t))^r .$$

It is easy to check that the kernel associated with the operator Q_r is

$$q_r(t,s) = r^2 q(t,s).$$

In formula (7) "almost sure" does not depend on B but it depends on r. Is it possible to make it independent from r? I shall give only partial answers to that question.

Theorem 1. *Suppose that $\sigma(\in M^+(T))$ is a Q-regular measure, that is*

$$\int_B Q_n(t)\sigma(dt) \to S(B) \quad in \quad L^1(\Omega) \qquad (n \to \infty)$$

for each Borelian B such that $\sigma(B) < \infty$. Then

$$\int_B Q_{r,n}(t)\sigma(dt) \to S_r(B) \quad in \quad L^1(\Omega) \qquad (n \to \infty)$$

uniformly with respect to r when $r \in [0,1]$, for each B such that $\sigma(B) < \infty$.

The proof follows from the inequality

$$E\left|\int_B (Q_{r,n+m}(t) - Q_{r,n}(t))\sigma(dt)\right| \leq E\left|\int_B (Q_{n+m}(t) - Q_n(t))\sigma(dt)\right| \tag{8}$$

when $0 < r < 1$, and (8) will be proved using the following notations:

$$Q_n = Q, \quad Q_{n+m} = PQ, \quad Q_{r,n} = Q', \quad Q_{r,n+m} = P'Q'.$$

Moreover we shall drop t, $\sigma(dt)$ and B in the integrals. What we have to prove is

$$E\left|\int (P'Q' - Q')\right| \leq E\left|\int (PQ - Q)\right|.$$

With obvious notations $p'(t,s) \leq p(t,s)$. Since P and P' are independent from Q', the key lemma, applied with $f(x) = \mid x - Q'^{-1} \int Q' \mid$ when Q' is fixed and $\sigma(dt)$ is replaced by $Q'(t)\sigma(dt)$, gives

$$E\left|\int (P'Q' - Q')\right| \leq E\left|\int (PQ' - Q')\right|. \tag{9}$$

Since

$$\begin{cases} q(t,s) = q'(t,s) + q''(t,s) \\ q''(t,s) = (1 - r^2)q(t,s) \end{cases}$$

We can decompose Q in the form $Q'Q''$ with $EQ'' = 1$ and Q'' independent from P and Q'. Let us write E'' for the expectation when P and Q' are fixed. Then

$$\left|\int (PQ' - Q')\right| \leq \left|\int E''((PQ' - Q')Q'')\right| \leq E''\left|\int (PQ' - Q')Q''\right|$$

and taking ordinary expectations now we obtain

$$E\left|\int (PQ' - Q')\right| \leq E|(PQ - Q)|. \tag{10}$$

(8) follows from (9) and (10) and Theorem 1 is proved.

Theorem 2. *Suppose that σ is a Q regular measure and moreover that*

$$E((Q\sigma)(K))^2 < \infty \text{ for some Borel set } K.$$

Then

$$\int_B Q_{r,n}(t)\sigma(dt) \to S_r(B) \quad in \quad L^2(\Omega) \quad and\ a.s. \quad (n \to \infty)$$

uniformly with respect to r when $r \in [0,a]$, whenever $a < 1$ for each Borel $B \subset K$.

Proof. We know that the assumption is equivalent with (5):

$$\iint_{K^2} \exp q(t,s)\sigma(dt)\sigma(ds) < \infty.$$

Moreover, from the L^2-martingale theory, it is equivalent with

$$\int_K Q_n(t)\sigma(dt) \to S(K) \quad \text{in} \quad L^2(\Omega).$$

Since (5) remains valid when $q(t,s)$ is changed into $r^2 q(t,s)$, we obtain immediately that

$$\int_K Q_{r,n}(t)\sigma(dt) \to S_r(K) \quad \text{in} \quad L^2(\Omega)$$

and the same holds when K is changed into $B(\subset K)$.

For the "almost sure" part of the conclusion we have to prove

$$\sup_{n,0<r<a} \int Q_{r,n} \in L^2(\Omega).$$

We shall make use of the inequality

$$\left|\sup_{0<r<a} f(r)\right|^2 \leq |f(0)|^2 + \int_0^a |f'(r)|^2 dr$$

for the function

$$f(r) = \int Q_{r,n}.$$

Let us compute $E\left(\left\|\int Q_{r,n}\right\|_{H_1}^2\right)$. For simplicity we assume $\sigma(K) = 1$ and we drop the index n in $Y_n(t)$, $q_n(t,s) = E\left(Y_n(t)Y_n(s)\right)$ and

$$Q_{r,n}(t) = \exp\left(rY_n(t) - \frac{1}{2}r^2 q_n(t,t)\right).$$

Moreover we drop t, $\sigma(dt)$ and K in the integrals when no confusion is possible. Then

$$\begin{aligned} E\left(\left\|\int Q_{r,n}\right\|_{H_1}^2\right) &= 1 + E\int_0^a \left(\int (Y - rq(t,t))Q_r\right)^2 dr \\ &= E\int_0^a \iint (Y(t) - rq(t,t))(Y(s) - rq(s,s))Q_r(t)Q_r(s)\sigma(dt)\sigma(ds)dr. \end{aligned}$$

Now we make use of the elementary equalities

$$\begin{aligned} A &= E\,\exp(uY(t)+vY(s)) = \exp\left(\frac{1}{2}u^2q(t,t)+\frac{1}{2}v^2q(s,s)+uvq(t,s)\right) \\ \frac{\partial A}{\partial u} &= E(Y(t)\exp(uY(t)+vY(s)) = (uq(t,t)+vq(t,s))A \\ \frac{\partial A}{\partial v} &= E(Y(s)\exp(uY(t)+vY(ts)) = (uq(t,s)+vq(s,s))A \\ \frac{\partial^2 A}{\partial u\partial v} &= E(Y(t)Y(s)\exp(uY(t)+vY(s))) \\ &= ((uq(t,t)+vq(t,s))(uq(t,s)+vq(s,s)+q(t,s))A. \end{aligned}$$

Hence one can check

$$\begin{aligned} &E\left((Y(t)-rq(t,t))(Y(s)-rq(s,s))Q_r(t)Q_r(s)\right) \\ &\qquad = \left(r^2(q(t,s))^2+q(t,s)\right)\exp\left(r^2q(t,s)\right) \end{aligned}$$

therefore

$$\begin{aligned} &E\left(\left\|\int Q_{r,n}\right\|_{H_1}^2\right) \\ &\qquad = 1+\iiint_0^a \left(r^2q_n^2(t,s)+q_n(t,s)\right)\exp\left(r^2q_n(t,s)\right)dr\,\sigma(dt)\sigma(ds) \end{aligned}$$

and, since our assumption is

$$\iint \exp q_n(t,s)\sigma(dt)\sigma(ds) = O(1) \qquad (n\to\infty),$$

we obtain

$$E\left(\left\|\int Q_{r,n}\right\|_{H_1}^2\right) = O(1). \tag{11}$$

The L^2 theory of martingales extends immediately to Hilbert space valued martingales. Since H_1 is a Hilbert space, (11) implies that $\int_K Q_{r,n}(t)\sigma(dt)$ converges a.s. to a limit in the space $H_1([0,1],dr)$, and the limit is identified as $S_r(K)$. The result for K implies the same for $B\subset K$. This is slightly stronger than Theorem 2. □

Theorems 1 and 2 leave a series of questions open.

1. Is it possible to define a multimeasure S_r $(=Q_r\sigma))$ $(0<r<1)$ almost surely whenever σ is Q-regular?
2. In the very particular case $T=\mathbb{R}^d$, $q(t,s)$ as in (4) with $u<2d$ and $\sigma\,\lambda$ (Lebesgue measure), is it possible to perform the multifractal analysis of S almost surely? The result is easy to guess:

$$D_\alpha = \left(\alpha-\frac{1}{2u}(\alpha-D)^2\right)^+ \qquad (D=d-\frac{u}{2})$$

(positive part of the parenthesis), in such a $0 \leq D_\alpha \leq \alpha$ for all $\alpha > 0$, $D_\alpha = D$ when $\alpha = D$, $\max_\alpha D = d = D_\beta$ with $\beta = D + u$ [9, 13].

3. In the case when T and $q(t,s)$ are as above but σ is no longer the Lebesgue measure, but σ is unidimensional, is it possible to perform the multifractal analysis of S a.s.?

4. When T has the doubling property and $q(t,s)$ is given by (3), for which σ is it possible to perform the multifractal analysis a.s.?

5. Give conditions on T, $q(t,s)$ and σ that insure the possibility of the almost sure multifractal analysis of S.

References

[1] J. Barral, *Moments, continuité et analyse multifractale des martingales de Mandelbrot*, Probability Theory and Related fields (to appear).

[2] F. Ben Nasr, *Analyse multifractale de mesures*, C. R. Acad. Sci. Paris, 1994.

[3] G. Brown, G. Michon, and J. Peyriere, *On the multifractal analysis of measures*, J. Stat. Phys., **66** (3/4) (1992), 775–790.

[4] A. Dvoretzky, *On covering a circle by randomly placed arcs*, Proc. Nat. Acad. Sci. USA, **42** (1956), 199–203.

[5] Fan Ai-Hua, *Sur les dimensions des mesures*, Studia Math., **111** (1) (1994), 1-17.

[6] Fan Ai-Hua, *Analyse multifractale de certains produits de Riesz avec phases*, C. R. Acad. Sci. Paris, **I-321** (1995), 399–404.

[7] Fan Ai-Hua, *Multifractal analysis of infinite products*, J. Stat. Phys., **80** (1997), 1313–1336.

[8] Y. Heurteaux, *Estimations de la dimension inférieure et de la dimension supérieure des mesures*, Ann. Inst. H. Poincaré, **34** (1998), 309–338.

[9] J.-P. Kahane, *Sur le chaos multiplicatif*, Ann. Sci. Math. Québec, **9** (1985), 105–150 et **10** (1986), 117–118.

[10] J.-P. Kahane, *Positive martingales and random measures*, Chinese Ann. Math., **8B1** (1987), 1–12.

[11] J.-P. Kahane, *Intervalles aléatoires et décomposition des mesures*, C. R. Acad. Sci. Paris, **304** (1987), 551–554.

[12] J.-P. Kahane, *Multiplications aléatoires et dimensions de Hausdorff*, Ann. Inst. H. Poincaré, Prob. & Stat., **23** (1987), 289–296.

[13] J.-P. Kahane, *Produits de poids aléatoires indépendants et applications*, in *Fractal geometry and analysis* (eds. J. Bélair and S. Dubuc), Kluwer Acad. Publ., **1991**, 277–324.

[14] J.-P. Kahane et Y. Katznelson, *Décomposition des mesures selon la dimension*, Colloq. Math., **58** (1990), 269–279.

[15] J.-P. Kahane et J. Peyriere, *Sur certaines martingales de Benoît Mandelbrot*, Advances in Math., **22** (1976), 131–145.

[16] B. Mandelbrot, *Possible refinement of the log-normal hypothesis concerning the distribution of energy dissipation in intermittent turbulence*, in *Statistical models and*

Turbulence. Symposium at U.C. San Diego 1971, Lecture Notes in Physics, Springer-Verlag, 1972, 333–351.

[17] B. Mandelbrot, *Intermittent turbulence in self similar cascades. Divergence of high moments and dimension of the carrier*, J. Fluid Mech., **62** (1974), 331–358.

[18] B. Mandelbrot, *Multiplications aléatoires itérées et distributions invariantes par moyenne pondérée aléatoire*, C. R. Acad. Sci. Paris, **278** (1974), 289–292 et 355–358.

[19] B. Mandelbrot, *Limit log-normal multifractal measures*, in *Frontiers of physics*, Landau memorial conference (ed. E. Gotsman), Tel-Aviv (1988), Pergamon, New York, 1989, 91–122.

[20] H. Mizumachi and H. Sato, *Absolute continuity of a random translation*, in *Trends in probability and related analysis* (ed. Shieh), 1997, 233–245.

[21] L. Olsen, *A multifractal formalism*, Adv. math., **116** (1995), 82–196.

[22] J. Peyriere, *Multifractal measures*, in *Probabilistic and stochastic methods in analysis, with applications*, ASI series, series C vol. 372, Kluwer Acad. Publ., 1992, 175–186.

[23] H. Sato and M. Tamashiro, *Multiplicative chaos and random translation*, Ann. Inst. H. Poincaré, Prob. & Stat., **30** (1994), 245–264.

Département de Mathématiques
Université Paris–Sud
Bât. 425
91405 Orsay France

Operator Theory:
Advances and Applications, Vol. 113

Some Remarks to Problems of Approximation with Prescribed Rate

S. Ya. Khavinson

Dedicated to the memory of S. A. Vinogradov

The paper consists of two sections. In §1 we establish two consequences of the well-known Bernstein theorem (see [1], [2]) ensuring the existence of an element with prescribed values of the best approximations. We examine the structure of the sets of all elements having common estimate for the rate of approximations.

In §2 we prove new modifications of the duality relations in approximation theory. Our modifications involve both the rate of the approximation and the magnitude of the coefficients of the approximating polynomials. The question about the magnitude of the coefficients arises in many approximation problems. The article [3] by Ph. Davis and Ky Fan was the first where a general approach to such problems was suggested; this approach is based on the Hahn–Banach theorem. We extended and developed this method in several papers. For the most detailed presentation, see [4], [5]. These papers also contain extensive bibliography. Our method is based on the duality theorem proved in [6]. However, in all the papers [3]–[6] the rate of approximation was not treated; attention was paid only to the coefficients and to approximability itself. The proof of our new duality relations consists in purposeful "identical transformations" (V. P. Havin's saying) applied to the fundamental duality relations of approximation theory. These "identical transformations" are not difficult, but they turn out to be useful, because we obtain the most convenient form for our problems. (The fundamental duality relations mentioned above belong to M. G. Krein and S. M. Nikolsky. These relations, again, have their origin in the Hahn–Banach theorem; see [7]–[10].)

1. Two consequences of Bernstein's theorem

Let X be a Banach space, and let

$$U_1 \subsetneqq U_2 \subsetneqq \cdots \subsetneqq U_n \subsetneqq \cdots \tag{1.1}$$

be an extending sequence of finite-dimensional linear subspaces in X such that $U = \bigcup_{n=1}^{\infty} U_n$ is everywhere dense in X. Let

$$\rho_1 \geq \rho_2 \geq \cdots \geq \rho_n \geq \cdots; \quad \rho_n > 0,\ n = 1, 2, \ldots; \quad \rho_n \to 0 \tag{1.2}$$

be a prescribed sequence of positive numbers. For $M > 0$, we denote by E_M the set

$$E_M = \{\omega \in X : d(\omega, U_n) \le M\rho_n,\ n = 1, 2, \dots\} \tag{1.3}$$

and by E the set

$$E = \bigcup_{M>0} E_M = \{\omega \in X : d(\omega, U_n) = O(\rho_n)\}. \tag{1.4}$$

Here $d(\omega, U_n)$ is the distance from ω to the subspace U_n.

The following important theorem is due to to S. N. Bernstein [1] (see also [2]).

For any sequence (1.2) *there exists an element* $\omega \in X$ *such that*

$$d(\omega, U_n) = \rho_n, \quad n = 1, 2, \dots . \tag{1.5}$$

From Bernstein's theorem we deduce two consequences.

Proposition 1. *For any* $M > 0$, *the set* E_M *is nonempty, closed, convex, and nowhere dense.*

Proof. The nonemptiness of E_M follows directly from Bernstein's theorem. The completeness and convexity of this set are evident. We prove that E_M is nowhere dense. Indeed, otherwise E_M contains a ball $S(x_0, r)$ (x_0 is the center of this ball, and $r > 0$ is the radius). If $x \in S(x_0, r)$, then $x - x_0 \in E_{2M}$. Hence, $E_{2M} \supseteq S(0, r)$. Therefore, $E \supseteq \bigcup_{k>0} kE_{2M} \supseteq \bigcup_{k>0} kS(0, r) = X$. We construct a sequence $\{s_n\}$ such that

$$\begin{gathered} s_1 \ge \dots \ge s_n \ge \cdots,\ s_n > 0,\ n = 1, \dots, \\ s_n \to 0 \quad \text{as} \quad n \to \infty,\ \frac{s_n}{\rho_n} \to 0 \quad \text{as} \quad n \to \infty. \end{gathered} \tag{1.6}$$

By Bernstein's theorem, there exists an element $\omega_0 \in X$ such that

$$d(\omega, U_n) = s_n, \quad n = 1, 2, \dots$$

From (1.6) it follows that $\omega_0 \notin E_M$ for every $M > 0$. Consequently, $\omega_0 \notin E = X$. This contradiction completes the proof. □

Proposition 2. i) *The set* E *is a linear subspace of* X; ii) *The set* E *is of type* F_σ *and of the first category (and therefore,* $E \neq X$); iii) *The set* $E \setminus U$ *is everywhere dense in* X.

Proof. Since statement i) is evident, and statement ii) follows from Proposition 1, we must prove only statement iii). Let $S(x_0, r)$ be an arbitrary ball. We show that $S(x_0, r)$ contains points belonging to $E \setminus U$. Since the subspace U is everywhere dense, we can assume that $x_0 \in U$. Then x_0 belongs to all U_n, $n \ge n_0$. There is a point $\omega \in X$ such that (1.5) is true for all n. The element

$$\omega_0 = x_0 + \frac{r'}{\|\omega\|}\omega, \quad 0 < r' < r,$$

belongs to the ball $S(x_0, r)$. Since

$$d(\omega, U_n) = \frac{r'}{\|\omega\|}\rho_n > 0, \quad n \geq n_0,$$

we have $\omega_0 \notin U$. We denote

$$d(\omega_0, U_n) = q_n, \quad n = 1, \ldots, n_0 - 1,$$

and select $M > 0$ such that

$$M \geq \frac{r'}{\|\omega\|}, \quad M \geq \frac{q_n}{\rho_n}; \quad n = 1, \ldots, n_0 - 1.$$

Then

$$d(\omega_0, U_n) \leq M\rho_n, \quad n = 1, 2, \ldots,$$

and $\omega_0 \in E_M \subset E$. Thus, $\omega_0 \in E \setminus U$, and Proposition 2 is proved. □

2. Duality relations involving both the rate of approximations and the magnitude of the coefficients of approximating polynomials

We select a basis in each of the subspaces U_j:

$$\text{in } U_1 : v_1, \ldots, v_{n_1}; \quad \text{in } U_j : v_{n_{j-1}+1}, \ldots, v_{n_j}; \ldots$$

Let $k_j = n_{j-1} + 1$; the dimension of U_j is $m_j = n_j - k_j + 1$. Any element $u_j \in U_j$ can be written in the form

$$u_j = \sum_{k_j}^{n_j} \lambda_i v_i, \quad \lambda_i \in T, \tag{2.1}$$

where $T = \mathbb{R}$ if X is a real space, and $T = \mathbb{C}$ if X is complex. For $N = 1, 2, \ldots$ we consider the space

$$T_N = T^{m_1} \times T^{m_2} \times \cdots \times T^{m_N}. \tag{2.2}$$

Thus, T_N consists of vectors

$$(\lambda) = (\lambda_1, \ldots, \lambda_{n_1}; \ldots; \lambda_{k_N}, \ldots, \lambda_{n_N}),\ \lambda_i \in T,\ i = 1, \ldots, n_N. \tag{2.3}$$

Suppose a norm Q is defined in T_N,

$$Q(\lambda) = Q(\lambda_1, \ldots, \lambda_{n_N}). \tag{2.4}$$

The dual space T_N^* consists of linear functionals, which are vectors

$$(\alpha) = (\alpha_1, \ldots, \alpha_{n_N}),\ \alpha_i \in T,\ i = 1, \ldots, n_N. \tag{2.5}$$

The functional $(\alpha) \in T_N^*$ acts on a vector $(\lambda) \in T_N$ by the formula

$$(\alpha)(\lambda) = \sum_{1}^{n_N} \alpha_i \lambda_i. \tag{2.6}$$

The norm on T_N^* dual to the norm Q on T_N will be denoted by

$$Q^*(\alpha_1, \ldots, \alpha_{n_N}). \tag{2.7}$$

Theorem 3 (on duality). *For any $\omega \in X$ we have*

$$\inf_{\lambda_1,\dots,\lambda_{n_N}} \left[\max_{1\le j\le N} \frac{\left\|\omega - \sum_{i=k_j}^{n_j} \lambda_i v_i\right\|}{\rho_j} + Q(\lambda_1,\dots,\lambda_{n_N}) \right] = \sup \left| \sum_{j=1}^{N} l_j(\omega) \right|. \quad (2.8)$$

Here the supremum on the right-hand side in (2.8) is taken over all collections of linear functionals $l_j \in X^*$, $j = 1,\dots,N$, satisfying

$$\sum_{1}^{N} \rho_j \|l_j\| \le 1, \quad (2.9)$$

$$Q^*\left(l_1(v_1),\dots,l_1(v_{n_1}); l_2(v_{k_2}),\dots,l_2(v_{n_2});\dots;l_N(v_{k_N}),\dots,l_N(v_{n_N})\right) \le 1. \quad (2.10)$$

The first term within the square brackets in (2.8) describes the rate (relative to the sequence (1.2)) of approximation of an element ω by the subspaces U_j, $j = 1,\dots,N$. The second term there takes the coefficients of the approximating combinations (2.1) into account.

Proof. Consider the space $Y = X^N$ formed by the vectors

$$y = (x_1,\dots,x_N), \quad x_j \in X, \quad j = 1,\dots,N. \quad (2.11)$$

We norm this space as follows:

$$\|y\| = \max_{1\le j\le N} \frac{\|x_j\|}{\rho_j}. \quad (2.12)$$

The space Y^* dual to Y consists of the collections $L = (l_1,\dots,l_N)$, with $l_j \in X^*$, $j = 1,\dots,N$. A functional $L \in Y^*$ acts on a vector $y \in Y$ by the formula

$$L(y) = \sum_{1}^{N} l_j(x_j), \quad (2.13)$$

and the norm of L looks like this:

$$\|L\| = \sum_{1}^{N} \rho_j \|l_j\|. \quad (2.14)$$

Finally, we consider the space

$$S = Y \times T_N \quad (2.15)$$

(T_N as in (2.2)). The space S consists of the elements s such that

$$\begin{gathered} s = (y,(\lambda)), \quad y \in Y, \quad (\lambda) \in T_N, \\ y = (x_1,\dots,x_N), \quad x_j \in X, \quad j = 1,\dots,N \\ (\lambda) = (\lambda_1,\dots,\lambda_{n_N}), \quad \lambda_i \in T, \quad i = 1,\dots,n_N. \end{gathered} \quad (2.16)$$

We equip S with a norm as follows:

$$\|s\| = \|y,(\lambda)\| = \|y\| + Q((\lambda)) = \max_{1\le j\le N} \frac{\|x_j\|}{\rho_j} + Q(\lambda_1,\dots,\lambda_{n_N}). \quad (2.17)$$

The dual space S^* is formed by the functionals Λ representable as couples:

$$\Lambda = \big(L, (\alpha)\big), \quad L \in Y^*, \quad (\alpha) \in T_N^*. \tag{2.18}$$

A pair Λ acts on an elements s of the space (2.15) by the formula

$$\Lambda(s) = \Lambda\big(y, (\lambda)\big) = L(y) + (\alpha)(\lambda) = \sum_1^N l_j(x_j) + \sum_1^{n_N} \alpha_i \lambda_i. \tag{2.19}$$

(We recall formulas (2.11), (2.13), (2.16), (2.5), (2.6).)

The norm of Λ looks like this:

$$\|\Lambda\| = \max\big(\|L\|, Q^*(\alpha_1, \dots, \alpha_{n_N})\big) = \max\left(\sum_{j=1}^N \rho_j \|l_j\|, Q^*(\alpha_1, \dots, \alpha_{n_N})\right) \tag{2.20}$$

(see (2.14)).

Now we introduce a subspace R of the space S,

$$R = \left\{\left(\sum_1^{n_1} \lambda_i v_i, \sum_{k_2}^{n_2} \lambda_i v_i, \dots, \sum_{k_N}^{n_N} \lambda_i v_i; \right.\right.$$
$$\left.\left. \lambda_1, \dots, \lambda_{n_1}, \lambda_{k_2}, \dots, \lambda_{n_2}, \dots, \lambda_{k_N}, \dots, \lambda_{n_N}\right)\right\}, \tag{2.21}$$

and consider the annihilator $R^\perp$ of R. If

$$\Lambda = \big(L, (\alpha)\big) = (l_1, \dots, l_N; \alpha_1, \dots, \alpha_{n_N}) \in R^\perp,$$

then

$$\sum_{i=k_j}^{n_j} [\lambda_i l_j(v_i) + \alpha_i \lambda_i] = 0, \quad j = 1, \dots, N, \quad \text{for any} \quad \lambda_1, \dots, \lambda_{n_N}.$$

This is equivalent to the identities

$$\alpha_i = -l_j(v_i), \quad k_j \le i \le n_j, \quad j = 1, \dots, N. \tag{2.22}$$

Conversely, the relation $\Lambda \in R^\perp$ follows from (2.22).

Let ω be an arbitrary element of X. Fixing an element $\Omega = (\omega, \dots, \omega) \in Y$, in S we consider the problem of the best approximation of the element

$$s_0 = (\Omega, \mathbb{O}) \tag{2.23}$$

by the elements of R. The fundamental Krein–Nikolsky duality relation (see [7]–[10]) takes the form

$$\inf_{s\in R} \|s_0 - s\| = \inf_{\lambda_1,\dots,\lambda_{n_N}} \left[\max_{1\le j\le N} \frac{\left\|\omega - \sum_{k_j}^{n_j} \lambda_i v_i\right\|}{\rho_j} + Q(\lambda_1,\dots,\lambda_{n_N}) \right]$$

$$= \max_{\substack{\Lambda\in R^\perp \\ \|\Lambda\|\le 1}} |\Lambda(\Omega,\mathbb{O})| = \max_{l_1,\dots,l_N} \left|\sum_1^N l_j(\omega)\right|. \tag{2.24}$$

Inequalities (2.9) and (2.10) follow from the conditions $\Lambda \in R^\perp$, $\|\Lambda\| \le 1$ and formulas (2.20), (2.22). This completes the proof of the theorem. □

We denote by $M_Q^N(\omega)$ the common value of the two extrema in (2.8) and assume that the norm Q is defined on all spaces T_N, $N = 1, 2, \dots$. Consider the quantity

$$M_Q(\omega) := \sup_N M_Q^N(\omega). \tag{2.25}$$

The condition

$$M_Q(\omega) < +\infty \tag{2.26}$$

means that $\omega \in E$ (see definition (1.4)) and that there are polynomials $u_j \in U_j$ approximating ω with the rate $O(\rho_j)$ and having coefficients bounded in the norm Q. Our theorem gives a duality test for such approximation.

If Q is identically zero, $Q \equiv 0$, then the condition $M_0(\omega) < \infty$ means that $\omega \in E$, and the condition $M_0(\omega) \le M$ means that $\omega \in E_M$. Our theorem gives duality tests for these membership relations. Now, inequality (2.10) means that $l_j \in U_j^\perp$, $j = 1, \dots, N$, because formally $Q^* = \infty$ if $l_j(v_i) \ne 0$. However, in this case it seems more convenient to use the Krein–Nikolsky duality directly for every $j = 1, \dots$.

Remarks 4. 1. *If we do not assume that the basis chosen in U_{n+1} is an extension of that in U_n, then the above notation only require obvious modifications.*

2. *The following situation is more convenient for estimating the coefficients. In every T^{m_j} we have the norm $q(\lambda_{k_j}, \dots, \lambda_{n_j})$, and*

$$Q(\lambda_1,\dots,\lambda_{n_N}) = \max_{1\le j\le N} q(\lambda_{k_j},\dots,\lambda_{n_j}).$$

In this case

$$Q^*(\lambda_1,\dots,\lambda_{n_N}) = \sum_1^N q^*(\alpha_{k_j},\dots,\alpha_{n_j}),$$

where q^* is the norm dual to the norm q.

In conclusion, we consider an example. Let $\{\delta_n\}$ be a sequence of positive numbers, and let Q be defined in the simplest way:

$$Q(\lambda_1,\dots,\lambda_{n_N}) = \max_{1\le j\le n_N} \left|\frac{\lambda_i}{\delta_i}\right|.$$

The dual norm is

$$Q^*(\alpha_1, \dots, \alpha_{n_N}) = \sum_1^{n_N} \delta_i |\alpha_i|.$$

Corollary 5. *An element ω admits approximation by the polynomials* (2.1) *with the rate $O(\rho_n)$ and with coefficients $\{\lambda_i\}$ such that*

$$\sup_i \left| \frac{\lambda_i}{\delta_i} \right| < +\infty \quad \textit{if and only if} \quad \sup \left| \sum_1^N l_j(\omega) \right| < +\infty.$$

Here the supremum is taken over all collections of linear functionals $l_j \in X^*$ satisfying (2.9) *and the condition*

$$\sum_1^N \sum_{i=k_j}^{n_j} \delta_i |l_j(v_i)| \le 1, \quad N = 1, \dots .$$

References

[1] S. N. Bernstein, *Sur le problème inverse de la théorie de la meilleur approximation des fonctions continues,* C. R. Ac. Sci, **206** (1938), 1520–1523.

[2] A. F. Timan, *Theory of Approximation of Functions of a Real Variable,* Pergamon Press, New York, 1963.

[3] Ph. Davis and Fan Ky, *Complete sequences and approximation in normed spaces,* Duke Math. J, **24**, no. 2 (1957), 183–192.

[4] S. Ya. Khavinson, *The concept of completeness involving the magnitude of the coefficients of the approximating functions,* Izvestia Akad. Nauk Armenia, Ser. Matematica, **6**, no. 2–3 (1971), 221–234.

[5] S. Ya. Khavinson, *On complete systems in Banach's space,* Izvestia Akad. Nauk Armenia, Ser. Matematica, **20**, no. 2 (1985), 89–111.

[6] S. Ya. Khavinson, *Two classes of extremal problems for polynomyals and moments,* Izvestia Akad. Nauk SSSR, Ser. Matematica, **25**, no. 4 (1961), 557–590.

[7] M. G. Krein, *The moment's problem in an abstract linear normed space,* Paper IV in the book: N. I. Akhiezer and M. G. Krein, *On some questions in the moment's theory,* Kharkov, 1938.

[8] S. M. Nikolsky, *The approximation of functions by trigonometrical polynomials in mean,* Izvestia Akad. Nauk SSSR, Ser. Matematica, **10**, no. 3 (1946), 207–256.

[9] N. P. Korneichuk, *Extremal problem of approximation theory,* Nauka, Moscow, 1976.

[10] S. Ya. Khavinson and E. Sh. Chatskaya, *The duality relations and tests for elements of best approximation,* Ingenerno–Stroitel'nyi Institut, Moscow, 1976.

Department of Higher Mathematics,
Moscow State Civil Engineering University,
Yaroslavskoe Shosse, 26
129337 Moscow, Russia

Operator Theory:
Advances and Applications, Vol. 113

Interpolation Involving Bounded Bianalytic Functions

Serguei Kislyakov

Dedicated to the memory of S. A. Vinogradov

Abstract. It is shown that in many cases the subspaces of bianalytic functions in various lattices of measurable functions on the torus $\mathbb{T}^2$ behave under real interpolation precisely as their parent lattices. Some applications to free interpolation by Fourier coefficients of bounded bianalytic functions are considered. (Note different meanings of the word "interpolation".)

1. Introduction

On the n-dimensional torus $\mathbb{T}^n$, consider the Hardy spaces

$$H^p(\mathbb{T}^n) = \overline{\operatorname{span}}_{L^p(\mathbb{T}^n)}\{z_1^{k_1}\dots z_n^{k_n}, k_1,\dots,k_n \in \mathbb{Z}_+\}$$

(for $p=\infty$ the closure is taken in the weak* topology of $L^\infty(\mathbb{T}^n)$). Are the interpolation properties of the scale $H^p(\mathbb{T}^n)$, $0<p\le\infty$, similar to those of the scale $L^p(\mathbb{T}^n)$?

Trivially, the answer is "yes" if we restrict ourselves to the interval $1<p<\infty$, because the orthogonal projection of $L^2(\mathbb{T}^n)$ onto $H^2(\mathbb{T}^n)$ is bounded for such p. Beyond this interval, we shall consider real interpolation only. Then, instead of specific interpolation theorems, it is natural to study the fundamental property of K-closedness. Let (X_0, X_1) be a compatible couple of quasi-Banach spaces, and let Y_0, Y_1 be closed subspaces of X_0, X_1, respectively. The couple (Y_0, Y_1) is said to be K-closed in (X_0, X_1) if for every vector $y \in Y_0+Y_1$ and every representation $y = x_0+x_1$ with $x_i \in X_i$ $(i=0,1)$ there exists another representation $y=y_0+y_1$, where $y_i \in Y_i$ and $\|y_i\|_{Y_i} \le C\|x_i\|_{X_i}, i=0,1$.

It is easily seen that K-closedness implies the formula

$$(Y_0,Y_1)_{\theta,q} = (Y_0+Y_1)\cap(X_0,X_1)_{\theta,q}$$

for the interpolation spaces of the real method. We refer the reader to the surveys [Ja, Kis2] for further discussion.

So, we are interested in the following question:

Supported in part by the Russian Foundation for Basic Research.

When is the couple $(H^{p_0}(\mathbb{T}^n), H^{p_1}(\mathbb{T}^n))$ K-closed in $(L^{p_0}(\mathbb{T}^n), L^{p_1}(\mathbb{T}^n))$?

In [Xu] it was proved that this is always true if $0 < p_0 < p_1 < \infty$. If one of the indices is infinite, the situation is much less clear: for $n \geq 3$ it is unknown whether the couple $(H^p(\mathbb{T}^n), H^\infty(\mathbb{T}^n))$ is K-closed. It has long been known that this couple is K-closed if $n = 1$ (many different proofs exist, see, e.g., the survey [Kis2]). For $n = 2$, we also have K-closedness, see [KX, Theorem 3].

We shall illustrate the importance of the case $p_1 = \infty$ by recalling some problems about "free interpolation" by Fourier coefficients. First, we quote the celebrated theorem of Kahane, Katznelson, and de Leeuw (see [Ka-K-dL]):

Let $\{e_k\}$ be a uniformly bounded orthogonal system on a probability space (Ω, μ). Then for every sequence $\{a_k\} \in l^2$ there is a function $G \in L^\infty(\mu)$ such that $|\langle G, e_k\rangle| = |\int G\bar{e}_k\, d\mu| \geq |a_k|$.

It is natural to consider a "holomorphic" version of this majorization problem. Specifically, let $\{a_k\}_{k\in\mathbb{Z}_+^n} \in l^2((\mathbb{Z}_+)^n)$. Does there exist a function $G \in H^\infty(\mathbb{T}^n)$ such that $|\widehat{G}(k)| \geq |a_k|$ for all $k \in (\mathbb{Z}_+)^n$?

It is known (see [Kis1]) that the following property would suffice to ensure a positive answer to this question: for every $\lambda > 0$, every function $f \in H^4(\mathbb{T}^n)$ with $\|f\|_4 = 1$ admits a decomposition $f = g + h$ with $g \in H^\infty(\mathbb{T}^n), h \in H^2(\mathbb{T}^n), |g| \leq \lambda, \|h\|_2 \leq \varepsilon(\lambda)$, where $\varepsilon(\lambda)$ behaves as a negative power of λ for $\lambda \geq 1$. We observe that, trivially, such a decomposition with g and h measurable (i.e., $g \in L^\infty(\mathbb{T}^n), h \in L^2(\mathbb{T}^n)$) exists (simply take the truncation of f at the level λ as g). So, if we know that the couple $(H^2(\mathbb{T}^n), H^\infty(\mathbb{T}^n))$ is K-closed in $(L^2(\mathbb{T}^n), L^\infty(\mathbb{T}^n))$, we can modify the above measurable decomposition to an analytic one, retaining the estimates. We see that the holomorphic majorization problem has a positive solution if $n = 1$ or $n = 2$, because in these dimensions we do have K-closedness, as mentioned above. Largely, this is the argument used in [Kis1], though K-closedness theorems in dimension 2 (and, probably, in dimension 1) involving H^∞ were not available at the time when that paper was written. However, the author knew a less refined statement of interpolatory nature, which sufficed.

For $n \geq 3$, the holomorphic majorization problem remains open, as does the problem of K-closedness if H^∞ is involved. Our aim in this paper is to present more information concerning dimension 2. It turns out that Theorem 3 in [KX] can be extended to many other couples of spaces of bianalytic functions. Though the main ideas will be the same as in the proof of that theorem, this extension is not immediate and, in some cases, requires much additional work.

As illustrations, we formulate two consequences of our general results.

Let Y be the subspace of bianalytic functions in $L^\infty(L^2) = L^\infty(\mathbb{T}; L^2(\mathbb{T}))$, i.e., $Y = \{ f \in L^\infty(L^2) : \hat{f}(k) = 0 \text{ for } k \notin \mathbb{Z}_+^2 \}$.

Theorem A. *The couple $(Y, H^\infty(\mathbb{T}^2))$ is K-closed in $(L^\infty(L^2), L^\infty(\mathbb{T}^2))$.*

Let a, b be two functions on $\mathbb{T}$ such that $0 \leq a \leq 1$, $0 \leq b \leq 1$. We define $L^\infty(a \otimes b) = \{ f : |f(\zeta, \omega)| \leq ca(\zeta)b(\omega) \}$ (the smallest c is declared to be the norm of f), and $H^\infty(a \otimes b) = H^\infty(\mathbb{T}^2) \cap L^\infty(a \otimes b)$ (which, in principle, may consist of 0 only).

Theorem B. *If* $\log a$, $\log b \in \mathrm{BMO}$, *then the couple* $(H^2(\mathbb{T}^2), H^\infty(a\otimes b))$ *is* K-*closed in* $(L^2(\mathbb{T}^2), L^\infty(a\otimes b))$.

From Theorem A it is easy to deduce the Bourgain theorem on free interpolation by lacunary one variable Fourier coefficients of bounded bianalytic functions. Let X be the space of all sequences $\{f_s\}_{s\geq 1}$, $f_s \in H^\infty(\mathbb{T})$, satisfying

$$\|f\| \stackrel{\text{def}}{=} \sup_z \Big(\sum |f_s(z)|^2\Big)^{1/2} < \infty.$$

We introduce an operator $\tau\colon H^\infty(\mathbb{T}^2)\to X$ by

$$\tau G = \Big\{\int_{\mathbb{T}} G(\,\cdot\,,\omega)\overline{\omega}^{2^n}\,dm(\omega)\Big\}$$

(m is normalized Lebesgue measure on $\mathbb{T}$).

Bourgain's Theorem (see [B]). *The operator* τ *is onto.*

Proof. Let $f = \{f_s\} \in X$ with $\|f\|_X = 1$. The bianalytic function $h(\zeta,\omega) = \sum_{s\geq 1} f_s(\zeta)\omega^{2^s}$ satisfies the equation $\tau h = f$, but, in general, is not bounded. However, $\|h\|_{L^\infty(L^4)} \leq c$ (c is a universal constant). So, fixing small $\varepsilon > 0$, we can find a constant λ (depending only on ε) such that $h = \varphi + \psi$ with measurable φ and ψ satisfying $|\varphi| \leq \lambda$, $\|\psi\|_{L^\infty(L^2)} \leq \varepsilon$ (truncation at the level λ does the job). By Theorem A, at the expense of changing λ and ε by $C\lambda$ and $C\varepsilon$ in the inequalities, we may assume that φ and ψ are bianalytic. Fixing $\varepsilon = (2C)^{-1}$, we see that $\|\tau\varphi - f\|_X \leq 1/2$ (and $|\varphi| \leq C\lambda$). Now, the passage to exact interpolation from this approximate one is standard. □

In fact, this proof is of about the same complexity level as Bourgain's original proof if we take the argument leading to Theorem A into account (even making all relevant simplifications in the latter compared to the general statements discussed below). However, it seems interesting to relate Bourgain's result with K-closedness.

Now, we give an application of Theorem B. This statement seems to be new.

Theorem C. *If* a *and* b *are as in Theorem* B, *then for every sequence* $\{a_k\} \in l^2(\mathbb{Z}^2_+)$ *there is a function* $G \in H^\infty(a\otimes b)$ *such that* $|\widehat{G}(k)| \geq |a_k|$, $k \in \mathbb{Z}^2_+$.

Proof. As was explained earlier (see again [Kis1]), it suffices to verify the following statement: for every $\lambda \geq 1$, every function $f \in H^4(\mathbb{T}^2)$ with $\|f\|_4 = 1$ admits a representation $f = g+h$, where g, h are bianalytic and $|g| \leq \lambda a(\zeta)b(\omega)$, $\|h\|_{L^2(\mathbb{T}^2)} \leq c\lambda^{-d}$ ($c, d > 0$ are some fixed constants). By Theorem B we must only ensure a measurable decomposition of this sort. Let $e = \{|f| \geq \lambda a(\zeta)b(\omega)\}$. The question reduces to estimating the integral $\iint_e |f|^2$ in terms of λ. Since $\log a$, $\log b \in \mathrm{BMO}$, the functions a^{-s}, b^{-s} are integrable for small s. Fixing such s, we write

$$\iint_e |f|^2 \leq |e|^{1/2}\|f\|_4^2 = |e|^{1/2}$$

and, for every $t < 4$,

$$|e| = \iint_e (a(\zeta)b(\omega))^t (a(\zeta)b(\omega))^{-t}$$

$$\leq \left(\iint_e [a(\zeta)b(\omega)]^4\right)^{t/4} \left(\iint_{\mathbb{T}\times\mathbb{T}} [a(\zeta)b(\omega)]^{-\frac{4t}{4-t}}\right)^{1-t/4}.$$

If we choose t such that $4t(4-t)^{-1} \leq s$, the second factor is finite. In the first factor, the integrand does not exceed $(|f|\lambda^{-1})^4$, so that we obtain $|e| \leq c\lambda^{-t}$. □

2. Preliminaries and statements

The general direction in which we shall advance is the investigation of the subspace of bianalytic functions in a "mixed norm space" of the form $X(L^p(\mathbb{T}; v))$, where X is some lattice of measurable functions on $\mathbb{T}$ and v is some weight. (We note that our definition of a space with mixed norm will yield $L^\infty(L^\infty) = L^\infty(\mathbb{T}^2)$.)

Let (S, ν) be a measure space. A *quasi-Banach lattice of measurable functions* on S is a space $(X, \|\cdot\|)$ consisting of measurable functions on S and complete in the quasinorm $\|\cdot\|$ such that if $x \in X$, y is measurable, and $|y| \leq x$, then $y \in X$ and $\|y\| \leq C\|x\|$.

Mostly, (S, ν) will coincide with $(\mathbb{T}, m)$, but sometimes we shall take $(S, \nu) = (\mathbb{T} \times \Omega, m \times \mu)$, where (Ω, μ) is some auxiliary measure space (the latter case includes the former if Ω is a singleton). For such S, a quasi-Banach lattice X of measurable functions is said to be BMO-regular if every $x \in X$ admits a majorant $u \geq |x|$ satisfying $\|u\|_X \leq C\|x\|$ and $\sup_\omega \|\log u(\,\cdot\,, \omega)\|_{\mathrm{BMO}} \leq C$ (C is a constant depending only on X).

We remind the reader that

$$\mathrm{BMO} = \mathrm{BMO}(\mathbb{T}) = \{\, f = \alpha + \mathcal{H}\beta : \alpha, \beta \in L^\infty(\mathbb{T}) \,\},$$

where $\mathcal{H}$ is the harmonic conjugation operator. As usual, the constants are disregarded, i.e., we put

$$\|f\|_{\mathrm{BMO}} = \inf\{\|\alpha\|_{L^\infty/\mathbb{C}} + \|\beta\|_{L^\infty/\mathbb{C}}\}$$

(over all representations of f as above).

We have already used the well-known fact that if $f \in \mathrm{BMO}$, then $\exp(\alpha|f|)$ is integrable for small α. Thus, any BMO-regular lattice X on $\mathbb{T}$ embeds in $L^p(\mathbb{T})$ for small p; we always assume that this embedding is continuous. Then the "analytic subspace" $X_A = X \cap H^r(\mathbb{T})$ is closed in X (the definition does not depend on $r \leq p$). It is possible to define X_A for more general X, but we do not dwell on this.

Many common spaces on $\mathbb{T}$ are BMO-regular. For instance, $L^p(w)$ is such if $0 < p \leq \infty$ and $\log w \in \mathrm{BMO}$. (It should be noted that for $p = \infty$ we adopt the definition $L^\infty(w) = \{\, g : |g| \leq cw \,\}$, with the natural norm.) Also, if for some α the operator $\mathcal{H}$ acts on $X^\alpha = \{\, y : |y|^\alpha \in X \,\}$ ($\|y\| = \||y|^\alpha\|_X^{1/\alpha}$), then X is BMO-regular. We refer the reader to [Kis2] for more discussion.

A version of our main result will deal with function spaces on $\mathbb{T}\times\Omega$, where Ω is a measure space of homogeneous type. This means that Ω is a (quasi)metric space with a measure μ satisfying the condition $\mu\{\rho(x,x_0)\le 2r\}\le c\mu\{\rho(x,x_0)\le r\}$, $r>0$. A *Calderón-Zygmund operator* (CZO) is an operator Q defined on $L^2(\Omega,\mu)$ and having a kernel K such that

$$\sup_{y,y_0}\int_{\rho(x,y_0)\ge 10\rho(y,y_0)}(|K(x,y)-K(x,y_0)|+|K(y,x)-K(y_0,x)|)\,d\mu(x)<\infty \quad (0)$$

(the *Hörmander condition*). That K is a kernel of Q means simply that for every bounded functions f,g supported on two balls at a positive distance from each other we have

$$\langle Qf,g\rangle=\int K(x,y)f(y)g(x)\,d\mu(x)\,d\mu(y).$$

It is well known that a CZO always acts on $L^p(\Omega)$, $1<p<\infty$, and is of weak type $(1,1)$.

Let Q be a *Calderón-Zygmund projection*, i.e., Q is a CZO and $Q^2=Q$ on $L^2(\Omega)$. We define

$$\mathcal{F}_p=\mathcal{F}_p^Q=\{\,f\in L^p:Qf=f\,\},\quad 1\le p<\infty;$$
$$\mathcal{F}_\infty^Q=(\mathcal{F}_1^{I-Q^*})^\perp$$

(duality is always bilinear, i.e., is given by the form $\int\varphi\psi$, and Q^* is understood accordingly). See [Kis2, §4] for more information about these spaces. If $\Omega=\mathbb{T}$ and Q is the Riesz projection ($Qf=\sum_{k\ge 0}\hat{f}(k)z^k$), then, of course, $\mathcal{F}_p=H^p(\mathbb{T})$, $1\le p\le\infty$.

We pass to functions of two variables. Let X,Y be quasi-Banach lattices of measurable functions on $\mathbb{T}$ and Ω, respectively. We define $X(Y)$ to be the space of all bimeasurable functions z on $\mathbb{T}\times\Omega$ for which the function $\zeta\mapsto\|z(\zeta,\cdot)\|_Y$ belongs to X (the X-norm of the latter function is declared to be the norm of z in $X(Y)$). Let a Calderón-Zygmund projection Q related to Ω be fixed, let $1\le p\le\infty$, and let X be a BMO-regular lattice on $\mathbb{T}$. Consider the embedding $X\subset L^r$ (with some small r) discussed earlier. We define $H^r(\mathcal{F}_p^Q)$ as the $L^r(\mathcal{F}_p^Q)$-closure of the set of $\mathcal{F}_p^Q$-valued analytic polynomials, and put

$$X_A(\mathcal{F}_p^Q)=X(L^p(\Omega))\cap H^r(\mathcal{F}_p^Q).$$

This space may be interpreted as the subspace of $X(L^p)$ consisting of the functions that belong to $\mathcal{F}_p^Q$ in the second variable and to X_A is the first.

Theorem 1. *If X and Y are* BMO*-regular lattices on $\mathbb{T}$ and $1<p<\infty$, then the couple $(X_A(\mathcal{F}_\infty^Q),Y_A(\mathcal{F}_p^Q))$ is K-closed in $(X(L^\infty),Y(L^p))$.*

Sometimes, this statement can be extended to $p=1$ (or even to $p<1$), but this requires other methods and "gluing" pieces of interpolation scales, as was done in [KX].

Also, we observe that Theorem 1 can be extended to weighted spaces. We restrict ourselves to the case where (Ω,μ) is something like $\mathbb{R}^d$ or $\mathbb{T}^d$ with Lebesgue measure. As usual in the case of weights, instead of (0) we impose the following condition on the kernel K of Q:

$$|K(y_0,x)-K(y,x)|,\ |K(x,y_0)-K(x,y)|\le c\frac{|y_0-y|^\alpha}{|y_0-x|^{d+\alpha}}$$
$$\text{if } |x-y_0|\ge 10|y-y_0|.$$

Next, let v be a weight on Ω satisfying the Muckenhoupt condition A_p; then the couple $\big(X_A(\mathcal{F}^Q_\infty), Y_A(\mathcal{F}^Q_p(v))\big)$ is K-closed in $\big(X(L^\infty), Y(L^p(v))\big)$. Here $\mathcal{F}^Q_p(v)\stackrel{\text{def}}{=} Q(L^p(v))$, and it is clear how to define the other objects involved that have not been defined earlier. Some hints to the proof will be given in §3.

However, in the principal case where $(\Omega,\mu)=(\mathbb{T},m)$ and Q is the Riesz projection (i.e., we return to bianalytic functions on the bidisc), it is possible to prove a much stronger weighted result. For a weight v on $\mathbb{T}$ with $\log v\in\mathrm{BMO}$ (no other conditions are imposed except $v\ge 0$), we put $H^p(v)=(L^p(v))_A$, $0<p\le\infty$.

Theorem 2. *Let u, v be nonnegative functions on $\mathbb{T}$ with $\log u$, $\log v\in\mathrm{BMO}$, let $1<p<\infty$, and let X, Y be BMO-regular lattices on $\mathbb{T}$. Then the couple $(X_A(H^\infty(u)), Y_A(H^p(v)))$ is K-closed in $(X(L^\infty(u)), Y(L^p(v)))$.*

Here the spaces of bianalytic functions $X_A(H^\infty(u))$, $Y(H^p(v))$ are defined as the intersections of $X(L^\infty(u))$, $Y(L^p(v))$ with $H^r(\mathbb{T}^2)$, where r is very small.

3. Proofs

First, we eliminate general lattices X and Y. Theorems 1 and 2 are consequences of the following fact.

Theorem 3. *The conclusions of Theorems 1 and 2 are valid if $X=L^\infty(w_0)$ and $Y=L^\infty(w_1)$, where $w_i\ge 0$ and $\log w_i\in\mathrm{BMO}$, $i=0,1$.*

Reduction will be shown for Theorem 1 only (the other case is similar). Let $f\in X_A(\mathcal{F}_\infty)+Y_A(\mathcal{F}_p)$ be represented in the form $f=g+h$ with $g\in X(L^\infty(\Omega))$, $h\in Y(L^p(\Omega))$. Let α and β be the norms of g and h in their respective spaces. We put

$$s_0(\zeta)=\operatorname{ess\,sup}_{\omega\in\Omega}|g(\zeta,\omega)|,\qquad s_1(\zeta)=\left(\int|h(\zeta,\omega)|^p\,d\mu(\omega)\right)^{1/p}$$

Then $\|s_0\|_X=\alpha$, $\|s_1\|_Y=\beta$ and, by BMO-regularity, we can find two weights $w_0\ge s_0$, $w_1\ge s_1$ with $\|w_0\|_X\le c\alpha$, $\|w_1\|_Y\le c\beta$, and $\|\log w_i\|_{\mathrm{BMO}}\le c$, $i=0,1$. We see that $f\in L^\infty(w_0)_A(\mathcal{F}_\infty)+L^\infty(w_1)_A(\mathcal{F}_p)$ and that $f=g+h$ is a decomposition of f with the summands belonging to $L^\infty(L^\infty(\Omega),w_0)$ and $L^\infty(L^p(\Omega),w_1)$, respectively. It is clear how to finish the reduction. □

Proof of Theorem 3 in the case related to Theorem 1. We need to prove the K-closedness of $(L^\infty_A(w_0)[\mathcal{F}_\infty], L^\infty_A(w_1)[\mathcal{F}_p])$ in $(L^\infty(L^\infty(\Omega),w_0), L^\infty(L^p(\Omega),w_1))$. Multiplying everything by a certain outer function on $\mathbb{T}$, we may assume that $w_0=1$, $w_1=b$, where $\log b\in\mathrm{BMO}$. Next, we use the duality lemma saying that a subcouple (E_0, E_1) of w^*-closed subspaces is K-closed in (F_0^*, F_1^*) if and only if the couple $(E_0^\perp, E_1^\perp)$ is K-closed in (F_0, F_1). See [P] or the survey [Kis2] for the proof.

We agree to denote by ζ and ω the variables running over $\mathbb{T}$ and Ω, and to indicate by $d\zeta$ and $d\omega$ integration over m and μ, respectively. Regardless of whether or not weights are involved, duality will be defined by the nonweighted bilinear form $\iint \varphi\psi \, d\zeta \, d\omega$.

So, in the preduals $L^1(\mathbb{T}\times\Omega)$, $L^1(L^q(\Omega);b)$ $(q^{-1}+p^{-1}=1)$, we consider the annihilators M and N of $L^\infty_A[\mathcal{F}^Q_\infty]$ and $L^\infty_A(b)[\mathcal{F}^Q_p]$. We must prove that the couple (M,N) is K-closed.

Lemma 1. *For $\varphi \in L^1(\mathbb{T}\times\Omega)$ (respectively, $\varphi \in L^1(L^q(\Omega);b)$), we have $\varphi \in M$ (respectively, $\varphi \in N$) if and only if $\bar{\zeta}Q^*(\varphi(\zeta,\,\cdot\,))$ is analytic in ζ.*

Proof. We include the argument, though it is standard. We treat the space M only. The case of N is similar and even simpler (in that case it is useful first to eliminate the weight b, multiplying everything by an appropriate outer function). So, let $\varphi \in L^1(\mathbb{T}\times\Omega)$, and let $\alpha_n(\omega) = \int \varphi(\zeta,\omega)\bar{\zeta}^n \, d\zeta$ be the (vector) Fourier coefficients of φ. Using Abel–Poisson means (in ζ), we easily see that $\bar{\zeta}Q^*(\varphi(\zeta,\,\cdot\,))$ is analytic in ζ if and only if $Q^*\alpha_n = 0$ for $n \le 0$. For any $f \in H^\infty(\mathbb{T})$, we put $\varphi^{(f)}(\omega) = \int \varphi(\zeta,\omega)f(\zeta)\,d\zeta$. Then

$$\varphi^{(f)} = \lim_{r\to 1-0} \sum_{n\le 0} \hat{f}(-n)\alpha_n r^{|n|} \quad \text{in } L^1(\Omega).$$

It follows that if $Q^*\alpha_n = 0$ for $n \le 0$, then $Q^*\varphi^{(f)} = 0$ (i.e., $\varphi^{(f)} \in \mathcal{F}_1^{I-Q^*}$) for all $f \in H^\infty(\mathbb{T})$. The converse is also true (put $f(\zeta) = \zeta^n$, $n = 0,1,2,\dots$). Recalling the definition of $\mathcal{F}^Q_\infty$, we conclude that $\bar{\zeta}Q^*(\varphi(\zeta,\,\cdot\,))$ is analytic in ζ if and only if

$$\iint \varphi(\zeta,\omega)f(\zeta)g(\omega)\,d\zeta\,d\omega = 0 \quad \text{for all } f \in H^\infty(\mathbb{T}),\ g \in \mathcal{F}^Q_\infty.$$

Now, the set $\{\, f(\zeta)g(\omega) : f \in H^\infty(\mathbb{T}),\ g \in \mathcal{F}^Q_\infty \,\}$ is ω^*-dense in $L^\infty_A(\mathcal{F}^Q_\infty)$. This easily follows from the definition of the latter space and the standard Hardy space theory. □

Remark 1. *If $(\Omega,\mu) = (\mathbb{T}, m)$, Q is the Riesz projection, and $b = 1$, then M and N consist of functions whose spectrum does not intersect the set $\{\,(i,j) \in \mathbb{Z}\times\mathbb{Z} : i,j \le 0\,\}$. (Also, it should be noted that in this case Q^* does not act as Q, because we use bilinear duality: we have $Q^*\psi = \overline{Q\overline{\psi}}$ on polynomials.) In the general case, this observation concerning the spectrum may serve only as a heuristic indication, but it is still useful.*

Now, let $f \in M + N$, and let $f = g + h$, where $g \in L^1(L^1(\Omega))$, $h \in L^1(L^q(\Omega), b)$. Till the end of the proof we shall denote by A and B the norms of g and h in their respective spaces. For each fixed $\zeta \in \mathbb{T}$, we do Calderón-Zygmund decomposition of $g(\zeta,\,\cdot\,)$ at a level $\lambda(\zeta) > 0$ with respect to the CZO Q^*. (We do not dwell on tiny bimeasurability problems, which may seem to arise.)

This procedure will result in two functions $g_0(\zeta,\,\cdot\,)$, $g_1(\zeta,\,\cdot\,)$ and a set $\Omega_\zeta \subset \Omega$ with the following properties:

$$
\begin{gathered}
g(\zeta,\,\cdot\,) = g_0(\zeta,\,\cdot\,) + g_1(\zeta,\,\cdot\,); \\
|g_0(\zeta,\,\cdot\,)| \le \lambda(\zeta); \\
\int |g_0(\zeta,\,\cdot\,)|\,d\omega,\ \int |g_1(\zeta,\,\cdot\,)|\,d\omega \le c \int |g(\zeta,\,\cdot\,)|\,d\omega; \\
|\Omega_\zeta| \le c\lambda(\zeta)^{-1} \int |g(\zeta,\,\cdot\,)|\,d\omega; \\
\int_{\Omega\setminus\Omega_\zeta} |(Q^* g_1)(\zeta,\,\cdot\,)|\,d\omega \le c \int |g(\zeta,\,\cdot\,)|\,d\omega.
\end{gathered}
\tag{1}
$$

The level function $\lambda(\zeta)$ is chosen as follows. We put $y(\zeta) = \left(\int_\Omega |h(\zeta,\cdot)|^q\,d\omega\right)^{1/q}$ and find a function z on $\mathbb{T}$ and a natural number $k \ge 2$ such that $z \ge y$, $|\mathcal{H}(z^{1/k})| \le cz^{1/k}$ (we recall that $\mathcal{H}$ is the harmonic conjugation operator), and

$$
\int zb\,d\zeta \le c \int yb\,d\zeta = cB. \tag{2}
$$

To see that such z exists, we first use the fact that the space $L^1(b)$ is BMO-regular (see [Kis2, Subsection 3.4]). So, there is a function $z \ge y$ with $\int zb\,d\zeta \le c\int yb\,d\zeta$ and $\|\log z\|_{\mathrm{BMO}} \le c$. Now, by [Kis2, Lemma 3.1], we can further modify z to ensure also the condition involving $\mathcal{H}$.

Once z is found, we put

$$
\lambda(\zeta) = z(\zeta)^p \Big/ \Big(\int |g(\zeta,\,\cdot\,)|\,d\omega\Big)^{p-1}.
$$

Then

$$
\Big(\int |g_0(\zeta,\,\cdot\,)|^q\,d\omega\Big)^{1/q} \le cz(\zeta). \tag{3}
$$

Indeed, by (1),

$$
\begin{aligned}
\int |g_0(\zeta,\,\cdot\,)|^q\,d\omega &\le \lambda(\zeta)^{q-1} \int |g_0(\zeta,\,\cdot\,)|\,d\omega \\
&\le c\frac{z(\zeta)^{p(q-1)}}{\Big(\int |g(\zeta,\,\cdot\,)|\,d\omega\Big)^{(p-1)(q-1)}} \int |g(\zeta,\,\cdot\,)|\,d\omega = cz(\zeta)^q.
\end{aligned}
$$

Now, we put

$$
u \stackrel{\mathrm{def}}{=} Q^* f = Q^* g_1 + Q^*(g_0 + h)
$$

(as usual, Q^* acts in the variable ω). By Lemma 1, $\bar{\zeta}u$ is analytic in ζ. Let $\Phi \in L^\infty(\mathbb{T}\times\Omega)$ be some function *analytic in the first variable.* We put $\alpha = \Phi u - Q^*(g_0 + h)$ and claim that, for some choice of Φ,

$$
f = [g_1 - \alpha] + [g_0 + h + \alpha]
$$

is the required decomposition of f. It will suffice to ensure the norm estimates of the summands. Indeed, the arguments below justify the following line:

$$Q^*[g_1 - \alpha] = Q^*g_1 - Q^*\Phi u + Q^*Q^*(g_0 + h) = Q^*f - Q^*\Phi u,$$

whence it easily follows that $g_1 - \alpha$ and $g_0 + h + \alpha$ satisfy the "spectral condition" formulated in Lemma 1, so that we have a decomposition in the sum $M + N$.

Now, by (1), (2), (3) we have

$$\iint |g_1| b\zeta \, d\omega \le cA$$

and

$$\int \left(\int |g_0 + h|^q \, d\omega \right)^{1/q} b \, d\zeta \le cB,$$

and we must prove that α admits similar estimates in both spaces. We have

$$\alpha = \Phi Q^* g_1 - (1 - \Phi) Q^*(g_0 + h).$$

We specify the choice of Φ. Besides k that arose when we chose z, we fix another integer s satisfying $ps^{-1} < 1$. Let $r = z^{1/k}$ (so that $|\mathcal{H}r| \le cr$), and let

$$\gamma(\zeta, \omega) = \max\{1, (\lambda(\zeta)^{-1} |(Q^* g_1)(\zeta, \omega)|)^{1/ks}\}.$$

Next, we put ($\mathcal{H}$ acts in the variable ζ, of course):

$$F = \frac{r + i\mathcal{H}r}{r\gamma + i\mathcal{H}(r\gamma)}, \qquad \Phi = 1 - (1 - F^{ks})^k.$$

We see that Φ is indeed analytic in ζ. Moreover,

$$|\Phi| \le c|F|^{ks} \le c(r\gamma)^{-ks}(r + |\mathcal{H}r|)^{ks} \le c'\gamma^{-ks},$$

whence it follows that Φ is uniformly bounded and

$$|\Phi||Q^* g_1| \le \lambda(\zeta). \tag{4}$$

Now, we recall that the estimates to be verified are

$$\int \left(\int |\alpha|^q \, d\omega \right)^{1/q} b \, d\zeta \le cB, \tag{5}$$

$$\iint |\alpha| \, d\omega \, d\zeta \le cA. \tag{6}$$

We check (5):

$$\int \left(\int |\alpha|^q \, d\omega \right)^{1/q} b \, d\zeta$$

$$\le c \left[\int \left(\int |\Phi Q^* g_1|^q \, d\omega \right)^{1/q} b \, d\zeta + \int \left(\int |1 - \Phi|^q |Q^*(g_0 + h)|^q \, d\omega \right)^{1/q} b \, d\zeta \right]$$

$$\stackrel{\text{def}}{=} c[I_1 + I_2].$$

In I_2, we replace $|1-\Phi|$ by a constant, then use the L^q-boundedness of Q^*, and then apply (3) and (2). This will result in $I_2 \le cB$. To estimate I_1, we analyze the integral $J(\zeta) = \int |\Phi Q^* g_1|^q \, d\omega$ (for each fixed ζ).

Introducing the distribution function

$$\rho(t) = \rho_\zeta(t) \stackrel{\text{def}}{=} |\{\, \omega : |\Phi Q^* g_1| > t \,\}|, \tag{7}$$

by (4) we see that $\rho(t) = 0$ for $t > \lambda(\zeta)$, and otherwise $\rho(t) \le ct^{-1} \int |g_1| \, d\omega \le c' t^{-1} \int |g| \, d\omega$ (by (1)). So,

$$J(\zeta) \le c \int_0^{\lambda(\zeta)} t^{q-1} \rho(t) \, dt \le c\lambda(\zeta)^{q-1} \int |g(\zeta, \cdot)| \, d\omega = Cz(\zeta)^q. \tag{8}$$

Now $I_1 = \int (J(\zeta))^{1/q} b(\zeta) \, d\zeta \le c \int zb \, d\zeta \le cB$ by (2).

We pass to the verification of (6). Again, we write

$$\begin{aligned}\iint |\alpha| \, d\omega \, d\zeta &\le C \Big[\iint |\Phi Q^* g_1| \, d\omega \, d\zeta + \iint |1-\Phi||Q^*(g_0 + h)| \, d\omega \, d\zeta \Big] \\ &\stackrel{\text{def}}{=} C(I_3 + I_4).\end{aligned}$$

To estimate I_3, we first treat the integral $J'(\zeta) = \int |\Phi Q^* g_1| \, d\omega$. Using (1) and the fact that the integrand does not exceed $\lambda(\zeta)$ (see (4)), we obtain

$$J'(\zeta) \le C \Big(\lambda(\zeta) |\Omega_\zeta| + \int_{\Omega \setminus \Omega_\zeta} |Qg_1| \, d\omega \Big) \le C \int |g(\zeta, \cdot)| \, d\omega, \tag{9}$$

whence $I_3 \le cA$. For I_4, using the L^q-boundedness of Q^*, we write

$$\begin{aligned} I_4 &\le \int \Big(\int |1-\Phi|^p \, d\omega \Big)^{1/p} \Big(\int |Q^*(g_0 + h)|^q \, d\omega \Big)^{1/q} d\zeta \\ &\le C \int \Big(\int |1-\Phi|^p \, d\omega \Big)^{1/p} z \, d\zeta, \end{aligned} \tag{10}$$

by (3) and the inequality $z(\zeta) \ge \big(\int |h(\zeta, \cdot)|^q \, d\omega \big)^{1/q}$. We have

$$|1-\Phi| \le c|1-F|^k, \qquad |1-F| \le \frac{\gamma-1}{\gamma} + \frac{|\mathcal{H}(r(\gamma-1))|}{r}; \tag{11}$$

the latter quantity is equal to $r^{-1}|\mathcal{H}(r(\gamma-1))|$ on the set where $\gamma = 1$ (i.e., where $|Q^* g_1| \le \lambda(\zeta)$). Thus,

$$\begin{aligned} I_4 \le C \Big[&\int |\{\gamma(\zeta, \cdot) > 1\}|^{1/p} z(\zeta) \, d\zeta \\ &+ \int \Big(\int |\mathcal{H}((\gamma-1)r)|^{kp} \, d\omega \Big)^{1/p} r^{-k} z \, d\zeta \Big]. \end{aligned} \tag{12}$$

Since $r^k = z$, the weight cancels in the second integral. Since $\mathcal{H}$ (acting in ζ) is bounded on $L^k(\mathbb{T}; L^{kp}(\Omega))$, we continue as follows:

$$\begin{aligned} I_4 &\le C\Big[\int \lambda(\zeta)^{-1/p}\Big(\int |g(\zeta,\,\cdot\,)|\,d\omega\Big)^{1/p} z(\zeta)\,d\zeta \\ &\quad + \int\Big(\int |(\gamma-1)r|^{kp}\,d\omega\Big)^{1/p} d\zeta\Big] \\ &= C\Big[\iint |g|\,d\omega\,d\zeta + \int z(\zeta)\lambda(\zeta)^{-1/s}\Big(\int_{\gamma>1} |Q^* g_1|^{p/s}\,d\omega\Big)^{1/p} d\zeta\Big]. \end{aligned} \tag{13}$$

The first term in brackets is already good. In the second one, we first estimate the inner integral for each fixed ζ, by using the fact that Q^* is of weak type (1,1) and $p/s < 1$:

$$\begin{aligned} \Big(\int_{\gamma>1} |Q^* g_1|^{p/s}\,d\omega\Big)^{1/p} &\le C\Big[|\{\,\omega : \gamma(\zeta,\omega) > 1\,\}|^{\frac{s-p}{p}} \int |g|\,d\omega\Big]^{1/s} \\ &\le C\Big(\lambda(\zeta)^{-1}\int |g|\,d\omega\Big)^{1/p-1/s}\Big(\int |g|\,d\omega\Big)^{1/s}. \end{aligned} \tag{14}$$

Thus,

$$I_4 \le C\Big(A + \int z(\zeta)\lambda(\zeta)^{-1/p}\Big(\int |g(\zeta,\,\cdot\,)|\,d\omega\Big)^{1/p} d\zeta\Big) \le CA, \tag{15}$$

as required. □

Now, we give a hint to the proof of the K-closedness of $(X_A(\mathcal{F}^Q_\infty), Y_A(\mathcal{F}^Q_p(v)))$ in $(X(L^\infty(\Omega)), Y(L^p(v)))$ under the condition $v \in A_p$. As before, we use duality to pass to a couple (M,N) of subspaces of $L^1(\mathbb{T}\times\Omega), L^1(L^q(\Omega,u),b)$, where $u \in A_q$. In a measurable representation $f = g+h$ of $f \in M+N$, we do Calderón-Zygmund decomposition, but this time of the function gu^{-1} with respect to the measure with density u. (Specifically, we examine the averages $\int_I |g|u/\int_I u$ over cubes I, etc.) This will lead to formulas like in (1), but with the second and the forth lines replaced by $|g_0(\zeta,\,\cdot\,)| \le \lambda(\zeta)u(\zeta)$ and $\int_{\Omega_\zeta} u(\,\cdot\,)\,d\omega \le C\lambda(\zeta)^{-1}\int |g(\zeta,\,\cdot\,)|\,d\omega$, respectively. The adjustment of the further details is left to the reader. The condition A_q is needed to ensure that the measure with density u satisfies a doubling condition, and that Q^* is bounded on $L^q(u)$.

Proof of Theorem 3 *in the case related to Theorem* 2. We retain the notation ζ, ω, $d\zeta$, $d\omega$. Now Q is the Riesz projection, acting in the variable ω.

We are dealing with the couple $(L^\infty(L^\infty(u); w_0), L^\infty(L^p(v); w_1))$, and the corresponding subspaces of bianalytic functions. Multiplication by appropriate outer functions first in ω, then in ζ reduces the problem to the case where $w_0 = 1$, $w_1 = b$, $u = a$, $v = a^{1-p}$, with $\log a, \log b \in \mathrm{BMO}$. Dualizing, we arrive at the couple $(L^1(L^1(a)), L^1(L^q(a), b))$, and the couple M, N of the corresponding annihilators (the remark after Lemma 1 should be kept in mind though, literally, it is not applicable).

Again, let $M + N \ni f = g + h$, where $g \in L^1(L^1(a))$, $h \in L^1(L^q(a), b)$. We denote $A = \|g\|_{L^1(L^1(a))}$, $B = \|h\|_{L^1(L^q(a),b)}$. The main innovation compared to the previous case is that we shall "kill" the "bad weight" a with the help of an analytic decomposition of unity subordinate to a. This is a system $\{\varphi_j\}_{j\in\mathbb{Z}} \subset H^\infty(\mathbb{T})$ satisfying

$$|\varphi_j|^{1/8}a \le c2^j, \quad \sum |\varphi_j|^{1/8}2^j \le ca, \qquad \sum |\varphi_j|^{1/8} \le c, \quad \sum \varphi_j = 1. \tag{16}$$

In [Kis3] it was proved that such a system exists if and only if $\log a \in \mathrm{BMO}$, which is our case. We retain the exponent 1/8 (which occurred in [Kis3]) to facilitate reference (a bigger exponent would suffice for what follows).

We factorize $\varphi_j = \theta_j\psi_j^8$ with θ_j inner and ψ_j outer. Next, we write $f\psi_j^4 = g\psi_j^4 + h\psi_j^4$ and do Calderón-Zygmund decomposition, without weight, for the function $(g\psi_j^4)(\zeta, \cdot)$ at a level $\lambda_j(\zeta)$; the CZO involved is Q^*. Some functions $g_0^j(\zeta, \cdot)$, $g_1^j(\zeta, \cdot)$ and some sets $\Omega_\zeta^j \subset \mathbb{T}$ will arise, such that, in place of (1), we shall have

$$\begin{gathered}(g\psi_j^4)(\zeta, \cdot) = g_0^j(\zeta, \cdot) + g_1^j(\zeta, \cdot);\\ |g_0^j(\zeta, \cdot)| \le \lambda_j(\zeta);\\ \int |g_0^j(\zeta, \cdot)|\, d\omega, \ \int |g_1^j(\zeta, \cdot)|\, d\omega \le c\int |(g\psi_j^4)(\zeta, \cdot)|\, d\omega;\\ |\Omega_\zeta^j| \le c\lambda_j(\zeta)^{-1}\int_{\mathbb{T}} |g(\zeta, \cdot)|\, d\omega;\\ \int_{\mathbb{T}\setminus\Omega_\zeta^j} |Q^*(g_1^j)(\zeta, \cdot)|\, d\omega \le c\int_{\mathbb{T}} |(g\psi_j^4)(\zeta, \cdot)|\, d\omega.\end{gathered} \tag{1'}$$

The $\lambda_j(\zeta)$ are chosen as follows. We put $y_j(\zeta) = \left(\int |(h\psi_j^4)(\zeta, \cdot)|^q\, d\omega\right)^{1/q}$. Then, by (16),

$$\sum_{j\in\mathbb{Z}} 2^j y_j(\zeta)^q \le c\int |h(\zeta, \cdot)|^q \sum_j 2^j|\psi_j(\cdot)|\, d\omega \le c\int |h(\zeta, \cdot)|^q a(\cdot)\, d\omega,$$

whence

$$\int \left(\sum 2^j y_j^q\right)^{1/q} b\, d\zeta \le cB.$$

The expression on the left in this inequality is a lattice norm on measurable functions on $\mathbb{T} \times \mathbb{Z}$ (i.e., we interpret the sequence $\{y_j\}$ as a single function of (ζ, j)). By [Kis2, Subsection 3.4], the corresponding lattice X is BMO-regular. Thus, we can find functions $z_j \geq y_j$ such that $\sup_j \|\log z_j\|_{\mathrm{BMO}} \leq c$ and

$$\int \Big(\sum 2^j z_j^q\Big)^{1/q} b\, d\zeta \leq c' B. \tag{2'}$$

After this we modify the z_j in accordance with Lemma 3.1 in [Kis2] (or, rather, with its proof) to ensure that $|\mathcal{H}(z_j^{1/k})| \leq c z_j^{1/k}$ for some $c > 0$ and some fixed integer $k \geq 2$ depending eventually only on the weights a and b, and the exponent q.

Now, we fix

$$\lambda_j(\zeta) = z_j(\zeta)^p \Big/ \Big(\int |(g\psi_j^4)(\zeta, \cdot)|\, d\omega\Big)^{p-1}.$$

As in (3), it is easily seen that

$$\Big(\int |g_0^j(\zeta, \cdot)|^q\, d\omega\Big)^{1/q} \leq c z_j(\zeta). \tag{3'}$$

Next, we put

$$u_j = Q^*(f\psi_j^4) = Q^*(g_1^j) + Q^*(g_0^j + h\psi_j^4).$$

As well as $\bar{\zeta} u$ in the previous proof, the functions $\bar{\zeta} u_j$ are analytic in ζ. We consider some $\Phi_j \in L^\infty(\mathbb{T} \times \mathbb{T})$ also analytic in ζ and put $\alpha_j = \Phi_j u_j - Q^*(g_0^j + h\psi_j^4)$. Then, clearly, in the representation

$$f\psi_j^4 = [g_1^j - \alpha_j] + [g_0^j + h\psi_j^4 + \alpha_j]$$

the spectra of both terms are contained in $\{(i,k) \in \mathbb{Z} \times \mathbb{Z} : i > 0 \text{ or } k > 0\}$ (this statement is not quite rigorous, but reflects the reality: application of Q^* (in ω) to each term yields a function analytic in ζ and vanishing at $\zeta = 0$). The required decomposition of f will look like this:

$$f = \sum \theta_j \psi_j^4 f \psi_j^4 = \Big[\sum \theta_j \psi_j^4 g_1^j - \Lambda\Big] + \Big[\sum \theta_j \psi_j^4 (g_0 + h\psi_j^4) + \Lambda\Big],$$

where $\Lambda = \sum \theta_j \psi_j^4 \alpha_j$. Again, after ensuring the due norm estimates of the summands $\sum \theta_j \psi_j^4 g_1^j - \Lambda$ and $\sum \theta_j \psi_j^4 (g_0 + h\psi_j^4) + \Lambda$, it is fairly easy to see that they are indeed in the annihilators M and N. To obtain the norm estimates in question, we need to choose the Φ_j. But first we treat the terms not involving Λ, merely to show how the weight a is killed and then reproduced. We only do the calculations for one of these two terms, the other one being treated similarly: by the convexity of the function $t \mapsto t^q$, and by the conditions $\sum |\psi_j| \leq c$, $|\psi_j| a \leq c 2^j$, we obtain

$$I \stackrel{\text{def}}{=} \int \Big(\int \Big|\sum \theta_j \psi_j^4 (g_0^j + h\psi_j^4)\Big|^q a\, d\omega\Big)^{1/q} b\, d\zeta$$
$$\leq c \int \Big(\sum 2^j \int |g_0^j + h\psi_j^4|^q\, d\omega\Big)^{1/q} b\, d\zeta.$$

By (3′), we see that

$$I \leq c' \int \Big(\sum 2^j z_j^q + \int |h|^q \sum 2^j |\psi_j| \, d\omega \Big)^{1/q} b,$$

which does not exceed $c''B$ by (2′) and the relation $\sum 2^j |\psi_j| \leq ca$.

We proceed to the choice of the Φ_j. Again, we fix $s \in \mathbb{N}$ such that $p/s < 1$. The corresponding formulas in the preceding proof will be modified as follows (as before, $\mathcal{H}$ acts in the variable ζ): $r_j = z_j^{1/k}$,

$$\gamma_j(\zeta,\omega) = \max\{1, (\lambda_j(\zeta)^{-1} |(Q^* g_1^j)(\zeta,\omega)|)^{1/ks}\},$$

$$F_j = \frac{r_j + i\mathcal{H} r_j}{r_j \gamma + i\mathcal{H}(r_j \gamma)}, \qquad \Phi_j = 1 - (1 - F_j^{ks})^k.$$

The Φ_j are analytic in ζ and $|\Phi_j| \leq c\gamma_j^{-ks}$, whence

$$|\Phi_j||Q^*(g_1^j)| \leq \lambda_j(\zeta). \tag{4′}$$

We must verify the inequalities

$$\int \Big(\int |\Lambda|^q a \, d\omega \Big)^{1/q} b \, d\zeta \leq cB, \tag{5′}$$

$$\iint |\Lambda| a \, d\omega \, d\zeta \leq cA. \tag{6′}$$

To check (5′), we write

$$\begin{aligned} \int \Big(\int |\Lambda|^q a \, d\omega \Big)^{1/q} b \, d\zeta &\leq c \Big[\int \Big(\sum 2^j \int |\Phi_j Q^* g_1^j|^q \, d\omega \Big)^{\frac{1}{q}} b \, d\zeta \\ &\quad + \int \Big(\sum 2^j \int |Q^*(g_0^j + h\psi_j^4)|^q \, d\omega \Big)^{\frac{1}{q}} b \, d\zeta \Big] \\ &\stackrel{\text{def}}{=} c[I_1 + I_2]. \end{aligned}$$

I_2 is treated easily (see the preceding proof). In I_1, we first estimate the inner integral $\int |Q^* g_1^j|^q \, d\omega$ for each fixed ζ, as in (7), (8). This will result in

$$I_1 \leq c \int \Big(\sum 2^j z_j^q \Big)^{1/q} b \, d\zeta,$$

which is dominated by cB (see (2′)).

To verify (6′), we write

$$\begin{aligned} \int |\Lambda| a \, d\omega \, d\zeta &\leq c \Big[\int \sum 2^j \int |\Phi_j Q^*(g_1^j)| \, d\omega \, d\zeta \\ &\quad + \int \sum 2^j \int |1 - \Phi_j||Q^*(g_0^j + h\psi_j^4)| \, d\omega \, d\zeta \Big] \\ &\stackrel{\text{def}}{=} c[I_3 + I_4]. \end{aligned}$$

Acting nearly as in (9), we see that $I_3 \le cA$. In place of (10), we obtain

$$I_4 \le c \int \sum 2^j \left(\int |1 - \Phi_j|^p \, d\omega \right)^{1/p} z_j \, d\zeta, \tag{10'}$$

and then, like in (11), (12), we arrive at

$$I_4 \le c \left(A + \int \sum 2^j z_j \lambda_j(\zeta)^{-1/s} \left(\int_{\gamma_j > 1} |Q^* g_1^j|^{p/s} \, d\omega \right)^{1/p} d\zeta \right). \tag{13'}$$

Each of the inner integrals (in ω) is estimated separately, as in (14), which will eventually lead to

$$I_4 \le c \left(A + \iint \sum 2^j |g \psi_j^4| \, d\omega \, d\zeta \right) \le c' A, \tag{15'}$$

again because $\sum 2^j |\psi_j| \le ca$. □

References

[B] J. Bourgain, *Bilinear forms on H^∞ and bounded bianalytic functions*, Trans. Amer. Math. Soc. **286** (1984), no. 1, 313–337.

[Ja] S. Janson, *Interpolation of subcouples and quotient couples*, Ark. Mat. **31** (1993), 307–338.

[Ka-K-dL] J.-P. Kahane, Y. Katznelson, and K. de Leeuw, *Sur les coéfficients de Fourier des fonctions continues*, C. R. Acad. Sci. Paris, **285** (1977), 1001–1004.

[Kis1] S. V. Kislyakov, *Fourier coefficients of boundary values of functions that are analytic in the disc and bidisc*, Trudy Math. Inst. Steklov, **155** (1981), 77–94.

[Kis2] S. V. Kislyakov, *Interpolation of H^p-spaces: some recent developments*, In: Function spaces, interpolation spaces, and related topics, Israel Mathematical Conference Proceedings, 13. Amer. Math. Soc., Providence, R.I., 1999.

[Kis3] S. V. Kislyakov, *Bourgain's analytic projection revisited*, Proc. Amer. Math. Soc. (to appear).

[KX] S. V. Kislyakov and Q. Xu, *Real interpolation and singular integrals*, St. Petersburg Math. J., **8** (1997), no. 4, 593–615.

[P] G. Pisier, *Interpolation between H^p spaces and noncommutative generalizations. I*, Pacific J. Math., **155** (1992), no. 2, 341–368.

[Xu] Q. Xu, *Some properties of the quotient space $L^1(\mathbb{T}^d)/H^1(\mathbb{D}^d)$*, Illinois J. Math., **37** (1993), no. 3, 437–454.

St. Petersburg Division
of the Steklov Mathematical Institute,
Fontanka 27,
191011 St. Petersburg, Russia
E-mail address: skis@pdmi.ras.ru

Operator Theory:
Advances and Applications, Vol. 113

Carleson's Interpolation Theorem Deduced from a Result of Pick

Paul Koosis

Dedicated to the memory of S. A. Vinogradov

The proof of Carleson's theorem to be given below was worked up in response to a particular teaching situation. About a year and a half ago, I was assigned our department's first (semester) course on functional analysis, normally devoted to the three principles of that subject and to some of their applications. My class consisted of quite good students with an adequate preparation in measure theory and integration but only a rather scanty knowlege of analytic functions; I desired nevetheless to show them how Carleson's result could be made to follow from the rudiments of the subject. The problem with this was that the students could not be expected to know anything about H_p spaces or their duality, and class time did not allow for any preliminary discussion of that material.

This circumstance led me to try to deduce Carleson's theorem from the one of Pick, which can be obtained as a fairly straightforward consequence of the Riesz extension theorem for positive linear functionals, itself an extremal variant of the Hahn–Banach theorem, one of the three principles of functional analysis. The attempt proved successful and its realization is described here.

1. Let us first consider the following problem: one is assigned distinct complex numbers $z_1, \dots, z_n$ having strictly positive imaginary part and others, $w_1, \dots, w_n$, with the same property and asked to determine whether or not there is a function $F(z)$, *analytic in* $\operatorname{Im} z > 0$ *and having positive imaginary part there*, such that

$$F(z_k) = w_k \quad \text{for} \quad k = 1, 2, \dots, n. \tag{1}$$

Already in 1916, Pick had found that *a necessary and sufficient condition for that is that the quadratic form*

$$\sum_{j=1}^{n} \sum_{k=1}^{n} \frac{w_j - \overline{w}_k}{z_j - \overline{z}_k} \alpha_j \overline{\alpha}_k \tag{2}$$

be positive definite.

Quite elementary proofs of this result are available; see for instance Ch. III, §3.3 of [1] or pp. 7–9 of [2]. Let us, however, sketch here its deduction from Riesz' extension theorem.

It is enough to show that positive definiteness of (2) is equivalent to the existence of a function $F(z)$ analytic for $\operatorname{Im} z > 0$, satisfying (1) and given by the formula

$$F(z) = az + b + \int_{-\infty}^{+\infty} \left(\frac{1}{t-z} - \frac{t}{t^2+1} \right) d\nu(t), \tag{3}$$

where $a \geq 0$, $b \in \mathbb{R}$, and ν is a positive Borel measure such that

$$\int_{-\infty}^{+\infty} \frac{d\nu(t)}{1+t^2} < \infty. \tag{4}$$

It is easy to check directly that if $F(z)$, from (3), satisfies (1), then (2) must be positive definite, so let us prove the *converse*. For that matter, we consider the one-point compactification $\mathbb{R}^* = \mathbb{R} \cup \{\infty\}$ of $\mathbb{R}$ and the vector space $\mathcal{E}$ of *real-valued* functions of the form $\varphi(t)/(1+t^2)$, where $\varphi \in C(\mathbb{R}^*)$ (sic). *Certain* linear combinations (not all!) of the functions $1/(t-z_k)$, $1/(t-\overline{z}_k)$, $k = 1, 2, \dots, n$, will lie in $\mathcal{E}$; these make up a subspace $\mathcal{L}$ of the latter. We define a linear form L on $\mathcal{L}$ by first putting $L(1/(t-z_k)) = w_k$, $L(1/(t-\overline{z}_k)) = \overline{w}_k$, $k = 1, 2, \dots, n$, then extending L to *all* the linear combinations of $1/(t-z_k)$, $1/(t-\overline{z}_k)$ by linearity, and finally *restricting* L to the subspace $\mathcal{L}$ of those linear combinations.

The form L is positive on $\mathcal{L}$. Indeed, any member of $\mathcal{L}$ can be expressed as $P(t)/Q(t)$, where $P(t)$ is a *real* polynomial in t of degree $\leq 2n-2$ and $Q(t) = \prod_{k=1}^{n}(t-z_k)(t-\overline{z}_k)$. If $P(t)/Q(t)$ is to be positive on $\mathbb{R}$, the roots of $P(t)$ must occur in conjugate pairs (or be of even order if they lie on $\mathbb{R}$), and thence it is easy to see that

$$\frac{P(t)}{Q(t)} = \left| \sum_{k=1}^{n} \frac{\alpha_k}{t-z_k} \right|^2$$

with certain constants α_k. The square on the right works out to

$$\sum_{j=1}^{n} \sum_{k=1}^{n} \frac{1}{z_j - \overline{z}_k} \left(\frac{1}{t-z_j} - \frac{1}{t-\overline{z}_k} \right) \alpha_j \overline{\alpha}_k,$$

and, applying L to this, we get just the form (2), which by hypothesis is positive! Therefore $L\left(P(t)/Q(t)\right) \geq 0$ if $P(t)/Q(t) \geq 0$ on $\mathbb{R}$.

Riesz' theorem (for which see, e.g., pp. 269–273 of [3]) will now yield a positive extension Λ of L to all of $\mathcal{E}$, *provided that any element* $\varphi(t)/(1+t^2)$ *of the latter is dominated by some member of* $\mathcal{L}$. But *that is so*; for the member of $\mathcal{L}$ we may take any sufficiently large constant multiple of $1/(t-z_1)(t-\overline{z}_1)$. Therefore Λ is available; it generates in obvious fashion a positive linear functional on $C(\mathbb{R}^*)$. Such a functional can always be represented as

$$\int_{\mathbb{R}^*} \varphi(t) d\mu(t)$$

with a finite Borel measure $\mu \geq 0$ on $\mathbb{R}^*$; on putting $\mu(\infty) = a$ and $d\mu(t) = d\nu(t)/(1+t^2)$ on $\mathbb{R}$, we get $\Lambda\left(\varphi(t)/(1+t^2)\right) = a\varphi(\infty) + \int_{-\infty}^{+\infty} (\varphi(t)/(1+t^2)) d\nu(t)$ for functions $\varphi(t)/(1+t^2)$ in $\mathcal{E}$; here the positive measure ν satisfies (4).

Apply the last relation to the particular members of $\mathcal{L}$ equal successively to $\operatorname{Im}(1/(t-z_k)) = \operatorname{Im} z_k/(t-z_k)(t-\overline{z}_k)$, $k = 1, 2, \ldots, n$, and to $\operatorname{Re}(1/(t-z_j) - 1/(t-z_k))$ with the different values of j and k; one finds that

$$w_j - w_k = a(z_j - z_k) + \int_{-\infty}^{+\infty} \left(\frac{1}{t-z_j} - \frac{1}{t-z_k} \right) d\nu(t)$$

for $j, k = 1, 2, \ldots, n$. It follows that the function $F(z)$ given by (3) will, for suitable choice of $b \in \mathbb{R}$, satisfy (1).

2. In the *Pick–Nevanlinna problem* one is given distinct points $z_1, \ldots, z_n$ in the open upper half-plane and complex numbers $s_1, \ldots, s_n$ with the $|s_k| < 1$, and asked whether or not there is a function $f(z)$, analytic for $\operatorname{Im} z > 0$ and in modulus < 1 there, such that

$$f(z_k) = s_k, \qquad k = 1, 2, \ldots, n. \tag{5}$$

This question can be reduced by a change of variable to the one just treated; putting

$$w_k = i\frac{1+s_k}{1-s_k}, \qquad k = 1, 2, \ldots, n,$$

we will have $\operatorname{Im} w_k > 0$, and an f with the desired properties will exist if and only if there is an $F(z)$, analytic in $\operatorname{Im} z > 0$ and with positive imaginary part there, such that (1) holds. The relation between F (if it exists) and f is expressed by the formula $F(z) = i\,(1+f(z))\,/\,(1-f(z))$.

In terms of the s_k, the form (2) works out to

$$\sum_{j=1}^{n}\sum_{k=1}^{n} \frac{2i(1-s_j\overline{s}_k)}{z_j - \overline{z}_k} \left(\frac{\alpha_j}{1-s_j}\right)\left(\frac{\overline{\alpha}_k}{1-\overline{s}_k}\right).$$

Making a further change of variable $\alpha_j/(1-s_j) = \gamma_j$, $j = 1, 2, \ldots, n$, we see by Pick's theorem that *a function $f(z)$ satisfying* (5) *and having the other desired properties exists if and only if the form*

$$\sum_{j=1}^{n}\sum_{k=1}^{n} i\frac{1-s_j\overline{s}_k}{z_j - \overline{z}_k}\gamma_j\overline{\gamma}_k \tag{6}$$

is positive definite.

3. Suppose now that we are given an infinite sequence of distinct points z_k in the open upper half-plane; we wish to know whether or not there is some $\eta > 0$ such that, for *any sequence* of values s_k with $|s_k| < \eta$ for $k = 1, 2, \ldots$, there is a function $f(z)$, *analytic for* $\operatorname{Im} z > 0$ *and in modulus* ≤ 1 *there, with* $f(z_k) = s_k$ *for all* k.

The answer is provided by Carleson's celebrated *interpolation theorem: such an $\eta > 0$ is forthcoming if and only if there is a $\delta > 0$ such that, for all n and k,*

$$\prod_{\substack{j \le n \\ j \ne k}} \left| \frac{z_k - z_j}{z_k - \overline{z}_j} \right| > \delta. \tag{7}$$

We are going to deduce this result from the one in §2.

It is, in the first place, only necessary to demonstrate the *sufficiency* of (7) (for $\{z_k\}$ to enjoy the property in question), because the *necessity* of that condition is established rather easily by use of certain test sequences $\{s_k\}$ and the maximum principle.

Granted (7) – for all n and k – then, we wish to find an $\eta > 0$ for which the above interpolation problem always has a positive solution. For that, it will be enough to get an $\eta > 0$ such that, for *any* n, there is an $f_n(z)$, analytic in the upper half-plane and in modulus < 1 there, with $f_n(z_k) = s_k$ for $k = 1, 2, \ldots, n$ as long as $|s_k| < \eta$, $k = 1, 2, \ldots, n$. If, indeed, we can *do* that, any *sequence* $\{s_k\}$ with $|s_k|$ always $< \eta$ will give us such functions $f_n(z)$ for $n = 1, 2, \ldots$, and a subsequence of *those* – they form a normal family – will converge in the upper half-plane to a function $f(z)$ having all the desired properties. The whole difficulty here is to find an $\eta > 0$ independent of n and, *without loss of generality*, < 1 which will work.

According to the result in §2, that requirement on η is *equivalent* to positive definiteness of the forms (6) for *all* n and *all choices* of the s_k with $|s_k| < \eta$.

It is perhaps not clear to begin with why (6) should be positive definite *even when all the s_j are zero. Then*, however, that is so because (6) reduces to the integral

$$\int_0^\infty \left| \sum_{j=1}^n \gamma_j e^{iz_j s} \right|^2 ds.$$

In the *general* situation, the sums

$$\sum_{j,k=1}^n \frac{i}{z_j - \overline{z}_k} \gamma_j \overline{\gamma}_k \qquad \text{and} \qquad \sum_{j,k=1}^n \frac{i}{z_j - \overline{z}_k} (s_j \gamma_j) \overline{(s_k \gamma_k)} \tag{8}$$

are *both positive.* Our procedure will be to obtain an *upper bound* on the *second* sum and a *lower bound* on the *first* one. The *former* will be seen to always be $\le$ the *latter* provided that all the $|s_k|$ are $<$ a suitable η which will depend (only) on the δ from (7). *Both bounds* will of course depend on the γ_k, and we first have to set up a yardstick involving those quantities to which the sums may be compared.

4. The two sums in (8) being of essentially the same form, it seems reasonable to try taking, as yardstick, *the average*

$$\left(\frac{1}{2\pi}\right)^n \underbrace{\int_0^{2\pi} \cdots \int_0^{2\pi}}_{n} \sum_{j,k=1}^{n} \frac{i}{z_j - \overline{z}_k} \left(e^{i\omega_j}\gamma_j\right) \overline{\left(e^{i\omega_k}\gamma_k\right)} \, d\omega_1 \ldots d\omega_n,$$

which no longer involves the arguments of the γ_j and works out to

$$\sum_{j=1}^{n} \frac{|\gamma_j|^2}{2 \operatorname{Im} z_j}.$$

Forgetting, for the moment, about the factors s_k in the *second sum* (8), we thus set out to compare

$$\sum_{j,k=1}^{n} \frac{i}{z_j - \overline{z}_k} \gamma_j \overline{\gamma}_k$$

with the last expression or, what comes to the same thing (put $\gamma_j/(\operatorname{Im} z_j)^{1/2} = \beta_j$), to compare the form

$$\sum_{j=1}^{n} \sum_{k=1}^{n} \frac{i(\operatorname{Im} z_j \operatorname{Im} z_k)^{1/2}}{z_j - \overline{z}_k} \beta_j \overline{\beta}_k \tag{9}$$

with $\sum_{j=1}^{n} |\beta_j|^2$, seeking first an *upper bound* on (9). It is claimed *that* (9) *is bounded above by a multiple, depending only on* δ, *of the last sum.*

5. We pass to the proof of the preceding statement. Write

$$A_{j,k} = \frac{i(\operatorname{Im} z_j \operatorname{Im} z_k)^{1/2}}{z_j - \overline{z}_k}, \tag{10}$$

and consider the $n \times n$ matrix

$$A = [A_{j,k}], \qquad j, k = 1, 2, \ldots, n.$$

According to the observation made near the end of §3, A is positive definite; *so, therefore, is the complex conjugate matrix*

$$\overline{A} = \left[\, \overline{A_{j,k}} \,\right].$$

For any complex n-vector $\beta = (\beta_1, \ldots, \beta_n)$ we therefore have (trick!)

$$\langle A\beta, \beta\rangle = \sum_{j,k=1}^{n} A_{j,k}\beta_j\overline{\beta}_k \le \langle (A + \overline{A})\beta, \beta\rangle = 2 \sum_{j,k=1}^{n} (\operatorname{Re} A_{j,k})\beta_j\overline{\beta}_k. \tag{11}$$

Since

$$\operatorname{Re} A_{j,k} = \frac{(\operatorname{Im} z_j + \operatorname{Im} z_k)(\operatorname{Im} z_j \operatorname{Im} z_k)^{1/2}}{|z_j - \overline{z}_k|^2}, \tag{12}$$

the matrix

$$\operatorname{Re} A = [\operatorname{Re} A_{j,k}], \qquad j, k = 1, 2, \ldots, n$$

has only positive entries, besides being symmetric and positive definite. By this observation we see, referring to (11), that

$$\langle A\beta, \beta\rangle \le 2\sum_{j,k=1}^{n} (\operatorname{Re} A_{j,k})|\beta_j||\beta_k| \le 4 \sum_{\operatorname{Im} z_j \le \operatorname{Im} z_k} (\operatorname{Re} A_{j,k})|\beta_j||\beta_k|, \qquad (13)$$

and our idea now is to apply the *Senichkin–Vinogradov test* to the *dissymmetrized* form standing on the right.

For that purpose, we put

$$B_{j,k} = \begin{cases} \operatorname{Re} A_{j,k}, & \operatorname{Im} z_j \le \operatorname{Im} z_k \\ 0 & \text{otherwise,} \end{cases} \qquad (14)$$

and also denote the vector $(|\beta_1|, \dots, |\beta_n|)$ by $|\beta|$. In terms of the matrix

$$B = [B_{j,k}], \qquad j, k = 1, 2, \dots, n,$$

(13) then reads

$$\langle A\beta, \beta\rangle \le 4\langle B|\beta|, |\beta|\rangle$$

with, by Schwarz' inequality,

$$\langle B|\beta|, |\beta|\rangle \le \sqrt{\langle B^*B|\beta|, |\beta|\rangle\langle|\beta|, |\beta|\rangle}. \qquad (15)$$

In the Senichkin–Vinogradov test, the procedure is to estimate *the entries* of B^*B in terms of those of B and then use this last relation.

By (12) and (14),

$$B_{j,k} \le \frac{2(\operatorname{Im} z_j)^{1/2}(\operatorname{Im} z_k)^{3/2}}{|z_j - \overline{z}_k|^2} \qquad \text{when} \quad \operatorname{Im} z_j \le \operatorname{Im} z_k,$$

$B_{j,k}$ being zero otherwise. For the (k, l)th entry $(B^*B)_{k,l}$ of B^*B we thus have

$$(B^*B)_{k,l} \le 4 \sum_{\substack{\operatorname{Im} z_j \le \operatorname{Im} z_k \\ \operatorname{Im} z_j \le \operatorname{Im} z_l}} \frac{(\operatorname{Im} z_k)^{3/2}\operatorname{Im} z_j (\operatorname{Im} z_l)^{3/2}}{|z_j - \overline{z}_k|^2|z_j - \overline{z}_l|^2}. \qquad (16)$$

Here, since $\operatorname{Im} z_j$ and $\operatorname{Im} z_k$ are > 0, we have (and this trick has been used by Vinogradov)

$$|z_k - \overline{z}_l| \le |z_k - z_j| + |z_j - \overline{z}_l| \le |\overline{z}_k - z_j| + |z_j - \overline{z}_l|.$$

Dividing both sides by $|z_j - \overline{z}_k||z_j - \overline{z}_l||z_k - \overline{z}_l|$, we get

$$\frac{1}{|z_j - \overline{z}_k||z_j - \overline{z}_l|} \le \frac{1}{|z_k - \overline{z}_l|}\left(\frac{1}{|z_j - \overline{z}_k|} + \frac{1}{|z_j - \overline{z}_l|}\right),$$

whence

$$\frac{1}{|z_j - \overline{z}_k|^2|z_j - \overline{z}_l|^2} \le \frac{2}{|z_k - \overline{z}_l|^2}\left(\frac{1}{|z_j - \overline{z}_k|^2} + \frac{1}{|z_j - \overline{z}_l|^2}\right).$$

Substituted into the right side of (16), this yields (and usually by far!)

$$(B^*B)_{k,l} \leq \frac{8(\operatorname{Im} z_k \operatorname{Im} z_l)^{1/2}}{|z_k - \overline{z}_l|^2} \left\{ \operatorname{Im} z_l \sum_{j=1}^{n} \frac{\operatorname{Im} z_j \operatorname{Im} z_k}{|z_j - \overline{z}_k|^2} + \operatorname{Im} z_k \sum_{j=1}^{n} \frac{\operatorname{Im} z_j \operatorname{Im} z_l}{|z_j - \overline{z}_l|^2} \right\}. \tag{17}$$

It is *only at this point* that we bring in the condition (7). According to *D. J. Newman's lemma* (see [2] or p. 201 of [4]), (7) implies the relation

$$\sum_{j=1}^{n} \frac{\operatorname{Im} z_j \operatorname{Im} z_k}{|z_j - \overline{z}_k|^2} \leq C_\delta = \frac{1}{4} + \frac{1}{2}\log\frac{1}{\delta}, \tag{18}$$

holding for every k. From (17) we thence get

$$(B^*B)_{k,l} \leq \frac{8(\operatorname{Im} z_k + \operatorname{Im} z_l)(\operatorname{Im} z_k \operatorname{Im} z_l)^{1/2}}{|z_k - \overline{z}_l|^2} \cdot C_\delta;$$

that is,

$$(B^*B)_{k,l} \leq 8C_\delta \max(B_{k,l}, B_{l,k}) \tag{19}$$

in view of (12) and (14).

Let us now plug (19) into the right side of (15). After squaring and cancellation, that gives us

$$\langle B|\beta|, |\beta|\rangle \leq 16C_\delta\langle|\beta|, |\beta|\rangle$$

so finally, by (13) and (14),

$$\langle A\beta, \beta\rangle \leq 64C_\delta \sum_{j=1}^{n} |\beta_j|^2, \tag{20}$$

as asserted at the end of §4.

We have completed our first task, which was to obtain an *upper bound* on the right-hand sum in (8). Replacing each β_j in (20) by $s_j\beta_j$ we have, in terms of the quantities $\beta_j = \gamma_j/\sqrt{\operatorname{Im} z_j}$, the

Lemma. *If* $|s_j| < \eta$ *for* $j = 1, 2, \ldots, n$ *and* (7) *holds, one has*

$$\sum_{j,k=1}^{n} \frac{i(\operatorname{Im} z_j \operatorname{Im} z_k)^{1/2}}{z_j - \overline{z}_k}(s_j\beta_j)\overline{(s_k\beta_k)} \leq 64C_\delta\eta^2 \sum_{j=1}^{n} |\beta_j|^2, \tag{21}$$

where $C_\delta = \frac{1}{4} + \frac{1}{2}\log(1/\delta)$.

6. Now we must find a *lower bound* on the *first* sum in (8). As in §§4, 5 we continue to work with the form (9) and the matrix A given by (10); our job is thus to find a lower bound on (9) in terms of $\sum_{j=1}^{n} |\beta_j|^2$. Since A is positive definite, this can be accomplished by getting an *upper bound* on $\langle A^{-1}\beta, \beta\rangle$ in terms of the last sum, and with that in mind we proceed to work out A^{-1} explicitly.

For that we use Cramer's rule, whose application requires the evaluation of certain determinants. Those are all of the form

$$\operatorname{Det}\left[\frac{1}{a_j + b_k}\right]$$

where j and k both range from 1 to $n-1$ or to n, and a formula going back to Cauchy (see Ch. I, §14 in [5]) says that this determinant is equal to

$$\prod_{j<k}(a_j - a_k)\cdot\prod_{r<s}(b_r - b_s)\Big/\prod_{p,q}(a_p + b_q)$$

Before attempting to find A^{-1}, it is better to first compute the inverse of $M = [M_{j,k}]$, $j,k = 1,2,\ldots,n$, where

$$M_{j,k} = \frac{1}{z_j - \overline{z}_k}. \tag{22}$$

For this matrix, the preceding formula yields

$$\operatorname{Det} M = \frac{\prod\limits_{r<s}(z_r - z_s)\prod\limits_{l<m}(\overline{z}_m - \overline{z}_l)}{\prod\limits_{p,q}(z_p - \overline{z}_q)}. \tag{23}$$

According to Cramer's rule, one obtains $(M^{-1})_{j,k}$, the (j,k)th entry of M^{-1}, by *first crossing out* the kth *row* (sic!) and jth *column* of M and evaluating the resulting $(n-1)\times(n-1)$ determinant, *then* dividing by $\operatorname{Det} M$, and *finally* multiplying by $(-1)^{j+k}$. Crossing out the kth row and jth column in M has the effect of eliminating z_k and $\overline{z}_j$, and aside from that the $(n-1)\times(n-1)$ determinant can be worked out by Cauchy's formula again. This leads to an expression like the right side of (23) save that, *in the numerator*, the factors

$$\prod_{r<k}(z_r - z_k)\prod_{s>k}(z_k - z_s)$$

and

$$\prod_{l<j}(\overline{z}_j - \overline{z}_l)\prod_{m>j}(\overline{z}_m - \overline{z}_j)$$

will be missing, while, *in the denominator*, the factors

$$\prod_{p}(z_p - \overline{z}_j) \qquad \text{and} \qquad \prod_{q\neq j}(z_k - \overline{z}_q)$$

will also be missing. Dividing the determinant thus found by $\operatorname{Det} M$ and then multiplying by $(-1)^{j+k}$, we get

$$(M^{-1})_{j,k} = \frac{\prod\limits_{q\neq j}(z_k - \overline{z}_q)\prod\limits_{p}(z_p - \overline{z}_j)}{\prod\limits_{r\neq k}(z_k - z_r)\prod\limits_{l\neq j}(\overline{z}_l - \overline{z}_j)}.$$

Here the right side can be rewritten as

$$(z_j - \overline{z}_j) \prod_{l \neq j} \left(\frac{z_l - \overline{z}_j}{\overline{z}_l - \overline{z}_j} \right) \cdot \frac{z_k - \overline{z}_k}{z_k - \overline{z}_j} \cdot \prod_{r \neq k} \left(\frac{z_k - \overline{z}_r}{z_k - z_r} \right).$$

(Equality of this with the preceding must be checked in two cases: when $j \neq k$ and when $j = k$.)

Let us introduce the partial Blaschke product whose modulus figures in (7):

$$B_k(z) = \prod_{j \leq n,\ j \neq k} \left(\frac{z - z_j}{z - \overline{z}_j} \right). \tag{24}$$

Then the last expression becomes

$$-4 \frac{\operatorname{Im} z_j}{\overline{B_j(z_j)}} \cdot \frac{1}{z_k - \overline{z}_j} \cdot \frac{\operatorname{Im} z_k}{B_k(z_k)}.$$

Referring to (22), we see that we have derived the formula

$$(M^{-1})_{j,k} = \frac{2 \operatorname{Im} z_j}{\overline{B_j(z_j)}} \cdot \overline{M_{j,k}} \cdot \frac{2 \operatorname{Im} z_k}{B_k(z_k)}. \tag{25}$$

Now it is easy to compute A^{-1}. By (10) and (22),

$$A = i \Delta M \Delta,$$

where Δ is the *diagonal matrix* with entries $(\operatorname{Im} z_1)^{1/2}, (\operatorname{Im} z_2)^{1/2}, \ldots, (\operatorname{Im} z_n)^{1/2}$. Thence,

$$A^{-1} = -i \Delta^{-1} M^{-1} \Delta^{-1}.$$

With (25), (22) and (10) this makes

$$(A^{-1})_{j,k} = \frac{2}{\overline{B_j(z_j)}} \cdot \overline{A_{j,k}} \cdot \frac{2}{B_k(z_k)},$$

i.e.,

$$A^{-1} = 4 \overline{D}\, \overline{A} D \tag{26}$$

with the *diagonal matrix*

$$D = \begin{pmatrix} 1/B_1(z_1) & & & 0 \\ & 1/B_2(z_2) & & \\ & & \ddots & \\ 0 & & & 1/B_n(z_n) \end{pmatrix}. \tag{27}$$

The lower bound spoken of at the begining of this § is now furnished by the

Lemma. *Provided that* (7) *holds, we have*

$$\langle A\beta, \beta \rangle \geq \frac{\delta^2}{256 C_\delta} \langle \beta, \beta \rangle, \tag{28}$$

where $C_\delta = \frac{1}{4} + \frac{1}{2} \log(1/\delta)$.

Proof. Since A is positive definite, $\langle A\beta,\beta\rangle \geq \lambda\langle\beta,\beta\rangle$ where λ is *the smallest proper value of* A and, at the same time, $1/\lambda$ is the *largest proper value* of A^{-1}. An *upper bound* on the latter will thus give us a *lower bound* on λ.

We have, from (26),

$$\langle A^{-1}\beta,\beta\rangle = 4\langle\overline{A}D\beta, D\beta\rangle = 4\overline{\langle A\overline{D\beta},\overline{D\beta}\rangle}.$$

By (20) and (27), the right-hand member is

$$\leq 256C_\delta\sum_{j=1}^{n}|\beta_j/B_j(z_j)|^2,$$

which, by (7), is $\leq (256C_\delta/\delta^2)\sum_{j=1}^{n}|\beta_j|^2$.

This makes $1/\lambda \leq 256C_\delta/\delta^2$ and hence $\lambda \geq \delta^2/256C_\delta$ with C_δ as in (18). We are done.

7. The lemmas at the end of §§5 and 6 yield, almost without further ado, the

Theorem. *Suppose that* (7) *holds for some* $\delta > 0$. *Then, if* $|s_k| < \delta/128C_\delta$ *for* $k = 1,2,\ldots,n$ *where* $C_\delta = \frac{1}{4}+\frac{1}{2}\log(1/\delta)$, *the form* (6) *is positive definite.*

Proof. Making the change of variable $\beta_j = \gamma_j/(\operatorname{Im} z_j)^{1/2}$ in (6), we see that we have to verify the inequality

$$\sum_{j,k=1}^{n} A_{j,k}\beta_j\overline{\beta}_k \geq \sum_{j,k=1}^{n} A_{j,k}(s_j\beta_j)\overline{(s_k\beta_k)}$$

with the $A_{j,k}$ given by (10).

Taking $\eta = \delta/128C_\delta$ in the lemma of §5, we see by that result that the right-hand sum is

$$\leq (\delta^2/256C_\delta)\sum_{j=1}^{n}|\beta_j|^2$$

when the s_k fulfill the condition in the hypothesis. At the same time, the left-hand sum is

$$\geq (\delta^2/256C_\delta)\sum_{j=1}^{n}|\beta_j|^2$$

by the lemma at the end of §6. This does it.

According to the discussion in §3, this theorem shows that (7) is *sufficient* for the interpolation property under consideration here to *hold* for the sequence $\{z_k\}$. If, namely, (7) *is fulfilled* (with a $\delta > 0$ and for all n, k), *there will be*, for any sequence $\{s_k\}$ with the $|s_k|$ all $< \delta/128C_\delta$, a function $f(z)$, analytic for $\operatorname{Im} z > 0$ and in modulus ≤ 1 there, satisfying $f(z_k) = s_k,\quad k = 1,2,\ldots$.

As already remarked in §3, *necessity of* (7) *is easy.* Carleson's interpolation theorem is thus proved.

8. Sequences $\{z_k\}$ enjoying the interpolation property just mentioned are called *interpolating sequences.* Another curious property of such sequences, which perhaps could not have been guessed at beforehand, is brought to light by the work of the preceding discussion: *for an interpolating sequence* $\{z_k\}$ *the forms*

$$\sum_{j,k=1}^{n} \frac{i(\operatorname{Im} z_j \operatorname{Im} z_k)^{1/2}}{z_j - \overline{z}_k} \beta_j \overline{\beta}_k \qquad \text{and} \qquad \sum_{j=1}^{n} |\beta_j|^2$$

are comparable, in a manner independent of n. That means that the first form is essentially "spherical" which is far from evident – it certainly doesn't *look* spherical!

That this new property *follows* from (7) is essentially what we proved in §§5 and 6. It also *implies* (7) because *it, by itself,* implies that $\{z_k\}$ *is* an interpolating sequence – see the proof of the theorem in §7! *Approximate sphericity* (independent of n) *of the first form is thus a necessary and sufficient condition for* $\{z_k\}$ *to be an interpolating sequence.*

The sphericity property has an equivalent formulation in terms of the normalized exponentials $(2 \operatorname{Im} z_k)^{1/2} e^{iz_k s}$ belonging to $L_2(0, \infty)$: *the integral*

$$\int_0^\infty \left| \sum_{k=1}^{n} \beta_k (2 \operatorname{Im} z_k)^{1/2} e^{iz_k s} \right|^2 ds$$

and the sum $\sum_{k=1}^{n} |\beta_k|^2$ *are comparable (in a manner independent of* n*) if and only if* $\{z_k\}$ *is an interpolating sequence.* Direct evaluation of the integral reduces this statement to the preceding one.

The second of these characterizations of interpolating sequences plays an important rôle in some work of Khrushchev, Nikolskii and Pavlov. For that, one may consult [7].

9. I remember having heard, and perhaps even having read (but *where?*), that Carleson's theorem "could not" be deduced from Pick's. Here, this seems to have been done.

However that may be, I am not entirely happy about having had to resort to the Senichkin–Vinogradov test in §5 in order to obtain the upper bound found there. The reason for this is that that test, *by itself,* already gives us Carleson's theorem if we refer to some simple properties of H_p spaces, including their duality. Even the computations made in the course of such a proof – Vinogradov's – are virtually identical to those in §5. For details, see Lecture VII in [6] and also Havin's discussion at the end of Appendix I in [4]. In the present treatment, Pick's theorem and the rather grubby linear algebra calculation in §6 have enabled us to avoid the use of H_p spaces.

Having once heard that Donald Marshall might also have deduced Carleson's theorem from Pick's, I described the above procedure to him in a letter. Later on Marshall told me in a telephone conversation that he and E. Amar (or perhaps it was just Amar) had once arrived at such a deduction, but that it involved Carleson measures and was different from the one given here. He said that a paper about this work had never been published, and offered to send me a letter with some details when he would find the time to write one. Up to now, I have not heard any more from him.

References

[1] N. I. Akhiezer, *The Classical Moment Problem*, Gos. Izd. Fiz.-Mat. Lit., Moscow, 1961 (Russian); Engl. transl.: N. I. Akhiezer, *The Classical Moment Problem*, Oliver & Boyd, Edinburgh, 1965.

[2] J. Garnett, *Bounded Analytic Functions*, Academic Press, New York, 1981.

[3] G. Choquet, *Lectures on Analysis*, volume II. Benjamin, New York, 1969.

[4] P. Koosis, *Introduction to H_p Spaces*, Second Edition, Cambridge University Press, 1998.

[5] N. I. Akhiezer, *Theory of Approximation*, Second edition, Nauka, Moscow, 1965 (Russian); English transl. of first edition, Ungar, New York, 1956; German transl.: N. I. Achieser, *Vorlesungen über Approximationstheorie*, Zweite Auflage, Akademie Verlag, Berlin, 1967.

[6] N. K. Nikolskii, *Treatise on the Shift Operator. Spectral Function Theory*, Springer, Berlin, 1986.

[7] S. V. Hruščev, N. K. Nikolskii, and B. S. Pavlov, *Unconditional bases of exponentials and of reproducing kernels*, in *Complex Analysis and Spectral Theory*, edited by V. P. Havin and N. K. Nikolskii, Lecture Notes in Mathematics, No. 864, Springer, Berlin, 1981, pp. 214–335.

Mathematics Department,
McGill University
805 Sherbrooke St. West
Montreal, Québec, H3A2K6
Canada

Operator Theory:
Advances and Applications, Vol. 113

$A^{-\alpha}$ Zero Sets: New Methods and Techniques

Boris Korenblum

Dedicated to the memory of S. A. Vinogradov

Abstract. We give a detailed presentation of the theory of $A^{-\alpha}$ zero sequences and of the underlying density concepts, based on new methods and techniques.

1. Introduction

In this paper we shall be concerned with *analytic functions of hyperbolic exponential type* in the unit disk $\mathbf{D}$, i.e., with functions f such that

$$T(f) = \limsup_{|z|\to 1-} \log|f(z)|/|\log(1-|z|)| < \infty. \tag{1.1}$$

If $f \not\equiv 0$ then $T(f) \geq 0$ and is called the *type of* f; if $T(f) = 0$, f is of *minimal type*; if for some $\alpha > 0$

$$\|f\|_{-\alpha} := \sup\{|f(z)|/(1-|z|)^{\alpha} : z \in \mathbf{D}\} < \infty \tag{1.2}$$

and $T(f) = \alpha$, we say that f is of *exact type* α. The Banach space of analytic functions f in $\mathbf{D}$ for which (1.2) holds is denoted $A^{-\alpha}$, with $\|\cdot\|_{-\alpha}$ as norm. Clearly, $A^{-\alpha}$ consists of 0, of all analytic f with $T(f) < \alpha$, and of all f of exact type α. We also consider the linear spaces

$$A_{+}^{-\alpha} = \{0\} \bigcup \{f : T(f) \leq \alpha\} = \bigcap_{\beta > \alpha} A^{-\beta},$$

$$A_{-}^{-\alpha} = \{0\} \bigcup \{f : T(f) < \alpha\} = \bigcup_{\beta < \alpha} A^{-\beta},$$

and

$$A^{-\infty} = \bigcup_{\alpha > 0} A^{-\alpha}.$$

A sequence $A = \{a_n\} \subset \mathbf{D}$, with no accumulation points in $\mathbf{D}$, is called an $A^{-\alpha}$ *zero sequence* (or zero set) if there is a function $f \in A^{-\alpha}$ with zeros precisely at A (the repeated values of a_n indicate the multiplicities of the zeros of f). The zero sets for $A_{+}^{-\alpha}$, $A_{-}^{-\alpha}$ and $A^{-\infty}$ are defined analogously.

In [1] the notion of *upper asymptotic* κ*-density* $D^{+}(A)$ was introduced (see also Section 2 below), and it was proved that $D^{+}(A) < \infty$ is a necessary and sufficient condition for A to be an $A^{-\infty}$ zero set. (Actually, a sharper result was proved: $D^{+}(A) < 2\alpha + \varepsilon$ is necessary and $D^{+}(A) < \frac{\alpha}{2} - \varepsilon$ is sufficient for A to be an $A^{-\alpha}$ zero set, $\forall \varepsilon > 0$.)

In [2] and [3] K. Seip was able to close the above gap by perfecting the method used in [1]; he proved that the condition $D^{+}(A) \leq \alpha$ is necessary and sufficient for A to be an $A_{+}^{-\alpha}$ zero sequence.

In [4], K. Seip obtained even sharper necessary or sufficient conditions for the $A^{-\alpha}$ zero sequences by modifying the underlying density concept. He also proved that the remaining "logarithmic" gap between the necessary and sufficient conditions is impossible to close in terms of that or similar notions of density.

The purpose in this paper is to give a self-sufficient presentation of the theory of $A^{-\alpha}$ zero sets and of the underlying density concepts, based on new methods and techniques. Instead of certain distortion estimates for conformal mappings used in [1–4], the method used in this paper for proving the necessary conditions is much simpler: a combination of "radial projection" of zeros and some elementary constructions of auxiliary harmonic functions. Besides its simplicity, the new method has the advantage of being applicable to similar potential-theory problems in higher dimensions.

As to the sufficient conditions, we by and large adhere to Seip's method of "oblique projection" of zeros [4] combined with some techniques from convex analysis first used in this context in [1].

2. Several notions of density

2.1. For a point $\zeta \in \partial\mathbf{D}$, we let S_ζ denote the *standard Stolz angle* with vertex at ζ and aperture $\pi/2$. Thus, S_ζ is the convex hull of the set

$$\{\zeta\} \cup \{z \in \mathbf{C} : |z| \leq 1/\sqrt{2}\}.$$

For a nonempty closed set $F \subsetneqq \partial\mathbf{D}$ with complementary arcs $\{I_n\}$ we define

$$\hat{\kappa}(F) = \sum_n \frac{|I_n|}{2\pi} \log \frac{2\pi e}{|I_n|}, \tag{2.1}$$

where $|I_n|$ is the length of I_n. The quantity $\hat{\kappa}(F)$ will be called the *Beurling–Carleson characteristic* of F (sometimes the term *entropy* is also used). We set $\hat{\kappa}(\varnothing) = 0$.

A closed set $F \subset \partial\mathbf{D}$ is called a Beurling-Carleson (B.-C.) set if F is nonempty, has Lebesgue measure 0, and $\hat{\kappa}(F) < \infty$. It is clear that for such sets $\hat{\kappa}(F) \geq 1$, with equality occurring only for one-point sets F. For much of our discussions we assume F to be a finite set.

In association with F we define a star-like region G_F as follows:

$$G_F = \bigcup \{S_\zeta : \zeta \in F\}. \tag{2.2}$$

For a sequence $A = \{a_n\}$ of (not necessarily distinct) points in $\mathbf{D}$ we form the "partial Blaschke sum"

$$\sigma(A,F) = \sum\{\frac{1}{2}(1-|a_n|^2) : a_n \in G_F\} \tag{2.3}$$

and the κ-density of A in G_F:

$$D(A,F) = \sigma(A,F)/\hat{\kappa}(F). \tag{2.4}$$

Definition 2.1. *Let F be finite subsets of $\partial\mathbf{D}$. We define*

$$D^+(A) = \limsup_{\hat{\kappa}(F)\to\infty} D(A,F). \tag{2.5}$$

$D^+(A)$ is called the upper asymptotic κ-density of A.

We proceed to give several equivalent definitions of the upper asymptotic κ-density. First, observe that replacing the standard Stolz angle S_ζ with a general Stolz angle $S_{\zeta,\alpha}$ having any fixed aperture $0 < \alpha < \pi$ and making the corresponding changes in the definitions of $G_F, \sigma(A,F)$, and $D^+(A)$ will not alter the quantity $D^+(A)$. What is somewhat surprising is that the angle α can be reduced to 0 with no effect on $D^+(A)$. More specifically, for a finite set $F \subset \partial\mathbf{D}$ and a sequence A we define

$$R_F = \bigcup\{r\zeta : 0 \le r < 1, \zeta \in F\} \tag{2.6}$$

and

$$\sigma'(A,F) = \sum\{\frac{1}{2}(1-|a_n|^2) : a_n \in R_F\}. \tag{2.7}$$

Then we have the following result.

Proposition 2.2.

$$D^+(A) = \limsup_{\hat{\kappa}(F)\to\infty} \frac{\sigma'(A,F)}{\hat{\kappa}(F)}. \tag{2.8}$$

To prove (2.8) we need yet another notion of density based on Carleson squares. Recall that for an arc $I \subset \partial\mathbf{D}$, $|I| < 1$, the associated Carleson square is

$$Q_I = \{z \in \mathbf{C} : 1-|I| < |z| < 1, \frac{z}{|z|} \in I\}.$$

For a finite set $F \subset \partial\mathbf{D}$ with complementary arcs $\{I_n\}$ we define

$$B_F = \mathbf{D} - \bigcup_n Q_{I_n}.$$

Then we arrive at another way of obtaining $D^+(A)$.

Proposition 2.3.

$$D^+(A) = \limsup_{\hat{\kappa}(F)\to\infty} \frac{\sigma''(A,F)}{\hat{\kappa}(F)}, \tag{2.9}$$

where

$$\sigma''(A,F)=\sum\{\frac{1}{2}(1-|a_n|^2):a_n\in B_F\}. \tag{2.10}$$

We proceed now to prove the two above propositions, that is, the equivalence of all three definitions of $D^+(A)$.

Note that the lim sup in each of these definitions will not change if we allow closed countable sets F of finite entropy as well.

We start with the proof of Proposition 2.3. We enlarge every finite set F by inserting in each complementary arc I_n of F some additional points accumulating at the end-points of I_n so that their distances from the nearest end-point of I_n form a geometric progression with some fixed ratio q. An elementary computation shows that the augmented set $F_1\supset F$ will have the property

$$\hat\kappa(F)<\hat\kappa(F_1)<\hat\kappa(F)+C$$

where C depends only on the ratio q. Also, we can choose q so that $B_F\subset G_F\subset B_{F_1}$, i.e.,

$$\sigma''(A,F)\le\sigma(A,F)\le\sigma''(A,F_1).$$

All this proves Proposition 2.3.

To prove Proposition 2.2, observe that $R_F\subset G_F$ and, thus, $\sigma'(A,F)\le\sigma(A,F)$, which implies

$$\limsup_{\hat\kappa(F)\to\infty}\frac{\sigma'(A,F)}{\hat\kappa(F)}\le\limsup_{\hat\kappa(F)\to\infty}\frac{\sigma(A,F)}{\hat\kappa(F)}.$$

The reverse inequality, by Proposition 2.3, is equivalent to

$$\limsup_{\hat\kappa(F)\to\infty}\frac{\sigma''(A,F)}{\hat\kappa(F)}\le\limsup_{\hat\kappa(F)\to\infty}\frac{\sigma'(A,F)}{\hat\kappa(F)}. \tag{2.11}$$

At first glance this looks highly improbable, since the sum defining $\sigma''(A,F)$ involves all points in B_F, and the sum defining σ' involves only the points lying on one of the radii from 0 to $\zeta\in F$. However, a more careful argument will prove (2.11). We may assume that the lim sup on the left-hand side of (2.11) is positive. Let K be a positive number less than this lim sup. This implies that there are finite subsets F of $\partial\mathbf{D}$ of arbitrary large $\hat\kappa(F)$ such that

$$\sigma''(A,F)=\sum\{\frac{1}{2}(1-|a_k|^2):a_k\in B_F\}<K\cdot\hat\kappa(F).$$

Until the end of the proof we shall assume that F satisfies this inequality. Let F_1 be the set F plus the radial projections $z/|z|$ of the points $z\in A\cap(B_F-R_F)$. Let k_n be the number of such radial projections (counting multiplicities) that lie on I_n. Observe now that the contribution to $\hat\kappa(F_1)$ from the complementary arcs of F_1 contained in I_n does not exceed the quantity

$$\frac{|I_n|}{2\pi}\left[\log\frac{2\pi e}{|I_n|}+\log(k_n+1)\right]$$

which corresponds to the case of k_n equidistant points of $(F_1 - F) \cap I_n$. Therefore $\hat{\kappa}(F_1) \le \hat{\kappa}(F) + r(F)$ with

$$r(F) = \frac{1}{2\pi} \sum_n |I_n| \log(k_n + 1).$$

We are going to show that

$$r(F) = o(\sigma''(A, F))$$

as $\hat{\kappa}(F) \to \infty$. In fact, pick a positive integer N and split the sum defining $r(F)$ in two:

$$\begin{aligned} r(F) &= \frac{1}{2\pi} \left[\sum_{k_n \le N} + \sum_{k_n > N} \right] |I_n| \log(k_n + 1) \\ &\le \log(N+1) + \frac{\log(N+1)}{N} \sum_{k_n > N} k_n |I_n| \\ &\le \log(N+1) + \frac{\log(N+1)}{N} \sigma''(A, F). \end{aligned}$$

Letting $\hat{\kappa}(F) \to \infty$ with $\sigma''(A, F) \to \infty$, first holding N constant and then letting $N \to \infty$, we obtain $r(F) = o(\sigma''(A, F))$. Consequently,

$$\hat{\kappa}(F_1) = \hat{\kappa}(F) + o(\sigma''(A, F))$$

as $\hat{\kappa}(F) \to \infty$. Since $\sigma'(A, F_1) \ge \sigma''(A, F)$, we get

$$\frac{\sigma'(A, F_1)}{\hat{\kappa}(F_1)} \ge \frac{\sigma''(A, F)}{\hat{\kappa}(F) + o(\sigma''(A, F))}$$

as $\hat{\kappa}(F) \to \infty$. This implies that

$$\lim_{\hat{\kappa}(F) \to \infty} \frac{\sigma'(A, F_1)}{\hat{\kappa}(F_1)} \ge K.$$

Since K can be arbitrarily close to the lim sup on the left-hand side of (2.11), and F_1 can be substituted for F on the right-hand side, Proposition 2.2 is proved.

2.2. Let $A = \{a_n\}$ be a sequence such that $m = D^+(A) < \infty$. As we know, that implies both

$$\sup\{\sigma'(A, F) - (m + \varepsilon)\hat{\kappa}(F)\} < +\infty$$

and

$$\sup\{\sigma(A, F) - (m + \varepsilon)\hat{\kappa}(F)\} < +\infty$$

for every $\varepsilon > 0$ (the supremum is taken over all finite $F \subset \partial\mathbf{D}$).

We shall need a similar comparison between σ' and σ for some other asymptotic restrictions on σ'.

Proposition 2.4. *Let*

$$\sup_F\{\sigma'(A,F) - m(\hat{\kappa}(F) + c\log\hat{\kappa}(F))\} < +\infty \tag{2.12}$$

with some $m > 0$ and $c > 0$. Then

$$\sup_F\{\sigma(A,F) - m[\hat{\kappa}(F) + (c+1)\log\hat{\kappa}(F)]\} < +\infty. \tag{2.13}$$

Proof. As in Subsection 2.1 we can show that (2.13) is equivalent to a similar estimate with σ replaced by σ'':

$$\sup_F\{\sigma''(A,F) - m[\hat{\kappa}(F) + (c+1)\log\hat{\kappa}(f)]\} < +\infty. \tag{2.14}$$

To derive (2.14) from (2.12), let $F = \{\zeta_n\}$, R_F, B_F, $\{I_n\}$ and $\{Q_{I_n}\}$ be as in Subsection 2.1. We can assume that there are no points from A on $R_F \cup \partial B_F$. Project all points from $A \cap B_F$ to $\partial\mathbf{D}$, and let $F' = \{\zeta = \frac{z}{|z|} : z \in A \cap B_F\}$ be the resulting set, with $k_n = \operatorname{card}(I_n \cap F')$. An elementary argument shows that

$$\hat{\kappa}(F') \le \hat{\kappa}(F) + \sum_n \frac{|I_n|}{2\pi}\log(k_n + 1), \tag{2.15}$$

with equality occurring if the k_n points of $F' \cap I_n$ divide I_n in $k_n + 1$ equal subarcs. On the other hand,

$$\sigma'(A,F') \ge \sigma''(A,F) \ge \sum_n |I_n| k_n. \tag{2.16}$$

The sum in (2.15) can be estimated using the concavity of $\log t$:

$$\sum_n \frac{|I_n|}{2\pi}\log(k_n+1) \le \log(1 + \frac{1}{2\pi}\sum_n k_n|I_n|) \le \log\sigma''(A,F) + O(1) \tag{2.17}$$

since $\sum_n \frac{1}{2\pi}|I_n| = 1$. Replace now F with F' in (2.12) and use (2.16) and (2.17) to get

$$\begin{aligned}\sigma''(A,F) &\le \sigma'(A,F') \le m(\hat{\kappa}(F') + c\log\hat{\kappa}(F')) + O(1)\\ &\le m[\hat{\kappa}(F) + \log\sigma''(A,F) + c\log(\hat{\kappa}(F) + \log\sigma''(A,F))] + O(1).\end{aligned}$$

From Subsection 2.1 we know that $\log\sigma''(A,F) \le \log\hat{\kappa}(F) + O(1)$, and thus

$$\sigma''(A,F) \le m[\hat{\kappa}(F) + (c+1)\log\hat{\kappa}(F)] + O(1),$$

which is equivalent to (2.14). □

3. $A^{-\alpha}$ zero sets: Necessary conditions

3.1. To establish necessary conditions for the $A^{-\alpha}$ zero sets we need the following balayage type estimate, which enables us to "sweep" the zeros of an analytic function f radially to the circumference $\partial\mathbf{D}$ and convert them into singular masses without increasing $|f|$ in a certain critical region.

Lemma 3.1. *Let S_1 be the standard Stolz angle at $\zeta = 1$. Then*

$$\left|\frac{a-z}{1-az}\right| \geq \exp\left\{(\log a)\frac{1-|z|^2}{|1-z|^2}\right\} \tag{3.1}$$

for all $0 < a < 1$ and all $z \in \mathbf{D} - S_1$.

Proof. Using the Cayley transform

$$w = K(z) = \frac{1+z}{1-z}$$

from $\mathbf{D}$ onto the half-plane $\mathbf{C}_+ = \{w = u + iv : u > 0\}$, we can rewrite the desired inequality as

$$\left|\frac{b-w}{b+w}\right| \geq \exp\left\{\left(\log\frac{b-1}{b+1}\right)u\right\}, \tag{3.2}$$

where $w \in \mathbf{C}_+ - K(S_1)$ and $b = (1+a)/(1-a) > 1$. We are going to show that (3.2) actually holds for w in a larger set $\mathbf{C}_+ - \Omega$ where

$$\Omega = \{w = u + iv : u > 1, |v| < u\}.$$

To see that Ω is smaller than $K(S_1)$ observe that $\partial K^{-1}(\Omega)$ consists of two orthogonal circles through 1 and -1 and an arc of the circle through 0 and 1 tangent to $\partial\mathbf{D}$. Then it is geometrically obvious that $K^{-1}(\Omega) \subset S_1$.

We now show that

$$\frac{1}{u}\log\frac{b^2+u^2+v^2+2bu}{b^2+u^2+v^2-2bu} < 2\log\frac{b+1}{b-1}$$

for $b > 1$ and $w = u + iv \in \mathbf{C}_+ - \Omega$. It is easy to check that the left-hand side above decreases, for any fixed u, as $|v|$ increases; and for $v = 0$ it is an increasing function of u. Thus the inequality above holds in the strip $0 \leq u \leq 1$ with equality attained at $u = 1$, $v = 0$. It remains to verify the case $|v| = u$;

$$\frac{1}{u}\log\frac{b^2+2u^2+2bu}{b^2+2u^2-2bu} \leq 2\log\frac{b+1}{b-1} \quad (u \geq 1).$$

Let $u = bt$. It suffices to show that

$$\frac{1}{t}\log\frac{1+2t^2+2t}{1+2t^2-2t} \leq 2b\log\frac{b+1}{b-1}$$

for $b > 1$ and $t > 0$. The right-hand side here is decreasing in b and tends to 4 as $b \to +\infty$. So it is enough to check that

$$\frac{1}{t}\log\frac{1+2t^2+2t}{1+2t^2-2t} \leq 4$$

for all $t > 0$. An easy computation shows that the function $f(t) = 4t + \log(1+2t^2-2t) - \log(1+2t^2+2t)$ has a positive derivative on $(0, +\infty)$: $f'(t) = 8(t^2+t^4)/(1+4t^4)$. Thus $f(t) > f(0) = 0$ for $t > 0$, and the proof of the lemma is complete. □

Corollary 3.2. *Suppose $f \in A^{-\alpha}$, and $\{z_1, \dots, z_n\}$ are some zeros of f, all of them different from 0. Let $F = \{\zeta_1, \dots, \zeta_n\}$, where $\zeta_k = z_k/|z_k|$ $(k = 1, \dots, n)$. Then*

$$\left|\frac{f(z)}{B(z)}\right| < \frac{\|f\|_{-\alpha}}{(1-|z|)^\alpha} \exp(P(z)), \quad z \in \mathbf{D} - G_F,$$

where $B(z)$ is the Blaschke product for $\{z_1, \dots, z_n\}$ and

$$P(z) = \sum_{k=1}^{n} (\log \frac{1}{|z_k|}) \frac{1-|z|^2}{|\zeta_k - z|^2}.$$

3.2. We need some estimates for several auxiliary harmonic functions. The notation $d_F(z)$ means the euclidean distance from $z \in \mathbf{C}$ to a set $F \subset \mathbf{C}$.

Lemma 3.3. *Let $F \subset \partial \mathbf{D}$ be a finite set whose complementary arcs are of length < 1. Then the harmonic function*

$$U_F(z) = \frac{1}{2\pi} \int_{\partial \mathbf{D}} \frac{1-|z|^2}{|\zeta - z|^2} \log \frac{1}{d_F(\zeta)} |d\zeta|, \quad z \in \mathbf{D},$$

satisfies

$$U_F(z) \geq \log \frac{1}{d_F(z)}, \quad z \in \mathbf{D}.$$

Proof. Since

$$\log \frac{1}{d_F(z)} = \max_{\zeta \in F} \log \frac{1}{|z - \zeta|},$$

we see that $\log \frac{1}{d_F(z)}$ is a positive subharmonic function in $\mathbf{D}$ whose boundary values equal those of $U_F(z)$. The desired result then follows. □

Next, let $0 < p < 1$ and $\zeta \in \partial \mathbf{D}$. Consider the harmonic function

$$V_{\zeta,p}(z) = \sec \frac{p\pi}{2} Re(1 - \bar{\zeta} z)^{-p}, \quad z \in \mathbf{D}. \tag{3.3}$$

Lemma 3.4. *Let p and ζ be as above, and let $0 < c < \frac{1}{4}$. Define the curve*

$$\gamma_{\zeta,p,c} = \{z \in \bar{\mathbf{D}} : 1 - |z|^2 = c|\zeta - z|^{2-p}\}.$$

Then

$$cV_{\zeta,p}(z) > \frac{1-|z|^2}{|\zeta - z|^2}$$

for all z in the region between $\partial \mathbf{D}$ and $\gamma_{\zeta,p,c}$.

Proof. If $1 - |z|^2 < c|\zeta - z|^{2-p}$ then

$$cV_{\zeta,p}(z) > c|1 - \bar{\zeta} z|^{-p} > \frac{1-|z|^2}{|\zeta - z|^2}.$$

□

Finally, for $F = \{\zeta_1, \dots, \zeta_n\} \subset \partial\mathbf{D}$ with complementary arcs $\{I_k\}_1^n$ satisfying $|I_k| < 1$, and for p and c as above, let

$$P(z) = \sum_{k=1}^{n} m_k \frac{1-|z|^2}{|\zeta_k - z|^2}, \quad z \in \mathbf{D}, \tag{3.4}$$

where $\{m_k\}$ are some nonnegative numbers. Define the curve

$$\gamma_{F,p,c} = \{z \in \bar{\mathbf{D}} : 1 - |z|^2 = c(d_F(z))^{2-p} \tag{3.5}$$

and the harmonic function

$$V_{F,p,P}(z) = \sec\frac{p\pi}{2} Re \sum_{k=1}^{h} m_k(1-\bar{\zeta}_k z)^{-p}, \quad z \in \mathbf{D}. \tag{3.6}$$

Lemma 3.5. *The inequality*

$$cV_{F,p,P}(z) > P(z) \tag{3.7}$$

holds for all z between $\partial\mathbf{D}$ and $\gamma_{F,p,c}$.

Proof. Apply Lemma 3.4 to each term of $P(z)$ and observe that the set of points z between $\partial\mathbf{D}$ and $\gamma_{F,p,c}$ is the intersection of the domains described in Lemma 3.4 for each ζ_k. □

3.3. A Jensen-type inequality. Let $0 \not\equiv f \in A^{-\alpha}$ have zeros (counting multiplicities) at $\{a_n\} = A$, with $0 \notin A$, and $F \subset \partial\mathbf{D}$ be an arbitrary finite set with complementary arcs of length < 1.

Set

$$T = \sum\{\log\frac{1}{|a_n|} : a_n \in R_F\} \tag{3.8}$$

(recall that R_F denotes the union of the radii from 0 to points $\zeta \in F$).

Theorem 3.6. *If $T > 4\alpha$ then*

$$\log|f(0)| \le \alpha[\hat{\kappa}(F) + \log\hat{\kappa}(F)] - T + \alpha\log T - \alpha(\log\alpha - 2) + \log\|f\|_{-\alpha}. \tag{3.9}$$

Remark. Inequality (3.9) implies that $T < \infty$ for every finite F.

Proof. We can assume A to be a finite sequence. By Corollary 3.2,

$$\log|f(z)/B(z)| \le \log\|f\|_{-\alpha} + \alpha\log\frac{1}{1-|z|} + P(z), \quad z \in \gamma_{F,p,c}, \tag{3.10}$$

where $B(z)$ is the Blaschke product for $A \cap R_F$, and

$$P(z) = \sum\{(\log\frac{1}{a_n})\frac{1-|z|^2}{|\zeta_n - z|^z} : a_n \in R_F\}, \quad \zeta_n = \frac{a_n}{|a_n|}$$

(see (3.5) for the definition of $\gamma_{F,p,c}$). We apply now Lemma 3.3 and Lemma 3.5 to obtain from (3.10)

$$\log|f(z)/B(z)| - \alpha(2-p)U_F(z) - \alpha\log\frac{1}{c} - cV_{F,p,P}(z) \le \log\|f\|_{-\alpha}, \quad z \in \gamma_{F,p,c}.$$

The left-hand side here is a subharmonic function in the region bounded by $\gamma_{F,p,c}$. By the maximum principle

$$\begin{aligned}
&\log|f(0)/B(0)| - \alpha(2-p)U_F(0) - \alpha\log\frac{1}{c} - cV_{F,p,P}(0)\\
&= \log|f(0)| + T - \alpha(2-p)\frac{1}{2\pi}\int_{\partial\mathbf{D}}\log\frac{1}{d_F(\zeta)}|d\zeta| - (c\sec\frac{p\pi}{2})T\\
&\leq \log\|f\|_{-\alpha} + \alpha\log\frac{1}{c}.
\end{aligned}$$

The integral above does not exceed $\hat{\kappa}(F)$, and $\sec\frac{p\pi}{2} < \frac{1}{1-p}$; thus,

$$\log|f(0)| \leq \alpha(2-p)\hat{\kappa}(F) + (\frac{c}{1-p} - 1)T + \alpha\log\frac{1}{c} + \log\|f\|_{-\alpha}.$$

To minimize the right-hand side we put $1-p = \frac{1}{\hat{\kappa}(F)}, c = \frac{\alpha}{T\hat{\kappa}(F)}$ and obtain (3.9). □

3.4. Two forms of a necessary condition

Theorem 3.7. *If $A = \{a_n\}$ is an $A^{-\alpha}$ zero sequence and $F \subset \partial\mathbf{D}$ are finite sets, then*

$$\sup_F\{\sigma'(A,F) - \alpha[\hat{\kappa}(F) + 2\log\hat{\kappa}(F)] < \infty. \tag{3.11}$$

Proof. Recall that

$$\sigma'(A,F) = \sum\frac{1}{2}(1-|a_n|^2) : a_n \in R_F\}.$$

Theorem 3.6 implies that

$$\sup_F\{\sigma'(A,F) - \alpha\log^+\sigma'(A,F) - \alpha[\hat{\kappa}(F) + \log\hat{\kappa}(F)]\} < \infty,$$

or, equivalently,

$$\sigma'(A,F) - \alpha\log^+\sigma'(A,F) < \alpha[\hat{\kappa}(F) + \log\hat{\kappa}(F)] + C$$

for some $C > 0$ and all finite sets $F \subset \partial\mathbf{D}$.

Taking the logarithm of both sides yields

$$\log^+[\sigma' - \alpha\log^+\sigma'(A,F)] < \log[\hat{\kappa}(F) + \log\hat{\kappa}(F)] + C_1$$

with some other constant C_1. Since $\log(x + \log x) - \log x \to 0$ and $\log x - \log(x - \alpha\log x) \to 0$ as $x \to +\infty$, we get

$$\log^+\sigma'(A,F) < \log\hat{\kappa}(F) + C_2$$

and

$$\sigma'(A,F) < \alpha[\hat{\kappa}(F) + 2\log\hat{\kappa}(F)] + C_3,$$

which is equivalent to (3.11). □

Theorem 3.8. *Under all the assumptions of Theorem 3.7 we have*

$$\sup_F\{\sigma(A,F)-\alpha[\hat{\kappa}(F)+3\log\hat{\kappa}(F)]\}<\infty. \tag{3.12}$$

Proof. (3.12) is a direct consequence of (3.11) and Proposition 2.4. □

Corollary 3.9. *If $A=\{a_n\}$ is an $A^{-\infty}$ zero sequence, then*
(i) $S(r)=\sum\{1-|a_n|:|a_n|<r\}=O(\log\frac{1}{1-r})$, $r\to 1^-$;
(ii) $\sum_n(1-|a_n|)/|\log(1-|a_n|)|^{1+\varepsilon}<\infty$ $(\varepsilon>0)$.

Proof. Taking $F=\{\exp(2k\pi i/N)\}_{k=1}^N$ in Theorem 3.8 and letting $N\to\infty$ yields (i). The following computation then proves (ii):

$$\begin{aligned}\sum_n(1-|a_n|)/|\log(1-|a_n|)|^{1+\varepsilon}&=\int_0^1\frac{dS(r)}{|\log(1-r)|^{1+\varepsilon}}\\&=(1+\varepsilon)\int_0^1\frac{S(r)dr}{(1-r)|\log(1-r)|^{2+\varepsilon}}<\infty.\end{aligned}$$

□

4. $A^{-\alpha}$ zero sets: A sufficient condition

4.1.

Theorem 4.1. $A=\{a_n\}_1^\infty\subset\mathbf{D}$ *is an $A^{-\alpha}$ zero sequence if*

$$m(A,\alpha)=\sup_F\{\sigma(A,F)-\alpha\hat{\kappa}(F)\}<\infty \tag{4.1}$$

(*the supremum is taken over all finite $F\subset\partial\mathbf{D}$, including $F=\phi$*).

To prove this theorem we need several auxiliary results.

Remark. As in Subsection 2.2, one can show that (4.1) is equivalent to a similar condition where $\sigma(a,F)$ is replaced by $\sigma''(A,F)$. However, replacing $\sigma(A,F)$ with $\sigma'(A,F)$ apparently renders (4.1) insufficient for $A^{-\alpha}$ zero sets; cf. [4]. Also, the standard Stolz angles involved in the definition of $\sigma(A,F)$ can be replaced by the Stolz angles with any fixed aperture – the resulting conditions are all equivalent.

4.2. Seip's "oblique projection" lemma. From now on S_ζ will denote a Stolz angle with vertex at $\zeta\in\partial\mathbf{D}$ and with an arbitrary but fixed aperture φ, $\frac{\pi}{2}\le\varphi<\pi$, i.e. the convex hull of $\{\zeta\}\cup\{|z|\le\sin\frac{\varphi}{2}\}$.

For an open arc $I\subset\partial\mathbf{D}$ with endpoints ω_1 and ω_2, we define the "tent" H_I as the component of $\mathbf{D}-G_{\{\omega_1,\omega_2\}}$ abutting on I. For a fixed $\zeta\in\mathbf{D}$ and arbitrary $\lambda\in\mathbf{D}-G_{\{\zeta,-\zeta\}}$ we define the "oblique projection" $\omega=\omega_\zeta(\lambda)\in\partial\mathbf{D}$ so that the shorter arc I between ω and ζ satisfies $\lambda\in\partial H_I$.

Lemma 4.2. *If $\pi - \varphi$ is sufficiently small then for all $z = t\zeta$, $0 < t < 1$, and $\lambda \in \mathbf{D} - G_{\{\zeta,-\zeta\}}$*

$$\log\left|\frac{\lambda - z}{1 - \bar{\lambda}z}\right| + \frac{1}{2}(1 - |z|^2)(1 - |\lambda|^2)/|1 - \overline{\omega}z|^2 \le 0, \tag{4.2}$$

where $\omega = \omega_\zeta(\lambda)$.

Note that (4.2) means that the modulus of the Blaschke factor $(\lambda - z)/(1 - \bar{\lambda}z)$ is dominated on $\{z = t\zeta : 0 < t < 1\}$ by the singular inner function $\exp\{-m(w + z)/(w - z)\}$, where $m = \frac{1}{2}(1 - |\lambda|^2)$.

Proof. Using the identity

$$1 - \left|\frac{\lambda - z}{1 - \bar{\lambda}z}\right|^2 = \frac{(1 - |\lambda|^2)(1 - |z|^2)}{|1 - \bar{\lambda}z|^2},$$

we can rewrite (4.2) as

$$\log(1 - 2ma_2) + 2ma_1 \le 0, \tag{4.3}$$

where

$$2m = 1 - |\lambda|^2, a_1 = \frac{1 - |z|^2}{|1 - \bar{z}\omega|^2}, a_2 = \frac{1 - |z|^2}{|1 - \bar{z}\lambda|^2}.$$

We have

$$\begin{aligned}\log(1 - 2ma_2) + 2ma_1 &= -\sum_{n=1}^{\infty}\frac{(2ma_2)^n}{n} + 2ma_1 \le 2m(a_1 - a_2) - \frac{(2ma_2)^2}{2(1 - ma_2)} \\ &= \frac{2m(a_1 - a_2 - ma_1a_2)}{1 - ma_2}.\end{aligned}$$

Therefore (4.3) will follow if we prove that $a_1 - a_2 - ma_1a_2 \le 0$, which is equivalent to $a_2^{-1} - a_1^{-1} \le m$, i.e.

$$\left|\frac{1}{\bar{z}} - \lambda\right|^2 - \left|\frac{1}{\bar{z}} - \omega\right|^2 \le \frac{1}{2}(1 - |\lambda|^2)\left(\frac{1}{|z|^2} - 1\right). \tag{4.4}$$

Let $\beta = |\mathrm{Arg}(\zeta_0\bar{\omega})|$, $\gamma = |\mathrm{Arg}(\zeta_0\lambda^{-1})|$. The definition of the oblique projection implies that $\frac{\beta}{2} \le \gamma \le \beta$ and $1 - |\lambda| < c(\pi - \varphi)(\beta - \gamma)$, where c is an absolute constant. Also,

$$\left|\frac{1}{\bar{z}} - \lambda\right|^2 = \frac{1}{|z|^2} + |\lambda|^2 - 2\frac{|\lambda|}{|z|}\cos\gamma, \qquad \left|\frac{1}{\bar{z}} - \omega\right|^2 = \frac{1}{|z|^2} + 1 - \frac{2}{|z|}\cos\beta.$$

Substituting this in (4.4) yields

$$\cos\beta - |\lambda|\cos\gamma \le \frac{1}{4}\left(|z| + \frac{1}{|z|}\right)(1 - |\lambda|^2).$$

Since $|z| + \frac{1}{|z|} > 2$ $(0 < |z| < 1)$ it is enough to prove that

$$\cos\beta - |\lambda|\cos\gamma \le \frac{1}{2}(1 - |\lambda|^2) \quad (\lambda \notin S_{\zeta_0}). \tag{4.5}$$

We can assume $\beta < \frac{\pi}{3}$ since otherwise (4.5) holds for all λ. Solving the quadratic inequality we get from (4.5):

$$0 \leq 1 - |\lambda| < (1 - \cos\gamma) + [(1 - \cos\gamma)^2 + 2\sin\frac{\beta - \gamma}{2}\sin\frac{\beta + \gamma}{2}]^{1/2}. \tag{4.6}$$

The right-hand side is greater than $(2/\pi)(\beta - \gamma)$. If $\pi - \varphi$ is small enough then $1 - |\lambda| < c(\pi - \varphi)(\beta - \gamma) < (2/\pi)(\beta - \gamma)$, which yields (4.6), (4.5) and (4.2). □

4.3. The use of Linear-Programming Technique. Suppose $A = \{a_n\} \subset \mathbf{D}$ is a finite sequence and w_0 a fixed point on $\partial\mathbf{D}$. Assume also that $\alpha > 0$ is fixed, as well as the aperture φ of all Stolz angles considered here.

Definition 4.3. *A nonnegative Borel measure μ on $\partial\mathbf{D}$ is (A, α, w_0)-admissible if*
(i) $\mu(\{w_0\}) = 0$;
(ii) *for each open arc $I \subset \partial\mathbf{D}$ with $w_0 \notin I$, the following inequality holds:*

$$0 \leq \mu(I) \leq \alpha\frac{|I|}{2\pi}\log\frac{2\pi e}{|I|} + \sum\{\frac{1}{2}(1 - |a_n|^2) : a_n \in H_I\}. \tag{4.7}$$

The set of all (A, α, w_0)-admissible measures will be denoted by $\mathcal{M}(A, \alpha, w_0)$, or just $\mathcal{M}$. (4.7) implies that $\mu(\{\zeta\}) = 0$ for any $\zeta \in \partial\mathbf{D}$ and not just for $\zeta = w_0$.

Lemma 4.4.

$$\sup\{\mu(\partial\mathbf{D}) : \mu \in \mathcal{M}\} = -m(A, \alpha, w_0) + \sum_n \frac{1}{2}(1 - |a_n|^2) \tag{4.8}$$

where

$$m(A, \alpha, w_0) = \sup\{\sigma(A, F) - \alpha\hat{\kappa}(F) : F \textit{ finite}, \ w_0 \in F\}, \tag{4.9}$$

and there is at least one maximal admissible measure μ_0 for which

$$\mu_0(\partial\mathbf{D}) = -m(A, \alpha, w_0) + \sum_n \frac{1}{2}(1 - |a_n|^2). \tag{4.10}$$

Proof. Without loss of generality we can assume that the circle $|z| = \sin\frac{\varphi}{2}$ associated with the Stolz angles S_ζ is free of points from A. Define a finite set F_0 consisting of w_0 and all those $\zeta \in \partial\mathbf{D}$ for which $(\partial S_\zeta) \cap A \neq \phi$, and let $\{I_k\}_1^N$ be the complementary arcs of F_0.

For every $\mu \in \mathcal{M}$ let $\tilde{\mu}$ denote the measure with constant density on each I_k and such that $\tilde{\mu}(I_k) = \mu(I_k)$, $\forall k$. We claim that $\mu \in \mathcal{M}$ implies $\tilde{\mu} \in \mathcal{M}$. In fact, if the endpoints e^{it_1} and e^{it_2} of some arc I are not in F_0 then the right-hand side of (4.7) is a locally continuous and concave function of t_1 and t_2. Therefore replacing μ by $\tilde{\mu}$ will not invalidate (4.7). For the same reason, to ascertain that a measure $\mu = \tilde{\mu}$ is in $\mathcal{M}$ it is enough to check (4.7) only for arcs I, $w_0 \notin I$, whose endpoints are in F_0. Each such measure is described by a vector $x = (x_1, \ldots, x_N)$ where $N = \text{card}F_0$ and $x_k = \mu(I_k)$. We are thus led to a standard problem from the optimization theory: maximize the functional $L(x) = x_1 + \cdots + x_N$ where the non-negative vector x satisfies $N(N+1)/2$ restrictions

$$x_k + x_{k+1} + \cdots + x_\ell \le a_{k\ell} \quad (1 \le k \le \ell \le N) \tag{4.11}$$

that correspond to (4.7) with arcs I having endpoints in F_0.

For the sake of convenience we introduce some terminology and notations. $\mathcal{J}$ stands for the set of all closed intervals $J = [k, \ell]$ whose endpoints are integers satisfying $1 \le k \le \ell \le N$. A system $\mathcal{P} = \{J_\nu\} = \{[k_\nu, \ell_\nu]\}$ of such intervals (repetitions are allowed) is called an *m-fold covering* of $\mathbf{N}_N = \{1, 2, \ldots, N\}$ if every $n \in \mathbf{N}_N$ belongs to exactly m intervals from $\mathcal{P}$ (if $m = 1$ we speak of a *simple* covering).

In terms of (4.11) the main assertion of Lemma 4.3 can be stated as follows:

If $\mathcal{B} \subset \mathbf{R}^N$ is the set of all nonnegative solutions $x = (x_1, \ldots, x_N)$ of (4.11), then

$$\max\{L(x) : x \in \mathcal{B}\} = \min\{\sum_\nu a_{k_\nu \ell_\nu} : \text{all simple coverings } \{[k_\nu, \ell_\nu]\} \text{ of } \mathbf{N}_N\}. \tag{4.12}$$

To prove (4.12) we need three elementary facts.

(i) *Every m-covering $\mathcal{P}$ of $\mathbf{N}_N$ is a union of m simple coverings.*

In fact, every interval $J = [k, \ell] \in \mathcal{P}$ with $\ell < N$ has the property that $\ell + 1$ is covered m times by $\mathcal{P} - \{J\}$ while ℓ is covered only $m - 1$ times. This is possible only if there is an interval in $\mathcal{P} - \{J\}$ whose left endpoint is $\ell + 1$. The rest is done by induction.

(ii) *For a system of linear algebraic equations with integer coefficients the set of rational solutions is dense in the set of all solutions.*

The proof is omitted.

(iii) *Assume (as we may) that $a_{k\ell} < a_{k_1\ell_1}$ whenever $[k, \ell]$ is strictly contained in $[k_1, \ell_1]$. Then among the points $x = (x_1, \ldots, x_N) \in B$ that give maximum value to $L(x)$ there is at least one with $x_j > 0$ for all $1 \le j \le N$.*

In fact, suppose $x_j = 0$ for $k < j < \ell$, but $x_k > 0$ and $x_\ell > 0$. Then we can choose $\varepsilon > 0$ sufficiently small so that the point $x' = (x_1, \ldots, x_{k-1}, x_k - \varepsilon(\ell - k - 1), \varepsilon, \varepsilon, \ldots, \varepsilon, x_\ell, \ldots, x_N)$ is in B and $L(x') = L(x)$.

Now we can prove (4.12). That the left-hand side of (4.12) cannot be greater than its right-hand side follows directly from (4.11). Conversely, the solution set $\mathcal{B}$ of (4.11) is a closed convex polyhedron in the positive cone of $\mathbf{R}^N$. The maximum L_0 of L on $\mathcal{B}$ determines a support hyperplane $x_1 + \cdots + x_N = L_0$ of $\mathcal{B}$. If $L_0 = a_{1N}$, then there is nothing to prove. If $L_0 < a_{1N}$ then we pick a vertex of $\mathcal{B}$ lying in that hyperplane and consider the faces of $\mathcal{B}$ that intersect at this vertex. The equation $x_1 + \cdots + x_N = L_0$ is then a positive linear combination with (as we may assume) rational coefficients of the equations of these faces. Multiplying by the common denominator q of these rational coefficients, we come to the conclusion that there is a q-covering $\mathcal{P}$ of $\mathbf{N}_N$ such that

$$\sum\{a_{k\ell} : [k,\ell] \in \mathcal{P}\} = qL_0 \tag{4.13}$$

(each $a_{k\ell}$ is repeated in accordance with the multiplicity of $[k,\ell]$ in $\mathcal{P}$). We then split $\mathcal{P}$ into q simple coverings, $\mathcal{P} = \mathcal{P}_1 \cup \cdots \cup \mathcal{P}_q$, and observe that for each $\mathcal{P}_j$ we have $\sum\{a_{k\ell} : [k,\ell] \in \mathcal{P}_j\} \geq L_0$. In view of (4.13) this can happen only if all the above inequalities are in fact equalities. □

4.4. Completion of the proof of Theorem 4.1. First observe that if an arc $I \subset \partial\mathbf{D}$ is partitioned into two arcs I_1 and I_2, then

$$\frac{|I|}{2\pi}\log\frac{2\pi e}{|I|} \leq \frac{|I_1|}{2\pi}\log\frac{2\pi e}{|I_1|} + \frac{|I_2|}{2\pi}\log\frac{2\pi e}{|I_2|} \leq \frac{|I|}{2\pi}\log\frac{2\pi e}{|I|} + \frac{\log 2}{2\pi}|I|.$$

This implies that if we add $\alpha(2\pi)^{-1}(\log 2)|I|$ to the right-hand side of (4.7) then the restriction $w_0 \notin I$ can be lifted.

Let $0 \notin A = \{a_n\}$ satisfy (4.1) and A_0 be a *finite* subsequence of A. Now we choose an arbitrary $w_0 \in \partial\mathbf{D}$, construct as in Lemma 4 a maximal (A_0, α, w_0)-admissible measure μ_0 and form the function

$$f_0(z) = B(z)\Phi(z),$$

where $B = B_{A_0}$ is the Blaschke product for A_0, and Φ is the outer function

$$\Phi(z) = \exp\{\int_{\partial\mathbf{D}} \frac{\zeta + z}{\zeta - z} d\mu_0(\zeta)\}.$$

We are going to obtain an upper estimate for $\|f_0\|_{-\alpha}$ and a lower estimate for $|f_0(0)|$, both independent of $A_0 \subset A$. To do this, fix $\zeta \in \partial\mathbf{D}$ and consider two subsequences of $A_0 : A_0' = A_0 \cap G_{\{\zeta_1,\zeta_2\}}$ and $A_0'' = A_0 - A_0'$. Let B_1 and B_2 be the Blaschke products for A_0' and A_0'', respectively. For each $a_n \in A_0''$, let $\omega_n = \omega_\zeta(a_n)$ be its oblique projection. Form an atomic measure τ by placing at each ω_n a mass $m_n = \frac{1}{2}(1 - |a_n|^2)$ and let

$$T(z) = \exp\{-\int_{\partial\mathbf{D}} \frac{\zeta + z}{\zeta - z} d\tau(\zeta)\}$$

and $\Psi = \Phi T$. The boundary measure for Ψ is $\mu_0 - \tau$; by the remark at the beginning of this section

$$(\mu_0 - \tau)(I) \leq \alpha\frac{|I|}{2\pi}\log\frac{2\pi e}{|I|} + \frac{\alpha\log 2}{2\pi}|I|$$

for every arc I of length $< \pi$ having ζ as one of its endpoints. Using partial integration we derive from this

$$|\Psi(z)| \leq C_\alpha(1 - |z|)^{-\alpha}, \quad z = t\zeta \quad (0 \leq t < 1),$$

with a constant C_α depending only on α. We now apply Lemma 4.2 to get $|B_2(z)| \leq |T(z)|$ for $z = t\zeta$ $(0 \leq t < 1)$. Since $|B(z)| \leq |B_2(z)|$, we obtain that $|f_0(z)| = |B(z)\Phi(z)| \leq |B_2(z)\Phi(z)| \leq |\Psi(z)| \leq C_\alpha(1 - |z|)^{-\alpha}$ $(z = t\zeta, 0 \leq t < 1)$. The point $\zeta \in \partial\mathbf{D}$ is arbitrary; therefore $\|f_0\|_{-\alpha} \leq C_2$.

To get a lower estimate for $|f_0(0)|$ we derive from (4.10):

$$\begin{aligned}\log|f_0(0)| &= -\sum\{\log\frac{1}{|a_n|} : a_n \in A_0\} + \mu_0(\partial\mathbf{D})\\ &= -\sum\{\frac{1}{2}(1-|a_n|^2) : a_n \in A_0\}\\ &\quad -\sum\{\log\frac{1}{|a_n|} - \frac{1}{2}(1-|a_n|^2) : a_n \in A_0\} + \mu_0(\partial\mathbf{D})\\ &\geq -m(A_0,\alpha,w_0) - \sum\{\log\frac{1}{|a_n|} - \frac{1}{2}(1-|a_n|^2) : a_n \in A_0\}\\ &\geq -m(A,\alpha) - \sum\{\log\frac{1}{|a_n|} - \frac{1}{2}(1-|a_n|^2) : a_n \in A\} > -\infty,\end{aligned}$$

because $0 \leq \log\frac{1}{|a_n|} - \frac{1}{2}(1-|a_n|^2) = O[(1-|a_n|)^2]$ and $\sum_n(1-|a_n|)^2 < \infty$. Now, take a nested sequence of finite subsets of A, $A_1 \subset A_2 \subset \cdots$, with $A = \bigcup_n A_n$, construct as above functions f_n for each A_n and note that $\{f_n\}$ is a normal family. Thus, there is a subsequence $\{f_{n_k}\}$ converging to an analytic function f uniformly on the compact subsets of $\mathbf{D}$; f is in $A^{-\alpha}$ and its zero sequence is A. □

Corollary 4.5. *The necessary and sufficient condition for A to be an $A_+^{-\alpha}$ zero sequence is $D^+(A) \leq \alpha$.*

References

[1] B. Korenblum, *An extension of the Nevanlinna theory*, Acta Math., **135** (1975), 187–219.

[2] K. Seip, *Beurling type density theorems in the unit disk*, Invent. Math., **113** (1993), 21–39.

[3] K. Seip, *On a theorem of Korenblum*, Ark. Mat., **32** (1994), 237–243.

[4] K. Seip, *On Korenblum's density condition for the zero sequences of $A^{-\infty}$*, J. d'Analyse Math., **67** (1995), 307–322.

Department of Mathematics and Statistics
SUNY
Albany, NY 1222-0001
USA

Operator Theory:
Advances and Applications, Vol. 113

Interpolation Sets for the Hölder Spaces of Functions Analytic in a Strip

A. M. Kotochigov

Dedicated to the memory of S. A. Vinogradov

Abstract. A full description of the free interpolation sets is obtained for the analytic Hölder classes in a strip in the complex plane; by a free interpolation set we mean a set such that any Hölder function on this set is the trace of some function belonging to the analytic Hölder class in the strip. A set is an interpolation set if and only if its inner part is sparse and in every boundary interval (of length less then 1) there is a "big" subinterval free from the points of the interpolation set.

1. Introduction

The main result of this paper consists in description of the closed subsets $\mathcal{E}$ of a strip such that any Hölder function on $\mathcal{E}$ can be extended to a function belonging to the corresponding analytic Hölder class in the strip. In this case we call $\mathcal{E}$ an interpolation set. The first result of this sort was a theorem proved by Dyn′kin in [1]. He solved this problem for the analytic Hölder classes on the disk. It turns out that a similar description of the interpolation sets is valid for the Lavrent′ev domains. This is a result of Kotochigov and Shirokov [2]. But if the domain is allowed to have cusp points on the boundary, the structure of the interpolation set may become more complicated. In [3], Kotochigov and Shirokov exemplified this phenomenon in the case of an outer cusp, and in [4] Kotochigov considered some specific domains with inner cusps. It turned out that the space of restrictions of the functions in question to $\mathcal{E}$ may have higher smoothness than the initial analytic space. Here we continue the study of this topic. We prove that the nature of the interpolation sets for the strip is very similar to that in the case of an outer cusp.

Given a set $\mathcal{E}$, we introduce the Hölder class $\Lambda^\alpha(\mathcal{E})$ $(0 < \alpha \leq 1)$ as the space of all functions $f : \mathcal{E} \to \mathbf{C}$ for which

$$\|f\| = \sup\{|f(z)| : z \in \mathcal{E}\} + \sup\left\{\left|\frac{f(z_1) - f(z_2)}{z_1 - z_2}\right| : z_1 \neq z_2, z_1, z_2 \in \mathcal{E}\right\} < \infty.$$

In this paper we deal with the strip

$$\mathcal{L} = \{z = x + iy : 0 < y < 1, -\infty < x < \infty\}$$

and with the corresponding analytic Hölder classes

$$\Lambda_a^\alpha(\mathcal{L}) = \{f \text{ is analytic in } \mathcal{L},\ f \in \Lambda^\alpha(\overline{\mathcal{L}})\}.$$

Throughout, we denote by C various constants in estimates.

A closed subset $\mathcal{E}$ of $\overline{\mathcal{L}}$ is called *a free interpolation set* for the pair $(\Lambda_a^\alpha(\mathcal{L}),\ \Lambda^\alpha(\mathcal{E}))$ if the space of restrictions $\{f|_{\mathcal{E}} : f \in \Lambda_a^\alpha(\mathcal{L})\}$ coincides with $\Lambda^\alpha(\mathcal{E})$.

We need the following terminology. A subset $\mathcal{E}$ of $\mathcal{L}$ is said to be *sparse* if there is a constant C_s such that

$$\operatorname{dist}(z, \partial\mathcal{L}) \le C_s \operatorname{dist}(z, \mathcal{E}\backslash\{z\}) \tag{1}$$

for any $z \in \mathcal{E}$.

A subset $\mathcal{E}$ of $\overline{\mathcal{L}}$ is said to be *locally porous* if there is a constant C_p such that every interval $I \subset \partial\mathcal{L}$ of length $|I| < 1$ contains a subinterval J with

$$|I| \le C_p|J|, \qquad \mathcal{E} \cap B_J = \varnothing. \tag{2}$$

Here B_J is the disk with diameter J.

The following theorem is our main result.

Theorem 1.1. *Let* $0 < \alpha < 1$, *and let* $\mathcal{E}$ *be a closed subset of* $\overline{\mathcal{L}}$. *Then* $\mathcal{E}$ *is in interpolation set for the pair* $(\Lambda_a^\alpha(\mathcal{L}),\ \Lambda^\alpha(\mathcal{E}))$ *if and only if the following two conditions are fulfilled:*

1) *the set* $\mathcal{E} \cap \mathcal{L}$ *is sparse, and*
2) $\mathcal{E}$ *is locally porous.*

Remark 1.2. *The theorem remains valid for* $\alpha = 1$, *provided that condition* 1 *is replaced by the requirement that* $\mathcal{E} \cap \mathcal{L}$ *be the union of two sparse sets.*

The method of the proof of Theorem 2.1 is similar to that used in [4]. However, some additional complications arise when we deal with the Cauchy integrals along $\partial\mathcal{L}$. To make sure that such integrals converge, we prefer to work with the subclass

$$\Lambda^{\alpha,0}(\mathcal{E}) = \left\{h :\ h(z) = \frac{f(z)}{z+i},\ f \in \Lambda^\alpha(\mathcal{E})\right\} \tag{3}$$

rather than with $\Lambda^\alpha(\mathcal{E})$.

2. Auxiliary statements

Proposition 2.1. *Let* $\mathcal{E} = \overline{\mathcal{E}} \subset \overline{\mathcal{L}}$. *If* $h \in \Lambda^{\alpha,0}(\mathcal{E})$, *then*

$$|h(z) - h(z_1)| \le c\frac{|z - z_1|^\alpha}{|z| + |z_1| + 1}, \qquad z, z_1 \in \mathcal{E}.$$

Proof. This follows from the inequality

$$|h(z) - h(z_1)| \le \frac{|f(z) - f(z_1)|}{|z+i|} + \frac{|f(z_1)||z - z_1|}{|z+i||z_1+i|} \le c||f||\frac{|z - z_0|^\alpha}{|z+i|}\left(1 + \frac{|z - z_1|}{|z_1+i|}\right).$$

□

Proposition 2.2. *If* $h \in \Lambda^{\alpha,0}(\mathcal{E})$, *then*

$$\int_{\partial\mathcal{L}} \frac{|h(z)||dz|}{|z - z_1|} \leq c\,||h||, \qquad z_1 \in \mathcal{L}.$$

Proof. Proposition 2.1 implies that $|h(z)| \leq c(|z| + 1)^{\alpha-1}$, and the convergence of the integral becomes obvious. □

In order to prove that some analytic function belongs to $\Lambda^\alpha_a(\mathcal{L})$ we shall use the analytic continuation of that function to some standard extension of the strip; such an extension is defined as follows.

Let $\mathcal{E} = \overline{\mathcal{E}} \subset \overline{\mathcal{L}}$, and let $\{l_n\}_n$ be the collection of all intervals complementary to the set $\mathcal{E} \cap \partial\mathcal{L}$ on $\partial\mathcal{L}$. We fix some number θ, $0 < \theta < \pi/2$. Let B_n be the disk such that the interval l_n is a chord of B_n, and the angle between this chord and the boundary of the disk is equal to θ. By the θ-extension of the strip we mean the domain

$$\mathcal{L}(\theta, \mathcal{E}) = \mathcal{L} \cup (\cup_n (B_n \cap \mathcal{L})). \tag{4}$$

Remark 2.3. *The definition of a sparse set and of a locally porous set can easily be carried over from the strip to the set* $\mathcal{L}(\theta, \mathcal{E})$. *Obviously, if* $\mathcal{E}$ *is sparse or locally porous relative to* $\mathcal{L}$, *than so is it for the set* $\mathcal{L}(\theta, \mathcal{E})$.

Now we describe an auxiliary function that is very important for our construction. This is a sort of an "analytic distance" to the set $\mathcal{E}$.

Proposition 2.4. *If* $\mathcal{E}$ *is a locally porous subset of* $\overline{\mathcal{L}}$, *then for any* $0 < \theta < \pi/2$ *on the strip* $\mathcal{L}(\theta, \mathcal{E})$ *there exists an outer function such that*

$$|g(z)| \asymp \operatorname{dist}(z, \mathcal{E}), \qquad z \in \mathcal{L}(\theta, \mathcal{E}).$$

Proof. Consider a conformal map of the strip $\mathcal{L}(\theta, \mathcal{E})$ onto the disk. An analysis of the distortions appearing under such a map (regarded as a perturbation of a map of the standard strip to the disk) shows that the image of $\mathcal{E}$ is a porous subset of the disk. (For the disk, porosity is equivalent to local porosity.) In Dyn'kin's paper [1] it was shown that in this case there exists a function whose modulus on the circle is equivalent to the distance from a point to the image of the set $\mathcal{E}$. The conformal transplantation of such a function to the strip $\mathcal{L}(\theta, \mathcal{E})$ gives a function g that we need. □

Proposition 2.5. *For any* θ_1, θ_2 *with* $0 < \theta_1 < \theta_2 < \pi/2$ *we have*

$$\operatorname{dist}(z, \mathcal{E}) \asymp \operatorname{dist}(z, \partial\mathcal{L}(\theta_2, \mathcal{E})), \qquad z \in \mathcal{L}(\theta_1, \mathcal{E}).$$

Moreover, any outer function constructed as in Proposition 3.4 *satisfies the estimate*

$$|g(z)| \asymp \operatorname{dist}(z, \mathcal{E}), \qquad z \in \mathcal{L}(\theta_1, \mathcal{E}). \tag{5}$$

Proof. For a strip, the first inequality is obvious. The properties of the function g show that the second inequality is simply a reformulation of the first. □

In what follows, we shall use a result (see [2]) concerning interpolation in the so-called *Lavrent'ev domains*; these are the domains such that for any two points z_1, z_2 on the boundary one of the arcs connecting them (we denote it by $J(z_1, z_2)$) satisfies the condition

$$|J(z_1, z_2)| \le C_L\ |z_1 - z_2|, \tag{6}$$

where C_L is a constant independent of z_1, z_2.

We introduce a partition of the strip into "squares" $\mathcal{L}(\theta, \mathcal{E})_n = \{z \in \mathcal{L}(\theta, \mathcal{E}) : n \le \mathrm{Re}(z) < n + 1\}$.

Remark 2.6. *All sets $\mathcal{L}(\theta, \mathcal{E})_n$ are Lavrent'ev domains; hence, the interpolation sets for each of them are precisely the sparse and porous sets.*

For an application of the Cauchy formula we need a standard extension of a function $f \in \Lambda^\alpha(\mathcal{E})$ to the entire complex plane. We shall modify the construction used in the paper [4], where the following assertion was proved.

Proposition 2.7. *Let $\mathcal{E}$ be a closed subset on the boundary of a Lavrent'ev domain $\mathcal{D}$, and let $\mathcal{E}$ satisfy the porosity condition. Then there exists a bounded linear operator S acting from $\Lambda^\alpha(\mathcal{E})$ to $\Lambda^\alpha(\mathbf{C})$ such that*

a) $Sf(z) = f(z)$ *for all* $z \in \mathcal{E}$;

b) $\left|\frac{\partial}{\partial \bar{z}} Sf(z)\right| \le c\ \mathrm{dist}^{\alpha-1}(z, \mathcal{E})$;

c) $\int_{\partial \mathcal{D}} \frac{Sf(z)\ dz}{g^\varepsilon(z)} = 0.$

Proposition 2.8. *For any θ with $0 < \theta < \pi/2$ there exists a unitary operator S acting from $\Lambda_0^\alpha(\mathcal{E})$ to $\Lambda_0^\alpha(\mathbf{C})$ such that any function h is taken to a function Sh that is infinitely differentiable off $\mathcal{E}$ and satisfies the conditions*

$$|Sh(z)| \le c\ ||h||(|z+1)^{\alpha-1}, \qquad |Sh(z) - Sh(z_1)| \le c\ \frac{|z - z_1|^\alpha}{|z| + |z_1| + 1}, \qquad z, z_1 \in \mathbf{C}, \tag{7}$$

$$\int_{\partial \mathcal{L}_n(\theta, \mathcal{E})} \frac{Sh(z)dz}{g^\varepsilon(z)} = 0. \tag{8}$$

Proof. Fixing some number n, we construct a linear extension of h from the set $\mathcal{E}$ to the complementary arcs of $\mathcal{E}$ on the boundary of the "square" $\mathcal{L}_n(\theta, \mathcal{E})$. This can be done by repeating the arguments of the proof of Proposition 3.7 in [4]. We do this for each "square"; then we check inequalities (7) for the function constructed. For z_1, $z_2 \in \mathcal{E}$, let $J(z_1, z_2)$ be a subarc of $\partial\mathcal{L}_n(\theta, \mathcal{E})$ connecting these points. If $z \in J(z_1, z_2)$, then

$$|S_0 h(z)| = \left| \frac{z - z_2}{z_1 - z_2} h(z_1) + \frac{z - z_1}{z_2 - z_1} h(z_2) \right| \le |h(z_1)| + |h(z_2)| \le c\ ||h||(|z| + 1)^{\alpha-1}.$$

An estimate for the difference of values may be obtained in a similar way. If z, $z^* \in \partial\mathcal{L}(\theta, \mathcal{E})$, and $J(z_1, z_2)$, $J(z_1^*, z_2^*)$ are the corresponding intervals in the complement, then

$$|S_0h(z) - S_0h(z^*)| \le |S_0h(z) - h(z_2)| + |h(z_2) - h(z_1^*)| + |h(z_1^*) - S_0h(z^*)| \le$$
$$\left|\frac{z - z_2}{z_1 - z_2}(h(z_2) - h(z_1)\right| + c||h||\frac{|z_2 - z_1^*|}{|z_2| + |z_1^*| + 1} + |\frac{z - z_1^*}{z_2^* - z_1^*}|h(z_1^*) - h(z_2^*)| \le$$
$$c\ ||h||\frac{|z - z^*|^\alpha}{|z| + |z^*| + 1}.$$

Here we have used the estimate proved in Proposition 2.1. Next, as in the proof of Proposition 3.7 (see [4]), on each of the contours $\partial\mathcal{L}_n(\theta, \mathcal{E})$ we choose a big interval of the complement. Such an interval does exist due to the porosity condition. Then we construct a correcting perturbation that does not affect the function Sh on the set $\mathcal{E}$ but allows us to gain identity (c) of the proposition by choosing a coefficient appropriately. The value of this coefficient is equivalent to

$$\left|\int_{\partial\mathcal{L}_n(\theta,\mathcal{E})} \frac{S_0h(z)}{g^\varepsilon(z)}dz\right| \le c\ ||h||\frac{1}{|n| + 1}$$

This estimate is based on an inequality for the function Sh obtained in the proof of Proposition 3.7. We get an extension of h to $\partial\mathcal{L}(\theta, \mathcal{E})$ that satisfies all requirements of the proposition. We define Sh to be zero off the strip $\{-1 < \mathrm{Im} z < 2\}$ and extend this function to the entire complex plane so as to obtain a function of class $\Lambda^{\alpha,0}(\mathbf{C})$. This is possible due to the theorem of N. A. Lebedev [5], and condition (b) will survive. □

The next statement, which is fundamental for many further estimates, was proved by E. M. Dyn′kin in [1].

Proposition 2.9. *Let J be an arc on the boundary of $\mathcal{L}(\theta, \mathcal{E})$, let $\mathcal{E} \cap J$ be a porous set, and let $\{l_n\}$ be the family of arcs complementary to $\mathcal{E}$ on $\partial\mathcal{L}(\theta, \mathcal{E})$. Then there exist positive constants β and C depending only on the constant occurring in the definition of porosity and such that*

$$\sum_{l_n \subset J} |l_n|^{1-\beta} \le C|J|^{1-\beta}.$$

Also, we need an estimate based on the preceding proposition.

Proposition 2.10. *Let an integer n, a real number γ, a point $z \in \mathcal{L}_n(\theta, \mathcal{E})$, and an arc $J \subset \partial\mathcal{L}(\theta, \mathcal{E})$ be fixed. Then for any σ with $\sigma < \beta$, where β is the constant occurring in Proposition 2.9, we have*

$$\int_{\partial\mathcal{L}_n(\theta,\mathcal{E})} \frac{|dw|}{|w - z|^\gamma \mathrm{dist}^\sigma(w, \mathcal{E})} \le c\ (\mathrm{dist}^{1-\gamma-\sigma}(z, \mathcal{E}) + |J|^{1-\gamma-\sigma}).$$

In conclusion, we establish an important auxiliary statement concerning estimates for the series of the form

$$S = \sum_{n \ne N_1, N_2} \frac{|n - N_1| + |n - N_2|}{(|n| + 1)|n - N_1|^2|n - N_2|^2}.$$

We put

$$l_0(n) = |n| + 1, \qquad l_1(n) = |N_1 - n|, \qquad l_2(n) = |N_2 - n|.$$

Consider the lower envelope of the functions l_0, l_1, l_2. First, we obtain a universal estimate for the parts of the sum (11) corresponding to the monotonicity intervals of this envelope.

Proposition 2.11. *Let $[a, b]$ be a monotonicity interval of the envelope mentioned above (a or b may by equal to infinity). Suppose also that one of the variables l_k equals 1 at one of the endpoints (let it be a) and that this variable does not exceed any of the remaining two variables on the entire interval $[a, b]$. Let A, B be the values of the remaining variables at a. Then*

$$S = \sum_{a<n\le b} \frac{|n - N_1| + |n - N_2|}{(|n| + 1)|n - N_1|^2 |n - N_2|^2} \le c\frac{A}{B}.$$

Proof. Without loss of generality we may assume that $a < b$. First we consider the case where $b = \infty$. Then all variables increase on the interval in question (their graphs are parallel). For getting estimates, it is only essential to know the location of l_0 relative to the remaining two variables, because we can interchange the latter. We analyze three possible cases. If l_0 is the smallest variable, then $a = 0$, and on the interval $0 < n < A$ one of the variables l_1, l_2 is equivalent to A and the other to B; on the interval $A < n \le B$ the smallest of l_1, l_2 is equivalent to l_0 and the remaining one to B; finally, on the semiaxis $B < n < \infty$ all variables are equivalent to l_0. Consequently,

$$S \le c\left(\sum_{0<n\le A} \frac{1}{nA^2B} + \sum_{A<n\le B} \frac{1}{n^3B} + \sum_{B<n<\infty} \frac{1}{n^4}\right) \le$$
$$c\left(\frac{1}{AB}\frac{\ln A}{A} + \frac{1}{A^2B} + \frac{1}{B^3}\right) \le \frac{c}{AB}. \tag{9}$$

If $l_1 \le l_0 \le l_2$ or $l_2 \le l_0 \le l_1$, then similar arguments involving equivalences and a shift of the smallest variable (so as to get summation starting with zero) allow us to obtain the estimate

$$S \le c\left(\sum_{0<k\le A} \frac{1}{Ak^2B} + \sum_{A<k\le B} \frac{1}{k^3B} + \sum_{B<k<\infty} \frac{1}{k^4}\right) \le c\left(\frac{1}{A^2B} + \frac{1}{A^2B} + \frac{1}{B^3}\right) \le \frac{c}{AB}. \tag{10}$$

Finally, if l_0 is the greatest variable, then, repeating the same arguments, we get an estimate like (10).

Now, let $b < \infty$; we show that the required estimate is a consequence of (9), (10). This is clear if all three variables simultaneously increase or decrease (then in the above estimates we can replace the infinite sums by suitable partial sums, which will only refine the estimate). Now, let L_0 be the smallest variable, and let the other variable L_1 be of a different character of monotonicity. Then it is easy to check that $2\min L_1(n) \geq \max L_1(n)$, whence $L_1(n) \asymp \min L_1(n)$. Now we can proceed as before, obtaining inequalities similar to (9), (10). □

Proposition 2.12. *Let N_1, N_2 be distinct integers; then*

$$S = \sum_{n \neq N_1, N_2} \frac{|n - N_1| + |n - N_2|}{(|n| + 1)|n - N_1|^2 |n - N_2|^2} \leq \frac{c}{|N_1| + |N_2|}.$$

Proof. In order to use Proposition 2.11, we must find the values of two variables at the point where the remaining one vanish, i.e., we must evaluate the quantities A and B. It suffices to list all these values:

$$|N_1|, \quad |N_2|, \quad |N_1| - |N_2|, \quad |N_1| + |N_2|.$$

For definiteness, suppose that $|N_1| < |N_2|$. Proposition 2.11 says that for any combination of such A and B, on each of the intervals where the lower envelope of the variables l_k is monotonic, the sum S is dominated by $|N_1|^{-1}(|N_2| - |N_1|)^{-1}$. It remains to observe that if $|N_2| - |N_1| < |N_2|/2$, then $|N_1| \asymp |N_2|$, otherwise $|N_2| - |N_1| \asymp |N_2|$. Hence, the following estimate is always true:

$$S(N_1, N_2) \leq \frac{c}{|N_1| + |N_2|}.$$ □

3. The necessity of the conditions of Theorem 1.1

Proposition 3.1. *Let $\mathcal{E} \subset \partial\mathcal{L}$ be an interpolation set for the couple of spaces $\Lambda^\alpha_a(\mathcal{L})$, $\Lambda^\alpha(\mathcal{E})$. Then $\mathcal{E}$ is sparse.*

Proof. Let $z_0 \in \mathcal{E} \cap \mathcal{L}$. Let K be a square centered at z_0 with one side lying on the boundary of the strip. Since any function of class $\Lambda^\alpha(\mathcal{E} \cap K)$ can be extended to the entire set $\mathcal{E}$ so as to get a function of class $\Lambda^\alpha(\mathcal{E})$ (see [5]), the set $\mathcal{E} \cap K$ is an interpolation set for the corresponding couple of spaces on the square, but the square is a Lavrent′ev domain (see (6)), and for such domains any interpolation set is known to be sparse (see [2]). □

Proposition 3.2. *Let $\mathcal{E} \subset \partial\mathcal{L}$ be an interpolation set for the couple $\Lambda^\alpha_a(\mathcal{L})$, $\Lambda^\alpha(\mathcal{E})$. Then $\mathcal{E}$ is a locally porous subset of the strip.*

Proof. We fix a segment $I \subset \partial\mathcal{L}$, $|I| < 1$. We must check condition (2). Arguing as in the preceding proposition, we can reduce the proof of (2) to the corresponding part of the proof of Theorem 1 in [2]. □

4. Construction of the extension operator

The sufficiency of the conditions of Theorem 1.1 will be proved if we construct an extension operator from the space $\Lambda^\alpha(\mathcal{E})$ to the space $\Lambda^\alpha_a(\mathcal{L})$. First, we construct such an extension operator in spaces of functions vanishing at infinity. Let $\alpha < \varepsilon <$

β, where β is the constant occurring in Proposition 2.9. We define an extension of a function h belonging to the space $\Lambda^{\alpha,0}(\mathcal{E})$ (see definition (3)) by the formula

$$H(z) = \frac{g^{\varepsilon}(z)}{2\pi i} \int_{\partial\mathcal{L}(\theta,\mathcal{E})} \frac{Sh(w)}{(w-z)} \frac{dw}{g^{\varepsilon}(w)}. \tag{11}$$

(The definition of the domain $\mathcal{L}(\theta,\mathcal{E})$ was given in (4).) Below we shall repeatedly use decomposition of such domains into the "squares" $\mathcal{L}(\theta,\mathcal{E})_n = \{z \in \mathcal{L}(\theta,\mathcal{E}) : n \le \mathrm{Re}(z) < n+1\}$.

Proposition 4.1. *If $\mathcal{E}$ satisfies condition 2 of Theorem 1.1, and $\alpha < \varepsilon < \beta$, then the integral (11) converges, and*

$$|H(z)| \le c\,||h||\,(|z|+1)^{\alpha-1}.$$

Proof. Using estimate (7) for Sh and the estimate for g proved in Proposition 2.5, we get the inequality

$$\left|\int_{\partial\mathcal{L}(\theta_2,\mathcal{E})} \frac{Sh(w)}{(w-z)} \frac{dw}{g^{\varepsilon}(w)}\right| \le c||h|| \int_{\partial\mathcal{L}(\theta_2,\mathcal{E})} \frac{1}{(|w|+1)^{1-\alpha}} \frac{1}{|w-z|} \frac{|dw|}{\mathrm{dist}^{\varepsilon}(w,\mathcal{E})}.$$

Let $z \in \mathcal{L}_N$ and $w \in \mathcal{L}_n$; then $|w|+1 \asymp |n|+1$, and for $n \ne N$ we have $|w-z| \asymp |n-N|$. To avoid unessential technical details, we assume that $N \ne 0$. We split the integral along the boundary into the integrals corresponding to the "squares". We have

$$\left|\int_{\partial\mathcal{L}(\theta_2,\mathcal{E})} \frac{Sh(w)}{(w-z)} \frac{dw}{g^{\varepsilon}(w)}\right| \le c||h|| \Bigg(\sum_{n\ne N} \frac{1}{|n|^{1-\alpha}|n-N|} \int_{\partial\mathcal{L}_n(\theta_2,\mathcal{E})} \frac{|dw|}{\mathrm{dist}^{\varepsilon}(w,\mathcal{E})} +$$

$$+ \frac{1}{|N|^{1-\alpha}} \int\limits_{\partial\mathcal{L}_N(\theta_2,\mathcal{E})} \frac{|dw|}{|w-z|\mathrm{dist}^{\varepsilon}(w,\mathcal{E})} \Bigg).$$

Applying Proposition 2.10 and choosing suitable values of the parameters, we get

$$\left|\int_{\partial\mathcal{L}(\theta_2,\mathcal{E})} \frac{Sh(w)}{(w+i)(w-z)} \frac{dw}{g^{\varepsilon}(w)}\right| \le$$

$$c\,||h||\left(\sum_{n\ne N} \frac{1}{|n|^{1-\alpha}|n-N|} + |N|^{1-\alpha} + |N|^{1-\alpha}\mathrm{dist}^{-\varepsilon}(z,\mathcal{E})\right) \le$$

$$c\,||h||\left(\frac{1}{N^{1-\alpha}} + \frac{\ln N}{N} + |N|^{1-\alpha}\mathrm{dist}^{-\varepsilon}(z,\mathcal{E})\right).$$

Hence, the integral converges for every fixed z; taking the assumption $|z| \asymp |N|$ into account, we obtain the desired estimate for the function H. □

Remark 4.2. *The convergence of the integral in (11) implies immediately that the function (11) solves the interpolation problem, i.e., $H(z) = h(z)$ for any $z \in \mathcal{E}$.*

Now we show that H satisfies the Hölder condition in the strip $\mathcal{L}$. We use the following simple estimate (see [6]).

Proposition 4.3. *If a function H is analytic in $\mathcal{L}$, and for every point $z \in \mathcal{L}$ and for the point $z_1 \in \mathcal{E}$ nearest to z we have*

$$|F(z) - F(z_1)| \leq c \operatorname{dist}^{\alpha}(z, \partial\mathcal{L}(\theta, \mathcal{E})),$$

then H satisfies the Hölder condition in $\mathcal{L}$.

Proposition 4.4. *If $\mathcal{E}$ satisfies condition 2 of Theorem 1.1, and if $\alpha < \varepsilon < \beta$, where β is as in Proposition 3.9, then the function H satisfies the Hölder condition in the strip $\mathcal{L}$, and, moreover,*

$$|H(z) - h(z_1)| \leq c \, ||h|| \frac{\operatorname{dist}^{\alpha}(z, \partial\mathcal{L}(\theta, \mathcal{E}))}{|z| + 1}, \tag{12}$$

where $z_1 \in \mathcal{E}$, $|z - z_1| = \operatorname{dist}(z, \mathcal{E})$.

Proof. By Proposition 5.3, it suffices to prove inequality (12). Let N be the index of the domain $\mathcal{L}_N$ that contains the point z; we rewrite the integral along the contour $\partial\mathcal{L}$ as the sum of the integrals along the contours $\partial\mathcal{L}_n$ (the supplementary integrals along the inner parts of the boundary arcs cancel each other). This yields

$$|H(z) - h(z_1)| =$$

$$\left| \frac{g^{\varepsilon}(z)}{2\pi} \sum_{n \neq N} \int_{\partial\mathcal{L}_n(\theta,\mathcal{E})} \frac{Sh(w)}{w - z} \frac{dw}{g^{\varepsilon}(w)} + \left(\int_{\partial\mathcal{L}_N(\theta,\mathcal{E})} \frac{Sh(w)}{w - z} \frac{dw}{g^{\varepsilon}(w)} - h(z_1) \right) \right| \leq$$

$$\left| \frac{g^{\varepsilon}(z)}{2\pi} \sum_{n \neq N} \int_{\partial\mathcal{L}_n(\theta,\mathcal{E})} \frac{Sh(w) - Sh(w_n)}{w - z} \frac{dw}{g^{\varepsilon}(w)} \right| +$$

$$\left| \frac{g^{\varepsilon}(z)}{2\pi} \int_{\partial\mathcal{L}_N(\theta,\mathcal{E})} \frac{Sh(w) - Sh(z_1)}{w - z} \frac{dw}{g^{\varepsilon}(w)} \right| =$$

$$I_1 + I_2. \tag{13}$$

Here the w_n are arbitrary points on the contours $\partial\mathcal{L}_n$. The replacement of $Sh(w)$ by $Sh(w) - Sh(w_n)$ does not change the integrals in the sum I_1, because the function $Sh(w_n) \cdot (w - z)^{-1} g^{-\varepsilon}(w)$ is analytic inside the contour of integration for $n \neq N$. As for I_2, we have used the fact that $z \in \mathcal{L}_N$, so that the Cauchy formula implies the identity $h(z_1) = g^{\varepsilon}(z)/(2\pi) \int_{\partial\mathcal{L}_N(\theta,\mathcal{E})} Sh(z_1)(w - z)^{-1} g^{-\varepsilon}(w) dw$. For estimating I_1, we insert the auxiliary term $1/(w_n - z)$, which is possible because

$$\int_{\partial\mathcal{L}_n(\theta,\mathcal{E})} (Sh(w) - Sh(w_n)) \frac{1}{w - z} \frac{dw}{g^{\varepsilon}(w)} =$$

$$\int_{\partial\mathcal{L}_n(\theta,\mathcal{E})} (Sh(w) - Sh(w_n)) \left(\frac{1}{w - z} - \frac{1}{w_n - z} \right) \frac{dw}{g^{\varepsilon}(w)}. \tag{14}$$

Relation (14) follows from formula (8), which ensures that

$$\int_{\partial\mathcal{L}_n(\theta,\mathcal{E})} \frac{1}{w_n - z} \frac{Sh(w) \, dw}{g^{\varepsilon}(w)} = 0,$$

and from the analyticity of the function $h_1(w) = h(w_n)(w_n - z)^{-1}g^{-\varepsilon}(w)$ inside the contour $\partial\mathcal{L}_n(\theta, \mathcal{E})$. Now we use (7), obtaining

$$|Sh(w) - Sh(w_n)| \le c\,||h||\,\frac{|w - w_n|^\alpha}{|w| + 1)}. \tag{15}$$

We finish the estimation of I_1 by performing the transformation (14) in each summand, using estimates (15), and taking into account the fact that for $w \in \partial\mathcal{L}_n(\theta, \mathcal{E})$ we have the equivalences $|w| \asymp n$, $|w - z| \asymp |n - N|$. This yields

$$I_1 \le c\,||h||\,|g^\varepsilon(z)| \sum_{n \ne N} \int_{\partial\mathcal{L}_n(\theta,\mathcal{E})} \frac{|w - w_n|^\alpha}{(|w| + 1)} \frac{|w - w_n|}{|w - z||w_n - z|} \frac{|dw|}{|g^\varepsilon(w)|} \le$$

$$c\,||h||\,|z - z_1|^\varepsilon \sum_{n \ne N} \frac{1}{|n||n - N|^2} \int_{\partial\mathcal{L}_n(\theta,\mathcal{E})} \frac{|dw|}{\operatorname{dist}^\varepsilon(w, \mathcal{E})}$$

$$\le c\,||h||\,\frac{1}{|N| + 1} \le c\,||h|| \frac{|z - z_1|^\varepsilon}{|z| + 1}$$

(we have used the inequality of Proposition 2.10). Now, the condition $\alpha < \varepsilon$ implies that I_1 satisfies the required estimate. Turning to the second summand in (13), we have

$$I_2 \le c\,||h||\,|z - z_1|^\varepsilon \int_{\partial\mathcal{L}_N(\theta,\mathcal{E})} \frac{|Sh(w) - Sh(z_1)|}{|w - z|} \frac{|dw|}{\operatorname{dist}^\varepsilon(w, \mathcal{E})} \le$$

$$c\,||h||\,|z - z_1|^\varepsilon \int_{\partial\mathcal{L}_N(\theta,\mathcal{E})} \frac{|w - z_1|^\alpha |dw|}{(|w| + 1)|w - z|\operatorname{dist}^\varepsilon(w, \mathcal{E})} \le c\,||h||\,\frac{|z - z_1|^\alpha}{|z| + 1}.$$

Here the estimate for the integral is based on Proposition 2.10 and can be obtained as in [4]. The inequalities established and Proposition 4.3 imply that $H \in \Lambda^\alpha_a(\mathcal{L})$. □

4.1. End of the proof of Theorem 1.1 in the case where $\alpha < \beta$

It remains to check the sufficiency of the conditions of the theorem. This can easily be done with the use of Proposition 4.4. We recall that this proposition is valid only for sufficiently small α. Let f be an arbitrary function of class $\Lambda^\alpha(\mathcal{E})$. Putting $h(z) = f(z)(z + i)^{-1}$, we construct an extension Sh of h to the entire complex plane in accordance with Proposition 2.8. Consider the function H defined by (11) and put $F(z) = (z + i)H(z)$. It is clear that $F(z) = f(z)$ for $z \in \mathcal{E}$. We check that F satisfies the Hölder condition by using the inequality proved in Proposition 4.4 (precisely here we need the assumption $\alpha < \beta$). We have

$$|F(z_1) - F(z_2)| = |(z_1 + 1)(H(z_1) - H(z_2)) + (z_1 - z_2)H(z_1)| \le$$

$$c||f|| \left(\frac{(|z_1| + 1)|z_1 - z_2|^\alpha}{|z_1| + |z_2| + 1} + |z_1 - z_2|^\alpha \frac{|z_1 - z_2|^{1-\alpha}|z_2|^\alpha}{|z_2| + 1} \right) \le c\,||f||\,|z_1 - z_2|^\alpha,$$

which proves the theorem for $\alpha < \beta$.

5. Construction of an auxiliary operator that reduces smoothness

As in [4], we construct an operator that reduces the smoothness of the initial function; then we use this operator to get the desired interpolation. Let $f \in \Lambda^\alpha(\mathcal{E})$. As before, for us it is more convenient to deal with the function $h(z) = f(z)(z+i)^{-1}$ and with its extension Sh constructed as in Proposition 2.8. We introduce an operator reducing smoothness by the formula

$$Qh(z) = \int_{\partial\mathcal{L}(\theta,\mathcal{E})} \frac{Sh(w) - Sh(z)}{w - z} \frac{dw}{g^\sigma(w)}, \qquad z \in \mathcal{E}. \tag{16}$$

Since the operator Q is defined via the extension of h to the entire complex plane, we can write

$$Qh(z) = \sum_n Q_n h(z), \qquad \text{where} \qquad Q_n h(z) = \int_{\partial\mathcal{L}_n(\theta,\mathcal{E})} \frac{Sh(w) - Sh(z)}{w - z} \frac{dw}{g^\sigma(w)}.$$

Proposition 2.2 shows that the above integral converges absolutely, so that the integrals along the new inner arcs cancel each other. We check the Hölder condition for a separate summand, considering various locations of the points involved with respect to the contour of integration.

Proposition 5.1 (two points lie on the contour). *Let $z_1, z_2 \in \partial\mathcal{L}_N(\theta,\mathcal{E})$, and let $\sigma < \alpha < \beta$, where β is as in Proposition 2.9. Then*

$$|Q_N h(z_1) - Q_N h(z_2)| \le c\,\|h\| \frac{|z_1 - z_2|^{\alpha-\sigma}}{|z_1| + |z_2| + 1}.$$

Proof. Consider the auxiliary arc $J = \partial\mathcal{L}_N(\theta,\mathcal{E}) \cap \{z : |z - (z_1+z_2)/2| < 2|z_1 - z_2|\}$ and its complement $\Gamma = \partial\mathcal{L}_N(\theta,\mathcal{E}) \setminus J$. We begin with splitting the integral into convenient fragments as follows:

$$|Q_N h(z_1) - Q_N h(z_2)| \le$$
$$\left|\int_J \frac{Sh(w) - Sh(z_1)}{w - z_1} \frac{dw}{g^\sigma(w)}\right| + \left|\int_J \frac{Sh(w) - Sh(z_2)}{w - z_2} \frac{dw}{g^\sigma(w)}\right| +$$
$$\left|\int_\Gamma \left(\frac{Sh(w) - Sh(z_1)}{w - z_1} - \frac{Sh(w) - Sh(z_2)}{w - z_2}\right) \frac{dw}{g^\sigma(w)}\right|. \tag{17}$$

The first two summands in (17) admit similar estimates due to inequality (7) and Proposition 2.10:

$$\left|\int_J \frac{Sh(w) - Sh(z_1)}{w - z_1} \frac{dw}{g^\sigma(w)}\right| \le$$
$$c\,\|f\| \int_J \frac{|dw|}{(|w| + |z_1| + 1)|w - z_1|^{1-\alpha} \mathrm{dist}^\sigma(w,\mathcal{E})} \le c\,\|f\| \frac{|z_1 - z_2|^{\alpha-\sigma}}{(|z_1| + 1)}.$$

For estimating the third summand we reshape the integrand to the form

$$\frac{Sh(w) - Sh(z_1)}{w - z_1} - \frac{Sh(w) - Sh(z_2)}{w - z_2} = \frac{Sh(z_2) - Sh(z_1)}{w - z_1} + (z_1 - z_2)\frac{Sh(w) - Sh(z_2)}{(w - z_1)(w - z_2)}. \tag{18}$$

After that, we act as in the first case:

$$\left|\int_\Gamma \left(\frac{Sh(w)-Sh(z_1)}{w-z_1} - \frac{Sh(w)-Sh(z_2)}{w-z_2}\right)\frac{dw}{g^\sigma(w)}\right| \le$$
$$c\,||f||\left(\frac{|z_1-z_2|^\alpha}{|z_2|+|z_1|+1}\int_\Gamma \frac{|dw|}{|w-z_1|\operatorname{dist}^\sigma(w,\mathcal{E})}\right)+$$
$$c\,||f||\left(|z_1-z_2|\int_\Gamma \frac{|w-z_2|^{\alpha-1}|dw|}{(|w|+|z_2|+1)|w-z_1|\operatorname{dist}^\sigma(w,\mathcal{E})}\right).$$

Now we apply Proposition 2.10, taking into account the fact that for the arc Γ we have the equivalence $|w-z_2| \asymp |w-z_1| \asymp \operatorname{dist}(z_1,\Gamma)$. Moreover, we may assume that $\operatorname{diam}\Gamma \ge |z_1-z_2|$, because otherwise we can put $J = \partial\mathcal{L}_N(\theta,\mathcal{E})$. As a result, we find

$$\left|\int_\Gamma \left(\frac{Sh(w)-Sh(z_1)}{w-z_1} - \frac{Sh(w)-Sh(z_2)}{w-z_2}\right)\frac{dw}{g^\sigma(w)}\right| \le$$
$$c\,||f||\left(\frac{|z_1-z_2|^\alpha}{|z_2|+|z_1|+1}(\operatorname{dist}^{-\sigma}(z_1,\Gamma)+\operatorname{diam}^{-\sigma}\Gamma)\right)+$$
$$c\,||f||\left(\frac{|z_1-z_2|}{|z_2|+|z_1|+1}(\operatorname{dist}^{\alpha-\sigma-1}(z_1,\Gamma)+\operatorname{diam}^{\alpha-\sigma-1}\Gamma)\right) \le$$
$$c\,||f|||z_1-z_2|^{\alpha-\sigma}.$$

Combining the estimates obtained for the summands, we arrive at inequality (17),

$$|Q_N h(z_1) - Q_N h(z_2)| \le c\,||h|||z_1-z_2|^{\alpha-\sigma}.$$

□

Remark 5.2. *Extending the squares $\mathcal{L}_n$ appropriately, in what follows we shall assume that*

$$|z-w| \asymp |N-n| \qquad \textit{for} \qquad z\in\mathcal{L}_N,\ w\in\mathcal{L}_n,\ N\ne n.$$

Proposition 5.3 (one point lies on the contour). *Let $z_k \in \partial\mathcal{L}_{N_k}(\theta,\mathcal{E})$, $k=1,2$, $N_1\ne N_2$, and let $\sigma<\alpha<\beta$, where β is as in Proposition 2.9. Then*

$$|Q_{N_2}h(z_1) - Q_{N_2}h(z_2)| \le c\,||f||\frac{|z_1-z_2|^{\alpha-\sigma}}{|z_1|+|z_2|+1}.$$

Proof. Modifying, if necessary, our partition of the strip into squares, we may assume that $|z_1-z_2|>1$. Hence, it suffices to prove the inequality

$$|Q_{N_2}h(z_1) - Q_{N_2}h(z_2)| \le c\,||f||\frac{1}{|z_1|+|z_2|+1}.$$

We show that such an inequality is fulfilled for any term of the difference in question, i.e., we prove that for any N and z we have

$$|Q_N h(z)| \le \frac{c\,||f||}{|N|+|z|+1}.$$

Arguing as in proof of Proposition 6.1, we obtain the first of the following inequalities:

$$|Q_N h(z)| \leq c\,||h|| \int_{\partial\mathcal{L}_N} \frac{|w-z|^{\alpha-1}|dw|}{(|w|+|z|+1)\,\mathrm{dist}^{\sigma}(w,\partial\mathcal{L}_N)} \leq$$

$$\frac{c\,||h||}{|N|+|z|+1} \int_{\partial\mathcal{L}_N} \frac{|w-z|^{\alpha-1}|dw|}{\mathrm{dist}^{\sigma}(w,\partial\mathcal{L}_N)} \leq \frac{c\,||h||}{|N|+|z|+1}.$$

The second inequality is based on the equivalence $|w| \asymp |N|$, on the estimate written in Proposition 2.10, on the equivalence $|\partial\mathcal{L}_N| \asymp 1$, and on the fact that if $\alpha < 1$, then the estimate mentioned above is valid even for $z \in \partial\mathcal{L}_N$. □

Proposition 5.4 (both points lie off the contour). *Let $z_k \in \partial\mathcal{L}_{N_k}(\theta,\mathcal{E})$, $k = 1,2$, and let $\sigma < \alpha < \beta$, where β is as in Proposition 2.9. Suppose that $n \neq N_1, N_2$. Then*

$$|Q_n h(z_1) - Q_n h(z_2)| \leq c\,||f||\,|z_1 - z_2| \frac{|n-N_1| + |n-N_2|}{(|n|+1)|n-N_1|^2|n-N_2|^2}.$$

Proof. Using (18), we rewrite the difference to be estimated as follows:

$$Q_n h(z_1) - Q_n h(z_2) =$$

$$(Sh(z_2) - Sh(z_1)) \int_{\partial\mathcal{L}_n(\theta,\mathcal{E})} \frac{dw}{(w-z_1)g^{\sigma}(w)} +$$

$$(z_1 - z_2) \int_{\partial\mathcal{L}_n(\theta,\mathcal{E})} \frac{(Sh(w) - Sh(z_2))dw}{(w-z_1)(w-z_2)g^{\sigma}(w)}.$$

By our assumptions, the point z_1 lies off the contour of integration; hence, the first integral vanishes. The same argument and identity (8) allow us to continue reshaping the second summand as follows:

$$Q_n h(z_1) - Q_n h(z_2) = (z_1 - z_2)\cdot$$

$$\int_{\partial\mathcal{L}_n} (Sh(w) - Sh(w_n)) \left(\frac{1}{(w-z_1)(w-z_2)} - \frac{1}{(w_n-z_1)(w_n-z_2)} \right) \frac{dw}{g^{\sigma}(w)} =$$

$$(z_1 - z_2) \int_{\partial\mathcal{L}_n} (Sh(w) - Sh(w_n)) \frac{(w_n - w)(w_n + w - z_1 - z_2)}{(w-z_1)(w-z_2)(w_n-z_1)(w_n-z_2)} \frac{dw}{g^{\sigma}(w)}.$$

Here the points $w_n \in \partial\mathcal{L}_n$ are chosen in such a way that $\mathrm{dist}(w_n, \partial\mathcal{L}_n) \asymp \mathrm{diam}\,\partial\mathcal{L}_n$ ($\asymp 1$). Under the assumptions of the proposition we have $|z_1 - z_2| \asymp |N_1 - N_2|$, $|w - z_k| \asymp |w_n - z_k| \asymp |N_k - n|$, $|w - w_n| \leq c$, and by Proposition 2.5 we have $|g(w)| \asymp \mathrm{dist}(w, \mathcal{E})$. This yields

$$|Q_n h(z_1) - Q_n h(z_2)| \leq$$

$$||h||\,|z_1 - z_2| \int_{\partial\mathcal{L}_n} \frac{|w-w_n|^{\alpha}}{|w|+|w_n|+1} \frac{|w-w_n|(|n-n_1|+|n-N_2|)}{|n-N_1|^2|n-N_2|^2} \frac{|dw|}{\mathrm{dist}^{\sigma}(w,\mathcal{E})} \leq$$

$$c\,||h||\,|z_1 - z_2| \frac{1}{|n|+1} \frac{|n-n_1|+|n-N_2|}{|n-N_1|^2|n-N_2|^2} \mathrm{diam}^{1-\sigma}\partial\mathcal{L}_n \leq$$

$$c\,||h||\,|z_1 - z_2| \frac{|n-N_1|+|n-N_2|}{(|n|+1)|n-N_1|^2|n-N_2|^2}.$$

□

Proposition 5.5. *Let $\mathcal{E}$ be a closed subset of the boundary of the strip $\mathcal{L}$ satisfying the assumptions of Theorem* 1.1, *and let $\sigma < \alpha$, $\sigma < \beta$, where β is as in Proposition* 2.9. *Then the operator Q is a continuous mapping from the space $\Lambda^{\alpha,0}(\mathcal{E})$ to the space $\Lambda^{\alpha-\sigma,0}(\mathcal{E})$.*

Proof. We estimate the difference $Qh(z_1) - Qh(z_2)$ for arbitrary points $z_1,\ z_2 \in \mathcal{E}$. Let $z_k \in \partial\mathcal{L}_{N_k},\ k = 1, 2$; we have

$$|Qh(z_1) - Qh(z_2)| \leq \sum_n |Q_n h(z_1) - Q_n h(z_2)|.$$

The summands with indices $n = N_1, N_2$ were estimated in Propositions 5.1 and 5.3, while the other summands can be estimated with the help of Proposition 6.4. The sum itself can be estimated by Proposition 2.12. Since $|z_k| + 1 \asymp |N_k| + 1$, we obtain the required estimate for the difference. □

Now, we introduce the following new version of the operator of interpolation:

$$Rh(z) = \frac{g^\sigma(z)}{2\pi i}\left(\int_{\partial\mathcal{L}(\theta_1,\mathcal{E})} \frac{Sh(w)dw}{(w-z)g^\sigma(w)} - H(z)\right).$$

Proposition 5.6. *Let $\theta_1 < \theta$, and let $\sigma < \alpha$, $\sigma + \alpha < 1$, $\sigma < \beta$, where β is as in Proposition* 3.9. *If $\mathcal{E}$ is an interpolation set for the couple $(\Lambda^\alpha_{\mathrm{a}}(\mathcal{L}(\theta,\mathcal{E})), \Lambda^\alpha(\mathcal{E}))$, then so is it for the couple $(\Lambda^{\alpha+\sigma}_{\mathrm{a}}(\mathcal{L}(\theta_1,\mathcal{E})), \Lambda^{\alpha+\sigma}(\mathcal{E}))$.*

Proof. Fixing a function $f \in \Lambda^{\alpha+\sigma}(\mathcal{E})$, we put $h(z) = f(z)(z+i)^{-1}$. By Proposition 5.5, we have $Qh \in \Lambda^{\alpha,0}(\mathcal{E})$. The assumptions of Proposition 5.6 ensure the existence of a function $H \in \Lambda^\alpha_{\mathrm{a}}(\mathcal{L}(\theta,\mathcal{E}))$ such that $H(z) = Qh(z)$ for any $z \in \mathcal{E}$. Proposition 4.4 implies the estimate

$$|H(z) - Qh(z_1)| \leq c\,||h||\,\frac{|z-z_1|^\alpha}{|z|+1}$$

for $z \in \mathcal{L}(\theta_1,\mathcal{E}),\ z_1 \in \mathcal{E},\ |z - z_1| = \operatorname{dist}(z,\mathcal{E})$. To finish the proof it suffices to show that for $z \in \mathcal{L}(\theta_1,\mathcal{E})$, $z_1 \in \mathcal{E}$, $|z - z_1| = \operatorname{dist}(z,\mathcal{E})$ we have

$$|Rh(z) - h(z_1)| \leq c\,\frac{|z-z_1|^{\alpha+\sigma}}{|z|+1}, \tag{19}$$

because Proposition 4.3 will then imply that $Rh \in \Lambda^\alpha_{\mathrm{a}}(\mathcal{L}(\theta,\mathcal{E}))$, and, on the other hand, the function Rh coincides with H on the set $\mathcal{E}$. Then, we can return to the function f arguing as in Subsection 4.1.

From the difference in question we extract the operator Q defined by (16); this operator will be estimated with the help of Proposition 4.4 and estimate (6). The remaining term will be handled in the spirit of Propositions 5.1–5.4. We have

$$Rh(z) - h(z_1) = \frac{g^\sigma(z)}{2\pi i}\left(\int_{\partial\mathcal{L}(\theta_1,\mathcal{E})} \frac{Sh(w) - h(z_1)}{(w-z)g^\sigma(w)}dw - H(z)\right)$$

$$
\begin{aligned}
&= \frac{g^\sigma(z)}{2\pi i}\left(\int_{\partial\mathcal{L}(\theta_1,\mathcal{E})} \frac{Sh(w)-h(z_1)}{(w-z_1)g^\sigma(w)}dw - H(z)\right) \\
&+ \frac{g^\sigma(z)}{2\pi i}\int_{\partial\mathcal{L}(\theta_1,\mathcal{E})} \frac{(z_1-z)(Sh(w)-h(z_1))}{(w-z)(w-z_1)g^\sigma(w)}dw = I_1 + I_2.
\end{aligned} \tag{20}
$$

Here the first equality follows from the Cauchy formula

$$
g^{-\sigma}(z) = (2\pi i)^{-1}\int_{\partial\mathcal{L}(\theta_1,\mathcal{E})} \frac{dw}{(w-z)g^\sigma(w)}.
$$

In order to estimate I_1, we note that the first term in I_1 is equal to $Qh(z_1)$, because $Qh(z_1) = H(z_1)$ by construction. Since $H \in \Lambda_a^\alpha(\mathcal{L}(\theta_1,\mathcal{E}))$, we have

$$
|I_1| = |\frac{g^\sigma(z)}{2\pi i}||H(z_1)-H(z)| \le c\,||h||\,\frac{|g^\sigma(z)||z-z_1|^\alpha}{|z|+1} \le c\,||h||\,\frac{|z-z_1|^{\alpha+\sigma}}{|z|+1}. \tag{21}
$$

For the second summand in (20) we can write

$$
|I_2| \le c\,|z-z_1|^\sigma \int_{\partial\mathcal{L}(\theta_1,\mathcal{E})} \frac{|z-z_1||Sh(w)-h(z_1)|}{|w-z||w-z_1||g^\sigma(w)|}|dw| \le
$$

$$
c\,||h||\,|z-z_1|^{1+\sigma}\sum_n \int_{\partial\mathcal{L}_n(\theta_1,\mathcal{E})} \frac{|w-z_1|^{\alpha+\sigma}|dw|}{(|w|+1)|w-z||w-z_1||g^\sigma(w)|}.
$$

Let $z \in \mathcal{L}_N(\theta_1,\mathcal{E})$. For the summands corresponding to $n \ne N$ we have $|w-z| \asymp |w-z_1| \asymp |N-n|$, $|w|+1 \asymp |n|+1$. Hence, it remains to estimate the integrals of the form $\int_{\partial\mathcal{L}_n(\theta_1,\mathcal{E})} \operatorname{dist}^{-\sigma(w)}(w,\mathcal{E})|dw|$. All these integrals are uniformly bounded by Proposition 2.10. The central term with $n = N$ needs some preliminary transformation. Let J be the arc on $\partial\mathcal{L}_N(\theta_1,\mathcal{E})$ centered at z_1 and such that $|J| = 2\operatorname{dist}(z,\mathcal{E})$. Now, for the central summand we can write

$$
\int_{\partial\mathcal{L}_N(\theta_1,\mathcal{E})} \frac{|w-z_1|^{\alpha+\sigma-1}|dw|}{(|N|+1)|w-z||w-z_1|\operatorname{dist}^\sigma(w,\mathcal{E})} \le
$$

$$
\int_J \frac{|w-z_1|^{\alpha+\sigma-1}|dw|}{(|N|+1)|z-z_1|\operatorname{dist}^\sigma(w,\mathcal{E})} + \int_{\partial\mathcal{L}_N(\theta_1,\mathcal{E})\setminus J} \frac{|w-z_1|^{\alpha+\sigma-2}|dw|}{(|N|+1)\operatorname{dist}^\sigma(w,\mathcal{E})}.
$$

Each of the above integrals is dominated, by Proposition 2.10, by the quantity $c\,|z-z_1|^{\alpha-1}(|z|+1)^{-1}$. Returning to I_2, we get

$$
|I_2| \le c\,||h||\,\frac{|z-z_1|^{\sigma+1}}{|z|+1}\left(|z-z_1|^{\alpha-1} + \sum_{n\ne N}|N-n|^{\alpha+\sigma-2}\right) \le c\,||h||\,\frac{|z-z_1|^{\alpha+\sigma}}{|z|+1}.
$$

Here we have used the condition $\alpha+\sigma < 1$ that ensures the convergence of the series. Combining the estimate obtained with (21), we get (19). □

5.1. Proof of the theorem in the case where $\mathcal{E} \subset \partial\mathcal{L}$

For any $\alpha \in (0,1)$ we can choose an integer M such that the number $\alpha_1 = \alpha - M\sigma$, with σ as in Proposition 5.6, satisfies the assumptions of Theorem 2.1; hence, for α_1 interpolation is possible. Consequently, by Proposition 6.6, it is also possible for the number $\alpha_1 + \sigma$. Continuing this procedure, we see that the number α also admits interpolation.

5.2. Proof of the theorem in the general case

We split the set $\mathcal{E}$ into two subsets: the set $\mathcal{E}_1$ of all interior points and the set $\mathcal{E}_2$ of all boundary points for our strip. In order to get interpolation on $\mathcal{E}_1$, we can repeat the arguments used in [2] in a similar situation.

6. Acknowledgment

The author thanks Prof. N. A. Shirokov and Prof. E. M. Dyn'kin for valuable discussions.

References

[1] Dyn'kin E. M., *Sets of free interpolation for Hölder classes,* Mat. Sb. **109** (1979), 107–128.

[2] Kotochigov A. M., Shirokov N. A., *Free interpolation for Hölder classes in Jordan domains*, Probl. Math. Analysis, SPbSU, SPb, **15** (1995), 109–138.

[3] Kotochigov A. M., Shirokov N. A., *Free interpolation for Hölder classes in quasiconformal domains*, Probl. Math. Analysis, SPbSU, SPb, **16** (1997), 145–166.

[4] Kotochigov A. M., Shirokov N. A., *Interpolation near a cusp point*, Algebra i Analiz, **10** (1998), 142–176.

[5] Lebedev N. A., *About extension of complex functions*, Vest. LGU, Ser. Mat. Mekh. Astronom. **4** (1963), 61–68.

[6] Shirokov N. A., *Free interpolation for the classes $C^A_{r<\omega}$*, Mat. Sb., **117** (1982), 337–358.

Department of Mathematics,
SPb Electrotechnical University,
Prof. Popova str. 5
Saint-Petersburg, 197376, Russia
E-mail address: kam@kam.sbp.su

Operator Theory:
Advances and Applications, Vol. 113

Scattering Problem with Physical Behavior of Scattering Matrix and Operator Relations

Pavel Kurasov and Boris Pavlov

Dedicated to the memory of S. A. Vinogradov

Abstract. Finite rank generalized singular perturbations of selfadjoint operators are studied. It is proved that the restriction of the arising operator relation to the subspace generated by the original Hilbert space is a selfadjoint operator (if the perturbation is singular). The relations with the corresponding scattering problem are investigated. In particular, the case where the scattering matrix has physical behavior at infinity is examined.

1. Introduction

It is known that Krein's formula for generalized resolvents of selfadjoint extensions of symmetric operators may produce not only the resolvent of an operator, but sometimes the resolvent of an operator relation, which corresponds to the singular choice of the unitary von Neumann parameter with an eigenvalue equal to one. The formal von Neumann construction gives an operator relation in this case, since the domain of it is not dense. This fact can be deduced from Krein's formula and the paper by M. Krasnoselskii [6]. See also the papers by S. Hassi, H. Langer, H. de Snoo, and B. Textorius [3, 5, 8] and by M. Malamud [9] where this question is discussed in more detail from the functional analytic point of view. The denominator of the additional term in Krein's formula for generalized resolvents grows linearly in this case. It is remarkable that precisely this property of Krein's formula is necessary for the scattering matrix to have physical behavior at infinity

$$S(k) \longrightarrow 1 \quad \text{as } k \to \infty$$

(see [11]). In this paper we show that the linear term in the asymptotics of the denominator can appear even if no operator relation occurs. Actually the operators (not operator relations) with exactly this property have been studied in a series of papers devoted to the investigation of Schrödinger operators with resonance interaction ("Inner structure") generated by a finite matrix ([1, 11, 12, 10]).

We study also the following question: Can operator relation occur if one considers generalizations of the resonance interactions involving symmetric operators with arbitrary finite deficiency indices? (Essentially, all operators considered earlier have deficiency indices (1,1).) It is proved that operator relations may occur, but are not interesting for applications, since the corresponding generalized resolvents just coincide with the generalized resolvents of the selfadjoint operators arising after restriction of the operator relations to the subspace of "physically realizable" elements, which belong to the linear hull of the elements generated by the "outer" space. Thus only the operator-part of the relation plays role in scattering processes.

The authors were supported by the Grant number 3368152 from the Marsden Fund of the Royal Society of New Zealand, by the system of Grants ESPRIT 21042, CTIAC, awarded by the Commission in Research and Development of European Community, and by the grant number 97-01-01149 of the Russian Foundation for Basic Research.

2. Residue of the resolvent at infinity

Let us consider first a bounded selfadjoint operator A acting in the Hilbert pace E. Assume that $N \subset E$, $\dim N = n$, is a subspace such that the angle between N and $\hat{N} \equiv \frac{A+iI}{A-iI} N$ is positive. This permits developing the construction of extensions, which is alternative to the classical von Neumann construction. This construction uses the *symmetric* Krein parameter for description of all selfadjoint extensions instead of the unitary parameter in von Neumann case. This alternative approach provides a straightforward proof of Krein's formula, [12]. We are going to explain below some details of this construction.

In what follows the subspaces $N, \hat{N}$ play roles of deficiency subspaces at the points $\lambda = i, -i$ respectively. It is clear that the Cayley transform of A transfers $E \ominus N$ into $E \ominus \hat{N}$ and N into $\hat{N}$. Redefining the Cayley transform on $N : U|_N \rightarrow U'|_N$ we get a new unitary operator U', which coincides with U on $E \ominus N$. If 1 is not an eigenvalue of it then the inverse Cayley transform $A' = i\frac{U'+1}{U'-1}$ exists and is a densely defined selfadjoint operator in E. If 1 is an eigenvalue, then the latter formula defines a selfadjoint operator relation. In both cases the corresponding resolvent is related to the resolvent of the corresponding unitary operator by the formula

$$R'_\lambda \equiv (A' - \lambda 1)^{-1} = \frac{1}{i-\lambda} \frac{U'-1}{U'-\zeta}, \ \zeta = \frac{\lambda + i}{\lambda - i},$$

and satisfies the Hilbert relation

$$R'_\lambda - R'_\mu = (\lambda - \mu) R'_\lambda R'_\mu.$$

In the singular case the residue of the resolvent at infinity is not equal to identity but the following equality holds

$$\frac{1}{2\pi i} \oint R'_\lambda d\lambda = -I + P_0,$$

where P_0 is the orthogonal projection onto the singular subspace $N_0 \subset N$ which corresponds to the singular eigenvalues $\zeta = 1$ of U' . Then the restriction of the relation A' to $E \ominus N_0$ is a usual selfadjoint operator.

The reduction of an operator relation to an operator can be done for finite rank perturbations of unbounded operators as well, with the only difference that 1 may be not an isolated eigenvalue of the Cayley transform U', but an embedded eigenvalue. Nevertheless the splitting of the corresponding singular subspace gives obviously a usual selfadjoint operator in the remaining part of the Hilbert space.

Consider now the case where the nonperturbed operator acts in an orthogonal sum $\mathcal{H}$ of two Hilbert spaces $\mathcal{H} = E \oplus H$ and is equal to the orthogonal sum of two operators $\mathcal{A} = A \oplus L$ acting in the spaces E and H respectively. We suppose that the first operator is bounded, the second one is unbounded. In the following section the role of A is played by a finite matrix, and the role of L is played by the Laplacian in $L_2(\mathbf{R}^3)$.

Let us assume that N_e and N_h are finite dimensional deficiency subspaces at the point $\lambda = i$ for the restricted operators A_0 and L_0 respectively. Then the sum $\mathcal{N} = N_e \oplus N_h$ is a deficiency subspace for the total restricted operator $\mathcal{A}_0 = A_0 \oplus L_0$ and $\mathrm{Dom}\,(\mathcal{A}_0) = \frac{I}{\mathcal{A}-iI}[\mathcal{H} \ominus \mathcal{N}]$. We assume that $\mathcal{N} \cap \mathrm{Dom}\,(L) = \varnothing$, so that the domain of the restricted operator L_0 is dense in H.

In what follows we need an expression for the singular subspace in terms of Krein's parameters. The Krein formula for the singular case has the same form as for the non-singular case (see [6], [10], [9]). Denoting by Γ any bounded selfadjoint operator in $\mathcal{N}$ (Krein's parameter) represented by the operator-matrix

$$\Gamma = \begin{pmatrix} \gamma_{ee} & \gamma_{eh} \\ \gamma_{he} & \gamma_{hh} \end{pmatrix},$$

we write this formula as

$$(\mathcal{A}_\Gamma - \lambda I)^{-1} = (\mathcal{A} - \lambda I)^{-1} + \frac{\mathcal{A} + iI}{\mathcal{A} - \lambda I} \frac{I}{\Gamma - \mathcal{Q}(\lambda)} P_{\mathcal{N}} \frac{\mathcal{A} - iI}{\mathcal{A} - \lambda I},$$

where

$$\mathcal{Q}(\lambda) = \begin{pmatrix} P_{N_e} \frac{I+\lambda A}{A-\lambda I} \mid_{N_e} & 0 \\ 0 & P_{N_h} \frac{I+\lambda L}{L-\lambda I} \mid_{N_h} \end{pmatrix} \equiv \begin{pmatrix} Q_A(\lambda) & 0 \\ 0 & Q_L(\lambda) \end{pmatrix}$$

and $P_{\mathcal{N}}$ is the orthogonal projection onto $\mathcal{N}$. Consider the case where the total operator $\mathcal{A}$ is bounded, i.e. where $\mathcal{A} = A$ and $\Gamma = \gamma_{ee}$. Then the residue of the resolvent at infinity can easily be calculated using the spectral representation of the finite dimensional operator $\Gamma + P_N A P_N$ acting in N with respect to the weighted scalar product

$$\langle\langle g, h \rangle\rangle \equiv \langle P_N (I + A^2) P_N g, h \rangle.$$

Lemma 2.1. *If the total original operator is bounded* ($\mathcal{A} = A$), *then the residue of the resolvent* $(\mathcal{A}_\Gamma - \lambda I)^{-1}$ *at infinity is equal to* $I - P_0$, *where*

$$P_0 * = \sum_{\alpha_s = 0} (A + iI) e_s \langle (A - iI) *, e_s \rangle,$$

and the e_s are the eigenvectors of the operator $\Gamma + P_N A|_N$ corresponding to the eigenvalue α_s and orthonormalized with respect to the weighted scalar product $\langle\!\langle \cdot,\cdot \rangle\!\rangle$.

Note that the sum in the latter formula is taken over all eigenvectors corresponding to the zero eigenvalue of the operator $\gamma_{ee} + P_{N_e} A|_{N_e} = \Gamma + P_{\mathcal{N}} A|_{\mathcal{N}}$.

Proof. Note that the additional term P_0 of the residue is determined by the second term in Krein's formula

$$P_0 = (A + iI) \lim_{\lambda\to\infty} \frac{1}{\lambda[\Gamma - P_N \frac{I+\lambda A}{A-\lambda I}|_N]} P_{N_e}(A - iI).$$

If the operator A is bounded, then the asymptotic representation for the inverse operator in the latter formula for large values of λ can be calculated via a perturbation procedure. Indeed, for a bounded operator A the denominator for large values of λ can be represented by the convergent series

$$\lambda[\Gamma + P_N A|_N] + P_N(1 + A^2)|_N + \sum_{l=1}^{\infty} \frac{P_N A^l (1 + A^2)|_N}{\lambda^l}.$$

The leading terms of this series can be inverted with the help of the spectral representation of $\Gamma + P_N A|_N$ in N with respect to the weighted scalar product $\langle\!\langle \cdot,\cdot \rangle\!\rangle$:

$$(\Gamma + P_N A) e_s = \alpha_s P_N (1 + A^2) e_s.$$

Assuming that the eigenvectors are orthonormalized $\langle\!\langle e_s, e_t \rangle\!\rangle = \delta_{st}$, we get for the inverse of the leading pair:

$$[\lambda\Gamma + \lambda P_N A|_N + P_N(1 + A^2)|_N]^{-1} f =$$
$$= \sum_s \frac{1}{1 + \lambda\alpha_s} e_s \langle f, e_s\rangle \equiv \mathcal{R}_0 f.$$

The term corresponding to the eigenvalue $\alpha_s = 0$ is constant, and the remaining terms tend to zero as λ goes to infinity. The operator $\mathcal{R}_0$ is bounded. Taking into account that the sum of the remaining terms of the series is small for large λ,

$$\sum_{l=1}^{\infty} \| \frac{P_N A^l (1 + A^2)|_N}{\lambda^l} \| = O(\frac{1}{\lambda}),$$

we conclude that

$$\frac{1}{\lambda[\Gamma - P_N \frac{I+\lambda A}{A-\lambda I}|_N]} = \mathcal{R}_0 [1 + O(\frac{1}{\lambda}) \mathcal{R}_0]^{-1},$$

and hence

$$
\begin{aligned}
P_0 f &= (A+iI)\lim_{\lambda\to\infty}\frac{1}{\lambda[\Gamma - P_N\frac{I+\lambda A}{A-\lambda I}|_N]}P_{N_e}(A-iI)f \\
&= \lim_{\lambda\to\infty}(A+iI)\mathcal{R}_0 P_{N_e}(A-iI)f \\
&= \sum_{\alpha_s=0}(A+iI)e_s\langle f,(A+iI)e_s\rangle .
\end{aligned}
$$

Notice that the dot product in the latter formula *is not weighted*, hence the operator P_0 is an orthogonal projection in E onto $(A+iI)N_0$. The Lemma is proved. □

In the case where the nontrivial unbounded component is present, the extensions could be parametrised by the boundary conditions imposed on the abstract boundary values $\xi_\pm$ of the elements (see [10]):

$$
u_{N_e} \sim \frac{A}{A-iI}\xi_+^e + \frac{1}{A-iI}\xi_-^e \in N_e + \hat{N}_e
$$

$$
h = h_0 + \frac{L}{L-iI}\xi_+^h + \frac{1}{L-iI}\xi_-^h \in H,
$$

$$
\begin{pmatrix} \xi_-^e \\ \xi_-^h \end{pmatrix} = \begin{pmatrix} \gamma_{ee} & \gamma_{eh} \\ \gamma_{he} & \gamma_{hh} \end{pmatrix}\begin{pmatrix} \xi_+^e \\ \xi_+^h \end{pmatrix}.
$$

Straightforward calculation gives the two-channel Krein formula

$$
(\mathcal{A}_\Gamma - \lambda I)^{-1} = \begin{pmatrix} \frac{I}{A-\lambda I} & 0 \\ 0 & \frac{I}{L-\lambda I} \end{pmatrix} +
$$

$$
\begin{pmatrix} \frac{A+iI}{A-\lambda I} & 0 \\ 0 & \frac{L+iI}{L-\lambda I} \end{pmatrix}\begin{pmatrix} \gamma_{ee} - P_{N_e}\frac{I+\lambda A}{A-\lambda I}|_{N_e} & \gamma_{eh} \\ \gamma_{he} & \gamma_{hh} - P_{N_h}\frac{I+\lambda L}{L-\lambda I}|_{N_h} \end{pmatrix}^{-1} P_{\mathcal{N}}
$$

$$
\begin{pmatrix} \frac{A+iI}{A-\lambda I} & 0 \\ 0 & \frac{L+iI}{L-\lambda I}. \end{pmatrix}.
$$

Calculation of the residue of the resolvent at infinity in this case obviously requires comparison of the behavior of the Q-operators $Q_A(\lambda)$ and $Q_L(\lambda)$ at infinity. Therefore, in the following section we are going to study the particular case where comparison can easily be carried out.

The general scattering problem can be studied using the technique developed in [1]. The case where the deficiency indices of the operator L_0 are equal to $(1,1)$ was investigated in detail [2, 7]. Physically relevant are the restrictions of the resolvent of the total operator $\mathcal{A}_\Gamma$ to the space H. These restrictions coincide with the generalized resolvents corresponding to the operator L_0. It has been proved that one can obtain an arbitrary generalized resolvent considering only the restrictions of the operator A with the deficiency indices $(1,1)$. Moreover, it has been proved that if both restricted operators A_0 and L_0 have deficiency indices $(1,1)$, and the boundary operator Γ is not singular ($\gamma_{eh} \neq 0 \neq \gamma_{he}$), then $\mathcal{A}_\Gamma$ is a selfadjoint operator (not an operator relation). In the exceptional case where $\gamma_{eh} = \gamma_{he} = 0$, $\mathcal{A}_\Gamma$ is equal to the orthogonal sum of a selfadjoint operator acting in H and, maybe, an operator relation in E. The latter case is not interesting, since the generalized resolvent in this case coincides with the resolvent of a selfadjoint operator acting in H. In the following section we show how to generalize the resolvent analysis to the case of an unbounded operator.

3. Resolvent analysis of a solvable model

Let us consider the selfadjoint operator $\mathcal{A} = A \oplus (-\Delta)$ acting in the orthogonal sum $E \oplus L_2(\mathbf{R}^3)$ with the finite-dimensional component E. Reducing the Laplacian to the domain containing all smooth square integrable functions vanishing at the origin we get the adjoint operator $-\Delta^+$ defined on locally $W_2^2(\mathbf{R}^3 \setminus \{0\})$ functions outside the origin with the asymptotic behavior

$$u(x) = \frac{a}{4\pi|x|} + b + o(1), \quad x \to 0.$$

The constants $a = a_u, b = b_u$ play the role of boundary data of the element u at the origin and the corresponding boundary form is equal to

$$\langle -\Delta^+ u, v\rangle - \langle u, -\Delta^+ v\rangle = b_u \bar{a}_v - a_u \bar{b}_v.$$

The boundary data and the boundary form are nontrivial only on the sum of deficiency subspaces of L_0:

$$M_h = N_h + \hat{N}_h = \bigvee \left(\frac{e^{ik|x|}}{4\pi|x|}, \frac{e^{-i\bar{k}|x|}}{4\pi|x|}\right), \; k^2 = i.$$

The finite-dimensional operator A is restricted $A \to A_0$ to the domain $\mathrm{Dom}(A_0) = \frac{1}{A-iI}(E \ominus N)$, N playing the role of the deficiency subspace at the spectral point $i, N = E \ominus (A_0 - iI)\mathrm{Dom}\,(A_0)$. If the angle between N_e and $\hat{N}_e$ is positive, then the elements ν of the orthogonal sum $M_e = N_e + \hat{N}_e$ can be characterized by the boundary values $\xi_\pm$:

$$\nu = \frac{A}{A-iI}\xi_+^\nu + \frac{1}{A-iI}\xi_-^\nu,$$

and the formal adjoint A_0^+ defined on M_e by the formulas

$$A_0^+\left(\frac{A}{A-iI}\xi_+^\nu\right) = -\frac{1}{A-iI}\xi_+^\nu, \qquad A_0^+\left(\frac{1}{A-iI}\xi_-^\nu\right) = \frac{A}{A-iI}\xi_-^\nu$$

has the boundary form

$$\langle A_0^+\nu, \mu\rangle_E - \langle \nu, A_0^+\mu\rangle_E = \langle \xi_-^\nu, \xi_+^\mu\rangle - \langle \xi_+^\nu, \xi_-^\mu\rangle.$$

The latter boundary form can be extended to the formal sum $\mathrm{Dom}\,(A_0) + M_e$.

The selfadjoint operator extensions and operator relations of the symmetric operator $A_0 \oplus L_0$ are parametrised by Lagrangian planes $\mathcal{L}$ of the total boundary form in $M_e \oplus M_h$, see [12]. In particular these planes $\mathcal{L}_\Gamma$ can be described by the boundary conditions defined by the bounded selfadjoint operators Γ connecting the boundary data:

$$\begin{pmatrix} \gamma_{ee} & \gamma_{eh} \\ \gamma_{he} & \gamma_{hh} \end{pmatrix} \begin{pmatrix} \xi_+ \\ a \end{pmatrix} = \begin{pmatrix} \xi_- \\ b \end{pmatrix},$$

as in [4]. One can easily check that the operator $\mathcal{A}|_{\mathcal{L}}$ defined on the Lagrangian plane $\mathcal{L}$ as a sum of the adjoint L_0^+ and the formal adjoint A_0^+ is symmetric on $\mathcal{L}$. Attaching to it the operators A_0 and L_0 we get a selfadjoint operator or operator relation in $E \oplus L_2(R_3)$.

Theorem 3.1. *The sum $\mathcal{D}_\Gamma = D(A_0) + \mathcal{L} + D(L_0)$ is direct in $E \oplus L_2(\mathbf{R}^3)$ and represents the domain of a selfadjoint operator or operator relation $\mathcal{A}_\Gamma$ defined by the corresponding Krein formula for the resolvent*

$$(\mathcal{A}_\Gamma - \lambda I)^{-1} = \begin{pmatrix} \frac{I}{A-\lambda I} & 0 \\ 0 & \int_{\mathbf{R}^3} \frac{e^{ik|x-y|}}{4\pi|x-y|} \cdot d^3y \end{pmatrix}$$
$$+ \begin{pmatrix} \frac{A+iI}{A-\lambda I} & 0 \\ 0 & \frac{e^{ik|x|}}{4\pi|x|} \end{pmatrix} \begin{pmatrix} \gamma_{ee} - P_N(\frac{I+\lambda A}{A-\lambda I}) & \gamma_{eh} \\ \gamma_{he} & \gamma_{hh} - \frac{ik}{4\pi} \end{pmatrix}^{-1}$$
$$\times \begin{pmatrix} P_{N_e}\frac{A-iI}{A-\lambda I} & 0 \\ 0 & \int_{\mathbf{R}^3} \frac{e^{ik|y|}}{4\pi|y|} \cdot d^3y \end{pmatrix},$$

where $k^2 = \lambda$.

The proof of this statement can be achieved by minor modification of one of standard proofs of Krein's formula that take into account the fact that the restricted operator is not densely defined (see [4], [12], [9]).

Our aim is the investigation of the asymptotic behavior of the resolvent at infinity and calculation of the corresponding residue. The central part in the calculation of the resolvent is the calculation of the inverse matrix in the second term of the expression for the generalized resolvent.

Lemma 3.2. *Let K denote the expression*

$$K \equiv \gamma_{hh} - \frac{ik}{4\pi} - \gamma_{he}\frac{I}{\gamma_{ee} - Q_0}\gamma_{eh}.$$

Then we have the following representation of the inverse matrix:

$$\begin{pmatrix} \gamma_{ee} - P_N(\frac{I+\lambda A}{A-\lambda I}) & \gamma_{eh} \\ \gamma_{he} & \gamma_{hh} - \frac{ik}{4\pi} \end{pmatrix}^{-1}$$
$$= \begin{pmatrix} \frac{1}{\gamma_{ee}-Q_0}[1 + \gamma_{eh}\frac{1}{\gamma_{ee}-Q_0}\gamma_{he}\frac{1}{K}\frac{1}{\gamma_{ee}-Q_0}] & -\frac{1}{\gamma_{ee}-Q_0}\gamma_{eh}\frac{1}{K} \\ -\frac{1}{K}\gamma_{he}\frac{1}{\gamma_{ee}-Q_0} & \frac{1}{K} \end{pmatrix}$$

The *Proof* is obtained by straightforward calculation.

Now we calculate the residue of the resolvent at infinity separately in the "internal" space E and in the "external" space $L_2(\mathbf{R}^3)$.

Let us consider the basis e_s in N introduced in the previous section. The vector $\gamma = \sum_{\alpha_s=0} (A+iI)e_s\langle\gamma_{eh}, e_s\rangle \in N_0$ is actually the orthogonal projection of $(A+iI)^{-1}\gamma_{eh}$ to $M_0 = (A+iI)N_0$. Denoting by γ_0 the corresponding normalized vector $\gamma_0 = \frac{\gamma}{|\gamma|}$, we form the orthogonal projection in E onto $M_\gamma \equiv M_0 \ominus \{\gamma\}$ as follows:

$$P_M * - \gamma_0\langle *, \gamma_0\rangle \equiv P_{M_\gamma} * .$$

Theorem 3.3. *The residue of the resolvent of the total operator $\mathcal{A}_\Gamma$ at infinity is equal to $I - P_{M_\gamma}$ where P_{M_γ} is the orthogonal projection onto M_γ in $E \oplus L_2(\mathbf{R}^3)$.*

Proof. Straightforward calculation of the expression K for large values of λ gives

$$\begin{aligned} K &\approx \gamma_{hh} - \frac{ik}{4\pi} - \sum_s \frac{\lambda}{1-\lambda\alpha_s}|\langle\gamma_{eh}, e_s\rangle|^2 \\ &= -\lambda \sum_{\alpha_s=0} |\langle\gamma_{eh}, e_s\rangle|^2 - \frac{ik}{4\pi} + \gamma_{hh} - \lambda \sum_{l\geq 1}\sum_{\alpha_s\neq 0} \frac{|\langle\gamma_{eh}, e_s\rangle|^2}{(-\alpha_s\lambda)^l} \\ &= -\lambda \sum_{\alpha_s=0} |\langle\gamma_{eh}, e_s\rangle|^2 + O(1). \end{aligned}$$

Then integration of the ee-element of the resolvent over a large circle gives precisely $I - P_{M_\gamma}$. Integration of the remaining elements of the resolvent applied to the lineal (dense in $L_2(\mathbf{R}^3)$) of all functions g_1 for which $\int \frac{g_1(x)}{|x|}dx = 0$ gives no contribution, since each of the elements contains the factor $\frac{e^{ik|x|}}{4\pi|x|}$; for instance the eh-element of the kernel of the additional term in Krein's formula gives the residue

$$\frac{\sum_{\alpha_s=0}(A+iI)e_s\langle\gamma_{eh}, e_s\rangle}{\sum_{\alpha_s=0}|\langle\gamma_{eh}, e_s\rangle|^2}\frac{1}{4\pi|x|},$$

which vanishes on this dense lineal. The theorem is proved. □

At the same time, straightforward calculation shows that any element ν in the subspace M_γ,

$$\nu = \begin{pmatrix} (A+iI)e \\ 0 \end{pmatrix}, \quad \text{where } e \in N_e,\ [\gamma_{ee} + P_N A|_N]e = 0,\ \langle\gamma_{eh}, e\rangle = 0,$$

satisfies the equation $(\mathcal{A}_\gamma - \lambda I)^{-1}\nu = 0$ for each λ. Indeed,

$$\begin{aligned} P_{N_e}\tfrac{A-i}{A-\lambda I}(A+iI)e &= P_{N_e}\tfrac{A^2+1}{A-\lambda} = \\ P_{N_e}Ae + P_{N_e}\tfrac{I+\lambda A}{A-\lambda I}e &= -\gamma_{ee}e + Q_0(\lambda)e. \end{aligned}$$

On the other hand

$$\begin{pmatrix} \gamma_{ee} - P_{N_e}(\frac{I+\lambda A}{A-\lambda I}) & \gamma_{eh} \\ \gamma_{he} & \gamma_{hh} - \frac{ik}{4\pi} \end{pmatrix}\begin{pmatrix} e \\ 0 \end{pmatrix} = \begin{pmatrix} (\gamma_{ee} - P_{N_e}(\frac{I+\lambda A}{A-\lambda I})\gamma_{eh})e \\ 0 \end{pmatrix},$$

since $\gamma_{he}e = \langle e, \gamma_{eh}\rangle = 0$. Then we have:

$$(\mathcal{A}_\gamma - \lambda I)^{-1}\nu = \left(\tfrac{A+iI}{A-\lambda I}e \right) - \left(\tfrac{A+iI}{A-\lambda I}e \right) = 0.$$

It follows that the space M_γ is contained in the kernel of the resolvent of $\mathcal{A}_\Gamma$.

In the case where the deficiency subspace N_e has dimension one and the parameters γ_{eh}, γ_{he} are not equal to zero, the subspace M_γ is trivial. It follows that in this case no operator relation occurs. If the dimension of the deficiency subspace N_e is greater than one, then $\mathcal{A}_\Gamma$ can be an operator relation. But the restriction of an operator relation to the orthogonal complement of M_γ in $E \oplus L_2(\mathbf{R}^3)$ is a selfadjoint operator. The resolvent of this new operator restricted to the space $L_2(\mathbf{R}^3)$ coincides with the restricted resolvent of the operator relation. Therefore general consideration can be confined to the case where no operator relation occurs, since only the generalized resolvent is interesting in applications.

4. Scattering Matrix

In our case the scattering amplitude is nontrivial in spherically-symmetric channel only. The corresponding scattering amplitude can be found from the condition that the Ansatz

$$\psi = \begin{pmatrix} \frac{A+iI}{A-\lambda I}\xi_+ \\ e^{ik\langle x,\omega\rangle} + f(k)\frac{e^{ik|x|}}{4\pi|x|} \end{pmatrix}$$

satisfies the homogeneous equation $(\mathcal{A}_\gamma - \lambda I)\psi = 0$ in the distributional sense, and the boundary conditions. This gives the following expressions for the scattering amplitude f and the boundary data ξ_+ in the internal channel:

$$f(k) = \frac{1}{-\gamma_{hh} + \gamma_{he}(\gamma_{ee} - Q_A(\lambda))^{-1}\gamma_{eh} + \frac{ik}{4\pi}}$$

$$\xi_+ = -(\gamma_{ee} - Q_A(\lambda))^{-1}\gamma_{eh} f(k).$$

It is obvious that under the condition that the orthogonal projection of $(A + iI)^{-1}\gamma_{eh}$ to $M_0 = (A+iI)N_0$ is not zero, that is the vector γ_{eh} is not orthogonal to the null-space of the operator $\gamma_{ee}+P_N A|_A$, the scattering matrix in the spherically-symmetric channel

$$S(k) = \frac{-\gamma_{hh} + \gamma_{he}(\gamma_{ee} - Q_A(\lambda))^{-1}\gamma_{eh} - \frac{ik}{4\pi}}{-\gamma_{hh} + \gamma_{he}(\gamma_{ee} - Q_A(\lambda))^{-1}\gamma_{eh} + \frac{ik}{4\pi}}$$

tends to 1 as $k^2 = \lambda \to \infty$. One can check by direct calculation that the inner components of scattered waves are orthogonal to the singular subspace of the operator relation.

We have shown that all model operators considered in the papers [1, 11, 12, 10] are really selfadjoint operators. These operators can model physically interesting problems only if the denominator in Krein's formula for the generalized resolvents has linear asymptotics at infinity. This situation has been usually related to the fact that the Q-function of the operator contains a linear term, and, therefore, to the operator relations. We have proved in this paper that this relationship is not inevitable.

The authors are grateful to Professor Malamud for fruitful and inspiring discussions.

References

[1] V. Adamyan, B. Pavlov, Zero-radius potentials and M.G. Krein's formula for generalized resolvents (in Russian), *Zap. Nauch. Sem. Leningrad. Otdel. Mat. Inst. Steklov. (LOMI)*, **149** (1986), *Issled. Linein. Oper. Teor. Functsii.* XV, 7–23.

[2] S. Albeverio, P. Kurasov, "Solvable Schrödinger type operators. Singular perturbations of differential operators", in preparation.

[3] E.A. Coddington and H. de Snoo, Positive selfadjoint extensions of positive symmetric subspaces, *Math. Z.* **159** (1978), 203–214.

[4] V.I. Gorbachuk, "Boundary problems for differential-operator equations", (in Russian) Kiev: Hauk. dumka, 1984, –284 p.

[5] S. Hassi, H. Langer, and H. de Snoo, Selfadjoint extensions for a class of symmetric operators with defect numbers $(1, 1)$, in: 15th OT Conf. Proc., Rom. Acad., Bucharest, 1995, pp. 115–145.

[6] M.A. Krasnoselskii, On selfadjoint extensions of Hermitian Operators, (in Russian), *Ukr. Mat. Journal* **1** (1949), 21–38

[7] P. Kurasov, B. Pavlov, in preparation, to be published in Proc. Conf. "Hyperfunctions, Operator and Dynamical Systems", Brussel, 1997.

[8] H. Langer, B. Textorius, On generalized resolvents and Q-functions of symmetric linear relations (subspaces) in Hilbert space, *Pacific J. Math.* **72** (1977), 135–165.

[9] M. M. Malamud, On formula of the generalized resolvents of a nondensely defined hermitian operators.(in Russian), *Ukr. Mat. Journal.* **44** (1992), 1658–1688. (English translation: *Sov.Math.*, Plenum Publ.Corp., 0041-5995/92/4412-1523, 1993, 1522–1546)

[10] B. Pavlov, A model of zero-radius potential with internal structure (in Russian), *Teor. Mat. Fiz.* **59** (1984), 345–353. (English translation: *Theoret. and Math. Phys.* **59** (1984), 544–550).

[11] B. S. Pavlov, Boundary values on thin manifolds and the semiboundedness of the three-body Schrödinger operator with point potential, *Math. USSR – Sbornik* **64** (1989), 161–175.

[12] B. S. Pavlov, The theory of the extensions and explicitly-soluble models, *Russian Mathematical Surveys* **42** (1987), 127–168.

Dept. of Math., Stockholm Univ.,
10691 Stockholm, SWEDEN;
Dept. of Math., The University of Auckland,
Private Bag 92019, Auckland, NEW ZEALAND
E-mail address: pak@matematik.su.se, pavlov@mat.auckland.ac.nz

Operator Theory:
Advances and Applications, Vol. 113

Spectra of Inner Functions and l^p-Multipliers

V. V. Lebedev

Dedicated to the memory of S. A. Vinogradov

Abstract. We give a survey of certain results on inner functions in the spaces $M_p^+(D)$ of l^p -multipliers of functions analytic in the open unit disc. A new result is proved: if the spectrum of an inner function I in its intersection with the boundary circle is of positive Lebesgue measure, then $I \notin M_p^+(D)$ for all $p \neq 2$.

We consider functions f analytic in the open unit disc $D = \{z \in \mathbb{C} : |z| < 1\}$ of the complex plane $\mathbb{C}$ and their Taylor series

$$f(z) = \sum_{n \geq 0} \hat{f}(n) z^n. \tag{1}$$

Let $A_p^+(D)$, $1 \leq p \leq \infty$, be the space of all functions (1) whose sequence of Taylor coefficients $\hat{f} = \{\hat{f}(n), n = 0, 1, \dots\}$ belongs to l^p, and let $\|f\|_{A_p^+(D)} = \|\hat{f}\|_{l^p}$. A function m analytic in D is called an l^p-multiplier if for every $f \in A_p^+(D)$ we have $m \cdot f \in A_p^+(D)$. The space of all such multipliers is denoted by $M_p^+(D)$. It is easy to see that $M_p^+(D)$ is a Banach algebra with respect to the norm

$$\|m\|_{M_p^+(D)} = \sup_{\|f\|_{A_p^+(D)} \leq 1} \|m \cdot f\|_{A_p^+(D)}$$

and the usual multiplication of functions.

It is well known that l^p-Fourier multipliers correspond to the bounded translation invariant operators in $l^p = l^p(\mathbb{Z})$ ($\mathbb{Z}$ is the set of integers). Similarly there is a natural relationship between the l^p-multipliers of analytic functions and the bounded right-shift invariant operators in $l^p = l^p(\mathbb{Z}_+)$ ($\mathbb{Z}_+$ is the set of non-negative integers); namely, every such operator Q is of the form $Q\hat{f} = \widehat{m \cdot f}$, $\hat{f} \in l^p$, where $m \in M_p^+(D)$. Note that there is no explicit characterization of the spaces $M_p^+(D)$ except for the following basic properties (see [12]): $M_p^+(D) = M_q^+(D)$ if $\frac{1}{p} + \frac{1}{q} = 1$, and

$$M_1^+(D) = M_\infty^+(D) = A_1^+(D) \subseteq M_p^+(D) \subseteq M_2^+(D) = H^\infty(D), \tag{2}$$

where $H^\infty(D)$ means the Hardy space.

The work was partially supported by RFBR, grant 98-01-00529.

The question we discuss in the present note is: Which inner functions belong to $M_p^+(D)$? We recall that a function I analytic in D is called *inner* if $|I(z)| \leq 1$, $z \in D$, and $|I(e^{it})| = 1$ almost everywhere. Taking (2) into account, one can see that the non-trivial case is that of $p \neq 2, 1, \infty$. We give a survey of some known results on the subject; an essentially new result is proved in Section 4, where we consider inner functions whose spectrum in the intersection with the boundary circle $\partial D = \{z \in \mathbb{C} : |z| = 1\}$ is of positive Lebesgue measure.

0. The relationship with l^p-Fourier multipliers

Identifying every function $m \in H^\infty(D)$ with its (non-tangential) boundary function $m^*(t) = m(e^{it})$, it is not difficult to see that $m \in M_p^+(D) \Leftrightarrow m^* \in M_p(\mathbb{T})$, where $M_p(\mathbb{T})$ is the space of l^p-Fourier multipliers on the circle $\mathbb{T} = \mathbb{R}/2\pi\mathbb{Z}$ ($\mathbb{R}$ is the real line). For general facts on $M_p(\mathbb{T})$ and relations between the spaces of multipliers $M_p(\Gamma)$ on the groups $\Gamma = \mathbb{T}$ and $\Gamma = \mathbb{R}$ see [7, 8].

1. Blaschke products with zeros accumulating to a single point

Every *Blaschke product*

$$B(z) = z^m \prod_{z_n \neq 0} \frac{-|z_n|}{z_n} \cdot \frac{z - z_n}{1 - \overline{z}_n z} \qquad (\{z_n\} \subset D, \ \sum_n (1 - |z_n|) < \infty)$$

is an inner function. The theorem below gives non-trivial examples of Blaschke products in $M_p^+(D)$, $p \neq 2$; it shows that if the zeros $\{z_n\}$ of a Blaschke product B tend to a single point quite rapidly, then $B \in M_p^+(D)$.

Theorem 1. *Let B be a Blaschke product with zeros $\{z_n\}$, $z_n \to 1$. If*

$$\sum_{n:|1-z_n|<\varepsilon} |1 - z_n| = O(\varepsilon), \qquad \varepsilon \to +0, \tag{3}$$

then $B \in \bigcap_{1<p<\infty} M_p^+(D)$.

With the additional assumption that the z_n tend to 1 non-tangentially, this theorem was firstly obtained by S. A. Vinogradov [4] (1974) as a consequence of his more general result on functions in $M_p^+(D)$ with one singular point: *if m is a bounded analytic function in the domain $G = \{z \in \mathbb{C} : |z| < r, \alpha < \arg(z-1) < 2\pi - \alpha\}$, where $r > 1$, $0 \leq \alpha < \pi/2$, then $m \in \bigcap_{1<p<\infty} M_p^+(D)$.* In the improved form presented here, Theorem 1 was found later (1980) by Vinogradov [1] and I. E. Verbitskiĭ [2] independently.

In [5] we showed that under some restrictions the converse to Theorem 1 is true:

Theorem 2. *Let B be a Blaschke product and let its zeroes $\{z_n\}$ tend to 1 non-tangentially remaining in the closed upper half-plane. Suppose that $B \in M_p^+(D)$ for some $p \neq 2$. Then we have* (3).

We do not know if the assumption of non-tangential convergence is essential in this statement. A question that seems more subtle is as to the extent to which the assumption of asymmetry $\operatorname{Im} z_n \geq 0$ is essential. It would be interesting to study Blaschke products with zeros placed symmetrically with respect to the real line. The possible mutual interference of zeros may produce some new effects in this case.

2. The limit sets of zeros of Blaschke products

Let us consider the following question: which sets on the circle ∂D can serve as the sets of limit points for zeros of a Blaschke product in $M_p^+(D)$, $p \neq 2$? Certainly, due to Theorem 1, any finite set has this property. We prove more:

Theorem 3 ([5]). *There exists a Blaschke product $B \in \bigcap_{1<p<\infty} M_p^+(D)$ such that the set of limit points of its zeros $\overline{\{z_n\}} \cap \partial D$ is a perfect set.*

We will give a sketch of the proof. Consider a closed set $F \subset \mathbb{R}$ of Lebesgue measure zero. Let $\{\Delta_n\}$ be the intervals complementary to F and S_n the operators defined by

$$(S_n g)\hat{} = 1_{\Delta_n} \cdot \hat{g},$$

where 1_{Δ_n} means the characteristic function of Δ_n ($\hat{}$ is the Fourier transform). Following [9] we say that F has property $LP(p)$ $(1 < p < \infty)$ if the corresponding Littlewood-Paley square function, given by

$$S(g) = \left(\sum_n |S_n g|^2\right)^{1/2},$$

satisfies

$$\|S(g)\|_{L^p(\mathbb{R})} \asymp \|g\|_{L^p(\mathbb{R})}.$$

These sets play remarkable role in the theory of the spaces of L^p- (or l^p-) Fourier multipliers. In particular, choosing arbitrarily some intervals Δ_{n_k}, $k = 1, 2, \dots$, among the intervals $\{\Delta_n\}$, for the characteristic function of their union we have

$$\|1_{\bigcup_k \Delta_{n_k}}\|_{M_p(\mathbb{R})} \leq \text{const}.$$

Moreover, any bounded measurable function m on $\mathbb{R}$ such that $\sup_n \operatorname{Var}_{\Delta_n} m < \infty$, where $\operatorname{Var}_{\Delta_n} m$ means the variation of m on Δ_n, belongs to $M_p(\mathbb{R})$ [9]. A classical example of a set that has property $LP(p)$ for all p, $1 < p < \infty$, is $\{\pm 2^k,\ k \in \mathbb{Z}\}$ [16]. We use the following fact [13]: *there is a perfect set F that has property $LP(p)$ for all p, $1 < p < \infty$.* The construction of this set is based on the work presented in [10]. Now, taking a portion of F, we may assume that F is contained in an interval of the length 2π. Let then

$$E = \{e^{it}, t \in F\}. \tag{4}$$

The theorem follows since for every closed nowhere dense set $E \subset \partial D$ we can construct a Blaschke product B with zeros accumulating to E so that the variations of B on the arcs complementary to E are uniformly bounded.

Certainly, for the same reason every set E of the form (4), where $F \subset (0, 2\pi)$ has property $LP(p)$, is the limit set of zeros of a Blaschke product $B \in M_p^+(D)$.

We do not know if there is a Blashke product $B \in M_p^+(D)$, $p \neq 2$, with zeros accumulating to an arbitrarily given closed set $E \subset \partial D$ of Lebesgue measure zero. The answer is unclear even if E is countable with a single limit point. (A negative answer seems more probable.) At the same time *if zeros of a Blaschke product B accumulate to a set of positive Lebesgue measure, then $B \notin M_p^+(D)$ for all $p \neq 2$.* This will be explained in Section 4.

We note also that the sets with property $LP(p)$ are quite rarefied. They are *porous* [18]. Here and below we call a set $E \subset \partial D$ porous if any arc $I \subseteq \partial D$ contains a subarc J of the length $|J| \geq \text{const} \cdot |I|$ such that $J \cap E = \varnothing$.

3. Singular inner functions

An inner function S is *singular* if it does not have zeros in D and $S(0) > 0$. Every singular inner function S is of the form

$$S(z) = \exp\Big(- \int_{\partial D} \frac{\xi + z}{\xi - z} d\mu(\xi) \Big),$$

where μ is a singular positive measure on the circle ∂D (μ is called the representing measure of S). The closed support $\operatorname{supp} \mu$ of the measure μ is the *spectrum* of S, it is the smallest (by inclusion) closed set $E \subseteq \mathbb{C}$ such that S can be analytically extended to $\mathbb{C} \backslash E$ [11].

It was proved by Verbitskiĭ [2] that *the simplest singular inner function*

$$S(z) = \exp\Big(- a\frac{1+z}{1-z} \Big), \qquad a > 0,$$

(the spectrum of this function is a single point $\{1\}$) *does not belong to $M_p^+(D)$, $p \neq 2$.* Are there any singular inner functions $S \not\equiv 1$ in $M_p^+(D)$, $p \neq 2$, at all? The question was posed by Vinogradov in [3]. We have not been able to answer this question, but obtained some partial results in [5]. Generally speaking if the spectrum of a singular inner function S is insufficiently massive, then $S \notin M_p^+(D)$ for all $p \neq 2$.

For a set $E \subseteq \partial D$ let $|E|$ denote its Lebesgue measure and $\overline{E}$ its closure. For an arc J let $d_E(J)$ mean the largest hole in J, i.e.

$$d_E(J) = \sup\{|\Delta| : \Delta \text{ is an arc}, \ \Delta \subseteq J, \ \Delta \cap E = \varnothing\}.$$

Following [15], we say that a point $\xi \in \partial D$ is a point of thickness of E if $d_E(J)/|J| \to 0$ as J contracts to ξ. By E^{th} we denote the set of all points of thickness of E.

Theorem 4. *Let E be a set on the circle ∂D and let $\overline{E^{th}} \neq E$. If S is a singular inner function whose spectrum is E, then $S \notin M_p^+(D)$ for all $p \neq 2$.*

Theorem 5. *Suppose that for every $\varepsilon > 0$ a (closed) set $E \subset \partial D$ can be covered with a family of arcs $\{J_n\}$ so that*

$$\sum_n |J_n|^4 / d_E(J_n)^3 < \varepsilon.$$

If S is a singular inner function whose spectrum is E, then $S \notin M_p^+(D)$ for all $p \neq 2$.

Note some corollaries. If a set $E \subset \partial D$ is porous, then $E^{th} = \varnothing$ and thus, if the spectrum of a singular inner function S ($S \not\equiv 1$) is porous, then $S \notin M_p^+(D)$, $p \neq 2$. The same conclusion is true if E is contained in a symmetric set of Hausdorff dimension less than 1, as well as if E satisfies the condition $|E_\delta| = o(\delta^{3/4})$, $\delta \to +0$, where E_δ is the δ-neighborhood of E.

We will briefly explain the reasons why the insufficient massiveness of the spectrum implies $S \notin M_p^+(D)$. Let $p > 2$. Consider an arc J, $J \cap E = \varnothing$. We may write $S(e^{it}) = e^{i\varphi(t)}$, where φ is a smooth function on J. Let $g = 1_J \cdot e^{-i\varphi}$, where 1_J is the characteristic function of J. Consider the Fourier transform of g

$$\hat{g}(k) = \int_J e^{-i(\varphi(t)+kt)} dt/2\pi, \qquad k \in \mathbb{Z}. \tag{5}$$

If E has large holes, then one can choose J so that a) $|J|$ is large; b) φ has large derivative of order $\nu \geq 2$ on J. The well-known van der Corput lemma allows us to estimate the integral (5), and, due to b), it is small uniformly with respect to k. So $\hat{g}$ is small and thus $\|\hat{g}\|_{l^p}$ is small as well ($p > 2$). At the same time, it is not difficult to show that $\|\hat{1}_J\|_{l^p} \asymp |J|^{1-\frac{1}{p}}$ and hence a) implies that $\|\widehat{e^{i\varphi}g}\|_{l^p} = \|\hat{1}_J\|_{l^p} \asymp |J|^{1-\frac{1}{p}}$ is large.

A similar method works for Blaschke products in Theorem 2; to prove it one needs to consider the arcs J chosen close to the point $\xi = 1$.

The approach described above uses only the real function technique. We hope that if one goes inside the disc, one will be able to obtain more.

We conclude this section with the following fact: *if the spectrum of a singular inner function S is of positive Lebesgue measure* (on $\mathbb{T}$), *then $S \notin M_p^+(D)$ for all $p \neq 2$.* A more general result is presented in the next section.

4. The continuity of l^p-multipliers

Let I be an inner function. The set of all points ξ such that $1/I$ cannot be analytically extended to a neighborhood of ξ is called the spectrum of I. We denote this set by $\sigma(I)$. It is well known that every inner function admits the factorization

$$I = \lambda \cdot B \cdot S, \tag{6}$$

where λ is a constant, $|\lambda| = 1$, B is a Blaschke product, and S is a singular inner function. Thus we have (see [17]) $\sigma(I) = \overline{\{z_n\}} \cup \operatorname{supp} \mu$ where $\{z_n\}$ are the zeros of the Blaschke factor B and μ is the representing measure of the singular factor S in (6).

We will consider the case where $|\sigma(I)\cap\partial D|>0$. To deal with this case we will need the following essential continuity property of Fourier multipliers, obtained in my joint work with A. Olevskiĭ [13] (see also [14]): *if $g\in M_p(\mathbb{T})$ for some $p\neq 2$, then g coincides almost everywhere with a function continuous almost everywhere.* Note that, on the other hand, any inner function I oscillates in every neighborhood U of a point $\xi\in\sigma(I)\cap\partial D$ so that the I-image of $U\cap D$ is dense in D [11]. In particular,

$$\liminf_{\substack{z\to e^{it}\\ z\in D}}|I(z)|=0, \qquad \limsup_{\substack{z\to e^{it}\\ z\in D}}|I(z)|=1 \tag{7}$$

for $e^{it}\in\sigma(I)\cap\partial D$. So, if $|\sigma(I)\cap\partial D|>0$, then the boundary function $I(e^{it})$ oscillates quite wildly. This naturally led us to conjecture in [6] (and also in [5]) that if $|\sigma(I)\cap\partial D|>0$, then $I\notin M_p^+(D)$, $p\neq 2$. We will prove more:

Theorem 6. *Let $m\in M_p^+(D)$ for some $p\neq 2$. Then*

$$\lim_{\substack{z\to e^{it}\\ z\in D}} m(z)$$

exists for almost every point $e^{it}\in\partial D$.

Proof. Recall that if $f\in L^\infty(\mathbb{T})$ and f is continuous at a point $t\in\mathbb{T}$, then for the Poisson integral $u(z)$, $z\in D$, of f we have

$$\lim_{\substack{z\to e^{it}\\ z\in D}} u(z)=f(t) \tag{8}$$

(see [11, Ch. I, Lemma 3.3]). Under the assumption of the theorem, $m\in H^\infty(D)$ (see (2)). Let $u(z)$ be the Poisson integral of the boundary function $m^*(t)=m(e^{it})$. By Fatou's theorem, $m(z)=u(z)$, $z\in D$ (see [11, Ch. I, Th. 5.3]). At the same time $m^*\in M_p(\mathbb{T})$ (see Section 0) and thus, according to the essential continuity property, there exists a function f continuous at almost every point $t\in\mathbb{T}$ such that $m^*(t)=f(t)$ almost everywhere. Certainly, $f\in L^\infty(\mathbb{T})$ and $u(z)$ is the Poisson integral of f. The theorem follows since at every point $t\in\mathbb{T}$ where f is continuous, we have (8). □

As the immediate consequence, we obtain

Theorem 7. *Let I be an inner function. If $|\sigma(I)\cap\partial D|>0$, then $S\notin M_p^+(D)$ for all $p\neq 2$.*

Proof. If $|\sigma(I)\cap\partial D|>0$, then the set of points $e^{it}\in\partial D$ where (7) is fulfilled is of positive Lebesgue measure. □

I am grateful to A. B. Aleksandrov for the help with the proof of the results presented in this section. The idea to use the Poisson integral belongs to him.

5. Rotation of zeros of a Blaschke product

The following result of Vinogradov [1] shows that generally there is no condition on the moduli of zeros of Blaschke products that could guarantee that a given Blaschke product belongs to M_p^+, $p \neq 2$.

Theorem 8. *Let B be a Blaschke product with zeros $\{z_n\}$. There exists a sequence $e^{i\alpha_n} \in \partial D$, $n = 1, 2, \ldots$, such that for the Blaschke product B' with zeros $\{z'_n = e^{i\alpha_n} z_n\}$ we have $B' \notin M_p^+(D)$ for all $p \neq 2$.*

This follows from Theorem 7, one should only choose $e^{i\alpha_n}$, $n = 1, 2, \ldots$, so that the sequence $z'_n = e^{i\alpha_n} z_n$ accumulate to a set of positive Lebesgue measure (say to the whole circle). We do not know what was the original proof of Vinogradov.

A natural question concerning Theorem 8 is whether, on the contrary, it is possible to improve any Blaschke product by rotation of its zeros. For instance, if B is a Blaschke product, is there some B' generated as above such that $B' \in M_p^+(D)$, $p \neq 2$? The question was posed by B. S. Kashin.

6. Right-shift invariant complemented subspaces in $l^p = l^p(\mathbb{Z}_+)$

We identify $l^p = l^p(\mathbb{Z}_+)$ with $A_p^+(D)$. So the right-shift invariant subspaces in $l^p(\mathbb{Z}_+)$ correspond to the subspaces in $A_p^+(D)$ invariant with respect to the operator $(Qf)(z) = zf(z)$. Consider only the case where $1 < p < 2$. Suppose that I is an inner function and $I \in M_p^+(D)$. Then the operator $(Pf)(z) = I(z)u(z)$, where u is the Poisson integral of $\overline{I(e^{it})} f(e^{it})$, is a bounded projection of $A_p^+(D)$ onto $I \cdot A_p^+(D)$. Thus $I \cdot A_p^+(D)$ is an invariant complemented subspace in $A_p^+(D)$. Does every invariant complemented subspace X in $A_p^+(D)$ have the form $X = I \cdot A_p^+(D)$, where I is an inner function, $I \in M_p^+(D)$? We do not know the answer. For $p = 2$ the answer is certainly positive (the Beurling theorem).

I would like to thank V. A. Ovsyannikov for financial help.

References

[1] S. A. Vinogradov, *Multiplicative properties of power series with coefficient sequence from l^p*, Dokl. Akad. Nauk SSSR, **254** (1980), no. 6; English transl. in Soviet Math. Dokl., **22** (1980), no. 2.

[2] I. E. Verbitskiĭ, *On multipliers of spaces l_A^p*, Functional analysis and its applications, **14** (1980), no. 3.

[3] S. A. Vinogradov, *Multiplicative properties of l_A^p*, In: Linear and Complex Analysis Problem Book, Lect. Notes in Math., No. 1034 (1984), 572–574.

[4] S. A. Vinogradov, *Multipliers of power series with the sequence of coefficients from l^p*, Zap. Nauch. Sem. Leningrad Otd. Mat. inst. Steklov (LOMI), **39** (1974), 30–39; English transl. in J. Soviet Math., **8** (1977), no. 1.

[5] V. V. Lebedev, *Inner functions and l^p-multipliers*, Functional analysis and its applications, **32** (1998), no. 4.

[6] V. V. Lebedev, *Spectra of inner functions and S. A. Vinogradov problem on l^p-multipliers*, Abstracts of 7-th Summer St. Petersburg Meeting in Mathematical Analysis, Euler Int. Math. Inst., St. Petersburg, June 17–20, 1998.

[7] K. De Leew, *On L^p -multipliers*, Ann. Math., **81** (1965), 364–379.

[8] M. Jodiet Jr., *Restrictions and extensions of Fourier multipliers*, Studia Math., **35** (1970), 215–226.

[9] P. Sjögren and P. Sjölin, *Littlewood-Paley decompositions and Fourier multipliers with singularities on certain sets*, Ann. Inst. Fourier, Grenoble 31 (1981), No. 1.

[10] K. Hare and I. Klemes, *On permutations of lacunary intervals*, Trans. Amer. Math. Soc., **347** (1995), no. 10.

[11] J. B. Garnet, *Bounded analytic functions*, Acad. Press, **1981**.

[12] N. K. Nikol'skiĭ, *On the spaces and algebras of Toeplitz matrices acting in l^p*, Sibirsk. Mat. Z., **7** (1966), no. 146; English transl. in Sibirian Math. J., **7** (1966).

[13] V. Lebedev and A. Olevskiĭ, *Bounded groups of translation invariant operators*, C. R. Acad. Sci. Paris, **322** (1996), Serie 1, 143–147.

[14] V. Lebedev and A. Olevskiĭ, *Idempotents in Fourier multipliers algebra*, Geometric and Functional Analysis (GAFA), **4** (1994), no. 5.

[15] N. K. Bari, *Trigonometric series*, Fizmatgiz, Moscow, **1961**; English transl., Vols 1, 2, Pergamon Press, Oxford, and Macmillan, New York, **1964**.

[16] E. M. Stein, *Singular integrals and differentiability properties of functions*, Princeton Univ. Press, Princeton, New Jersey, **1970**.

[17] N. K. Nikolski, *Lectures on the shift operator*, Nauka, Moscow, **1980**; English transl., *Treatise on the shift operator, spectral function theory*, Springer, Berlin and New York, **1984**.

[18] V. V. Lebedev, *On Littlewood-Paley sets* (in preparation).

Moscow State Institute of Electronics and Mathematics (MGIEM),
Dept. of Math. Analysis,
Bol'shoĭ Trekhsvyatitel'skii 3/12,
Moscow, 109028,
Russia
E-mail address: matan@miem.edu.ru

Operator Theory:
Advances and Applications, Vol. 113

The Theorem on Three Spheres for Harmonic Differential Forms

Eugenia Malinnikova

Dedicated to the memory of S. A. Vinogradov

Abstract. We study a generalization of the Hadamard theorem on three circles to harmonic differential forms. An inequality for the L^2-norms of a harmonic form over concentric spheres is proved. Also, we obtain an estimate for the L^∞-norms.

1. Introduction

Let u be a harmonic differential form on the *spherical shell* $\{r_1 \le |x| \le r_2\} \subset \mathbb{R}^n$. By $\|u\|_R$ we denote the L^2-norm of the form u over the sphere $\{|x| = R\}$, i.e., $\|u\|_R^2 = R^{1-n} \int_S |u(Rx')|^2 ds(x')$. In the first part of the paper we prove that

$$\|u\|_r \le \|u\|_{r_1}^{\alpha} \|u\|_{r_2}^{1-\alpha}, \tag{1}$$

where $r_1 < r < r_2$ and $\alpha = \log(r_2/r)/\log(r_2/r_1)$, i.e., $r = r_1^\alpha r_2^{1-\alpha}$. In order to obtain this inequality, we represent the norm $\|u\|_r^2$ as a series $\sum_{-\infty}^{\infty} c_k r^k$ with non-negative coefficients c_k. Then we give another proof of (1), using direct calculations of the derivatives of $\|u\|_r$ with respect to r. In the second part of the paper we employ (1) to obtain a rather rough estimate for the L^∞-norms.

Many results of this type have been proved for solutions of second order elliptic equations. We mention here the recent paper by Brummelhuis [B], the paper concerning harmonic functions by Korevaar and Meyers [KM], and the paper by Garofalo and Lin [GL]. In all these papers, inequalities similar to (1) and the corresponding inequalities for the L^∞-norms were considered for the functions satisfying a certain elliptic equation near the entire ball $\{|x| \le r_2\}$, while the classical Hadamard theorem is true for the functions analytic on the closed annulus $\{r_1 \le |x| \le r_2\}$. Note that the harmonic differential forms present a natural generalization of analytic functions. For a detailed discussion of this issue, see the paper [PH] by Presa Sagué and Havin.

Supported in part by the RFBR grant 96-01-00541 and by Norwegian Research Council project no. 118861/410.

Finally, we mention the paper [JP] by Janson and Peetre and the later paper [PS] by Peetre and Sjölin, where the three-line theorem was proved in the more general setting of the Clifford analysis.

The author is grateful to V. P. Havin for constant scientific supervision and for fruitful discussions of the content of this paper.

We use the usual notation d, $*$, δ, and Δ for the standard operators in $\mathbb{R}^n$. We denote by d_S, $*_S$, δ_S, and Δ_S the corresponding operators on the unit sphere $\mathbb{S}^{n-1}$. A differential form with C^1-smooth coefficients is said to be *harmonic* on an open set $O \subset \mathbb{R}^n$ if $du = 0$ and $\delta u = 0$ on O. It is easy to check that the coefficients of a harmonic form are harmonic functions, so that they are C^∞-smooth.

For $x \in \mathbb{R}^n$, we write $x = rx'$, where $r = \sqrt{x_1^2 + \ldots + x_n^2}$, $x' \in \mathbb{S}^{n-1}$. By $S(r)$ we denote the sphere $\{|x| = r\}$. A p-form u defined on a spherical shell can be uniquely decomposed into a sum $u(x) = u_t(x) + u_n(x) \wedge dr$, where $(u_t)_r(x') = u_t(rx')$ and $(u_n)_r(x') = u_n(rx')$ are forms on the unit sphere of degree p and $p-1$, respectively. We say that u_t is the *tangent* part of u and u_n is the *normal* part of u.

2. L^2-inequality

Let q be a positive integer, $0 < q < n-1$. We denote by L_q^2 the space of square integrable q-forms on the unit sphere $\mathbb{S}^{n-1}$. Consider the Laplace–Beltrami operator Δ_S on L_q^2. By the Hodge theorem, we have $L_q^2 = H_1^q + H_2^q$, where H_1^q (H_2^q) is the closure of the space of exact (coexact) forms in L_q^2. Observe that the spaces H_1^q and H_2^q are invariant under the action of the operator Δ_S. We shall use the following facts about the eigenvalues and eigenforms of the operator Δ_S (see [IK] and [P]). The eigenvalues of the Laplace–Beltrami operator Δ_S on the space of q-forms on the unit sphere are $\lambda_k = -(k+q)(n-q+k)$, $k = 0, 1, \ldots,$ and $\mu_k = -(k+q+1)(n-q+k-1)$, $k = 0, 1, \ldots,$ each eigenvalue being of finite multiplicity. The eigenforms corresponding to λ_k are closed, and the eigenforms corresponding to μ_k are coclosed. In L_q^2 we can choose a basis consisting of eigenforms of Δ_S. If $\{\beta_{kl}\}$, $k = 0, 1, \ldots,$ $1 \le l \le N_k$, is an orthonormal basis of the space H_2^q such that $\Delta_S \beta_{kl} = \mu_k \beta_{kl}$, then $\{|\mu_k|^{-1/2} d_S \beta_{kl}\}$, $k = 0, 1, \ldots,$ $1 \le l \le N_k$, is an orthonormal basis of the space H_1^{q+1}, and $\Delta_S(d_S\beta_{kl}) = \mu_k(d_S\beta_{kl})$ (here N_k is the multiplicity of the eigenvalue μ_k).

Let u be a harmonic p-form on O, where O is an open subset of $\mathbb{R}^n$ containing the spherical shell $K = \{r_1 \le |x| \le r_2\}$. Consider the tangent and the normal part of u, $u = u_t + u_n \wedge dr$. Suppose that $\{\beta_{kl}\}$ is a basis of the space H_2^{p-1} such that $\Delta_S \beta_{kl} = -(k+p)(n+k-p)\beta_{kl}$; then the forms u_t and u_n have the following decompositions in L_p^2 and in L_{p-1}^2, respectively.

Lemma 2.1. *There exist real numbers a_{kl} and b_{kl}, $k = 0, 1, \dots,\ 1 \le l \le N_k$, such that*

$$u_t(x) = (-1)^{p+1} \sum_{k,l} (a_{kl} r^{k+p} + b_{kl} r^{-n-k+p}) d_S \beta_{kl}(x'),$$
$$u_n(x) = \sum_{k,l} ((k+p) a_{kl} r^{k+p-1} - (n+k-p) b_{kl} r^{-n-k+p-1}) \beta_{kl}(x'). \tag{2}$$

The series on the right-hand sides are convergent in $L^2_p(S(r))$ and $L^2_{p-1}(S(r))$, respectively, for $r_1 \le r \le r_2$.

Proof. We rewrite the equations $du = 0$ and $\delta u = 0$ in terms of the forms u_t and u_n. Evidently, the equation $du = 0$ is equivalent to the system

$$0 = (du)_t = d_S u_t, \qquad 0 = (du)_n = d_S u_n + (-1)^p \partial u_t / \partial r. \tag{3}$$

Direct calculation shows that for any q-form ω we have $(*\omega)_n = r^{n-2q-1} *_S \omega_t$ and $(*\omega)_t = (-1)^{n-q} r^{n-2q+1} *_S \omega_n$. Hence, the equation $\delta u = 0$ can be written as follows:

$$0 = (\delta u)_t = r^{-2} \delta_S u_t + (-1)^p (n - 2p + 1) r^{-1} u_n + (-1)^p \partial u_n / \partial r, \tag{4}$$
$$0 = (\delta u)_n = r^{-2} \delta_S u_n. \tag{5}$$

Combined with (5), the first identity in (3) implies that $(u_t)_r \in H^p_1$ and $(u_n)_r \in H^{p-1}_2$. Thus, we can decompose the forms u_t and u_n into the series

$$u_t(x) = \sum_{k,l} f_{kl}(r) d_S \beta_{kl}(x'), \quad u_n(x) = \sum_{k,l} g_{kl}(r) \beta_{kl}(x'). \tag{6}$$

Obviously, the functions f_{kl} and g_{kl} are C^∞-smooth on $[r_1, r_2]$. We use the following notation for the scalar product in L^2_q. If $\phi, \psi \in L^2_q$, then

$$\langle \phi, \psi \rangle = \int_S \phi \wedge * \psi.$$

Clearly, $\langle \phi, \psi \rangle = \langle \psi, \phi \rangle$. The definition of the operator δ_S shows that $\langle d_S w, v \rangle = \langle w, \delta_S v \rangle$. Multiplying in L^2_p the second identity in (3) by $d_S \beta_{kl}$, we get

$$0 = \langle d_S u_n, d_S \beta_{kl} \rangle + \langle (-1)^p \partial u_t / \partial r, d_S \beta_{kl} \rangle = \langle u_n, \delta_S d_S \beta_{kl} \rangle + (-1)^p \partial / \partial r \langle u_t, d_S \beta_{kl} \rangle.$$

By (6), this yields

$$g_{kl}(r) = (-1)^{p+1} f'_{kl}(r). \tag{7}$$

Next, we multiply (4) by β_{kl}, obtaining

$$\begin{aligned} 0 &= \langle \delta_S u_t, \beta_{kl} \rangle + (-1)^p (n - 2p + 1) r \langle u_n, \beta_{kl} \rangle + (-1)^p r^2 \langle \partial u_n \partial r, \beta_{kl} \rangle \\ &= (k+p)(n+k-p) f_{kl}(r) + (-1)^p \left((n - 2p + 1) r g_{kl} + r^2 g'_{kl} \right). \end{aligned}$$

Substituting $(-1)^{p+1} f'_{kl}$ for g_{kl} in the above formula, we conclude that the function $f_{kl}(r)$ satisfies the differential equation

$$(k+p)(n+k-p) f(r) - (n - 2p + 1) r f'(r) - r^2 f''(r) = 0.$$

Then

$$f_{kl}(r) = (-1)^{p+1}(a_{kl}r^{k+p} + b_{kl}r^{-n-k+p}), \quad \text{where} \quad a_{kl}, b_{kl} \in \mathbb{R}. \tag{8}$$

Finally, (6), (7), and (8) imply (2). □

Theorem 2.2. *Suppose that u is a p-form continuous on $\{r_1 \le |x| \le r_2\}$ and harmonic in $\{r_1 < |x| < r_2\}$; then inequality* (1) *is fulfilled.*

Proof #1. The norm $\|u\|_r$ is continuous with respect to r on $[r_1, r_2]$. Therefore, in the proof of (1) we may assume that u is harmonic near the set $\{r_1 \le |x| \le r_2\}$. Then we can use Lemma 2.1. In order to compute the L^2-norm of the form u, we separate the normal and the tangent parts of u. Namely,

$$r^{1+2p-n}|u(rx')|^2 = r^{1+2p-n} * (u(rx') \wedge *u(rx'))$$
$$= *(u_t(rx') \wedge *_S u_t(rx') \wedge dr + r^2 u_n(rx') \wedge *_S u_n(rx') \wedge dr).$$

Let ω be a differential form on the unit sphere. We put $\|\omega\|^2 = \int_S \omega \wedge *_S\omega$. We recall that $\|u\|_R^2 = R^{1-n}\int_S |u(Rx')|^2 ds(x')$. Thus, we have

$$\|u\|_r^2 = r^{-2p}\|(u_t)_r\|^2 + r^{-2p+2}\|(u_n)_r\|^2.$$

Since the $d_S\beta_{kl}$ are mutually orthogonal in L_p^2, the β_{kl} are mutually orthogonal in L_{p-1}^2, and $\|d_S\beta_{kl}\|^2 = (k+p)(n+k-p)$, $\|\beta_{kl}\|^2 = 1$, from (2) we deduce that

$$\|(u_t)_r\|_S^2 = \sum_{k,l}(k+p)(n+k-p)(a_{kl}r^{k+p} + b_{kl}r^{-n-k+p})^2,$$

$$\|(u_n)_r\|_S^2 = \sum_{k,l}\left((k+p)a_{kl}r^{k+p-1} - (n+k-p)b_{kl}r^{-n-k+p-1}\right)^2.$$

Consequently,

$$\|u\|_r^2 = \sum_{k,l}((k+p)(n+2k)a_{kl}^2 r^{2k} + (n+k-p)(n+2k)b_{kl}^2 r^{-2n-2k}).$$

Now, it suffices to observe that the right-hand side of the above identity is a sum of logarithmically convex positive functions of the form $c_m r^m$. Hence, $\|u\|_r^2$ is also logarithmically convex, which proves inequality (1). □

Proof #2. Put $q(r) = r^{2p}\|u\|_r^2$. Inequality (1) means that q satisfies the following differential inequality:

$$rq''(r)q(r) + q'(r)q(r) - r(q'(r))^2 \ge 0 \tag{9}$$

((9) expresses the logarithmic convexity of q). Now we calculate the left-hand side of (9) with the help of (3) and (4). First, we note that

$$q(r) = \|(u_t)_r\|^2 + r^2\|(u_n)_r\|^2 = \langle u_t, u_t\rangle + r^2\langle u_n, u_n\rangle.$$

We omit the index r, having in mind that all forms actually depend on r, and we use f' to denote the differentiation of f with respect to r. We have

$$\begin{aligned} q'(r) = 2\langle u_t', u_t\rangle + 2r\langle u_n, u_n\rangle + 2\langle u_n', u_n\rangle = (-1)^{p+1}2\langle d_S u_n, u_t\rangle \\ +2r\langle u_n, u_n\rangle + (-1)^{p+1}2\langle \delta_S u_t, u_n\rangle - 2(n-2p+1)r\langle u_n, u_n\rangle \\ = (-1)^{p+1}4\langle d_S u_n, u_t\rangle - 2(n-2p)r\langle u_n, u_n\rangle. \end{aligned} \tag{10}$$

In the same way we calculate the second derivative:

$$\begin{aligned} q''(r) =& (-1)^{p+1}4\langle u_n', \delta_S u_t\rangle + (-1)^{p+1}4\langle d_S u_n, u_t'\rangle - 2(n-2p)\langle u_n, u_n\rangle \\ &+4(n-2p)r\langle u_n', u_n\rangle = 4\langle \delta_S u_t, \delta_S u_t\rangle + (-1)^p 4(n-2p+1)\langle u_n, \delta_S u_t\rangle \\ &+4\langle d_S u_n, d_S u_n\rangle + (-1)^p 4(n-2p)\langle \delta_S u_t, u_n\rangle - 2(n-2p)\langle u_n, u_n\rangle \\ &+4(n-2p)(n-2p+1)\langle u_n, u_n\rangle = 4\langle \delta_S u_t, \delta_S u_t\rangle + 4\langle d_S u_n, d_S u_n\rangle \\ &+(-1)^p 4(2n-4p+1)\langle \delta_S u_t, u_n\rangle + 2(n-2p)(2n-4p+1)\langle u_n, u_n\rangle. \end{aligned}$$

After trivial calculations, we get

$$\begin{aligned} A :=& \frac{r}{4}\left(rq''(r)q(r) + q'(r)q(r) - r(q'(r))^2\right) = r^2(n-2p)^2\|u_n\|^2\|u_t\|^2 \\ &+2(-1)^p(n-2p)r\left(\|u_t\|^2 - r^2\|u_n\|^2\right)\langle \delta_S u_t, u_n\rangle + r^2\|d_S u_n\|^2\|u_t\|^2 \\ &+r^2\|\delta_S u_t\|^2\|u_n\|^2 + r^4\| d_S u_n\|^2\|u_n\|^2 + \|\delta_S u_t\|^2\|u_t\|^2 - 4r\langle \delta_S u_t, u_n\rangle^2. \end{aligned}$$

Now we rearrange the terms of the sum above to show that A is positive:

$$\begin{aligned} A =& \left(r(n-2p)\|u_t\|\,\|u_n\| + (-1)^p\frac{\|u_t\|^2 - r^2\|u_n\|^2}{\|u_t\|\,\|u_n\|}\langle d_S u_n, u_t\rangle\right)^2 \\ &+\frac{\| u_t\|^2}{\|u_n\|^2}\left(\|\delta_S u_t\|^2\|u_n\|^2 - \langle \delta_S u_t, u_n\rangle^2\right) \\ &+r^4\frac{\|u_n\|^2}{\|u_t\|^2}\left(\|d_S u_n\|^2\|u_t\|^2 - \langle d_S u_n, u_t\rangle\right)^2 \\ &+r^2\left(\|d_S u_n\|^2\|u_t\|^2 + \|\delta_S u_t\|^2\|u_n\|^2 - 2\langle \delta_S u_t, u_n\rangle\langle d_S u_n, u_t\rangle\right). \end{aligned} \tag{11}$$

Each summand on the right in (11) is indeed positive. The first is a perfect square, the second and the third are positive thanks to the Cauchy inequality, namely,

$$\langle \delta_S u_t, u_n\rangle \le \|\delta_S u_t\|\,\|u_n\| \qquad \text{and} \qquad \langle d_S u_n, u_t\rangle \le \|d_S u_n\|\,\|u_t\|.$$

To show that the last term in (11) is also positive, we can use the Cauchy inequality twice:

$$\begin{aligned} \|d_S u_n\|^2\|u_t\|^2 + \|\delta_S u_t\|^2\|u_n\|^2 &\ge 2\,\|d_S u_n\|\,\|u_t\|\,\|\delta_S u_t\|\,\|u_n\| \\ &\ge 2\langle d_S u_n, u_t\rangle\langle \delta_S u_t, u_n\rangle. \end{aligned}$$

□

3. An estimate for the L^∞-norms

Using the L^2-inequality proved in the preceding section, we obtain an estimate for the L^∞-norms. This may be regarded as an illustration for "propagation of smallness", but the inequality we prove is rather rough. Our proof is similar to that in [KM]. Also, we wish to emphasize that in the case of an analytic function u the function $\log|u|$ is subharmonic, and the maximum principle implies the theorem on three circles for the sup-norms. However, no explicit form of a differential equation for $\log|u|$ is known for harmonic forms in $\mathbb{R}^n$, $n > 2$, although such an equation is known to exist (for the case of harmonic vector fields, see, e.g., [GH]).

For a form u defined near the sphere $S(r)$, we put $\|u\|_{r,\infty} = \sup_{|x|=r}|u(x)|$.

Theorem 3.1. *Let r be a fixed number in (r_1, r_2). Then for any p-form u continuous on $\{r_1 \le |x| \le r_2\}$ and harmonic on $\{r_1 < |x| < r_2\}$ we have*

$$\|u\|_{r,\infty} \le \max\{\ \|u\|^{\beta}_{r_1,\infty}\|u\|^{1-\beta}_{r_2,\infty},\ \|u\|^{\gamma}_{r_1,\infty}\|u\|^{1-\gamma}_{r_2,\infty}\ \} \tag{12}$$

with some constants $\beta = \beta(r_1, r, r_2)$ and $\gamma = \gamma(r_1, r, r_2)$, $0 < \beta < \gamma < 1$,.

Proof. Fixing a point $z \in S(r)$, we put $f(y) := *(u(y) \wedge *u(z))$. Then f is a harmonic function, $|f(y)| \le |u(y)||u(z)|$ for any $y \in \{r_1 \le |x| \le r_2\}$, and $|f(z)| = |u(z)|^2$. We choose $r_3 \in (r, r_2)$ and represent $f(z)$ by the Poisson integral of its boundary values on the spheres $S(r_1)$ and $S(r_3)$. Let $p(x, y)$ denote the Poisson kernel for the Laplace operator on the spherical shell $K_1 = \{r_1 \le |x| \le r_3\}$. Then

$$f(z) = \int_{S(r_1)} p(z, y) f(y) ds(y) + \int_{S(r_3)} p(z, y) f(y) ds(y). \tag{13}$$

The first integral can easily be estimated. Obviously, $p(z, y)$ is positive, and we have $\int_{S(r_1)} p(z, y) ds(y) \le \int_{\partial K_1} p(z, y) ds(y) = 1$. Hence,

$$\int_{S(r_1)} p(x, y) f(y) ds(y) \le sup_{|y|=r_1}|f(y)| \le |u(z)|\,\|u\|_{r_1,\infty}. \tag{14}$$

To estimate the second integral in (13) we apply the Cauchy–Schwartz inequality:

$$\int_{S(r_3)} p(z, y) f(y) ds(y) \le \left(\int_{S(r_3)} f(y)^2 ds(y) \int_{S(r_3)} p(z, y)^2 ds(y)\right)^{1/2}$$
$$\le r_3^{(n-1)/2}\|u\|_{r_3}|u(z)| \left(\max_{|y|=r_3} p(z, y)\right)^{1/2}.$$

We put $C_1 = r_3^{(n-1)/2}\left(\max_{|y|=r_3} p(z, y)\right)^{1/2}$ and $\beta_1 = \log(r_2/r_3)/\log(r_2/r_1)$. Now we use (1), having in mind that $\|u\|_\rho \le \|u\|_{\rho,\infty}$. This yields

$$\int_{S(r_3)} p(z, y) f(y) ds(y) \le C_1|u(z)|\|u\|^{\beta_1}_{r_1,\infty}\|u\|^{1-\beta_1}_{r_2,\infty}. \tag{15}$$

Since $f(z) = |u(z)|^2$, from (13), (14), and (15) we get

$$|u(z)| \le \|u\|_{r_1,\infty} + C_1 \|u\|_{r_1,\infty}^{\beta_1} \|u\|_{r_2,\infty}^{1-\beta_1}. \tag{16}$$

The same argument shows that

$$|u(z)| \le \|u\|_{r_2,\infty} + C_2 \|u\|_{r_1,\infty}^{\gamma_1} \|u\|_{r_2,\infty}^{1-\gamma_1}, \tag{17}$$

where $0 < \beta_1 < \alpha < \gamma_1 < 1$, and $\alpha = \log(r_2/r)/\log(r_2/r_1)$. The maximum principle applied to the harmonic function f implies yet another inequality for $|u(z)|$, namely,

$$f(z) \le \frac{r^{2-n} - r_2^{2-n}}{r_1^{2-n} - r_2^{2-n}} \|f\|_{r_1,\infty} + \frac{r_1^{2-n} - r^{2-n}}{r_1^{2-n} - r_2^{2-n}} \|f\|_{r_2,\infty}.$$

Dividing both sides by $|u(z)|$ and denoting the first ratio on the right by c, we obtain

$$|u(z)| \le c\|u\|_{r_1,\infty} + (1-c)\|u\|_{r_2,\infty}. \tag{18}$$

First, we assume that $\|u\|_{r_1,\infty} \le \|u\|_{r_2,\infty}$ and compare estimates (16) and (18). Let t_1 be the smallest positive solution of the equation $t + C_1 t^{\beta_1} = ct + 1 - c$. We choose β such that $t_1^\beta = ct_1 + 1 - c$. It can easily be verified that $t^\beta \ge \min\{t + C_1 t^{\beta_1}, ct + 1 - c\}$ for $0 \le t \le 1$. Thus, we have

$$|u(z)| \le \|u\|_{r_1,\infty}^{\beta} \|u\|_{r_2,\infty}^{1-\beta} \qquad \text{whenever} \qquad \|u\|_{r_1,\infty} \le \|u\|_{r_2,\infty}.$$

Now, suppose that $\|u\|_{r_1,\infty} \ge \|u\|_{r_2,\infty}$. In this case we use (17) and (18). Let t_2 be the greatest solution of the equation $1 + C_2 t^{\gamma_1} = ct + 1 - c$. Then we introduce γ by $t_2^\gamma = ct_2 + 1 - c$, and claim that $t^\gamma \ge \min\{1 + C_2 t^{\gamma_1}, ct + 1 - c\}$ for $t > 1$. Consequently,

$$|u(z)| \le \|u\|_{r_1,\infty}^{\gamma} \|u\|_{r_2,\infty}^{1-\gamma} \qquad \text{whenever} \qquad \|u\|_{r_1,\infty} \ge \|u\|_{r_2,\infty}.$$

This completes the proof. □

References

[B] R. Brummelhuis, *Three-spheres theorem for second order elliptic equations*, Journ. d'Anal. Math., **65** (1995), 179–206.

[GH] M. Giesecke and V. P. Havin, *The length of a harmonic vector*, Vestnik Leningr. Univ., Mat. Mekh. Astron. (1987), no. 3, 33–38.

[GL] N. Garofalo and F. Lin, *Monotonicity properties of variational integrals, A_p weights and unique continuation*, Indiana Univ. Math. Journ., **35** (1986), no. 2, 245–268.

[H] V. P. Havin, *On the modulus of a harmonic vector field*, Potential Theory – ICPT 94, Edited by Král, Lukeš, Netuka and Veselý, Walter de Gruyter, (Berlin – New York), 1996, 481–483.

[IK] I. Iwasake and K. Katase, *On the spectra of Laplace operator on* $\Lambda^*(\mathbf{S}^n)$, Proc. Japan Acad., Ser. A Math. Sci., **55** (1979), no. 4, 141–145.

[JP] S. Janson and J. Peetre, *Harmonic interpolation*, Interpolation spaces and allied topics in Analysis (Lund, 1983), Lecture Notes in Math., **1070**, Springer, (Berlin-New York), 1984, 92–124.

[KM] J. Korevaar and J. L. H. Meyers, *Logarithmic convexity for supremum norms of harmonic functions*, Bull. London Math. Soc., **26** (1994), 353–362.

[P] L. Paquet, *Méthode de séparation des variables et calcul du spectre d'opérateurs sur les formes différentielles*, Bull. Sci. Math. (2), **105** (1981), no. 1, 85–112.

[PH] A. Presa Sagué and V. P. Havin, *Uniform approximation by harmonic differential forms in Euclidean space*, Algebra i Analiz, **7** 1995, no. 6, 104–152.

[PS] J. Peetre and P. Sjölin, *Three-line theorem and Clifford analysis*, Complex Variables Theory Appl., **19** (1992), no. 3, 151–163.

Department of Mathematics and Mechanics,
St.Petersburg State University,
2, Bibliotechnaya pl.,
Stary Peterhof 198904 St. Petersburg, Russia
E-mail address: eugenia@dbs.spb.ru

Operator Theory:
Advances and Applications, Vol. 113

Traces and Extensions of Multipliers in Pairs of Sobolev Spaces

V. Maz′ya and T. Shaposhnikova

Dedicated to the memory of S. A. Vinogradov

Abstract. It is well known that the space $W_p^{l-1/p}(\mathbf{R}^{n-1})$ with integer l is the space of traces on $\mathbf{R}^{n-1}$ of functions in the Sobolev space $W_p^l(\mathbf{R}_+^n)$, where $\mathbf{R}_+^n = \{(x,y) : x \in \mathbf{R}^{n-1},\ y > 0\}$. We show that a similar result holds for spaces of pointwise multipliers acting in a pair of Sobolev spaces. Namely, we prove that the traces on R^n of functions in the multiplier space $M(W_p^m(\mathbf{R}_+^n) \to W_p^l(\mathbf{R}_+^n))$ form the space $M(W_p^{m-1/p}(\mathbf{R}^{n-1}) \to W_p^{l-1/p}(\mathbf{R}^{n-1}))$, and that there exists a linear continuous extension operator which maps $M(W_p^{m-1/p}(\mathbf{R}^{n-1}) \to W_p^{l-1/p}(\mathbf{R}^{n-1}))$ to $M(W_p^m(\mathbf{R}_+^n) \to W_p^l(\mathbf{R}_+^n))$. We apply this result to the Dirichlet problem for the Laplace equation in the half-space.

1. Introduction

1.1. Function spaces

Let U be an arbitrary function in the Sobolev space $W_p^m(\mathbf{R}_+^n)$, where $\mathbf{R}_+^n = \{(x,y) : x \in \mathbf{R}^{n-1},\ y > 0\}$ and m is a positive integer. It is well known that U has a trace u on the hyperplane $\mathbf{R}^{n-1}$ which belongs to the fractional Sobolev space $W_p^{m-1/p}(\mathbf{R}^{n-1})$ endowed with the norm

$$\left(\|D_{p,m-1/p}\, u; \mathbf{R}^{n-1}\|_{L_p}^p + \|u; \mathbf{R}^{n-1}\|_{L_p}^p \right)^{1/p}.$$

Here

$$(D_{p,m-1/p}\, u)(x) = \left(\int_{\mathbf{R}^{n-1}} |\nabla_{m-1} u(x+h) - \nabla_{m-1} u(x)|^p |h|^{-n-p} dh \right)^{1/p}$$

with ∇_k denoting the gradient of order k. The trace u admits the estimate

$$\|u; \mathbf{R}^{n-1}\|_{W_p^{m-1/p}} \le c \|U; \mathbf{R}_+^n\|_{W_p^m} \tag{1}$$

and there exists a linear bounded extension operator which maps $W_p^{m-1/p}(\mathbf{R}^{n-1})$ to $W_p^m(\mathbf{R}_+^n)$.

In (1) and elsewhere $c, c_1, c_2, \dots$ are positive constants depending on the parameters m, p, n and the like. We use the equivalence relation $a \sim b$ if there exist c_1, c_2 such that $c_1 a \le b \le c_2 a$. In what follows $B_r^{(n-1)}(x)$ is the ball in $\mathbf{R}^{n-1}$ with radius r and center at $x \in \mathbf{R}^{n-1}$. If x is the origin, we write simply $B_r^{(n-1)}$.

By a multiplier acting from one Banach function space S_1 into another S_2 we call a function γ such that $\gamma u \in S_2$ for any $u \in S_1$. By $M(S_1 \to S_2)$ we denote the space of multipliers $\gamma : S_1 \to S_2$ with the norm

$$||\gamma||_{M(S_1 \to S_2)} = \sup\{||\gamma u||_{S_2} : \ ||u||_{S_1} \le 1\}.$$

We write MS instead of $M(S \to S)$, where S is a Banach function space. We shall use the same notation both for spaces of scalar and vector-valued multipliers.

We shall also need the spaces S_{loc} and S_{unif} of functions on $\mathbf{R}^{n-1}$ defined as follows. By S_{loc} we denote the space

$$\{u : \ \eta u \in S \ \text{ for all } \ \eta \in C_0^\infty(\mathbf{R}^{n-1})\}$$

and by S_{unif} we mean the space

$$\{u : \ \sup_\xi \ ||\eta_\xi u||_S < \infty\},$$

where $\eta_\xi(x) = \eta(x - \xi)$, $\eta \in C_0^\infty(\mathbf{R}^{n-1})$, $\eta = 1$ on $B_1^{(n-1)}$. The norm in S_{unif} is introduced by

$$||u||_{S_{unif}} = \sup_\xi ||\eta_\xi u||_S.$$

The following description of the space $M(W_p^{m-1/p}(\mathbf{R}^{n-1}) \to W_p^{l-1/p}(\mathbf{R}^{n-1}))$ is given in [1], Sec. 3.2.

Proposition 1.1.

(i) *Suppose* $pm > n$, $m \ge l$, *and* $1 < p < \infty$. *Then* $\gamma \in M(W_p^{m-1/p}(\mathbf{R}^{n-1}) \to W_p^{l-1/p}(\mathbf{R}^{n-1}))$ *if and only if* $\gamma \in W_{p,unif}^{l-1/p}(\mathbf{R}^{n-1})$. *The equivalence relation*

$$||\gamma; \mathbf{R}^{n-1}||_{M(W_p^{m-1/p} \to W_p^{l-1/p})} \sim ||\gamma; \mathbf{R}^{n-1}||_{W_{p,unif}^{l-1/p}}.$$

is valid.

(ii) *Suppose* $pm \le n$, $m > l$, *and* $1 < p < \infty$. *Then* $\gamma \in M(W_p^{m-1/p}(\mathbf{R}^{n-1}) \to W_p^{l-1/p}(\mathbf{R}^{n-1}))$ *if and only if* $\gamma \in (W_{p,loc}^{l-1/p} \cap L_{1,unif})(\mathbf{R}^{n-1})$ *and*

$$D_{p,l-1/p}\gamma \in M(W_p^{m-1/p}(\mathbf{R}^{n-1}) \to L_p(\mathbf{R}^{n-1})).$$

The equivalence relation

$$\begin{aligned} &||\gamma; \mathbf{R}^{n-1}||_{M(W_p^{m-1/p} \to W_p^{l-1/p})} \\ &\sim ||D_{p,l-1/p}\,\gamma; \mathbf{R}^{n-1}||_{M(W_p^{m-1/p} \to L_p)} + ||\gamma; \mathbf{R}^{n-1}||_{L_{1,unif}} \end{aligned} \tag{2}$$

is valid. If $m = l$, *one should replace here* $||\gamma; \mathbf{R}^{n-1}||_{L_{1,unif}}$ *by* $||\gamma; \mathbf{R}^{n-1}||_{L_\infty}$. *One can also add the sum*

$$\sum_{j=1}^{l-1} ||D_{p,l-j-1/p}\gamma; \mathbf{R}^{n-1}||_{M(W_p^{m-j-1/p} \to L_p)} + \sum_{j=1}^{l} ||\nabla_{l-j}\gamma; \mathbf{R}^{n-1}||_{M(W_p^{m-j} \to L_p)}$$

to the right-hand side of (2).

The space $M(W_p^{k-1/p}(\mathbf{R}^{n-1}) \to L_p(\mathbf{R}^{n-1}))$, where k is a positive integer, can be characterized as follows.

Proposition 1.2. ([2], see also [1], Sec. 3.2.) *Let* $pk \le n$, $1 < p < \infty$. *Then* $\gamma \in M(W_p^{k-1/p}(\mathbf{R}^{n-1}) \to L_p(\mathbf{R}^{n-1}))$ *if and only if*

$$\sup_{\substack{e \subset \mathbf{R}^{n-1} \\ \mathrm{diam}(e) \le 1}} \frac{||\gamma; e||_{L_p}}{(\mathrm{cap}_{p,k-1/p}(e))^{1/p}} < \infty,$$

where $\mathrm{cap}_{p,k-1/p}(e)$ *is the capacity of a compact set* $e \subset \mathbf{R}^{n-1}$ *defined by*

$$\mathrm{cap}_{p,k-1/p}(e) = \inf \{||u; \mathbf{R}^{n-1}||^p_{W_p^{k-1/p}} : \ u \in C_0^\infty(\mathbf{R}^{n-1}), \ u \ge 1 \ \text{on} \ e\}.$$

Furthermore,

$$||\gamma; \mathbf{R}^{n-1}||_{M(W_p^{k-1/p} \to L_p)} \sim \sup_{\substack{e \subset \mathbf{R}^{n-1} \\ \mathrm{diam}(e) \le 1}} \frac{||\gamma; e||_{L_p}}{(\mathrm{cap}_{p,k-1/p}(e))^{1/p}}.$$

This equivalence relation also holds without the restriction $\mathrm{diam}(e) \le 1$.

Other characterizations of the space $M(W_p^{k-1/p}(\mathbf{R}^{n-1}) \to L_p(\mathbf{R}^{n-1}))$ can be found in [3] and [4].

The space $M(W_p^m(\mathbf{R}_+^n) \to W_p^l(\mathbf{R}_+^n))$ with integer m and l can be described analogously.

Proposition 1.3. ([1], Ch. 6)

(i) *Let* $pm > n$, $m \ge l$, $1 < p < \infty$. *Then* $\gamma \in M(W_p^m(\mathbf{R}_+^n) \to W_p^l(\mathbf{R}_+^n))$ *if and only if* $\gamma \in W^l_{p,unif}(\mathbf{R}_+^n)$, *where* $W^l_{p,unif}(\mathbf{R}_+^n)$ *is the space with the norm*

$$\sup_{z \in \overline{\mathbf{R}_+^n}} ||\eta_z u; \ \mathbf{R}_+^n||_{W_p^l},$$

with $\eta_z(w) = \eta(w - z)$, $\eta \in C_0^\infty(\overline{\mathbf{R}_+^n})$, $\eta = 1$ *on the unit ball centered at the origin. The equivalence relation*

$$||\gamma; \mathbf{R}_+^n||_{M(W_p^m \to W_p^l)} \sim ||\gamma; \mathbf{R}_+^n||_{W^l_{p,unif}} \tag{3}$$

holds.

(ii) *Let* $pm \le n$, $m \ge l$, $1 < p < \infty$. *Then* $\gamma \in M(W_p^m(\mathbf{R}_+^n) \to W_p^l(\mathbf{R}_+^n))$ *if and only if* $\gamma \in (W^l_{p,loc} \cap L_{1,unif})(\mathbf{R}_+^n)$, *and* $\nabla_l \gamma \in M(W_p^m(\mathbf{R}_+^n) \to L_p(\mathbf{R}_+^n))$. *The norm in* $M(W_p^m(\mathbf{R}_+^n) \to W_p^l(\mathbf{R}_+^n))$ *satisfies*

$$||\gamma; \mathbf{R}_+^n||_{M(W_p^m \to W_p^l)} \sim ||\nabla_l \gamma; \mathbf{R}_+^n||_{M(W_p^m \to L_p)} + ||\gamma; \mathbf{R}_+^n||_{L_{1,unif}}.$$

If $m = l$, one should replace here $||\gamma; \mathbf{R}^n_+||_{L_{1,unif}}$ by $||\gamma; \mathbf{R}^n_+||_{L_\infty}$. Furthermore, the equivalence relation

$$||\gamma; \mathbf{R}^n_+||_{M(W^m_p \to W^l_p)} \sim \sum_{j=0}^{l} ||\nabla_{l-j}\gamma; \mathbf{R}^n_+||_{M(W^{m-j}_p \to L_p)}$$

is valid.

1.2. Main results

In [1] we showed that $MW^{m-1/p}_p(\mathbf{R}^{n-1})$ coincides with the space of traces on $\mathbf{R}^{n-1}$ of functions from $MW^m_p(\mathbf{R}^n_+)$. (In other words, the multipliers of traces are the traces of multipliers.) In the present article we extend this assertion to the multiplier space $M(W^m_p(\mathbf{R}^n_+) \to W^l_p(\mathbf{R}^n_+))$. We restrict ourselves to the case $pm \le n$, since this result follows directly from (3) for $pm > n$. In Section 3 we prove the following result.

Theorem 1.4. *Let m and l be integers, $m \ge l \ge 1$, $1 < p < \infty$, and $pm \le n$.*

(i) *If*

$$\Gamma \in M(W^m_p(\mathbf{R}^n_+) \to W^l_p(\mathbf{R}^n_+))$$

and γ is the trace of Γ on $\mathbf{R}^{n-1}$, then

$$\gamma \in M(W^{m-1/p}_p(\mathbf{R}^{n-1}) \to W^{l-1/p}_p(\mathbf{R}^{n-1}))$$

and the estimate

$$||\gamma; \mathbf{R}^{n-1}||_{M(W^{m-1/p}_p \to W^{l-1/p}_p)} \le c||\Gamma; \mathbf{R}^n_+||_{M(W^m_p \to W^l_p)} \tag{4}$$

holds.

(ii) *Let T be the extension operator defined by*

$$(T\gamma)(z) = \int_{\mathbf{R}^{n-1}} \zeta(\xi)\gamma(x + y\xi)d\xi,$$

where $\zeta \in (W^{l+1}_{1,loc} \cap L_1)(\mathbf{R}^{n-1})$ and

$$\int_{\mathbf{R}^{n-1}} \zeta(\xi)d\xi = 1.$$

Suppose that

$$\int_{\mathbf{R}^{n-1}} (|\xi|^{l-1/p} + |\xi|^{l-m}) \sum_{i=0}^{l+1} \sup_{\partial B^{(n-1)}_{|\xi|}} |\nabla_i\zeta(\xi)|(1 + |\xi|)^i d\xi < \infty. \tag{5}$$

Then the operator T :

$$M(W^{m-1/p}_p(\mathbf{R}^{n-1}) \to W^{l-1/p}_p(\mathbf{R}^{n-1})) \to M(W^m_p(\mathbf{R}^n_+) \to W^l_p(\mathbf{R}^n_+))$$

is bounded.

In Section 4 we give a description of the multiplier space $M(w_p^{m-1/p}(\mathbf{R}^{n-1}) \to w_p^{l-1/p}(\mathbf{R}^{n-1}))$, where $mp < n$ and $w_p^{k-1/p}(\mathbf{R}^{n-1})$ is the completion of $C_0^\infty(\mathbf{R}^{n-1})$ in the norm

$$||u;\mathbf{R}^{n-1}||_{w_p^{k-1/p}} = ||D_{p,k-1-1/p}u;\mathbf{R}^{n-1}||_{L_p}.$$

Furthermore, from Theorem 1.4 we deduce a similar result for the multiplier space $M(w_p^{m-1/p}(\mathbf{R}^{n-1}) \to w_p^{l-1/p}(\mathbf{R}^{n-1}))$ (see Theorem 4.4). The role of $W_p^k(\mathbf{R}_+^n)$ is played by the completion of $C_0^\infty(\overline{\mathbf{R}_+^n})$ in the norm $||\nabla_k u;\mathbf{R}_+^n||_{L_p}$, which will be denoted by $w_p^k(\mathbf{R}_+^n)$.

In the last Section 5 we give the following application of Theorems 1.4 and 4.4.

Theorem 1.5.

(i) *Let*

$$\phi \in M(W_p^{m-1/p}(\mathbf{R}^{n-1}) \to W_p^{l-1/p}(\mathbf{R}^{n-1})),$$

where m, l are integers, $m \geq l \geq 1$, $p \in (1,\infty)$, $mp \leq n$. Then the Dirichlet problem

$$\Delta u = 0 \quad on\ \mathbf{R}_+^n, \quad u\big|_{\mathbf{R}^{n-1}} = \phi \tag{6}$$

has a unique weak solution u in the space $M(W_p^m(\mathbf{R}_+^n) \to W_p^l(\mathbf{R}_+^n))$. (The inclusion $u \in W_{p,unif}^1(\mathbf{R}_+^n)$ implies the uniqueness of u.)

(ii) *A similar result with $mp < n$ and W replaced by w is valid.*

2. Lemmas on the extension operator T

2.1. Pointwise estimate

In the proof of Theorem 1.4 we shall need the following estimate for the function $T\gamma$ and its derivatives.

Lemma 2.1. *Let $\gamma \in M(W_p^{m-l}(\mathbf{R}^{n-1}) \to L_p(\mathbf{R}^{n-1}))$, where m and l are integers, $m \geq l \geq 1$ $pm \leq n$, $1 < p < \infty$. Then*

$$\sum_{j=0}^{l} y^j |\nabla_j(T\gamma(z))| \leq cK(1+y^{l-m})||\gamma;\mathbf{R}^{n-1}||_{M(W_p^{m-l}\to L_p)}. \tag{7}$$

where K equals the integral in (5).

Proof. Let $\nu = (\nu_1, \dots, \nu_{n-1})$ be a multiindex of order $|\nu|$. We have

$$\begin{aligned} |\nabla_j(T\gamma)(z)| &= \sum_{i+|\nu|=j} \Big|\frac{\partial^i}{\partial y^i}(y^{1-n-|\nu|}\int_{\mathbf{R}^{n-1}}(D^\nu\zeta)(\frac{\xi-x}{y})\gamma(\xi)d\xi)\Big| \\ &\leq cy^{1-n-j}\sum_{i+|\nu|=j}\sum_{k=0}^{i}\int_{\mathbf{R}^{n-1}}|\nabla_{|\nu|+k}\zeta(\frac{\xi-x}{y})|(1+(\frac{|\xi-x|}{y})^k)|\gamma(\xi)|d\xi. \end{aligned} \tag{8}$$

We introduce the function

$$\Psi_j(t) = \sum_{i=0}^{j} |\nabla_i\zeta(t)|\,(1+|t|)^i. \tag{9}$$

The right-hand side of the inequality (8), written in spherical coordinates, equals

$$cy^{-j}\int_0^\infty \int_{\partial B_1^{(n-1)}} \Psi_j(\rho\theta)|\gamma(\rho y\theta+x)|\rho^{n-2}d\theta d\rho \tag{10}$$

which, due to the estimate

$$\Psi_j(\rho\theta) \le \int_\rho^{2\rho}\big(|\nabla\Psi_j(s\theta)| + s^{-1}\Psi_j(s\theta)\big)ds,$$

does not exceed

$$cy^{-j}\int_0^\infty \int_{\partial B_1^{(n-1)}} |\gamma(\rho y\theta+x)| \int_\rho^{2\rho}\big(|\nabla\Psi_j(s\theta)|+s^{-1}\Psi_j(s\theta)\big)ds\rho^{n-2}d\theta d\rho. \tag{11}$$

Let

$$Q(s) = \sup_{\theta\in\partial B_1^{(n-1)}}\big(|\nabla\Psi_j(s\theta)|+s^{-1}\Psi_j(s\theta)\big),$$

then (11) is majorized by

$$cy^{-j}\int_0^\infty Q(s)\int_{s/2}^{s}\int_{\partial B_1^{(n-1)}}|\gamma(\rho y\theta+x)|d\theta\rho^{n-2}d\rho ds.$$

Hence,

$$y^j\,|\nabla_j T\gamma(z)| \le cy^{-n+1}\int_0^\infty Q(s)\int_{sy/2}^{sy}\int_{\partial B_1^{(n-1)}}|\gamma(\tau\theta+x)|d\theta\tau^{n-2}d\tau ds. \tag{12}$$

We represent the integral over $(0,\infty)$ as the sum

$$\int_0^{1/y}\dots ds + \int_{1/y}^\infty \dots ds.$$

By Hölder's inequality,

$$\int_0^{1/y}\dots ds \le c\int_0^{1/y} Q(s)(sy)^{(n-1)(p-1)/p}||\gamma; B_{sy}^{(n-1)}(x)||_{L_p}ds$$
$$\le c\int_0^{1/y} Q(s)(sy)^{l-m+1}ds \sup_{\substack{r\in(0,1)\\ x\in\mathbf{R}^{n-1}}} r^{m-l-(n-1)/p}\,||\gamma; B_r^{(n-1)}(x)||_{L_p}.$$

Hence and by the relation

$$\mathrm{cap}_{p,m-l}(B_r^{(n-1)}(x)) \sim r^{n-1-(m-l)p} \quad \text{for} \quad 0<r<1, \;\; p(m-l)<n-1,$$

we obtain

$$\int_0^{1/y}\dots ds \le cy^{l-m+n-1}\int_0^{1/y} Q(s)s^{l-m+n-1}ds \sup_e \frac{||\gamma; e||_{L_p}}{(\mathrm{cap}_{p,m-l}(e))^{1/p}},$$

where supremum is taken over all compact sets $e \in \mathbf{R}^{n-1}$. Proposition 1.2 enables one to replace the last supremum by $\|\gamma; \mathbf{R}^{n-1}\|_{M(W_p^{m-l} \to L_p)}$. Since

$$\mathrm{cap}_{p,m-l}(B_r^{(n-1)}(x)) \sim r^{n-1} \text{ for } r > 1,$$

the same argument gives the estimate

$$\int_{1/y}^{\infty} \ldots ds \le c \int_{1/y}^{\infty} Q(s)(sy)^{n-1} \sup_{\substack{r>1 \\ x \in \mathbf{R}^{n-1}}} r^{-(n-1)/p} \|\gamma; B_r^{(n-1)}(x)\|_{L_p}$$

$$\le c y^{n-1} \int_{1/y}^{\infty} Q(s) s^{n-1} ds \|\gamma; \mathbf{R}^{n-1}\|_{M(W_p^{m-l} \to L_p)}.$$

Hence, the right-hand side in (12) is dominated by

$$c\|\gamma; \mathbf{R}^{n-1}\|_{M(W_p^{m-l} \to L_p)} (1 + y^{l-m}) \left\{ \int_0^{1/y} Q(s) s^{l-m+n-1} ds + \int_{1/y}^{\infty} Q(s) s^{n-1} ds \right\}.$$

The sum in curly brackets does not exceed

$$\int_0^{\infty} (1 + s^{l-m}) \big(s \sup_{\theta \in \partial B_1^{(n-1)}} |\nabla \Psi_j(s\theta)| + \sup_{\theta \in \partial B_1} \Psi_j(s\theta)\big) s^{n-2} ds$$

$$\le c \int_{\mathbf{R}^{n-1}} (1 + |\xi|^{l-m}) \sum_{k=0}^{j+1} \sup_{\partial B_{|\xi|}^{(n-1)}} |\nabla_k \zeta(\xi)| (1 + |\xi|)^k d\xi \le cK.$$

Hence,

$$y^j |\nabla_j (T\gamma)(z)| \le cK \|\gamma; \mathbf{R}^{n-1}\|_{M(W_p^{m-l} \to L_p)} (1 + y^{l-m}).$$

The proof is complete. □

2.2. Weighted L_p-estimates

In the following lemma and elsewhere we use the same constant K as in Lemma 2.1.

Lemma 2.2. *Let $\gamma \in M(W_p^{m-l}(\mathbf{R}^{n-1}) \to L_p(\mathbf{R}^{n-1}))$, where m and l are integers, $m > l \ge 1$, $pm \le n$, $1 < p < \infty$. Then, for any integer $k = j, \ldots, m - l + j - 1$,*

$$\int_0^1 y^{pk} |\nabla_j (T\gamma)(z)|^p dy$$
$$\le cK^p \|\gamma; \mathbf{R}^{n-1}\|^{p\frac{k-j+1/p}{m-l}}_{M(W_p^{m-l} \to L_p)} [(M\gamma)(x)]^{p(1-\frac{k-j+1/p}{m-l})}, \tag{13}$$

where M is the Hardy-Littlewood maximal operator in $\mathbf{R}^{n-1}$.

Proof. Let δ be a number in (0,1] which will be chosen later. We set

$$\int_0^1 y^{pk} |\nabla_j (T\gamma)(z)|^p dy = \int_0^{\delta} \ldots dy + \int_{\delta}^1 \ldots dy.$$

By (12) and the definition of the operator M,

$$\int_0^\delta \dots dy \le c \int_0^\delta y^{p(k-j-n+1)} \Big(\int_0^\infty Q(s) \int_{sy/2}^{sy} \int_{\partial B_1} |\gamma(\tau\theta + x)| d\theta \tau^{n-2} d\tau ds \Big)^p dy$$

$$\le c[(M\gamma)(x)]^p \, \Big(\int_0^\infty Q(s) s^{n-1} ds \Big)^p \int_0^\delta y^{p(k-j)} dy.$$

Clearly,

$$\int_0^\infty Q(s) s^{n-1} ds \ \le \ \int_0^\infty \big(s \sup_{\theta \in \partial B_1^{(n-1)}} |\nabla \Psi_j(s\theta)| + \sup_{\theta \in \partial B_1^{(n-1)}} \Psi_j(s\theta)\big) s^{n-2} ds \le cK,$$

where Ψ_j is the function defined by (9). Hence,

$$\int_0^\delta \dots dy \ \le \ cK^p [(M\gamma)(x)]^p \delta^{p(k-j)+1}. \tag{14}$$

By Lemma 2.1,

$$\int_\delta^1 \dots dy \le \ cK^p \, ||\gamma; \mathbf{R}^{n-1}||^p_{M(W_p^{m-l} \to L_p)} \delta^{1-p(m-l+j-k)}. \tag{15}$$

Suppose that

$$(M\gamma)(x) \ \le \ ||\gamma; \mathbf{R}^{n-1}||_{M(W_p^{m-l} \to L_p)}.$$

Then (13) follows from (14) with $\delta = 1$. Now let

$$(M\gamma)(x) \ > \ ||\gamma; \mathbf{R}^{n-1}||_{M(W_p^{m-l} \to L_p)}.$$

We set

$$\delta \ = \ \Big(\frac{||\gamma; \mathbf{R}^{n-1}||_{M(W_p^{m-l} \to L_p)}}{(M\gamma)(x)} \Big)^{1/(m-l)}$$

in (14) and (15) and sum up the estimates obtained. The proof is complete. □

We shall need the following auxiliary assertion proved in [1], Sec. 5.1.2, 5.1.4.

Lemma 2.3. *Let j and r be integers, $r \le j \le l$. Then*

$$\Big(\int_0^\infty y^{p(j-r)} \, |\nabla_j (T\gamma)(z)|^p dy \Big)^{1/p} \ \le \ cK (D_{p,r-1/p}\gamma)(x).$$

3. Proof of Theorem 1.4

(i) By setting ΓU and γu instead of U and u in (1) we arrive at the estimate

$$||\gamma u;\mathbf{R}^{n-1}||_{W_p^{l-1/p}} \le ||\Gamma;\mathbf{R}^n_+||_{M(W_p^m\to W_p^l)}||U;\mathbf{R}^n_+||_{W_p^m}.$$

Minimizing the right-hand side over all extensions U of u we obtain

$$||\gamma u;\mathbf{R}^{n-1}||_{W_p^{l-1/p}} \le ||\Gamma;\mathbf{R}^n_+||_{M(W_p^m\to W_p^l)}||u;\mathbf{R}^{n-1}||_{W_p^{m-1/p}}$$

which gives (4).

(ii) Let $U\in W_p^m(\mathbf{R}^n_+)$ and let $\gamma\in M(W_p^{m-1/p}(\mathbf{R}^{n-1})\to W_p^{l-1/p}(\mathbf{R}^{n-1}))$. We have

$$||UT\gamma;\mathbf{R}^n_+||^p_{W_p^l} \le c\int_{\mathbf{R}^n_+}\Big(\sum_{j=0}^{l}|\nabla_j(T\gamma)|^p|\nabla_{l-j}U|^p+|T\gamma|^p|U|^p\Big)dz$$

$$=c\Big(\int_{0<y<1}\dots dz+\int_{y>1}\dots dz\Big). \tag{16}$$

By Lemma 2.1,

$$\sum_{j=1}^{l}|\nabla_j(T\gamma(z))| \le cK||\gamma;\mathbf{R}^{n-1}||_{M(W_p^{m-l}\to L_p)}$$

for $y>1$ and hence,

$$\int_{y>1}\dots dz \le cK^p||\gamma;\mathbf{R}^{n-1}||^p_{M(W_p^{m-l}\to L_p)}\,||U;\mathbf{R}^n_+||^p_{W_p^l}.$$

This together with Proposition 1.1 yields

$$\int_{y>1}\dots dz \le cK^p||\gamma;\mathbf{R}^{n-1}||^p_{M(W_p^{m-l/p}\to W_p^{l-1/p})}\,||U;\mathbf{R}^n_+||^p_{W_p^m}. \tag{17}$$

Now we put

$$\mathcal{R}_{l-j}U(z)=\nabla_{l-j}U(z)-\sum_{k=0}^{m-l+j-1}\frac{\partial^k}{\partial y^k}\nabla_{l-j}U(x,0)\frac{y^k}{k!},$$

$$\mathcal{R}U(z)=U(z)-\sum_{k=0}^{m-l-1}\frac{\partial^k}{\partial y^k}U(x,0)\frac{y^k}{k!}.$$

Then

$$\int_{0<y<1} \dots dz$$

$$\leq c \int_{0<y<1} \Big(\sum_{j=0}^{l} |\nabla_j (T\gamma)(z)|^p \, |\mathcal{R}_{l-j} U(z)|^p + |(T\gamma)(z)|^p \, |\mathcal{R} U(z)|^p \Big) dz$$

$$+ c \int_{0<y<1} \Big(\sum_{j=0}^{l} \sum_{k=0}^{m-l+j-1} |\nabla_j (T\gamma)(z)|^p |\nabla_{l-j+k} U(x,0)|^p y^{pk} \tag{18}$$

$$+ |T\gamma(z)|^p \sum_{k=0}^{m-l-1} |\nabla_k U(x,0)|^p y^{pk} \Big) dz.$$

By Lemma 2.1, the first integral on the right in (18) is not greater than

$$cK^p ||\gamma; \mathbf{R}^{n-1}||^p_{M(W_p^{m-l} \to L_p)} \sum_{j=0}^{l} \int_{0<y<1} |y|^{p(l-m-j)} |\mathcal{R}_{l-j} U(z)|^p dz. \tag{19}$$

Since

$$\mathcal{R}_{l-j} U(z) = \frac{y^{m-l+j}}{(m-l+j-1)!} \int_0^1 \frac{\partial^{m-l+j}}{\partial y^{m-l+j}} \nabla_{l-j} U(x, ty)(1-t)^{m-l+j-1} dt,$$

we have the estimate

$$\int_{0<y<1} y^{p(l-m-j)} |\mathcal{R}_{l-j} U(z)|^p dz \ \leq \ c \int_{0<y<1} \Big(\int_0^1 |\nabla_m U(x,ty)| dt \Big)^p dz.$$

Hence and by Minkowsky's inequality, (19) is majorized by

$$cK^p \; ||\gamma; \mathbf{R}^{n-1}||^p_{M(W_p^{m-l} \to L_p)} \; ||U; \mathbf{R}^n_+||^p_{W_p^m}.$$

Similarly,

$$\int_{0<y<1} |(T\gamma)(z)|^p |\mathcal{R} U(z)|^p dz \ \leq \ cK^p ||\gamma; \mathbf{R}^{n-1}||^p_{M(W_p^{m-l} \to L_p)} ||U; \mathbf{R}^n_+||^p_{W_p^{m-l}}.$$

Hence and by Proposition 1.1 we find that the first integral in the right-hand side of (18) does not exceed

$$cK^p ||\gamma; \mathbf{R}^{n-1}||^p_{M(W_p^{m-l/p} \to W_p^{l-1/p})} ||U; \mathbf{R}^n_+||^p_{W_p^m}.$$

Now we pass to the estimate of the second integral in the right-hand side of (18). Consider first the terms with $k = j, \dots, m-l+j-1$ (there are none if $m = l$). Applying Lemma 2.2 and putting $\lambda = m-l+j-k-1/p$, $\mu = m-l$, we have

$$\int_{0<y<1} y^{pk} |\nabla_j (T\gamma)(z)|^p |\nabla_{l-j+k} U(x,0)|^p dz$$

$$\leq cK^p ||\gamma; \mathbf{R}^{n-1}||^{p(\mu-\lambda)/\mu}_{M(W_p^{m-l} \to L_p)} \int_{\mathbf{R}^{n-1}} \big(M\gamma(x) \big)^{p\lambda/\mu} |\nabla_{l-j+k} U(x,0)|^p dx\,,$$

with the last integral dominated by

$$||(M\gamma)^{p\lambda/\mu};\mathbf{R}^{n-1}||^p_{M(W^\lambda_p\to L_p)}||\nabla_{l-j+k}U(\cdot,0);\mathbf{R}^{n-1}||^p_{W^\lambda_p}.$$

Using Lemma 3.2.1/3 from [1] we obtain

$$||(M\gamma)^{\lambda/\mu};\mathbf{R}^{n-1}||_{M(W^\lambda_p\to L_p)}\le c||(M\gamma);\mathbf{R}^{n-1}||^{\lambda/\mu}_{M(W^{m-l}_p\to L_p)}. \tag{20}$$

By Verbitsky's theorem on the boundedness of the maximal operator M in the space $M(W^{m-l}_p(\mathbf{R}^{n-1})\to L_p(\mathbf{R}^{n-1}))$ (see [1], Sec. 2.6.3) one can replace $M\gamma$ by γ in the right-hand side. Therefore,

$$\begin{aligned}&\int_{0<y<1} y^{pk}|\nabla_j(T\gamma)(z)|^p|\nabla_{l-j+k}U(x,0)|^p dz\\ &\le cK^p||\gamma;\mathbf{R}^{n-1}||^p_{M(W^{m-l}_p\to L_p)}||U;\mathbf{R}^n_+||^p_{W^m_p}\\ &\le cK^p||\gamma;\mathbf{R}^{n-1}||^p_{M(W^{m-1/p}_p\to W^{l-1/p}_p)}||U;\mathbf{R}^n_+||^p_{W^m_p}\end{aligned}$$

with the second inequality following from Proposition 1.1. The same argument shows that for $k=0,\dots,m-l-1$,

$$\begin{aligned}&\int_{0<y<1} y^{pk}|\nabla_j(T\gamma)(z)|^p|\nabla_k U(x,0)|^p dz\\ &\le cK^p||\gamma;\mathbf{R}^{n-1}||^{\frac{pk+1}{m-l}}_{M(W^{m-l}_p\to L_p)}\int_{\mathbf{R}^{n-1}}(M\gamma(x))^{\frac{p(m-l-k)-1}{m-l}}|\nabla_k U(x,0)|^p dx\\ &\le cK^p||\gamma;\mathbf{R}^{n-1}||^{\frac{pk+1}{m-l}}_{M(W^{m-l}_p\to L_p)}||(M\gamma)^{\frac{p(m-l-k)-1}{p(m-l)}};\mathbf{R}^{n-1}||^p_{M(W^{m-l-k-1/p}_p\to L_p)}\\ &\times||\nabla_k U(\cdot,0);\mathbf{R}^{n-1}||^p_{W^{m-l-k-1/p}_p}.\end{aligned}$$

Hence, by (20) with $\lambda=m-l-k-1/p,\quad \mu=m-l$ and by the boundedness of the operator M in the space $M(W^{m-l}_p(\mathbf{R}^{n-1})\to L_p(\mathbf{R}^{n-1}))$, we arrive at the estimate

$$\begin{aligned}&\int_{0<y<1} y^{pk}|\nabla_j(T\gamma)(z)|^p|\nabla_k U(x,0)|^p dz\\ &\le cK^p||\gamma;\mathbf{R}^{n-1}||^p_{M(W^{m-l}_p\to L_p)}||U;\mathbf{R}^n_+||^p_{W^{m-l}_p}\end{aligned}$$

where the right-hand side is dominated by

$$cK^p||\gamma;\mathbf{R}^{n-1}||^p_{M(W^{m-1/p}_p\to W^{l-1/p}_p)}||U;\mathbf{R}^n_+||^p_{W^m_p}$$

due to Proposition 1.1.

It remains to estimate the integral

$$\int_{0<y<1} y^{pk}|\nabla_j(T\gamma)(z)|^p|\nabla_{l-j+k}U(x,0)|^p dz$$

for $k=0,\dots,j-1$. By Lemma 2.3, this integral does not exceed

$$cK^p\int_{\mathbf{R}^{n-1}}(D_{p,j-k-1/p}\gamma(x))^p|\nabla_{l-j+k}U(x,0)|^p dx. \tag{21}$$

By using Propositions 1.2 and 1.1 we obtain that (21) is not greater than

$$cK^p\|D_{p,j-k-1/p}\gamma;\mathbf{R}^{n-1}\|^p_{M(W_p^{m-l+j-k-1/p}\to L_p)}\|\nabla_{l-j+k}U(\cdot,0);\mathbf{R}^{n-1}\|^p_{W_p^{m-l+j-k-1/p}}$$
$$\le cK^p\|\gamma;\mathbf{R}^{n-1}\|^p_{M(W_p^{m-1/p}\to W_p^{l-1/p})}\|U;\mathbf{R}^n_+\|^p_{W_p^m}.$$

Thus we arrive at the estimate

$$\int_{0<y<1}\dots dz\le cK^p\|\gamma;\mathbf{R}^{n-1}\|^p_{M(W_p^{m-1/p}\to W_p^{l-1/p})}\|U;\mathbf{R}^n_+\|^p_{W_p^m}$$

which together with (17) and (16) completes the proof.

4. The space $M(w_p^{m-1/p}(\mathbf{R}^{n-1})\to w_p^{l-1/p}(\mathbf{R}^{n-1}))$

In this section we assume that $mp<n$.

Lemma 4.1.

(i) *The inequality*

$$\|\gamma;\mathbf{R}^{n-1}\|_{M(W_p^{m-1/p}\to L_p)}\le\|\gamma;\mathbf{R}^{n-1}\|_{M(w_p^{m-1/p}\to L_p)}\tag{22}$$

is valid.

(ii) *Let $\gamma_\rho(x)=\gamma(x/\rho)$ with $\rho>0$ and let $\gamma\in M(w_p^{m-1/p}(\mathbf{R}^{n-1})\to L_p(\mathbf{R}^{n-1}))$. Then*

$$\lim_{\rho\to0}\|\rho^{-m}\,\gamma_\rho;\mathbf{R}^{n-1}\|_{M(W_p^{m-1/p}\to L_p)}=\|\gamma;\mathbf{R}^{n-1}\|_{M(w_p^{m-1/p}\to L_p)}.\tag{23}$$

(iii) *The function γ satisfies*

$$\|\gamma;\mathbf{R}^{n-1}\|_{M(w_p^{m-1/p}\to L_p)}\ge c\sup_{\substack{x\in\mathbf{R}^{n-1}\\ r>0}}r^{m-n/p}\|\gamma;B_r^{(n-1)}(x)\|_{L_p}.\tag{24}$$

Proof. (i) We have

$$\|\gamma;\mathbf{R}^{n-1}\|_{M(W_p^{m-1/p}\to L_p)}=\sup_{u\in C_0^\infty(\mathbf{R}^{n-1})}\frac{\|\gamma u;\mathbf{R}^{n-1}\|_{L_p}}{\|u;\mathbf{R}^{n-1}\|_{W_p^{m-1/p}}}\tag{25}$$

and (22) follows from the inequality

$$\|u;\mathbf{R}^{n-1}\|_{W_p^{m-1/p}}\ge\|u;\mathbf{R}^{n-1}\|_{w_p^{m-1/p}}.$$

(ii) Clearly,

$$\|\rho^{-m}\,\gamma_\rho;\mathbf{R}^{n-1}\|_{M(W_p^{m-1/p}\to L_p)}=\sup_{U\in C_0^\infty(\mathbf{R}^{n-1})}\frac{\|\rho^{-m}\,\gamma_\rho U;\mathbf{R}^{n-1}\|_{L_p}}{\|U;\mathbf{R}^{n-1}\|_{W_p^{m-1/p}}}$$
$$=\sup_{u\in C_0^\infty(\mathbf{R}^{n-1})}\frac{\|\rho^{-m}\,\gamma_\rho u_\rho;\mathbf{R}^{n-1}\|_{L_p}}{(\|D_{p,m-1/p}u_\rho;\mathbf{R}^{n-1}\|^p_{L_p}+\|u_\rho;\mathbf{R}^{n-1}\|^p_{L_p})^{1/p}}.$$

The right-hand side majorizes

$$\frac{\|\gamma u;\mathbf{R}^{n-1}\|_{L_p}}{(\|D_{p,m-1/p}u;\mathbf{R}^{n-1}\|_{L_p}^p+\rho^{mp-1}\|u;\mathbf{R}^{n-1}\|_{L_p}^p)^{1/p}}.$$

Hence

$$\liminf_{\rho\to 0}\|\rho^{-m}\,\gamma_\rho;\mathbf{R}^{n-1}\|_{M(W_p^{m-1/p}\to L_p)}\geq\|\gamma;\mathbf{R}^{n-1}\|_{M(w_p^{m-1/p}\to L_p)}. \tag{26}$$

Noting that

$$\|\rho^{-m}\,\gamma_\rho;\mathbf{R}^{n-1}\|_{M(w_p^{m-1/p}\to L_p)}=\|\gamma;\mathbf{R}^{n-1}\|_{M(w_p^{m-1/p}\to L_p)}$$

and using (22) we obtain

$$\limsup_{\rho\to 0}\|\rho^{-m}\,\gamma_\rho;\mathbf{R}^{n-1}\|_{M(W_p^{m-1/p}\to L_p)}\leq\|\gamma;\mathbf{R}^{n-1}\|_{M(w_p^{m-1/p}\to L_p)}. \tag{27}$$

The result follows by combining (26) and (27).

(iii) Let $\eta\in C_0^\infty(B_2^{(n-1)})$, $\eta=1$ on $B_1^{(n-1)}$. The estimate (24) follows from (25) by choosing the test function $u(\xi)=\eta((\xi-x)/r)$. □

Lemma 4.2.

(i) *Let $m\geq l$ and let $\gamma\in M(w_p^{m-1/p}(\mathbf{R}^{n-1})\to w_p^{l-1/p}(\mathbf{R}^{n-1}))$. Then*

$$\|\gamma;\mathbf{R}^{n-1}\|_{M(W_p^{m-1/p}\to W_p^{l-1/p})}\leq c\|\gamma;\mathbf{R}^{n-1}\|_{M(w_p^{m-1/p}\to w_p^{l-1/p})}. \tag{28}$$

(ii) *The inequality*

$$\liminf_{\rho\to 0}\|\rho^{l-m}\,\gamma_\rho;\mathbf{R}^{n-1}\|_{M(W_p^{m-1/p}\to W_p^{l-1/p})}\geq\|\gamma;\mathbf{R}^{n-1}\|_{M(w_p^{m-1/p}\to w_p^{l-1/p})} \tag{29}$$

is valid.

Proof. (i) Let η be the same as in the proof of Lemma 4.1 (iii), and let $\eta_x(\xi)=\eta(x-\xi)$. It is well known that

$$\|\gamma;\mathbf{R}^{n-1}\|_{M(W_p^{m-1/p}\to W_p^{l-1/p})}\leq c\sup_{x\in\mathbf{R}^{n-1}}\|\eta_x\gamma;\mathbf{R}^{n-1}\|_{M(W_p^{m-1/p}\to W_p^{l-1/p})}$$

(see [1], p.131). The norm in the right-hand side is equal to

$$\sup_{u\in C_0^\infty(\mathbf{R}^{n-1})}\frac{\|\eta_x\gamma u;\mathbf{R}^{n-1}\|_{W_p^{l-1/p}}}{\|u;\mathbf{R}^{n-1}\|_{W_p^{m-1/p}}}.$$

Since $w_p^{l-1/p}(\mathbf{R}^{n-1})\subset L_{p(n-1)/(n-lp)}(\mathbf{R}^{n-1})$, the norm in the numerator is equivalent to the norm $\|\eta_x\gamma u;\mathbf{R}^{n-1}\|_{w_p^{l-1/p}}$. Hence

$$\|\gamma;\mathbf{R}^{n-1}\|_{M(W_p^{m-1/p}\to W_p^{l-1/p})}\leq c\sup_{x,u}\frac{\|\eta_x\gamma u;\mathbf{R}^{n-1}\|_{w_p^{l-1/p}}}{\|u;\mathbf{R}^{n-1}\|_{W_p^{m-1/p}}}$$

$$\leq c\|\gamma;\mathbf{R}^{n-1}\|_{M(w_p^{m-1/p}\to w_p^{l-1/p})}\sup_{x,u}\frac{\|\eta_x u;\mathbf{R}^{n-1}\|_{w_p^{m-1/p}}}{\|u;\mathbf{R}^{n-1}\|_{W_p^{m-1/p}}}.$$

The result follows.

(ii) Obviously, for any $u \in C_0^\infty(\mathbf{R}^{n-1})$,

$$\|\rho^{l-m}\,\gamma_\rho; \mathbf{R}^{n-1}\|_{M(W_p^{m-1/p}\to W_p^{l-1/p})} \geq \frac{\|\rho^{l-m}\,\gamma_\rho u_\rho; \mathbf{R}^{n-1}\|_{W_p^{l-1/p}}}{\|u_\rho; \mathbf{R}^{n-1}\|_{W_p^{m-1/p}}}$$

$$= \Big(\frac{\|D_{p,l-1/p}(\gamma u); \mathbf{R}^{n-1}\|_{L_p}^p + \rho^{pl-1}\|\gamma u; \mathbf{R}^{n-1}\|_{L_p}^p}{\|D_{p,m-1/p}u; \mathbf{R}^{n-1}\|_{L_p}^p + \rho^{pm-1}\|u; \mathbf{R}^{n-1}\|_{L_p}^p}\Big)^{1/p}.$$

Passing to the limit as $\rho \to 0$ we complete the proof. □

Now we give a description of the space $M(w_p^{m-1/p}(\mathbf{R}^{n-1}) \to w_p^{l-1/p}(\mathbf{R}^{n-1}))$ which follows essentially from Proposition 1.1.

Proposition 4.3. *Let $mp < n$, $m \geq l$, $1 < p < \infty$. Then $\gamma \in M(w_p^{m-1/p}(\mathbf{R}^{n-1}) \to w_p^{l-1/p}(\mathbf{R}^{n-1}))$ if and only if $\gamma \in w_{p,loc}^{l-1/p}(\mathbf{R}^{n-1})$,*

$$D_{p,l-1/p}\gamma \in M(w_p^{m-1/p}(\mathbf{R}^{n-1}) \to L_p(\mathbf{R}^{n-1})),$$

and

$$\begin{aligned} &\gamma \in L_\infty(\mathbf{R}^{n-1}) && \text{for} \quad m = l,\\ &\lim_{r\to\infty} r^{1-n}\|\gamma; B_r^{(n-1)}\|_{L_1} = 0 && \text{for} \quad m > l. \end{aligned} \tag{30}$$

The norm in the space $M(w_p^{m-1/p}(\mathbf{R}^{n-1}) \to w_p^{l-1/p}(\mathbf{R}^{n-1}))$, $m > l$, is subject to the equivalence relation

$$\|\gamma; \mathbf{R}^{n-1}\|_{M(w_p^{m-1/p}\to w_p^{l-1/p})} \sim \|D_{p,l-1/p}\gamma; \mathbf{R}^{n-1}\|_{M(w_p^{m-1/p}\to L_p)}. \tag{31}$$

The norm $\|\gamma; \mathbf{R}^{n-1}\|_{L_\infty}$ should be added to the right-hand side of this relation if $m = l$.

The equivalence relation

$$\begin{aligned} \|\gamma; \mathbf{R}^{n-1}\|_{M(w_p^{m-1/p}\to w_p^{l-1/p})} &\sim \sum_{j=0}^{l-1} \|D_{p,l-j-1/p}\gamma; \mathbf{R}^{n-1}\|_{M(w_p^{m-j-1/p}\to L_p)}\\ &\quad + \sum_{j=1}^{l} \|\nabla_{l-j}\gamma; \mathbf{R}^{n-1}\|_{M(w_p^{m-j}\to L_p)} \end{aligned} \tag{32}$$

also holds.

Proof. By Proposition 1.1, the analogue of (32), with W instead of w, is valid and we replace γ by $\rho^{l-m}\gamma_\rho$ in it. The relation (32) follows from Lemmas 4.1 and 4.2 as $\rho \to 0$. We put $\rho^{l-m}\gamma_\rho$ as γ in (2) to obtain

$$\begin{aligned} \|\rho^{l-m}\gamma_\rho; \mathbf{R}^{n-1}\|_{M(W_p^{m-1/p}\to W_p^{l-1/p})} &\leq c\big(\|D_{p,l-1/p}(\rho^{l-m}\gamma_\rho); \mathbf{R}^{n-1}\|_{M(W_p^{m-1/p}\to L_p)}\\ &\quad + \sup_{x,R} R^{m-l-n+1}\|\rho^{l-m}\gamma_\rho; B_R^{(n-1)}(x)\|_{L_1}\big). \end{aligned}$$

Since the second term on the right is equal to

$$\sup_{x,r} r^{m-l-n+1} \|\gamma; B_r^{(n-1)}(x)\|_{L_1},$$

and the first term tends to

$$\|D_{p,l-1/p}\gamma; \mathbf{R}^{n-1}\|_{M(w_p^{m-1/p} \to L_p)}$$

as $\rho \to 0$ by (23), the reference to (29) gives

$$\begin{aligned} &\|\gamma; \mathbf{R}^{n-1}\|_{M(w_p^{m-1/p} \to w_p^{l-1/p})} \\ &\le c\big(\|D_{p,l-1/p}\gamma; \mathbf{R}^{n-1}\|_{M(w_p^{m-1/p} \to L_p)} + \sup_{x,r} r^{m-l-n+1} \|\gamma; B_r^{(n-1)}(x)\|_{L_1}\big). \end{aligned} \tag{33}$$

It remains to remove the second term on the right in the case $m > l$. We shall make use of the following inequality

$$|\gamma(\xi)| \le c[(D_{p,l-1/p}\gamma)(\xi)]^{\frac{p(m-l)}{mp-1}} \big(\sup_{\substack{r>0,\\ x \in \mathbf{R}^{n-1}}} r^{m-n/p} \|D_{p,l-1/p}\gamma; B_r^{(n-1)}(x)\|_{L_p}\big)^{\frac{pl-1}{pm-1}}. \tag{34}$$

Its proof is the same as that of Verbitsky's Lemma 2.6.1/3 in [1]. Unlike this lemma, (34) does not contain the additional term $\|\gamma; \mathbf{R}^{n-1}\|_{L_{1,unif}}$ in the right-hand side due to the condition (30). Integrating (34) over an arbitrary ball $B_r^{(n-1)}(x)$, we arrive at

$$\begin{aligned} r^{m-l-n+1} \|\gamma; B_r^{(n-1)}(x)\|_{L_1} &\le c \sup_{\substack{x \in \mathbf{R}^{n-1},\\ r>0}} r^{m-n/p} \|D_{p,l-1/p}\gamma; B_r^{(n-1)}(x)\|_{L_p} \\ &\le c\|D_{p,l-1/p}\gamma; \mathbf{R}^{n-1}\|_{M(w_p^{m-1/p} \to L_p)}. \end{aligned}$$

(The last inequality is a consequence of (24).) This along with (33) and (32) leads to (31). □

Now we can pass to an analogue of Theorem 1.4 for the multiplier space $M(w_p^{m-1/p}(\mathbf{R}^{n-1}) \to w_p^{l-1/p}(\mathbf{R}^{n-1}))$.

Theorem 4.4. *Let m and l be integers, $m \ge l \ge 1$, $1 < p < \infty$, $pm < n$.*

(i) *If*

$$\Gamma \in M(w_p^m(\mathbf{R}_+^n) \to w_p^l(\mathbf{R}_+^n))$$

and γ is the trace of Γ on $\mathbf{R}^{n-1}$, then

$$\gamma \in M(w_p^{m-1/p}(\mathbf{R}^{n-1}) \to w_p^{l-1/p}(\mathbf{R}^{n-1}))$$

and the estimate

$$\|\gamma; \mathbf{R}^{n-1}\|_{M(w_p^{m-1/p} \to w_p^{l-1/p})} \le c\|\Gamma; \mathbf{R}_+^n\|_{M(w_p^m \to w_p^l)}$$

is valid.

(ii) *The operator T introduced in Theorem 1.4 continuously maps the space $M(w_p^{m-1/p}(\mathbf{R}^{n-1}) \to w_p^{l-1/p}(\mathbf{R}^{n-1}))$ to $M(w_p^m(\mathbf{R}_+^n) \to w_p^l(\mathbf{R}_+^n))$.*

Proof. (i) It is sufficient to change W for w in the proof of Theorem 1.4 (i).
(ii) Let $\gamma \in M(w_p^{m-1/p}(\mathbf{R}^{n-1}) \to w_p^{l-1/p}(\mathbf{R}^{n-1}))$. From Theorem 4.4 (ii) and Lemma 4.2 (i) the inequality

$$||UT\gamma; \mathbf{R}^n_+||_{w_p^l} \le cK||\gamma; \mathbf{R}^{n-1}||_{M(w_p^{m-1/p} \to w_p^{l-1/p})}||U; \mathbf{R}^n_+||_{W_p^m}$$

follows. Replacing here $\gamma(x)$ by $\rho^{l-m}\gamma(x/\rho)$ and putting $\rho^{m-n/p}U(z/\rho)$ instead of $U(z)$ we note that the left-hand side and the norm of γ on the right remain intact. Hence

$$||UT\gamma; \mathbf{R}^n_+||_{w_p^l}$$
$$\le cK||\gamma; \mathbf{R}^{n-1}||_{M(w_p^{m-1/p} \to w_p^{l-1/p})}(||\nabla_m U; \mathbf{R}^n_+||_{L_p} + \rho^m||U; \mathbf{R}^n_+||_{L_p})$$

Passing to the limit as $\rho \to 0$ we complete the proof. □

5. Proof of Theorem 1.5

(i) Let $\phi \in M(W_p^{m-1/p}(\mathbf{R}^{n-1}) \to W_p^{1-1/p}(\mathbf{R}^{n-1}))$. Since the Poisson kernel $cy(|x|^2+y^2)^{-n/2}$ satisfies (5) with $l = 1$ then Theorem 1.4 implies that the problem (6) has a solution u in the space $M(W_p^m(\mathbf{R}^n_+) \to W_p^1(\mathbf{R}^n_+))$, $m \ge 1$.

Suppose Theorem has been proved for $l = 1, \dots, \mathcal{L} - 1$, where $\mathcal{L} \ge 2$. Let

$$\phi \in M(W_p^{m-1/p}(\mathbf{R}^{n-1}) \to W_p^{\mathcal{L}-1/p}(\mathbf{R}^{n-1})), \quad m \ge \mathcal{L}.$$

By Proposition 1.1,

$$\phi \in M(W_p^{m-\mathcal{L}+1-1/p}(\mathbf{R}^{n-1}) \to W_p^{1-1/p}(\mathbf{R}^{n-1})).$$

Hence

$$u \in M(W_p^{m-\mathcal{L}+1}(\mathbf{R}^n_+) \to W_p^1(\mathbf{R}^n_+)),$$

and $u \in W^1_{p,unif}(\mathbf{R}^n_+)$ by Proposition 1.3.

Using Proposition 1.1 once more, we obtain

$$\frac{\partial \phi}{\partial x_k} \in M(W_p^{m-1/p}(\mathbf{R}^{n-1}) \to W_p^{\mathcal{L}-1-1/p}(\mathbf{R}^{n-1}))$$

for $k = 1, \dots, n-1$. Then, by the induction hypothesis,

$$\frac{\partial u}{\partial x_k} \in M(W_p^m(\mathbf{R}^n_+) \to W_p^{\mathcal{L}-1}(\mathbf{R}^n_+)).$$

Hence and by Proposition 1.3,

$$\nabla_{\mathcal{L}-1} \frac{\partial u}{\partial x_k} \in M(W_p^m(\mathbf{R}^n_+) \to L_p(\mathbf{R}^n_+)). \tag{35}$$

Using the harmonicity of u and (35) we find

$$\frac{\partial^{\mathcal{L}} u}{\partial y^{\mathcal{L}}} = -\frac{\partial^{\mathcal{L}-2}(\Delta_x u)}{\partial y^{\mathcal{L}-2}} \in M(W_p^m(\mathbf{R}^n_+) \to L_p(\mathbf{R}^n_+))$$

which together with (35) implies

$$\nabla_{\mathcal{L}} u \in M(W_p^m(\mathbf{R}_+^n) \to L_p(\mathbf{R}_+^n)).$$

It remains to refer to Proposition 1.3.

(ii) The proof is the same, with Theorem 4.4 playing the role of Theorem 1.4.

Acknowledgement. It is our pleasure to thank I. Verbitsky for fruitful comments.

References

[1] V. G. Maz'ya and T. O. Shaposhnikova, *Theory of multipliers in spaces of differentiable functions*, Pitman, **1985**.

[2] V. G. Maz'ya, *Summability, with respect to an arbitrary measure, of functions from S. L. Sobolev–L. N. Slobodetskiĭ spaces* (Russian), Investigations on linear operators and the theory of functions, IX, Zap. Nauchn. Sem. Leningrad. Otdel. Mat. Inst. Steklov (LOMI) **92** (1979), 192–202.

[3] R. Kerman and E. T. Sawyer, *The trace inequality and eigenvalue estimates for Schrödinger operators*, Ann. Inst. Fourier (Grenoble), **36** (1986), 207–228.

[4] V. G. Maz'ya and I. E. Verbitsky, *Capacitary inequalities for fractional integrals, with applications to partial differential equations and Sobolev multipliers*, Ark. Mat. **33** (1995), no. 1, 81–115.

Department of Mathematics,
Linköping University,
S-581 83 Linköping, Sweden
E-mail address: vlmaz@mai.liu.se, tasha@mai.liu.se

Operator Theory:
Advances and Applications, Vol. 113

Complete Version of Turan's Lemma for Trigonometric Polynomials on the Unit Circumference

Fedor L. Nazarov

Dedicated to the memory of S. A. Vinogradov

Abstract. We prove that for every trigonometric polynomial $p(z)$ in n characters, every measurable subset E of the unit circumference $\mathbb{T}$ of measure $\mu(E) \geqslant \frac{1}{3}$, and every $q \in [0, 2]$, the inequality

$$\|p\|_{L^q(\mathbb{T})} \leqslant e^{A(n-1)\mu(\mathbb{T}\setminus E)} \|p\|_{L^q(E)}$$

is true with some absolute constant $A > 0$. The result is sharp up to the numerical value of A.

Notation

Everywhere below we shall use the following notation.

$\mathbb{Z}$	is the set of all integers;
$\mathbb{N}$	is the set of all positive integers;
$\mathbb{R}$	is the real line;
$\mathbb{C}$	is the complex plane;
$\mathbb{T}$	is the unit circumference $\{z \in \mathbb{C} : \|z\| = 1\}$;
$\widetilde{E}$	is the complement $\mathbb{T} \setminus E$ of a set $E \subset \mathbb{T}$;
μ	is the Haar measure on $\mathbb{T}$, i.e., the Lebesgue measure normalized by the condition $\mu(\mathbb{T}) = 1$;
$\|E\|$	is the length of a set $E \subset \mathbb{R}$;
$\\|f\\|_{L^q(H)}$	is the L^q-norm of a function $f : \mathbb{T} \to \mathbb{C}$ restricted to the subset H of $\mathbb{T}$, i.e., $\\|f\\|_{L^q(H)}$ is equal

to $\left(\frac{1}{\mu(H)}\int_H |f|^q d\mu\right)^{1/q}$ if $0 < q < +\infty$,

to $\sup_{z\in H} |f(z)|$ if $q = +\infty$,

and to $\exp\left\{\frac{1}{\mu(H)}\int_H \log|f|\, d\mu\right\}$ if $q = 0$.

§0. Introduction

Let $p(z) = \sum_{k=1}^{n} c_k z^{m_k}$ ($c_k \in \mathbb{C}$, $m_1 < \cdots < m_n \in \mathbb{Z}$) be a trigonometric polynomial in n characters regarded as a function on the unit circumference $\mathbb{T}$. The classical Turan lemma claims that

$$\|p\|_{L^\infty(\mathbb{T})} \leqslant \left\{\frac{4e}{\mu(E)}\right\}^{n-1} \|p\|_{L^\infty(E)}$$

for every arc $E \subset \mathbb{T}$. In [1] it was shown that this estimate remains valid for an arbitrary measurable set $E \subset \mathbb{T}$ if we replace $4e$ by 14 in the numerator.

Recall that the decreasing rearrangement Φ of the function $|p|$ is defined as the (unique) continuous decreasing function acting from $[0,1]$ to $[0,+\infty)$ and such that $\mu\{z \in \mathbb{T} : |p(z)| > y\} = |\{t \in [0,1] : \Phi(t) > y\}|$ for every $y > 0$. Turan's lemma for arbitrary measurable sets can now be restated as follows:

$$\Phi(t) \geqslant \left(\frac{1-t}{14}\right)^{n-1} \Phi(0) \qquad (0 \leqslant t \leqslant 1).$$

Therefore, for each set $E \subset \mathbb{T}$,

$$\|p\|_{L^0(E)} \geqslant \exp\left\{\frac{1}{\mu(E)} \int_{1-\mu(E)}^{1} \log \Phi(t)\, dt\right\}$$
$$\geqslant \left(\frac{\mu(E)}{14e}\right)^{n-1} \Phi(0) = \left(\frac{\mu(E)}{14e}\right)^{n-1} \|p\|_{L^\infty(\mathbb{T})},$$

whence we conclude immediately that

$$\|p\|_{L^q(\mathbb{T})} \leqslant \left\{\frac{14e}{\mu(E)}\right\}^{n-1} \|p\|_{L^q(E)}$$

for every $q \in [0,+\infty]$. The example of the polynomial $p(z) = (z-1)^{n-1}$ and a small arc $E \subset \mathbb{T}$ centered at 1 shows that this estimate cannot be significantly improved for the sets E of small measure even if we put the $L^0(\mathbb{T})$-norm on the left-hand side and the $L^\infty(E)$-norm on the right-hand side. On the other hand, for sets E close to the entire unit circumference $\mathbb{T}$, one could expect some inequality of the kind $\|p\|_{L^q(\mathbb{T})} \leqslant B(\mu(E))^{n-1}\|p\|_{L^q(E)}$, where $B(\mu) \to 1$ as $\mu \to 1-$. Meanwhile, the value $B(\mu) = \frac{4e}{\mu}$ given by Turan's lemma remains very far from 1 even for the entire unit circumference, and this drawback cannot be removed by a minor modification of any proof known so far.

Our main aim in this paper is to confirm that some estimate of the above kind is indeed true. Namely, we shall prove the following

Theorem. *For every $0 \leqslant q \leqslant 2$ and every set $E \subset \mathbb{T}$ of measure $\mu(E) \geqslant \frac{1}{3}$, we have*

$$\|p\|_{L^q(\mathbb{T})} \leqslant e^{A(n-1)\mu(\widetilde{E})} \|p\|_{L^q(E)} \qquad (*)$$

with some constant $A > 0$ independent of q, E, and n.

It is not difficult (but quite boring, so that I am not going to perform it here) to show that there exists an absolute constant $a > 0$ such that for every $q \in [0, 2]$, every $n \in \mathbb{N}$, and every arc $E \subset \mathbb{T}$ of measure at least $\frac{1}{3}$ centered at -1, one of the polynomials $p_{m,r}(z) \stackrel{\text{def}}{=} \left(\frac{z^m - 1}{z - 1}\right)^r$ $(m, r \in \mathbb{N};\ mr \leqslant n)$ satisfies

$$\|p\|_{L^q(\mathbb{T})} \geqslant e^{a(n-1)\mu(\widetilde{E})} \|p\|_{L^q(E)}.$$

So, estimate $(*)$ is sharp up to the numerical value of the absolute constant involved.

§1. The proof of $(*)$ for $q = 2$

Lemma 1. *For any four real numbers $0 \leqslant a \leqslant A$ and $0 \leqslant b \leqslant B$, we have*

$$A + B - 2\sqrt{ab} \geqslant \left(\frac{1}{a + A} + \frac{1}{b + B}\right)\sqrt{(A^2 - a^2)(B^2 - b^2)}.$$

Proof. Substituting $A = x(1 + \alpha)$, $a = x(1 - \alpha)$, $B = y(1 + \beta)$, $b = y(1 - \beta)$, we obtain the inequality

$$x(1 + \alpha) + y(1 + \beta) \geqslant 2\sqrt{xy(1 - \alpha)(1 - \beta)} + 2(x + y)\sqrt{\alpha\beta}$$

to prove. But

$$x(1 - \beta) + y(1 - \alpha) \geqslant 2\sqrt{xy(1 - \alpha)(1 - \beta)},$$
$$(x + y)(\alpha + \beta) \geqslant 2(x + y)\sqrt{\alpha\beta},$$

and it only remains to add these estimates. □

Lemma 2. *Let $f, g : \mathbb{T} \to \mathbb{C}$. Assume that f is even and g is odd, i.e., that $f(-z) = f(z)$, $g(-z) = -g(z)$ for every $z \in \mathbb{T}$. For every set $H \subset \mathbb{T}$, put $F(H) \stackrel{def}{=} \int_H |f|^2 d\mu$, $G(H) \stackrel{def}{=} \int_H |g|^2 d\mu$. For every set $E \subset \mathbb{T}$, put $E^+ \stackrel{def}{=} E \cup (-E)$, $E^- \stackrel{def}{=} E \cap (-E)$. Then for the function $p = f + g$ we have*

$$\int_E |p|^2 d\mu \geqslant \sqrt{\frac{F(E^+)}{F(\mathbb{T})}\frac{F(E^-)}{F(\mathbb{T})}\frac{G(E^+)}{G(\mathbb{T})}\frac{G(E^-)}{G(\mathbb{T})}} \int_{\mathbb{T}} |p|^2 d\mu.$$

Proof. Let E^* be the "antisymmetric" part of E, i.e., $E^* \stackrel{\text{def}}{=} \{z \in E : -z \notin E\}$. Since both $|f|^2$ and $|g|^2$ are even functions, we get $F(E^\pm) = F(E) \pm F(E^*)$, $G(E^\pm) = G(E) \pm F(E^*)$, whence

$$\sqrt{\frac{F(E^+)}{F(\mathbb{T})}\frac{F(E^-)}{F(\mathbb{T})}\frac{G(E^+)}{G(\mathbb{T})}\frac{G(E^-)}{G(\mathbb{T})}} = \frac{\sqrt{(F(E)^2 - F(E^*)^2)(G(E)^2 - G(E^*)^2)}}{F(\mathbb{T})G(\mathbb{T})}.$$

We also have

$$\int_E |p|^2 d\mu = F(E) + G(E) + 2Re \int_E f\bar{g}\, d\mu$$
$$= F(E) + G(E) + 2Re \int_{E^*} f\bar{g}\, d\mu \geqslant F(E) + G(E) - 2\sqrt{F(E^*)G(E^*)},$$

because $f\bar{g}$ is an odd function, so that its integral over the symmetric part of E vanishes. Finally, $\int_{\mathbb{T}} |p|^2 d\mu = F(\mathbb{T}) + G(\mathbb{T})$. Thus, it suffices to show that

$$F(E) + G(E) - 2\sqrt{F(E^*)G(E^*)}$$
$$\geqslant \left(\frac{1}{F(\mathbb{T})} + \frac{1}{G(\mathbb{T})}\right)\sqrt{(F(E)^2 - F(E^*)^2)(G(E)^2 - G(E^*)^2)}.$$

Observe that $F(\mathbb{T}) \geqslant F(E^+) = F(E) + F(E^*)$ and $G(\mathbb{T}) \geqslant G(E^+) = G(E) + G(E^*)$. Substituting these estimates in denominators, we see that Lemma 2 is a direct consequence of Lemma 1 with $A = F(E)$, $a = F(E^*)$, $B = G(E)$, and $b = G(E^*)$. □

Now, we prove the inequality

$$\int_E |p|^2 d\mu \geqslant e^{-An\mu(\widetilde{E})} \int_{\mathbb{T}} |p|^2 d\mu \qquad (**)$$

by induction on n (as before, we assume that p is a trigonometric polynomial in n characters and $\mu(E) \geqslant \frac{1}{3}$). Note that estimate $(*)$ with $q = 2$ is equivalent to the inequality

$$\int_E |p|^2 d\mu \geqslant \mu(E) e^{-2A(n-1)\mu(\widetilde{E})} \int_{\mathbb{T}} |p|^2 d\mu,$$

which is always weaker than $(**)$ except for the trivial case where $n = 1$.

Induction base. If $n = 1$, we have

$$\int_E |p|^2 d\mu = \mu(E) \int_{\mathbb{T}} |p|^2 d\mu \geqslant e^{-An\mu(\widetilde{E})} \int_{\mathbb{T}} |p|^2 d\mu$$

for every set E of measure $\mu(E) \geqslant \frac{1}{3}$, because $t \geqslant e^{-A(1-t)}$ for $t \in [\frac{1}{3}, 1]$, provided that $A \geqslant 3$, say.

If $\frac{1}{3} \leqslant \mu(E) \leqslant \frac{2}{3}$, the estimate for "small" sets cited above implies that

$$\int_E |p|^2 d\mu \geqslant \mu(E) \left\{\frac{\mu(E)}{14e}\right\}^{2(n-1)} \int_{\mathbb{T}} |p|^2 d\mu$$
$$\geqslant \left\{\frac{\mu(E)}{14e}\right\}^{2n} \int_{\mathbb{T}} |p|^2 d\mu \geqslant e^{-An\mu(\widetilde{E})} \int_{\mathbb{T}} |p|^2 d\mu$$

for every $n \in \mathbb{N}$, because $\left\{\frac{t}{14e}\right\}^{2n} \geqslant e^{-An(1-t)}$ for $\frac{1}{3} \leqslant t \leqslant \frac{2}{3}$, provided that $A \geqslant 30$, say.

Induction step.

Let p be a polynomial in $n > 1$ characters and let E be a subset of $\mathbb{T}$ of measure $\mu(E) \geqslant \frac{2}{3}$. We split p into the sum of its even and odd parts $f(z) \stackrel{\text{def}}{=} \frac{1}{2}(p(z)+p(-z))$ and $g(z) \stackrel{\text{def}}{=} \frac{1}{2}(g(z) - g(-z))$.

Without loss of generality, we may assume that both f and g do not vanish identically. Indeed, the "worst" set E, i.e., the set that delivers the least value to the integral $\int_E |p|^2 d\mu$, is a level set of $|p|$. Therefore, if p contains only odd or only even powers, this set must be symmetric together with $|p|$. But then we can merely take z^2 as a new variable.

Note that both f and g are again trigonometric polynomials. Let n_1 and n_2 be the number of characters in f and g, respectively. Clearly, $n_1 + n_2 = n$. In the notation of Lemma 2, we get

$$\int_E |p|^2 d\mu \geqslant \sqrt{\frac{F(E^+)}{F(\mathbb{T})}\frac{F(E^-)}{F(\mathbb{T})}\frac{G(E^+)}{G(\mathbb{T})}\frac{G(E^-)}{G(\mathbb{T})}} \int_{\mathbb{T}} |p|^2 d\mu.$$

Observe that both n_1 and n_2 are strictly less than n, while our assumption $\mu(E) \geqslant \frac{2}{3}$ implies that $\mu(E^+) \geqslant \mu(E^-) \geqslant \frac{1}{3}$. Therefore, we may apply the induction hypothesis to each polynomial f and g and each set E^+ and E^-, obtaining

$$\begin{aligned}
&\sqrt{\frac{F(E^+)}{F(\mathbb{T})}\frac{F(E^-)}{F(\mathbb{T})}\frac{G(E^+)}{G(\mathbb{T})}\frac{G(E^-)}{G(\mathbb{T})}} \\
&\qquad \geqslant \sqrt{e^{-An_1\mu(\mathbb{T}\setminus E^+)}e^{-An_1\mu(\mathbb{T}\setminus E^-)}e^{-An_2\mu(\mathbb{T}\setminus E^+)}e^{-An_2\mu(\mathbb{T}\setminus E^-)}} \\
&\qquad\qquad = \sqrt{e^{-A(n_1+n_2)(\mu(\mathbb{T}\setminus E^+)+\mu(\mathbb{T}\setminus E^-))}} = e^{-An\mu(\widetilde{E})}.
\end{aligned}$$

This completes the proof of inequality $(**)$.

§2. The case where $0 \leqslant q < 2$

The L^2-estimates for small and large sets can be united in one inequality:

$$\|p\|_{L^2(E)} \geqslant \mu(E)^{A(n-1)} \|p\|_{L^2(\mathbb{T})} \qquad (***)$$

with some (large) absolute constant A (from the above argument it follows that $A = 30$ fits).

We are going to show that

$$\|p\|_{L^q(E)} \geqslant \mu(E)^{2A(n-1)} \|p\|_{L^q(\mathbb{T})}$$

for every $0 < q < 2$ (the case of the L^0-norm can be settled by letting $q \to 0$). For $\mu(E) \geqslant \frac{1}{3}$ this inequality is equivalent to $(*)$ up to the absolute constant involved. To proceed, we need the following classical lemma on convex functions.

Lemma 3. *Let $F, G : [0,1] \to [0,+\infty)$ be two continuously differentiable decreasing convex functions such that $F(1) = G(1)$ and $F(t) \geqslant G(t)$ for every $t \in [0,1]$. Then for every $\beta \in (0,1]$ and for every $s \in [0,1]$,*

$$\int_s^1 |F'(t)|^\beta dt \geqslant \int_s^1 |G'(t)|^\beta dt.$$

Now, let Φ be the decreasing rearrangement of $|p|$. We must prove that

$$\int_\delta^1 \Phi(t)^q dt \geqslant (1-\delta)^{2Aq(n-1)+1} \int_0^1 \Phi(t)^q dt,$$

or, equivalently,

$$\int_\delta^1 \Phi(t)^q dt \geqslant \frac{(1-\delta)^{2Aq(n-1)+1}}{1-(1-\delta)^{2Aq(n-1)+1}} \int_0^\delta \Phi(t)^q dt \qquad (0<\delta<1).$$

Let $F(s) \stackrel{\text{def}}{=} \int_s^1 \Phi(t)^2 dt$. Obviously, F is a convex decreasing function on $[0,1]$. Since multiplication of the polynomial p by a constant factor changes nothing, there is no loss of generality in assuming that $\int_{\mathbb{T}} |p|^2 d\mu = 1$, i.e., $F(0) = 1$. Then, estimate $(***)$ can be rewritten as $F(t) \geqslant (1-t)^{2A(n-1)+1} \stackrel{\text{def}}{=} G(t)$. Using the Hölder inequality, we get

$$\int_0^\delta \Phi(t)^q dt \leqslant \delta \left\{ \frac{1}{\delta} \int_0^\delta \Phi(t)^2 dt \right\}^{q/2}$$
$$= \delta \left\{ \frac{1}{\delta}(1-F(\delta)) \right\}^{q/2} \leqslant \delta \left\{ \frac{1}{\delta}(1-(1-\delta)^{2A(n-1)+1}) \right\}^{q/2}.$$

On the other hand, from Lemma 3 it follows that

$$\int_\delta^1 \Phi(t)^q dt = \int_\delta^1 |F'(t)|^{q/2}$$
$$\geqslant \int_\delta^1 |G'(t)|^{q/2} = \{2A(n-1)+1\}^{q/2} \frac{1}{Aq(n-1)+1} (1-\delta)^{Aq(n-1)+1}.$$

So, it suffices to show that

$$\{2A(n-1)+1\}^{q/2} \frac{1}{Aq(n-1)+1} (1-\delta)^{Aq(n-1)+1}$$
$$\geqslant \frac{(1-\delta)^{2Aq(n-1)+1}}{1-(1-\delta)^{2Aq(n-1)+1}} \delta \left\{ \frac{1}{\delta}(1-(1-\delta)^{2A(n-1)+1}) \right\}^{q/2}.$$

Since

$$\frac{1}{\delta}(1-(1-\delta)^{2A(n-1)+1}) \leqslant 2A(n-1)+1,$$

it suffices to check that

$$\frac{1}{Aq(n-1)+1}(1-\delta)^{Aq(n-1)+1} \geqslant \frac{(1-\delta)^{2Aq(n-1)+1}}{1-(1-\delta)^{2Aq(n-1)+1}} \delta,$$

or, equivalently,

$$1-(1-\delta)^{2Aq(n-1)+1} \geqslant (Aq(n-1)+1)\delta(1-\delta)^{Aq(n-1)}.$$

Denoting $r \stackrel{\text{def}}{=} Aq(n-1)$, we get the inequality $1-(1-\delta)^{2r+1} \geqslant (r+1)\delta(1-\delta)^r$ to prove. But the left-hand side equals

$$(2r+1)\int_0^\delta (1-t)^{2r}dt = (2r+1)\int_0^{\delta/2} \{(1-t)^{2r}+(1-\delta+t)^{2r}\}$$
$$\geqslant (2r+1)2\int_0^{\delta/2} \{(1-t)(1-\delta+t)\}^r \geqslant (2r+1)\delta(1-\delta)^r,$$

because $(1-t)(1-\delta+t) \geqslant 1-\delta$ for every $t \in [0, \frac{\delta}{2}]$. This is even stronger than the inequality required, and we are done.

§3. Concluding remarks

First, we recall that both the classical Turan lemma and its generalization for "small" sets remain valid for the exponential polynomials $p(t) = \sum_{k=1}^n c_k e^{i\lambda_k t}$ ($c_k \in \mathbb{C}$, $\lambda_k \in \mathbb{R}$), with no *a priori* separation condition imposed on the λ's. Namely, for every interval $I \subset \mathbb{R}$, every measurable set $E \subset I$, and every $q \in [0, +\infty]$, we have

$$\|p\|_{L^q(I)} \leqslant \left\{\frac{A|I|}{|E|}\right\}^{n-1} \|p\|_{L^q(E)}$$

with $A = 314e$, say. On the other hand, the argument in §1 depends heavily on the group structure on $\mathbb{T}$, and I have no idea of how to get a good estimate for arbitrary exponential polynomials and for the sets E close to the entire interval I. The second remark concerns the case where $q > 2$.

If the set $E \subset \mathbb{T}$ is not "too large", namely, if $\mu(E) \leqslant 1 - \frac{\log n}{n}$, then, with the help of the trivial estimate $\|p\|_{L^\infty(\mathbb{T})} \leqslant \sqrt{n}\,\|p\|_{L^2(\mathbb{T})}$, inequality $(*)$ can be proved for all $q > 2$ (with $A+1$ in place of A).

If the measure of $\widetilde{E}$ is essentially smaller than $\frac{\log n}{n}$, inequality $(*)$ may fail for $q > 2$. In this case I cannot prove estimates as sharp as those for $0 \leqslant q \leqslant 2$. However, I can completely answer the question as to how large the measure of a set $E \subset \mathbb{T}$ should be to ensure that the norms $\|p\|_{L^q(I)}$ and $\|p\|_{L^q(E)}$ are equivalent for all trigonometric polynomials $p(z)$ in n characters. The answer is that for $2 < q < +\infty$ we must require that $\mu \leqslant \dfrac{\text{Const}}{n^{q/2}}$, and for $q = +\infty$ this condition should be replaced by $\mu \leqslant e^{-\text{const}\, n}$.

References

[1] F. L. Nazarov, *Local estimates for exponential polynomials and their applications to inequalities of the uncertainty principle type*, Algebra i Analiz, **5** (1993), no. 4, 3–66.

F.L. Nazarov
Department of Mathematics,
Michigan State University
East Lansing, MI
48824, USA
E-mail address: fedja@math.msu.edu

Operator Theory:
Advances and Applications, Vol. 113

Ball, Haagerup, and Distribution Functions

Fedor L. Nazarov, Anatoliy N. Podkorytov

Dedicated to the memory of S. A. Vinogradov

Abstract. We discuss an elementary trick that sometimes yields simple proofs of integral inequalities of the kind $\int_{\mathcal{X}}(f^s - g^s)\,d\mu \geqslant 0$. We use this trick to obtain "computation-free" proofs of two famous theorems: Ball's theorem on the sections of a cube and Haagerup's theorem on the sharp constants in the Khinchin inequality for Rademacher functions.

Introduction

This paper does not contain any new results. Its only aim is to acquaint the reader with a completely elementary trick that often allows one to prove the integral inequality $\Delta(s) = \int_{\mathcal{X}} (f^s - g^s)\,d\mu \geqslant 0$ for $s > s_0$, if one knows that it holds for $s = s_0$ (in particular, if $f, g \in L^s(\mathcal{X}, \mu)$, we are interested in the preservation of the inequality between the L^s-norms of the functions f and g as s increases). Such estimates have various applications in analysis, especially in Banach space geometry.

In this paper we will use our trick to get rid of ugly numerical computations in the proofs of two famous results. The first chapter is devoted to the beautiful theorem of Keith Ball [1] that claims that the maximal $(n-1)$-dimensional volume of a section of the n-dimensional unit cube by a hyperplane is $\sqrt{2}$, i.e., that the maximal section is just the product of the diagonal of the two-dimensional unit square and the $(n-2)$-dimensional unit cube. The second chapter deals with the problem of finding the sharp constants in the Khinchin inequality for Rademacher functions, which was first solved by Uffe Haagerup [5]. We believe that in both cases[1] our proofs are much simpler (both technically and ideologically), shorter, and more elegant than the original ones.

Our main idea will be to use distribution functions, the tool that proved extremely helpful in the probability theory. In this way we shall present a simple sufficient condition for "almost monotonicity" of the integral $\Delta(s)$. More precisely, our condition will allow us to conclude that the product $s^{-1}Q^s\Delta(s)$ is monotone increasing for some $Q > 0$.

[1]To be honest, we will prove only one half of Haagerup's theorem in this paper. The second half can be obtained by our methods as well, but the proof becomes more complicated and less transparent. So, after hesitating for a while, we finally decided not to include it.

Chapter I. Ball's theorem on the sections of the unit cube

Let $t \in \mathbb{R}$. Let $a = (a_1, \ldots, a_n)$ be a unit vector in $\mathbb{R}^n$ $(n > 1)$. By $\sigma(a, t)$ we will denote the $(n-1)$-dimensional volume of the section of the cube $\left[-\frac{1}{2}, \frac{1}{2}\right]^n$ by the hyperplane $\{x \in \mathbb{R}^n : (x, a) = t\}$.

Theorem (Ball). *For any $t \in \mathbb{R}$ and for any unit vector $a \in \mathbb{R}^n$, one has*

$$\sigma(a, t) \leqslant \sqrt{2}.$$

Equality is attained only if $t = 0$ and the vector a has only two non-zero coordinates, each having absolute value $\frac{1}{\sqrt{2}}$, i.e., when the section is a product of the cube $\left[-\frac{1}{2}, \frac{1}{2}\right]^{n-2}$ and the diagonal of the square $\left[-\frac{1}{2}, \frac{1}{2}\right]^2$.

Among other things, this theorem implies that in Euclidean space of sufficiently large dimension all central sections of the ball of n-dimensional volume 1 have greater $(n-1)$-dimensional volumes than the corresponding sections of the unit cube. To prove this (rather counter-intuitive) fact, it is enough to observe that the $(n-1)$-dimensional volumes of the central sections of the ball of n-dimensional volume 1 tend to $\sqrt{e}$ as $n \to +\infty$. Actually, this phenomenon occurs starting with $n = 10$ (see [2] for details).

Thus, Ball's theorem provides the simplest counterexample to the famous conjecture of Busemann and Petty [3]. (They conjectured that of two convex bodies in $\mathbb{R}^n$ symmetric with respect to the origin, the one with greater $(n-1)$-dimensional volumes of central sections always has greater n-dimensional volume). The statement of the conjecture is obvious in $\mathbb{R}^2$. Recently it has been proved for $\mathbb{R}^3$ (see [4]). For $5 \leqslant n \leqslant 9$ there are counterexamples (see [6]), though not quite as simple as a ball and a cube. The case $n = 4$ remains controversial to the best of our knowledge. The latest rumor we heard was that the previously claimed counterexample turned out to be faulty and that a proof of the conjecture was found instead. We leave it to the reader to ferret out the truth.

Proof. Fix a unit vector $a \in \mathbb{R}^n$. If at least one coordinate a_j of the vector a vanishes, we can reduce the dimension of the problem. So it is enough to consider the case when all $a_j \neq 0$. Since the cube $\left[-\frac{1}{2}, \frac{1}{2}\right]^n$ is symmetric with respect to each coordinate hyperplane, we may also assume below that $a_j > 0$ for each j.

We shall split the proof of Ball's theorem into several steps.

Step 1. One of the coordinates a_j is large

Observe that $\sigma(a, t) \leqslant \frac{1}{a_j}$ for each $j = 1, \ldots, n$. Indeed, the volume of a section of the unit cube by a hyperplane orthogonal to a equals the volume of its projection to the coordinate hyperplane orthogonal to the j-th basic unit vector divided by the cosine of the angle between the vector a and that basic unit vector, i.e., by a_j. It remains to note that the projection of *the whole unit cube* to any of the coordinate hyperplanes has volume 1. Thus $\sigma(a, t) < \sqrt{2}$ if one of the coordinates a_j is greater than $\frac{1}{\sqrt{2}}$.

From now on, we shall assume that $a_j \leqslant \frac{1}{\sqrt{2}}$ for all $j = 1, \ldots, n$.

Step 2. Integral representation for $\sigma(a,t)$
Consider the function $S(t) := \sigma(a,t)$. For every $u \in \mathbb{R}$, we have

$$\widehat{S}(u) = \int_{\mathbb{R}} \sigma(a,t)e^{-2\pi iut}dt = \int_{[-\frac{1}{2},\frac{1}{2}]^n} e^{-2\pi iu(x,a)}\,dx$$
$$= \prod_{j=1}^{n} \int_{-\frac{1}{2}}^{\frac{1}{2}} e^{-2\pi iua_jx_j}\,dx_j = \prod_{j=1}^{n} \frac{\sin \pi a_j u}{\pi a_j u}.$$

Taking the inverse Fourier transform, we get

$$\sigma(a,t) = S(t) = \int_{-\infty}^{\infty} \widehat{S}(u)e^{2\pi iut}du = \int_{-\infty}^{\infty} e^{2\pi iut} \prod_{j=1}^{n} \frac{\sin \pi a_j u}{\pi a_j u} du.$$

From this representation one can derive that the naive conjecture that the section of the largest volume is the central section orthogonal to the main diagonal is false. Indeed, the volume of this section (corresponding to $t = 0$, $a_j = \frac{1}{\sqrt{n}}$) equals

$$\int_{\mathbb{R}} \left(\frac{\sin \frac{\pi u}{\sqrt{n}}}{\frac{\pi u}{\sqrt{n}}} \right)^n du = \sqrt{n} \int_{\mathbb{R}} \left(\frac{\sin \pi v}{\pi v} \right)^n dv$$

Using the Laplace asymptotic formula, we find that this volume tends to $\sqrt{\frac{6}{\pi}} < \sqrt{2}$ as $n \to \infty$.

Step 3. Reduction to Ball's integral inequality
Since $a_1^2 + \cdots + a_n^2 = 1$, the integral representation of $\sigma(a,t)$ together with the Hölder inequality implies

$$\sigma(a,t) \leqslant \int\limits_{\mathbb{R}} \prod_{j=1}^{n} \left| \frac{\sin \pi a_j u}{\pi a_j u} \right| du \leqslant \prod_{j=1}^{n} \left(\int\limits_{\mathbb{R}} \left| \frac{\sin \pi a_j u}{\pi a_j u} \right|^{1/a_j^2} du \right)^{a_j^2}.$$

Thus, we need only show that every integral $\int_{\mathbb{R}} \left| \frac{\sin \pi a_j u}{\pi a_j u} \right|^{1/a_j^2} du$ does not exceed $\sqrt{2}$ and that equality is possible only when $a_j = \frac{1}{\sqrt{2}}$. Taking into account that $\frac{1}{a_j^2} \geqslant 2$ for every j and making the change of variable $u \to a_j u$, we see that it would suffice to prove the following **Ball's integral inequality**:

$$\int_{\mathbb{R}} \left| \frac{\sin \pi x}{\pi x} \right|^s dx < \sqrt{\frac{2}{s}} \qquad \text{for every } s > 2$$

(for $s = 2$, it turns into identity).

Ball's original proof of this inequality, though quite straightforward, involved a lot of unpalatable numerical computations some of which were almost impossible to verify without a calculator. Below we shall use a completely different approach.

Step 4. Lemma on distribution functions

Let $(\mathcal{X}, \mu)$ be a measure space. For a non-negative function $f : \mathcal{X} \to [0, +\infty)$, its **distribution function** $F(y)$, $y > 0$ is defined by

$$F(y) := \mu\{x \in \mathcal{X} \,:\, f(x) > y\}.$$

Clearly, F is a non-increasing function on $(0, +\infty)$.

Lemma on distribution functions. *Let f and g be any two nonnegative measurable functions on a measure space $(\mathcal{X}, \mu)$. Let F and G be their distribution functions. Assume that both $F(y)$ and $G(y)$ are finite for every $y > 0$. Assume also that at some point y_0 the difference $F - G$ changes sign from $-$ to $+$, i.e., $F(y) \leqslant G(y)$ for all $y \in (0, y_0)$ and $F(y) \geqslant G(y)$ for all $y > y_0$.*

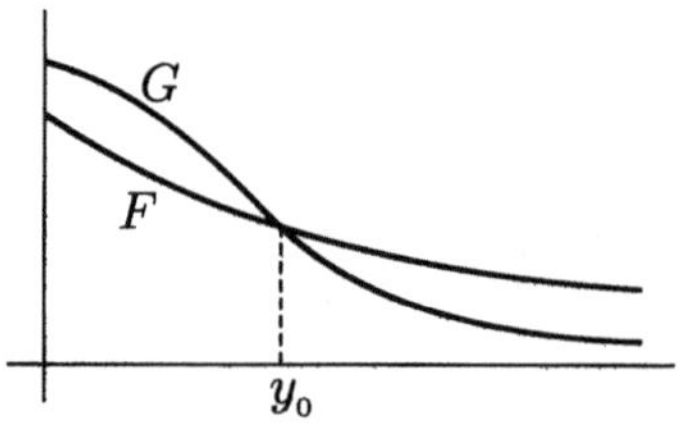

Let $S = \{s > 0 \,:\, f^s - g^s \in L^1(\mathcal{X}, \mu)\}$. Then the function

$$\varphi(s) = \frac{1}{s y_0^s} \int_{\mathcal{X}} (f^s - g^s)\, d\mu$$

is monotone increasing on S. In particular, if $\int_{\mathcal{X}} (f^{s_0} - g^{s_0}) d\mu = 0$, then $\int_{\mathcal{X}} (f^s - g^s) d\mu \geqslant 0$ for each $s > s_0$. The equality may hold only if the functions F and G coincide.

Remark. *Note that we do not demand the finiteness of the integrals $\int_{\mathcal{X}} f^s\, d\mu$ and $\int_{\mathcal{X}} g^s\, d\mu$ in the definition of the set S. Only the difference $f^s - g^s$ is required to be Lebesgue integrable with respect to the measure μ. Though this observation is of little importance now, it will play crucial role in Chapter II.*

Proof. Observe first of all that if the difference $f - g$ of two nonnegative functions f and g is Lebesgue integrable and both their distribution functions $F(y)$ and $G(y)$ are finite for all $y > 0$, then

$$\int_{\mathcal{X}} (f - g)\, d\mu = \int_0^\infty (F(y) - G(y))\, dy.$$

Indeed, let $h(x) = \min(f(x), g(x))$. The distribution function $H(y)$ of the function h does not exceed $F(y)$ and therefore is finite for every $y > 0$. Also, the (non-negative) difference $f - h$ does not exceed $|f - g|$, and thereby is Lebesgue integrable. Now, applying the Fubini theorem to the characteristic function of the set $\{(x,y) \in \mathcal{X} \times (0,+\infty) : h(x) \leqslant y < f(x)\}$, we obtain

$$0 \leqslant \int_{\mathcal{X}} (f-h)\,d\mu = \int_0^\infty (F(y) - H(y))\,dy < +\infty.$$

Analogously,

$$0 \leqslant \int_{\mathcal{X}} (g-h)\,d\mu = \int_0^\infty (G(y) - H(y))\,dy < +\infty.$$

Subtracting the second identity from the first, we obtain the desired statement.

Since for $s > 0$ the distribution function of, say, f^s equals $F(y^{1/s})$, we get

$$\int_{\mathcal{X}} (f^s - g^s)\,d\mu = \int_0^\infty \left(F(y^{1/s}) - G(y^{1/s})\right) dy = s\int_0^\infty y^{s-1}\,(F(y) - G(y))\,dy.$$

Therefore for $s > s_0$ $(s, s_0 \in S)$, we have

$$\varphi(s) - \varphi(s_0) = \frac{1}{y_0}\int_0^\infty \left(\left(\frac{y}{y_0}\right)^{s-1} - \left(\frac{y}{y_0}\right)^{s_0-1}\right)(F(y) - G(y))\,dy \geqslant 0,$$

because both factors in the integrand simultaneously change their signs at y_0. □

Step 5. Proof of Ball's integral inequality

Using the Euler-Poisson integral $\int_{\mathbb{R}} e^{-\pi x^2}dx = 1$, we can rewrite Ball's inequality as $\int_0^\infty g^s < \int_0^\infty f^s$, where $f(x) = e^{-\pi x^2/2}$, $g(x) = \left|\frac{\sin \pi x}{\pi x}\right|$. We need to prove this inequality for all powers $s > 2$. Since for $s = 2$ we have an identity, it is enough to show that the corresponding distribution functions F and G satisfy the condition of the lemma from Step 4 (with $\mathcal{X} = (0,\infty)$, $d\mu = dx$, $s_0 = 2$). Observe that both functions f and g are strictly less than 1 on the whole ray $(0,+\infty)$ and thereby $F(y) = G(y) = 0$ for each $y \geqslant 1$. So, we may restrict ourselves to the values $y \in (0,1)$.

Since f is monotone decreasing on $(0,\infty)$, we have

$$F(y) = f^{-1}(y) = \sqrt{\frac{2}{\pi}\ln\frac{1}{y}}.$$

To estimate $G(y)$, consider the quantities $y_m = \max_{[m,m+1]} g$, $m \in \mathbb{N}$ (the graph of $g(x)$ for $x \geqslant 1$ looks as a sequence of decreasing humps and y_m is just the height of the m-th hump in this sequence – see the picture below). Obviously, $\frac{1}{\pi(m+\frac{1}{2})} < y_m < \frac{1}{\pi m}$.

For $x \in (0,1)$, we have

$$g(x) = \prod_{k=1}^\infty \left(1 - \frac{x^2}{k^2}\right) \leqslant \prod_{k=1}^\infty e^{-x^2/k^2} = e^{-(\pi x)^2/6} < e^{-\pi x^2/2} = f(x).$$

Therefore, for all $y \in (y_1, 1)$,

$$G(y) = \mu\{x \in (0,1) : g(x) > y\} < \mu\{x \in (0,1) : f(x) > y\} \leqslant F(y),$$

i.e., the difference $F - G$ is positive on the interval $(y_1, 1)$.

The difference $F-G$ must change sign *at least* once on $(0,+\infty)$ because $\int_0^\infty 2y(F(y)-G(y))\,dy = \int_0^\infty (f(x)^2-g(x)^2)\,dx = 0$. To prove that the change of sign occurs *only* once, it now suffices to show that $F-G$ is monotone increasing on $(0,y_1)$. Since both F and G are continuous monotone decreasing functions, it is enough to prove that $|G'(y)| > |F'(y)|$ on each interval (y_{m+1}, y_m), $m \in \mathbb{N}$.

For $y \in (0,y_1)$, the value $G(y)$ can be visualized as the sum of the lengths of the thick black intervals on the picture below.

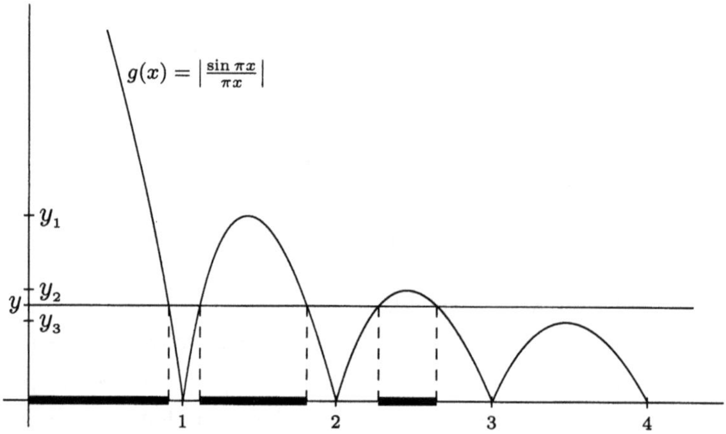

$G(y)$ equals the total length of the thick black intervals.

It is clear from this picture that for each $y \in (0,y_1)$, $y \neq y_j$, the function G is differentiable and

$$|G'(y)| = \sum_{x>0\,:\,g(x)=y} \frac{1}{|g'(x)|}.$$

When $y \in (y_{m+1}, y_m)$, the equation $g(x) = y$ has one root on the interval $(0,1)$ and two roots on each interval $(k,k+1)$, $k = 1,\dots,m$. Let us estimate $|g'(x)|$ from above at these roots.

- If the root $x \in (k,k+1)$, $k \geqslant 1$, then

$$|g'(x)| = \left|\frac{\cos \pi x}{x} - \frac{\sin \pi x}{\pi x^2}\right| \leqslant \frac{1}{x}\left(1 + \frac{|\sin \pi(x-k)|}{\pi x}\right) \leqslant \frac{1}{x}\left(1+\frac{\pi(x-k)}{\pi k}\right) = \frac{1}{k};$$

- If the root $x \in (0,1)$, then

$$|g'(x)| = \frac{\sin \pi x - \pi x \cos \pi x}{\pi x^2} = \frac{1}{\pi x^2}\int_0^{\pi x} t \sin t\,dt \leqslant \frac{1}{\pi x^2}\int_0^{\pi x} t\,dt = \frac{\pi}{2}.$$

Thus for every $y \in (y_{m+1}, y_m)$, we have

$$|G'(y)| \geqslant \frac{2}{\pi} + 2\sum_{k=1}^{m} k = \frac{2}{\pi} + m + m^2.$$

Recalling that $F(y) = \sqrt{\frac{2}{\pi}\ln\frac{1}{y}}$, we obtain

$$\left|\frac{G'(y)}{F'(y)}\right| = |G'(y)|\, y\sqrt{2\pi\ln\frac{1}{y}} \geqslant \left(\frac{2}{\pi} + m + m^2\right) y\sqrt{2\pi\ln\frac{1}{y}}.$$

Since $y_1 < \frac{1}{\pi} < \frac{1}{\sqrt{e}}$, the function $y\sqrt{2\pi\ln\frac{1}{y}}$ is monotone increasing on $(0, y_1)$. Recalling that $y > y_{m+1} > \dfrac{1}{\pi(m+\frac{3}{2})}$, we get

$$\left|\frac{G'(y)}{F'(y)}\right| > \frac{\frac{2}{\pi} + m + m^2}{\frac{3}{2} + m}\sqrt{\frac{2}{\pi}\ln\pi(m+\frac{3}{2})} > \sqrt{\frac{2}{\pi}\ln\frac{5\pi}{2}}$$

($m \geqslant 1$ and $\frac{2}{\pi} + m^2 > \frac{1}{2} + 1 = \frac{3}{2}$).

It remains to note that the right hand side is greater than 1 because $\ln 5x > x$ for every $x \in [1,2]$, and, in particular, for $x = \frac{\pi}{2}$ (since the function $\ln 5x$ is concave, it is enough to check this estimate at the ends of the interval, where it is obvious).

Ball's inequality is thus completely proved. □

Chapter II. Haagerup's theorem on sharp constants in the Khinchin inequality

Recall that the Rademacher functions $r_k(t)$ ($k = 1, 2, \ldots$, $t \in [0,1]$) are defined by

$$r_k(t) = \operatorname{sign}\sin(2^k\pi t)$$

From the probabilistic point of view, they are just independent random variables assuming values $+1$ and -1 with probability $\frac{1}{2}$ each. One of the most famous results on Rademacher functions is the **Khinchin inequality**:

$$A_p\left(\sum_{k=1}^{n} a_k^2\right)^{p/2} \leqslant R_p(a) := \int_0^1 \left|\sum_{k=1}^{n} a_k r_k(t)\right|^p dt \leqslant B_p\left(\sum_{k=1}^{n} a_k^2\right)^{p/2}$$

for any real coefficient vector $a = (a_1, \ldots, a_n)$ and for any $0 < p < +\infty$. Uffe Haagerup found the sharp values of the constants A_p and B_p in this inequality for all powers $p > 0$ (see [5]).

Formulation of Haagerup's theorem

To find the best constants A_p and B_p in the Khinchin inequality is the same as to solve the extremal problems

$$\int_0^1 \left|\sum_{k=1}^{n} a_k r_k(t)\right|^p dt \to \min, \qquad \sum_k a_k^2 = 1$$

and

$$\int_0^1 \left|\sum_{k=1}^{n} a_k r_k(t)\right|^p dt \to \max, \qquad \sum_k a_k^2 = 1$$

respectively. There are three "natural" linear combinations $\sum_k a_k r_k(t)$ to look at:

- $n = 1$, $a_1 = 1$ (a single Rademacher function). In this case $R_p(a) = 1$.
- $n = 2$, $a_1 = a_2 = \frac{1}{\sqrt{2}}$ (the sum or the difference of two Rademacher functions). In this case $R_p(a) = 2^{\frac{p-2}{2}}$.
- $n \to +\infty$, $a_1 = \cdots = a_n = \frac{1}{\sqrt{n}}$. In this case it is more convenient to accept the probabilistic point of view and to conclude, using the central limit theorem, that we are actually talking about the normal distribution. Hence, the corresponding (limit) value of $R_p(a)$ equals

$$\frac{1}{\sqrt{2\pi}} \int_{\mathbb{R}} |x|^p e^{-\frac{x^2}{2}} dx = \sqrt{\frac{2}{\pi}} \int_0^\infty x^p e^{-\frac{x^2}{2}} dx$$
$$= 2^{\frac{p-1}{2}} \sqrt{\frac{2}{\pi}} \int_0^\infty y^{\frac{p-1}{2}} e^{-y} dy = \frac{2^{\frac{p}{2}}}{\sqrt{\pi}} \Gamma\left(\frac{p+1}{2}\right) = 2^{\frac{p-2}{2}} \frac{\Gamma\left(\frac{p+1}{2}\right)}{\Gamma\left(\frac{3}{2}\right)}$$

(the statement of the central limit theorem for bounded symmetric random variables is strong enough to ensure convergence of moments of any order).

Now Haagerup's theorem can be stated in a very simple way:

Theorem (Haagerup). *For each $p \in (0, +\infty)$, the extremum in either of the extremal problems is one of these three values.*

What does this mean in terms of formulae? First of all, note that $A_p = 1$ for $p \geqslant 2$ and $B_p = 1$ for $p \leqslant 2$ (this is just the Hölder inequality). So, the only non-trivial tasks are to find A_p for $p < 2$ and to find B_p for $p > 2$. Haagerup's theorem claims that in these cases the extreme value is delivered by either the sum of two Rademacher functions or by the normal distribution, i.e.,

$$A_p = 2^{\frac{p-2}{2}} \min\left(1, \frac{\Gamma(\frac{p+1}{2})}{\Gamma(\frac{3}{2})}\right); \qquad B_p = 2^{\frac{p-2}{2}} \max\left(1, \frac{\Gamma(\frac{p+1}{2})}{\Gamma(\frac{3}{2})}\right).$$

Using the alternative representation for the Γ-function

$$\Gamma(q) = \int_0^\infty t^q e^{-t} \frac{dt}{t} = \lim_{n\to\infty} \int_0^n t^q \left(1 - \frac{t}{n}\right)^n \frac{dt}{t} = \lim_{n\to\infty} \frac{n^q\, n!}{q(q+1)\dots(q+n)},$$

we conclude that the second derivative $\frac{d^2}{dq^2} \ln \Gamma(q) = \sum_{k=0}^{\infty} \frac{1}{(q+k)^2}$ of the function $\ln \Gamma(q)$ is *positive and monotone decreasing* on $(0, +\infty)$. Since $\Gamma(1) = \Gamma(2) = 1$, we can deduce from this that the function $\ln \Gamma(q)$ is convex on $(0, +\infty)$ and attains its minimum on the interval $[1, 2]$ somewhere to the left from its middle point $\frac{3}{2}$. In particular, this implies that the function Γ is monotone increasing on $[\frac{3}{2}, +\infty)$. So for $p > 2$, the maximum of $R_p(a)$ is delivered by the normal distribution and

$$B_p = 2^{\frac{p-2}{2}} \frac{\Gamma(\frac{p+1}{2})}{\Gamma(\frac{3}{2})}.$$

The case where $p < 2$ is slightly more complicated because there exists $q_0 \in (1, \frac{3}{2})$ such that $\Gamma(q) > \Gamma(\frac{3}{2})$ for $0 < q < q_0$, $\Gamma(q_0) = \Gamma(\frac{3}{2})$, and $\Gamma(q) < \Gamma(\frac{3}{2})$ for $q_0 < q < \frac{3}{2}$. Therefore

$$A_p = \begin{cases} 2^{\frac{p-2}{2}} & \text{if } 0 < p \leqslant p_0; \\ 2^{\frac{p-2}{2}} \frac{\Gamma(\frac{p+1}{2})}{\Gamma(\frac{3}{2})} & \text{if } p_0 \leqslant p \leqslant 2 \end{cases}$$

where $p_0 = 2q_0 - 1 \in (1, 2)$ is the only nontrivial positive root of the equation $\Gamma(\frac{p+1}{2}) = \Gamma(\frac{3}{2})$.

Our aim in this chapter will be to give a (relatively) simple proof of one half of Haagerup's theorem. Namely, we shall consider the case $p < 2$ (or, which is quite the same, shall find the sharp value of A_p for all $p > 0$). Haagerup also considers the cases $p < 2$ and $p > 2$ separately and it is almost impossible to decide for which of them his original proof is more painstaking.

Proof. Again, we will split the proof into several steps.

Step 1. Reduction of the case of $0 < p < p_0$ to the case of $p = p_0$

Assume that we already know that $A_{p_0} = 2^{\frac{p_0-2}{2}}$, i.e., that $R_p(a) \geqslant 2^{\frac{p_0-2}{2}}$ for any $a = (a_1, \dots, a_n) \in \mathbb{R}^n$ satisfying $\sum_k a_k^2 = 1$. Let now $0 < p < p_0$. Then, using the Hölder inequality

$$R_{p_0}(a) \leqslant R_p(a)^{\frac{2-p_0}{2-p}} R_2(a)^{\frac{p_0-p}{2-p}}$$

and taking into account that $R_2(a) = \sum_k a_k^2 = 1$, we conclude that

$$R_p(a) \geqslant \left(2^{\frac{p_0-2}{2}}\right)^{\frac{2-p}{2-p_0}} = 2^{\frac{p-2}{2}},$$

so $A_p = 2^{\frac{p-2}{2}}$ for all $0 < p < p_0$.

Thus we need only consider the powers $p_0 \leqslant p \leqslant 2$.

Step 2. Integral representation for $R_p(a)$

Let $0 < p < 2$. Then for every $s \in \mathbb{R}$,

$$|s|^p = C_p \int_0^\infty \frac{1 - \cos us}{u^{p+1}} \, du$$

where the constant $C_p > 0$ depends on p only. Plugging $s = \sum_1^n a_k r_k(t)$ into this identity and integrating over $t \in [0, 1]$, we find

$$\int_0^1 \left| \sum_1^n a_k r_k(t) \right|^p dt = C_p \int_0^\infty \frac{1 - \int_0^1 \cos\left(u \sum_1^n a_k r_k(t)\right) dt}{u^{p+1}} \, du.$$

Observing that

$$\int_0^1 \cos\left(u \sum_1^n a_k r_k(t)\right) = \operatorname{Re} \int_0^1 e^{iu \sum_1^n a_k r_k(t)} = \prod_1^n \cos a_k u,$$

we finally arrive at **Haagerup's integral representation**:

$$R_p(a) = \int_0^1 \left| \sum_1^n a_k r_k(t) \right|^p dt = C_p \int_0^\infty \frac{1 - \prod_1^n \cos a_k u}{u^{p+1}}\, du.$$

Step 3. The case of small coefficients: Reduction to Haagerup's integral inequality Let us again assume that $\sum_j a_k^2 = 1$. Young's inequality yields

$$\prod_1^n \cos a_k u \leqslant \prod_1^n |\cos a_k u| \leqslant \sum_1^n a_k^2 |\cos a_k u|^{1/a_k^2} = 1 - \sum_1^n a_k^2 (1 - |\cos a_k u|^{1/a_k^2})$$

and therefore

$$\int_0^1 \left| \sum_1^n a_k r_k(t) \right|^p dt \geqslant \sum_1^n a_k^2\, \mathcal{I}_p\left(\frac{1}{a_k^2}\right) \qquad (*)$$

where

$$\mathcal{I}_p(s) = C_p \int_0^\infty \left(1 - \left|\cos \frac{u}{\sqrt{s}}\right|^s\right) \frac{du}{u^{p+1}} \qquad (s > 0).$$

The value of the integral $\mathcal{I}_p(s)$ can be computed in two cases:

- **Case 1:** $s = 2$. It follows immediately from Haagerup's integral representation for $n = 2$, $a_1 = a_2 = \frac{1}{\sqrt{2}}$ that

$$\mathcal{I}_p(2) = 2^{\frac{p-2}{2}}.$$

- **Case 2:** $s = \infty$. The trivial estimate $\ln \cos\theta = O(\theta^2)$ $(\theta \to 0)$ implies that $1 - \left|\cos \frac{u}{\sqrt{s}}\right|^s \leqslant 1 - e^{-\operatorname{Const} u^2} \leqslant \operatorname{Const} u^2$ for $0 < u < 1$ and sufficiently large s. Since this difference also never exceeds 1, we may apply Lebesgue's dominated convergence theorem and conclude that the limit $\lim_{s\to\infty} \mathcal{I}_p(s)$ does exist and equals

$$\mathcal{I}_p(\infty) = C_p \int_0^\infty \left(1 - e^{-\frac{u^2}{2}}\right) \frac{du}{u^{p+1}}.$$

On the other hand, for every *even* $n \in \mathbb{N}$, Haagerup's integral representation with $a_1 = \cdots = a_n = \frac{1}{\sqrt{n}}$ yields

$$R_p(\tfrac{1}{\sqrt{n}}, \ldots, \tfrac{1}{\sqrt{n}}) = \mathcal{I}_p(n).$$

As we have seen above, the left hand side tends to $2^{\frac{p-2}{2}} \frac{\Gamma(\frac{p+1}{2})}{\Gamma(\frac{3}{2})}$ (which is our hypothetical value for A_p) as $n \to \infty$. Thus we also have the identity

$$\mathcal{I}_p(\infty) = 2^{\frac{p-2}{2}} \frac{\Gamma(\frac{p+1}{2})}{\Gamma(\frac{3}{2})}.$$

Recall that $\mathcal{I}_p(2) \geqslant \mathcal{I}_p(\infty)$ for all $p \in [p_0, 2)$. If we could only show that $\mathcal{I}_p(s) \geqslant \mathcal{I}_p(\infty)$ for all other $s \in [1, +\infty)$ as well, then the inequality (*) would yield

$$\int_0^1 \left|\sum_1^n a_k r_k(t)\right|^p dt \geqslant \left(\sum_1^n a_k^2\right) \mathcal{I}_p(\infty) = \mathcal{I}_p(\infty) = 2^{\frac{p-2}{2}} \frac{\Gamma\left(\frac{p+1}{2}\right)}{\Gamma\left(\frac{3}{2}\right)},$$

proving Haagerup's theorem. Note now that

$$\begin{aligned}\mathcal{I}_p(s) - \mathcal{I}_p(\infty) &= C_p \int_0^\infty \left(1 - \left|\cos \frac{u}{\sqrt{s}}\right|^s\right) \frac{du}{u^{p+1}} - C_p \int_0^\infty \left(1 - e^{-\frac{u^2}{2}}\right) \frac{du}{u^{p+1}} \\ &\overset{(u=x\sqrt{s})}{=} C_p s^{\frac{p}{2}} \int_0^\infty \left(e^{-s\frac{x^2}{2}} - |\cos x|^s\right) \frac{dx}{x^{p+1}}.\end{aligned}$$

So, the desired estimate $\mathcal{I}_p(s) \geqslant \mathcal{I}_p(\infty)$ is equivalent to **Haagerup's integral inequality**

$$H(p, s) := \int_0^\infty \left(e^{-s\frac{x^2}{2}} - |\cos x|^s\right) \frac{dx}{x^{p+1}} \geqslant 0,$$

which again is an inequality of the kind $\int_{\mathcal{X}} (f^s - g^s)\, d\mu \geqslant 0$, only now with $\mathcal{X} = (0, +\infty)$, $f(x) = e^{-x^2/2}$, $g(x) = |\cos x|$, and $d\mu = \frac{dx}{x^{p+1}}$.

Unfortunately, there is no hope to prove the inequality $H(p, s) \geqslant 0$ for every $p \in [p_0, 2)$, $s \geqslant 1$ by our distribution function technique. Indeed, since $H(p_0, 2) = 0$, such a proof of the estimate $H(p_0, s) > 0$ for $s > 2$ would automatically be a proof of the estimate $H(p_0, s) < 0$ for $s < 2$ and vice versa. Fortunately, it is still possible to show that $H(p, s) \geqslant 0$ for $s \geqslant 2$, which is enough to establish Haagerup's theorem for the case when all the coefficients a_k do not exceed $\frac{1}{\sqrt{2}}$.

Step 4. Modified lemma on distribution functions and the proof of Haagerup's integral inequality

Our lemma on distribution functions from Chapter I cannot be applied directly now because the corresponding distribution functions $F(y)$ and $G(y)$ are infinite for small values of y. Nevertheless, the *modified distribution functions*

$$F_*(y) = \mu\{x \in \mathcal{X} : f(x) < y\} \quad \text{and} \quad G_*(y) = \mu\{x \in \mathcal{X} : g(x) < y\}$$

are finite for all $0 < y < 1$, so we may use the following

Modified lemma on distribution functions. *Let $Y > 0$. Let $f, g : \mathcal{X} \to [0, Y]$ be any two measurable functions on a measure space $(\mathcal{X}, \mu)$. Let F_* and G_* be their modified distribution functions. Assume that both $F_*(y)$ and $G_*(y)$ are finite for every $y \in (0, Y)$. Assume also that at some point y_0 the difference $F_* - G_*$ changes* *sign from + to −, i.e., $F_*(y) \geqslant G_*(y)$ for all $y \in (0, y_0)$ and $F_*(y) \leqslant G_*(y)$ for all $y \in (y_0, Y)$.*

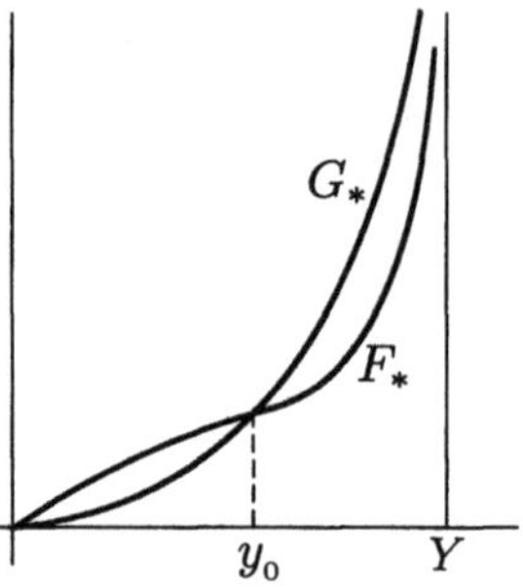

Let again $S = \{s > 0 : f^s - g^s \in L^1(\mathcal{X}, \mu)\}$. Then the function

$$\varphi(s) = \frac{1}{s y_0^s} \int_{\mathcal{X}} (f^s - g^s)\, d\mu$$

is monotone increasing on S. In particular, if $\int_{\mathcal{X}} (f^{s_0} - g^{s_0}) d\mu \geqslant 0$, then $\int_{\mathcal{X}} (f^s - g^s) d\mu \geqslant 0$ for each $s > s_0$.

The proof is essentially the same as in Chapter I.

Taking $\mathcal{X}$, μ, f, g as above and putting $Y = 1$, $s_0 = 2$, we see that to prove Haagerup's integral inequality in the desired range $p_0 \leqslant p \leqslant 2$, $s \geqslant 2$, we need only show that the corresponding modified distribution functions F_* and G_* satisfy the conditions of the modified lemma. These functions are not hard to compute: for $y \in (0, 1)$, we have

$$F_*(y) = \mu\{x > 0 : e^{-\frac{x^2}{2}} < y\} = \mu\left(\sqrt{2\ln\tfrac{1}{y}}, +\infty\right) = \int_{\sqrt{2\ln(1/y)}}^{+\infty} \frac{dx}{x^{p+1}} = \frac{1}{p}(2\ln\tfrac{1}{y})^{-p/2},$$

$$G_*(y) = \mu\{x > 0 : |\cos x| < y\} = \mu \bigcup_{k=0}^{\infty} (\pi k + \arccos y, \pi k + \pi - \arccos y)$$

$$= \sum_{k=0}^{\infty} \int_{\pi k + \arccos y}^{\pi k + \pi - \arccos y} \frac{dx}{x^{p+1}} = \frac{1}{p} \sum_{k=0}^{\infty} \left(\frac{1}{(\pi k + \arccos y)^p} - \frac{1}{(\pi k + \pi - \arccos y)^p} \right).$$

Make (decreasing) change of variable $y = \cos t$, $t \in (0, \frac{\pi}{2})$. Then our aim is to show that the difference $L(t)^{-p/2} - S(t)$, where

$$L(t) = (pF_*(\cos t))^{-2/p} = 2\ln\frac{1}{\cos t},$$

$$S(t) = pG_*(\cos t) = \frac{1}{t^p} - \frac{1}{(\pi - t)^p} + \frac{1}{(\pi + t)^p} - \frac{1}{(2\pi - t)^p} + \frac{1}{(2\pi + t)^p} - \dots,$$

changes its sign only once on the interval $(0, \frac{\pi}{2})$, this change being from $-$ to $+$, or, which is the same, that the difference

$$\Phi(t) = L(t) - S(t)^{-p/2}$$

changes its sign only once on the interval $(0, \frac{\pi}{2})$, this change being from $+$ to $-$.

The behavior of $\Phi(t)$ on the interval $(0, \frac{\pi}{2})$ is quite complicated: there is no obvious monotonicity or concavity that would allow us to give a simple proof of the desired property by *real variable methods*. Nevertheless our functions are simpler than the distribution functions from the proof of Ball's integral inequality in another respect: they both can be extended analytically to the complex plane. So, it seems reasonable to try complex variable methods. We will prove the following

Lemma on zeroes. *Let $1 < p < 2$. Then the analytic extension $\Phi(z)$ of the difference $\Phi(t)$ to the vertical strip $\{0 < \operatorname{Re} z < \frac{\pi}{2}\}$ has exactly one zero in that strip. This zero z_0 is real and the sign of $\Phi(t)$ changes from $+$ to $-$ at z_0.*

Proof. Since the minuend $L(t)$ and the subtrahend $S(t)^{-2/p}$ involve logarithms and fractional powers, some comment is due on how the corresponding branches in their extensions $L(z)$ and $S(z)^{-2/p}$ must be chosen. Since our complex-valued functions $L(z)$ and $S(z)$ of complex variable z are going to be extensions of the real-valued functions $L(t)$ and $S(t)$ of real variable t, we should have the symmetry $L(\overline{z}) = \overline{L(z)}$ and $S(\overline{z}) = \overline{S(z)}$. So, it is enough to trace the branches in the upper half-strip, i.e., when $\operatorname{Im} z \geqslant 0$.

For every such z, we have $\operatorname{Im} \cos z = (e^{-\operatorname{Im} z} - e^{\operatorname{Im} z}) \sin \operatorname{Re} z < 0$. So, $\cos z$ stays in the lower half-plane and, thereby, $\frac{1}{\cos z}$ stays in the upper half-plane as z runs over the upper half-strip. Besides, the values of $L(z)$ on the interval $(0, \frac{\pi}{2})$ should be real. Thus, the branch of logarithm we need to choose is the one with the imaginary part between 0 and π.

Let us now turn to $S(z)$. Note first of all that for $p > 1$, the series in the definition of $S(z)$ converges absolutely for every z in the upper half-strip. Now observe that for every z in the upper half-strip, all the complex numbers z, $\pi + z$, $2\pi + z$, $\dots$ belong to the right upper quadrant $\{\operatorname{Re} \zeta > 0, \operatorname{Im} \zeta \geqslant 0\}$. The branch of the function $\zeta \to \zeta^{-p}$ that is positive on the positive real semiaxis sends this quadrant to the angle $\Delta_p := \{-\frac{\pi p}{2} < \arg \zeta \leqslant 0\}$ (since $p < 2$, this angle is a subset of the lower half-plane). On the other hand, the numbers $\pi - z$, $2\pi - z$, $\dots$ belong to the right lower quadrant, which is sent by the mapping $\zeta \to -\zeta^{-p}$ to the same angle Δ_p. So, the whole sum $S(z)$ stays in this angle when z runs over the upper half-strip and the branch of the power $-2/p$ in the subtrahend $S(z)^{-2/p}$ should be the one that sends the angle Δ_p to the upper half-plane (which is the "natural" branch in this angle).

Note now that the functions $L(z)$ and $S(z)$ are analytic everywhere in the closed strip $\{0 \leqslant \operatorname{Re} z \leqslant \frac{\pi}{2}\}$ except two singular points 0 and $\frac{\pi}{2}$. Let us look at what happens when z approaches these singularities.

When $z \to 0$, we have $L(z) = 2\ln\frac{1}{\cos z} = z^2 + \frac{1}{6}z^4 + o(z^4)$ while $S(z) = z^{-p} + O(z) = z^{-p}(1 + o(z^2))$ because $p > 1$. Thus

$$\Phi(z) = z^2 + \frac{z^4}{6} + o(z^4) - z^2(1 + o(z^2)) = \frac{z^4}{6} + o(z^4) \qquad \text{as } z \to 0.$$

Note now that $\frac{1}{\cos z} = \frac{1}{\sin(\frac{\pi}{2} - z)} = O\Big(\frac{1}{|\frac{\pi}{2} - z|}\Big)$ and $L(z) = O\Big(\ln\frac{1}{|\frac{\pi}{2} - z|}\Big)$ as $z \to \frac{\pi}{2}$. On the other hand, $\frac{\pi}{2}$ is a simple zero of the function $S(z)$ and therefore $S(z) = D\,(\frac{\pi}{2} - z)(1 + o(1))$ where $D = -S'(\frac{\pi}{2}) = 2p\left[\left(\frac{2}{\pi}\right)^{p+1} + \left(\frac{2}{3\pi}\right)^{p+1} + \dots\right] > 0$. Thus

$$\Phi(z) \approx -D^{-2/p}\left(\frac{\pi}{2} - z\right)^{-2/p} \qquad \text{as } z \to \frac{\pi}{2}.$$

From these asymptotic relations, we can conclude that the function $\Phi(t)$ is positive near 0 and negative near $\frac{\pi}{2}$. So it has at least one root on the interval $(0, \frac{\pi}{2})$ and, if this root is unique, changes its sign from $+$ to $-$ there.

Now let us estimate the increment of the argument of the function $\Phi(z)$ over the boundary of the region Ω, the upper half Ω_+ of which is shown on the picture below.

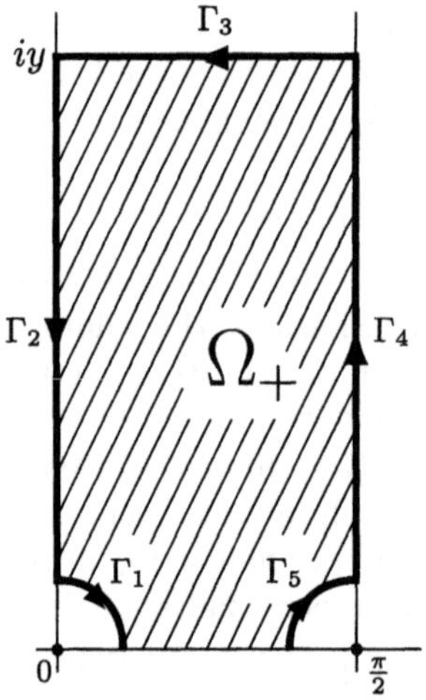

Note that $\Delta_{\partial\Omega} \arg\Phi = 2\Delta_{\partial\Omega_+} \arg\Phi$ due to the symmetry $\Phi(\overline{z}) = \overline{\Phi(z)}$. The boundary $\partial\Omega_+$ consists of 5 parts $\Gamma_1, \dots, \Gamma_5$. We assume that the radii of the circle arcs Γ_1 and Γ_5 are very small and the height y of Γ_3 is very large.

Estimates for Γ_1 and Γ_5. From the asymptotic relations above, we immediately deduce that the function $\Phi(z)$ does not vanish on Γ_1 and Γ_5 and

$$\Delta_{\Gamma_1} \arg\Phi \approx -2\pi\,; \qquad \Delta_{\Gamma_5} \arg\Phi \approx \frac{\pi}{p}.$$

ESTIMATE FOR Γ_3. Clearly,

$$L(z) = 2\ln\frac{1}{\cos z} = 2\ln\frac{2}{e^{iz}+e^{-iz}} = 2iz + O(1) \quad \text{as } \operatorname{Im} z \to +\infty,\ z \in \Omega_+.$$

As to $S(z)$, we have

$$\operatorname{Im}\frac{1}{z^p},\ \operatorname{Im}\frac{1}{(\pi+z)^p}, \dots < 0 \quad \text{and} \quad \operatorname{Im}\frac{1}{(\pi-z)^p},\ \operatorname{Im}\frac{1}{(2\pi-z)^p}, \dots > 0$$

for every $z \in \Omega_+$. What is more, the absolute values of the imaginary parts of the first $\operatorname{Im} z \approx |z|$ terms in the series for $S(z)$ are not less than $C|z|^{-p}$ with some absolute $C > 0$. Hence, $|S(z)| \geqslant |\operatorname{Im} S(z)| \geqslant \text{Const}\,|z|^{1-p}$ and, thereby,

$$S(z)^{-2/p} = O(|z|^{2(p-1)/p}) = o(|z|)$$

on Γ_3 (recall that $p < 2$). Thus $\Phi(z) \approx 2iz$ on Γ_3, so Φ does not vanish on Γ_3 and

$$\Delta_{\Gamma_3} \arg\Phi \approx 0.$$

ESTIMATE FOR Γ_2. For every $z = it$, $t > 0$ on the vertical side Γ_2, we have $L(z) = 2\ln\frac{2}{e^{-t}+e^{t}} \in \mathbb{R}$. Representing the function $S(z)$ as

$$S(z) = S(it) = \frac{1}{(it)^p} - \left[\frac{1}{(\pi-it)^p} - \frac{1}{(\pi+it)^p}\right] - \left[\frac{1}{(2\pi-it)^p} - \frac{1}{(2\pi+it)^p}\right] - \dots$$

and observing that every difference in brackets is a "positive imaginary" number, we arrive at the following picture:

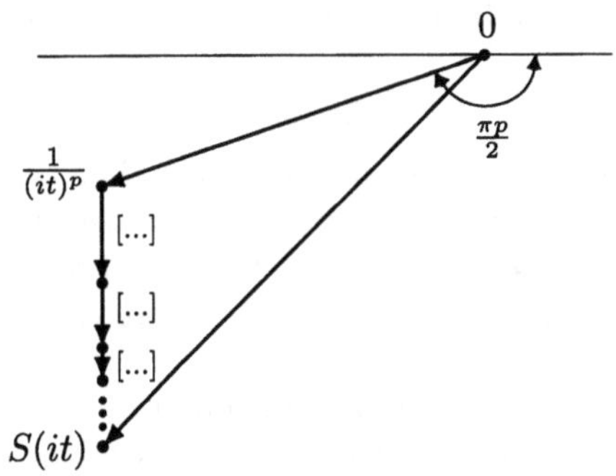

Hence $0 < \arg\left[S(it)^{-2/p}\right] < \pi$, i.e., $\operatorname{Im}\left[S(it)^{-2/p}\right] > 0$. Therefore $\operatorname{Im}\Phi(it) < 0$ for all $t > 0$, or, in other words, $\Phi(z)$ stays in the (open) lower half-plane as z runs over Γ_2. Therefore Φ does not vanish on Γ_2 and

$$\Delta_{\Gamma_2} \arg\Phi \leqslant \pi.$$

ESTIMATE FOR Γ_4. This is the most interesting part of the contour $\partial\Omega_+$. On Γ_4 we have $z = \frac{\pi}{2} + it,\ t > 0$. Hence

$$L(z) = 2\ln\frac{2}{ie^{-t} - ie^{t}} = 2\ln i + 2\ln\frac{2}{e^t - e^{-t}} = \pi i + 2\ln\frac{2}{e^t - e^{-t}}.$$

Meanwhile,

$$S(z) = \left[\frac{1}{(\frac{\pi}{2} + it)^p} - \frac{1}{(\frac{\pi}{2} - it)^p}\right] + \left[\frac{1}{(\frac{3\pi}{2} + it)^p} - \frac{1}{(\frac{3\pi}{2} - it)^p}\right] + \dots.$$

Every difference in brackets is a "negative imaginary" number. Thus $S(z) \in -i\mathbb{R}_+$ and thereby $S(z)^{-2/p} \in e^{i\pi/p}\,\mathbb{R}_+$. So, when t runs up from 0 to $+\infty$, the minuend $L(\frac{\pi}{2} + it)$ runs over the horizontal line $\operatorname{Im} z = \pi$ from right to left, while the subtrahend $S(\frac{\pi}{2} + it)^{-2/p}$ wanders over the ray $\arg z = \frac{\pi}{p}$ (see the picture below).

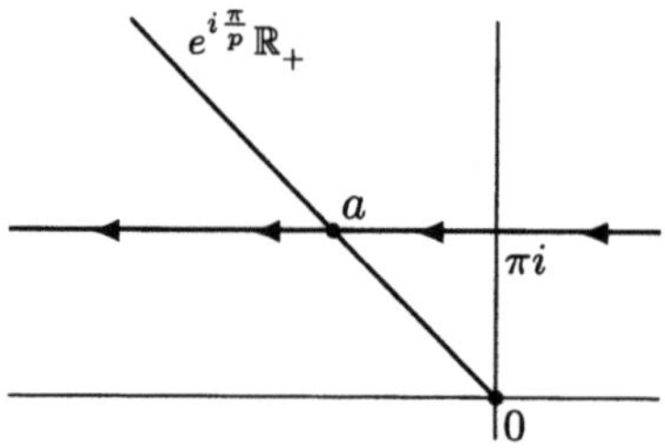

As long as $L(\frac{\pi}{2} + it)$ remains to the right from this ray, the difference $\Phi = L - S^{-2/p}$ stays in the "upper right" half-plane bounded by the line $e^{i\pi/p}\,\mathbb{R}$. So, the corresponding increment of $\arg\Phi$ does not exceed π. Once $L(\frac{\pi}{2} + it)$ has crossed this ray, it never leaves the "left lower" half-plane bounded by the same line and again $\arg\Phi$ cannot change by more than π. Thus, if Φ does not vanish on Γ_4, we have

$$\Delta_{\Gamma_4}\Phi \leqslant 2\pi.$$

Bringing these estimates together and recalling that $\Delta_{\partial\Omega}\arg\Phi = 2\,\Delta_{\partial\Omega_+}\arg\Phi$, we see that in this case the total change of $\arg\Phi$ over the boundary $\partial\Omega$ does not exceed ("up to epsilon") $2(-2\pi + \pi + 0 + 2\pi + \frac{\pi}{p}) = 2\pi(1 + \frac{1}{p})$. For $p > 1$, this quantity is strictly less than 4π and thereby there can be no second root of Φ in the strip $0 < \operatorname{Re} z < \frac{\pi}{2}$.

Suppose now that Φ vanishes at some point $z_0 \in \Gamma_4$. Replace a small interval of Γ_4 centered at z_0 by the clockwise oriented semicircle bypassing z_0 from the left. Then we shall get an extra *negative* change of argument by π (if the root is simple) or more (if it is multiple) when going over this semicircle. And the more roots we have and the higher their multiplicity is, the more we gain on the negative side! [2] So if Φ had any zeroes on Γ_4, we would end up with even better estimate

[2] Actually this last remark is empty because from what we said above, it is clear that Φ can vanish on Γ_4 only at the point where $L = a$ and only if $S(\frac{\pi}{2}+it)^{-2/p}$ reaches the point a simultaneously

$\Delta_{\partial\Omega} \arg \Phi < 2\pi$, which would prohibite Φ to have roots in the strip at all! But we know that at least one such root exists, ruling this possibility out.

Haagerup's integral inequality is thus completely proved. □

Step 5. What to do if one of the coefficients a_k is large?

From now on, it will be more convenient to replace our usual normalization condition $\sum_{k=1}^n a_k^2 = 1$ by the condition $a_1 = 1$. Our aim is to prove the inequality

$$R_p(a) = \int_0^1 \left| \sum_{k=1}^n a_k r_k(t) \right|^p dt \geqslant A_p(1 + a_2^2 + \cdots + a_n^2)^{p/2} =: A_p\, \varphi_p(a_2^2 + \cdots + a_n^2)$$

with $A_p = 2^{\frac{p-2}{2}} \frac{\Gamma(\frac{p+1}{2})}{\Gamma(\frac{3}{2})}$ for all $p \in [p_0, 2)$, $a = (a_1, \ldots, a_n) \in \mathbb{R}^n$. (Here and below $\varphi_p(x)$ stands for $(1+x)^{p/2}$.)

THE IDEA OF THE PROOF We already know that the desired inequality holds when $\max_k a_k^2 \leqslant \frac{1}{2} \sum_k a_k^2$ (this is just the result of Step 4, only slightly reformulated). How to get the desired estimate in full generality? We will try to use the almost obvious formula

$$R_p(a) = \frac{R_p(a^+) + R_p(a^-)}{2}$$

where $a^\pm = (a_1, \ldots, a_{n-2}, a_{n-1} \pm a_n)$ $(n \geqslant 3)$ to reduce the bad case when one of the coefficients is large to the case when the number of coefficients is $n-1$ instead of n. Formally that means that we are going to use induction on n. Since for $n \geqslant 3$ we have the identity

$$a_2^2 + \cdots + a_n^2 = \frac{1}{2} \left\{ \left[(a_2^-)^2 + \cdots + (a_{n-1}^-)^2 \right] + \left[(a_2^+)^2 + \cdots + (a_{n-1}^+)^2 \right] \right\},$$

the induction step would be fairly easy *if the function* $\varphi(x) = (1+x)^{p/2}$ *were convex*. Then, denoting for brevity $a_2^2 + \cdots + a_n^2$ by x and $(a_2^\pm)^2 + \cdots + (a_{n-1}^\pm)^2$ by $x^\pm$, we could make the induction step in just one line:

$$R_p(a) = \frac{R_p(a^+) + R_p(a^-)}{2} \overset{\substack{\text{induction}\\ \text{assumption}}}{\geqslant} A_p \frac{\varphi_p(x^+) + \varphi_p(x^-)}{2}$$
$$\overset{\text{convexity}}{\geqslant} A_p\, \varphi_p\left(\frac{x^+ + x^-}{2} \right) = A_p \varphi_p(x).$$

Unfortunately there is one small, but annoying obstacle on this otherwise excellent way: for $p < 2$ the function $\varphi_p(x)$ is *concave*! When proving inequalities by induction, the standard trick that allows us to get rid of such "small but annoying" obstacles is to strengthen the statement. In our case it would be ideal to find

with $L(\frac{\pi}{2} + it)$. Also, this root must be simple because the speed at which $L(\frac{\pi}{2} + it)$ runs over the line $\mathbb{R} + i\pi$ never vanishes.

a convex function $\Phi_p(x)$ that is pointwise not less than $\varphi_p(x)$ and to prove the desired estimate with Φ_p in place of φ_p, i.e., to prove the estimate

$$R_p(a) \geqslant A_p\, \Phi_p(a_2^2 + \cdots + a_n^2). \tag{**}$$

Unfortunately, there is no finite function $\Phi_p(x)$ dominating $\varphi_p(x)$ and convex on the whole ray $[0, +\infty)$ either. But we do not actually need that much: we can always reshuffle the coefficients $a_1, \ldots, a_n$ in the original inequality so that $a_1 = 1$ is the largest of them and then just to refer to the result of Step 4 when there is no large coefficient, i.e., when $x \geqslant 1$. So, it seems that we only need to change our function $\varphi_p(x)$ on the interval $[0, 1]$. It is really the case and the function that will work is

$$\Phi_p(x) = \begin{cases} \varphi_p(x), & \text{if } x \geqslant 1; \\ 2\varphi_p(1) - \varphi_p(2 - x), & \text{if } 0 \leqslant x \leqslant 1. \end{cases}$$

In other words, Φ_p is the function whose graph is obtained by replacing the part of the graph of φ_p on $[0, 1]$ by the symmetric image of the graph of φ_p on $[1, 2]$ with respect to the point $(1, \varphi_p(1)) = (1, 2^{p/2})$ (see the picture below).

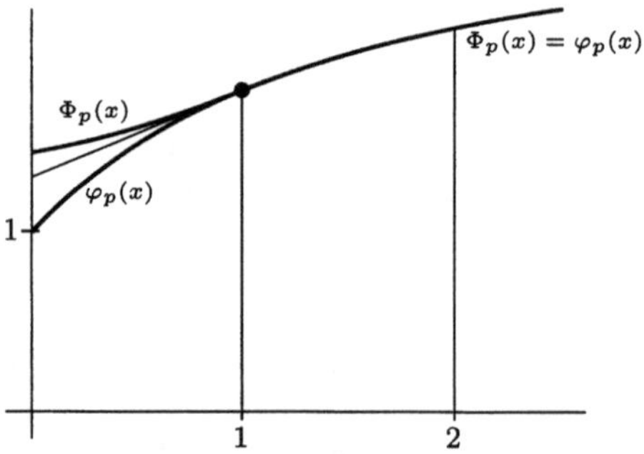

On the interval $[0, 1)$, the graphs of the functions Φ_p and φ_p are separated by their common tangent line at the point $(1, \varphi_p(1))$ and therefore $\Phi_p(x) > \varphi_p(x)$ when $0 \leqslant x < 1$, and $\Phi_p(x) = \varphi_p(x)$ when $x \geqslant 1$. Now we are ready to make

THE INDUCTION STEP. Observe first of all that

$$\Phi_p\left(\frac{x' + x''}{2}\right) \leqslant \frac{\Phi_p(x') + \Phi_p(x'')}{2} \quad \text{whenever} \quad x', x'' \geqslant 0 \text{ and } \frac{x' + x''}{2} \leqslant 1.$$

(This property is a little bit stronger than just convexity on $[0, 1]$, because we do not assume that $\max\{x', x''\} \leqslant 1$. It was the desire to have this "extended convexity" that prevented us from just replacing the part of the graph of φ_p on $[0, 1]$ by the corresponding piece of the tangent line.)

Indeed, we may assume without loss of generality that $x' < x''$. If $x'' \leqslant 1$, the desired inequality follows immediately from the convexity of Φ_p on $[0, 1]$. Suppose now that $x'' > 1$. Notice that

$$\frac{\partial}{\partial x'}\Phi_p\left(\frac{x'+x''}{2}\right) = \frac{1}{2}\left.\frac{d\Phi_p}{dx}\right|_{x=\frac{x'+x''}{2}} \geqslant \frac{1}{2}\left.\frac{d\Phi_p}{dx}\right|_{x=x'} = \frac{\partial}{\partial x'}\frac{\Phi_p(x')+\Phi_p(x'')}{2}$$

whenever $x' < x''$, $\frac{x'+x''}{2} \leqslant 1$, because the derivative Φ_p' is monotone increasing on the interval $[0,1]$. Therefore the function $\Phi_p(\frac{x'+x''}{2})$ increases in x' faster then $\frac{\Phi_p(x')+\Phi_p(x'')}{2}$, and it is enough to check our inequality for the maximal possible value of x', i.e., for $x' = 2 - x''$. But in this case our inequality turns into identity due to the definition of Φ_p.

Now suppose that $n \geqslant 3$ and that the estimate $(**)$ is already established for linear combinations of $n-1$ Rademacher functions. Let $a = (a_1, a_2, \dots, a_n) \in \mathbb{R}^n$, $a_1 = 1$. Denote as before $x = a_2^2 + \dots + a_n^2$. Three cases are possible:

- $a_1 = 1$ is not the largest coefficient. Then $x > 1$ and $(**)$ coincides with the homogeneous estimate $R_p(a) \geqslant A_p(a_1^2 + \dots + a_n^2)^{p/2}$. Rearranging the coefficients in the order of decreasing absolute value (this doesn't change either part of the homogeneous estimate) and renormalizing so that the first coefficient is again 1, we shall arrive at some new version of the inequality $(**)$ in which $a_1 = 1$ is the largest coefficient. This new version is not weaker than the homogeneous estimate, i.e., not weaker than the original version we wanted to prove.
- $a_1 = 1$ is the largest coefficient, $x \geqslant 1$. In this case, $(**)$ still coincides with the homogeneous estimate and $\max_k a_k^2 = 1 \leqslant \frac{1}{2}(1+x) = \frac{1}{2}\sum_k a_k^2$. So we may just refer to Step 4.
- $a_1 = 1$ is the largest coefficient, $x < 1$. Then (in the same notation as above)

$$\begin{aligned} R_p(a) = \frac{R_p(a^+)+R_p(a^-)}{2} &\overset{\substack{\text{induction}\\ \text{assumption}}}{\geqslant} A_p\frac{\Phi_p(x^+)+\Phi_p(x^-)}{2} \\ &\geqslant A_p\Phi_p\left(\frac{x^+ + x^-}{2}\right) = A_p\Phi_p(x) \end{aligned}$$

(because $x^\pm \geqslant 0$ and $\frac{x^- + x^+}{2} = x < 1$).

So it remains only to prove

THE INDUCTION BASE $n = 2$. We need to prove the inequality

$$R_p(1, a_2) \geqslant A_p\Phi_p(a_2^2).$$

Again we may assume that $a_1 = 1$ is the largest of the two coefficients, i.e., $a_2^2 = x \leqslant 1$ (otherwise, after interchanging a_1 and a_2 and renormalizing, we shall obtain a stronger inequality to prove). Computing the left hand side and recalling the definition of $\Phi_p(x)$ and our hypothetical formula for A_p, we can rewrite it as

$$\frac{(1+\sqrt{x})^p + (1-\sqrt{x})^p}{2} \geqslant 2^{\frac{p-2}{2}}\frac{\Gamma(\frac{p+1}{2})}{\Gamma(\frac{3}{2})}\left[2\cdot 2^{p/2} - (3-x)^{p/2}\right] \qquad (0 \leqslant x \leqslant 1).$$

The ratio $\frac{\Gamma(\frac{p+1}{2})}{\Gamma(\frac{3}{2})}$ does not exceed 1 for all $p \in [p_0, 2)$, so it is enough to prove the estimate

$$\frac{(1+\sqrt{x})^p + (1-\sqrt{x})^p}{2} \geqslant 2^{\frac{p-2}{2}} \left[2 \cdot 2^{p/2} - (3-x)^{p/2}\right] = 2^{p-1}\left[2 - \left(\frac{3-x}{2}\right)^{p/2}\right].$$

At first glance, this strengthening of the statement to prove may seem completely unmotivated, but if we recall that for $p = p_0$ this ratio equals 1 and that we do not know (or, at least, do not want to know) what p_0 is, it becomes quite natural. Dividing both parts by 2^{p-1} and transferring $\left(\frac{3-x}{2}\right)^{p/2}$ to the left hand side, we can restate our estimate as

$$\left(\frac{1+\sqrt{x}}{2}\right)^p + \left(\frac{1-\sqrt{x}}{2}\right)^p + \left(\frac{3-x}{2}\right)^{p/2} \leqslant 2.$$

Now denote

$$a = \frac{1+\sqrt{x}}{2}, \qquad b = \frac{1-\sqrt{x}}{2}.$$

Taking into account that $1 - x = 4ab$, we can rewrite our inequality as

$$a^p + b^p + (1+2ab)^{p/2} \leqslant 2 \qquad (p \leqslant 2;\ a, b \geqslant 0;\ a+b=1).$$

Since for $p = 2$ we have an identity and since the left hand side is *convex* in p, it is sufficient to show that

$$\left.\frac{d}{dp}\right|_{p=2} \left[a^p + b^p + (1+2ab)^{p/2}\right] \leqslant 0,$$

or, which is the same, that

$$a^2 \ln a + b^2 \ln b + \frac{1}{2}(1+2ab)\ln(1+2ab) \leqslant 0.$$

Note now that $\frac{d^2}{dt^2}[(1+t)\ln(1+t)] = \frac{1}{1+t} \leqslant 1$ for every $t \geqslant 0$. Thereby, using the second order Taylor formula at the point $t_0 = 0$, we obtain $(1+t)\ln(1+t) \leqslant t + \frac{1}{2}t^2$. In particular,

$$\frac{1}{2}(1+2ab)\ln(1+2ab) \leqslant \frac{1}{2}\left(2ab + \frac{1}{2}(2ab)^2\right) = ab + (ab)^2.$$

On the other hand, $a^2 \ln a = a^2 \ln(1-b) = -a^2\left(b + \frac{b^2}{2} + \dots\right) \leqslant -a^2\left(b + \frac{b^2}{2}\right)$ and, similarly, $b^2 \ln b \leqslant -b^2\left(a + \frac{a^2}{2}\right)$. Thus

$$a^2 \ln a + b^2 \ln b \leqslant -a^2\left(b + \frac{b^2}{2}\right) - b^2\left(a + \frac{a^2}{2}\right) = -ab(a+b) - (ab)^2 = -ab - (ab)^2.$$

Haagerup's theorem (case $p < 2$) is thus completely proved. □

References

[1] K. Ball, *Cube slicing in* $\mathbb{R}^n$, Proc. Amer. Math. Soc., **97 (3)** (1986), 465–473.

[2] K. Ball, *Some remarks on the geometry of convex sets,* Geometric aspects of functional analysis (1986/87), 251–260, Lecture Notes in Math., **1317**, Springer, Berlin-New York (1988).

[3] H. Busemann, C. M. Petty, *Problems on convex bodies,* Math. Scand., **4** (1956), 88–94.

[4] R. J. Gardiner, *A positive answer to the Busemann-Petty problem in three dimensions,* Ann. of Math. (2) **140** (1994), no. 2, 435–447.

[5] U. Haagerup, *The best constants in the Khintchine inequality,* Studia Math., **70 (3)** (1982), 231–283.

[6] G. Y. Zhang, *Intersection bodies and the Busemann-Petty inequalities in* $\mathbb{R}^4$, Ann. of Math. (2) **140** (1994), no. 2, 331–346.

F. L. Nazarov
Department of Mathematics,
Michigan State University
East Lansing, MI
48824, USA
E-mail address: `fedja@math.msu.edu`

A. N. Podkorytov
Department of Mathematics,
St. Petersburg State University
Bibliotechnaya pl. 2
Peterhof, St. Petersburg
198904, Russia
E-mail address: `pan@pan.usr.pu.ru`

Operator Theory:
Advances and Applications, Vol. 113

Carleson Measures of Bergman Spaces in Domains with Nonsmooth Boundary

Vladimir L. Oleinik

Dedicated to the memory of S. A. Vinogradov

Abstract. For a wide class of domains, we characterize the bounded identity operator from the weighted Bergman space of harmonic (analytic) functions $L^p_h(L^p_a)$ to the Lebesgue space $L^q(\mu)$ in terms of Carleson-type conditions.

1. Introduction

This paper is a continuation and extension of some topics discussed in [2], [3], [7], and [9].

In order to formulate our result we need some notation. Let G be an open set in $\mathbf{R}^n$, $G \neq \mathbf{R}^n$, $n \geq 2$, and let m be Lebesgue measure on G; for $x \in \mathbf{R}^n$ we set $d(x) = d(x, \partial G) = d_x$, $B(x) = \{y \in \mathbf{R}^n : |y - x| < d(x)/2\}$.

In what follows, we shall consider a special class $E(c, d_0)$, $c \in (0, 1/2]$, $0 < d_0 \leq +\infty$, of domains and the corresponding number $\beta_0 = \beta_0(G) < 0$ (see Definition 2.3), as well as a class $S(A, G)$ of functions on G, $A \geq 1$, which contains, e.g., all subharmonic functions (see Definition 2.9).

Our main theorem is as follows.

Theorem 1.1. *Let $c \in (0, 1/2]$, $0 < d_0 \leq \infty$, $A \geq 1$, $0 < p \leq +\infty$, and $0 < q < +\infty$. The following conditions imposed on open sets $G \in \mathbf{R}^n$ $(G \neq \mathbf{R}^n)$ of class $E(c, d_0)$, on positive Borel measures μ on G with $\mu(\{x \in G : 3d(x) \geq d_0\}) < \infty$, and on numbers $\beta \in (\beta_0(G), \infty)$ are equivalent:*

(1) $\mu(B(x))d_x^{-n-(\beta q/p)} \in L^r(G)$ *with* $r = p/(p-q)$ *if* $p > q$, *and* $\mu(B(x))d_x^{-(n+\beta)q/p} \in L^\infty(G)$ *if* $p \leq q$;

(2) *there exists a constant $C > 0$ such that for every harmonic function f in G (for every $f \in S(A, G)$) we have*

$$\left(\int_G |f|^q d\mu\right)^{1/q} \leq C \left(\int_G |f(x)|^p d_x^\beta dm(x)\right)^{1/p};$$

(3) *if $n = 2$, there exists a constant $C > 0$ such that the inequality in (2) is fulfilled for every analytic function f in G.*

Special cases of Theorem 1.1 were proved in [8] ($p \le q$, $\beta \ge 0$, $A = 1$, G is a bounded open set with smooth boundary), in [5] ($q < p$, $\beta = 0$, G is the open unit disk, f is an analytic function), in [6] ($q < p \le 2$, $\beta \ge 0$, $A = 1$, G is a bounded open set satisfying the "cone" condition), and in [9] ($p \le q$ or $q < p \le 2$, $G \in E(c, d_0)$, $\beta > \beta_0(G)$). Note that the implications (1) $\Rightarrow$ (2) and (1) $\Rightarrow$ (3) were proved in [9] for every open set G, $G \ne \mathbf{R}^n$, and every $\beta \in \mathbf{R}$. Moreover, the implication (2) $\Rightarrow$ (3) is obvious. Thus we only must prove the implications (2) $\Rightarrow$ (1) and (3) $\Rightarrow$ (1) for $p > 2$, $q < p$. The main ingredient of the proof of these implications is the use of a certain Khintchin type inequality. This method was suggested by L. Carleson [1] for $p = 2$, $q = 1$ and was used later by the author [6] in the case where $q < p \le 2$ (see also [9]).

In Section 2 we collect the necessary information on the Whitney coverings and present several conditions equivalent to condition (1) of Theorem 1.1. In Section 3 we prove the main lemmas. The necessity of condition (1) is proved in Section 4.

This research was partly supported by the International Soros Science Education Program (ISSEP, d98-387).

2. Preliminary results

For an open set $G \subset \mathbf{R}^n$ ($n \ge 2$) with $G \ne \mathbf{R}^n$ we denote $d_x = d(x) = d(x, \partial G)$. For $x \in \mathbf{R}^n$ and $r > 0$, let $B(x,r) = \{y \in \mathbf{R}^n : |x - y| < r\}$, $B_r = B(0,r)$, and $B = B(0,1)$. We call $B(x) = B(x, d(x)/2)$ the Whitney ball centered at x.

Lemma 2.1 (see [9], Lemma 2.5). *Let $a \in (0,1)$. For every open set $G \subset \mathbf{R}^n$, $G \ne \mathbf{R}^n$, there exists a sequence of distinct points $\bar{x} = \{x_k\}$ in G such that*

(1) *$B(x_k, ad(x_k)/3)$, $k = 1, 2, \ldots$, are disjoint,*

(2) *$G = \bigcup_{k=1}^{\infty} B(x_k, ad(x_k))$,*

(3) *there exists a positive number N such that every $x \in G$ belongs to at most N balls of the sequence $\{B(x_k, ad(x_k))\}$. We can choose $N < 2 \cdot 20^n \cdot (a(1-a))^{-n} + 1$.*

Definition 2.2. *Let $a \in (0,1)$, let $G \subset \mathbf{R}^n$ be an open set, and let $\bar{x} = \{x_k\}$ be a sequence as in Lemma 2.1. The covering $\{B(x_k, ad(x_k)) : k = 1, 2, \ldots\}$ of G is called an a-covering of G and is denoted by $\mathcal{B}_a(\bar{x})$. We call $\mathcal{B}(\bar{x}) = \mathcal{B}_{1/2}(\bar{x})$ the Whitney covering of G.*

Definition 2.3. *An open set G in $\mathbf{R}^n$ is said to be of class $E(c, d_0)$ ($I(c, d_0)$), $c \in (0, 1/2]$, $0 < d_0 \le +\infty$, if for all $z \in \partial G$ and all $r \in (0, d_0)$ there exists $x \in \mathbf{R}^n \setminus \overline{G}$ ($x \in G$) with $B(x, cr) \subset (\mathbf{R}^n \setminus \overline{G}) \bigcap B(z, r)$ (with $B(x, cr) \subset G \bigcap B(z, r)$). For $G \in E(c, d_0) \bigcup I(c, d_0)$, we denote by $\beta_0 = \beta_0(G)$ the negative constant $\log_2(1 - k^{-n})$ with $k = (9c + 2)/(2c)$.*

The next two lemmas explain the meaning of the number $\beta_0(G)$.

Lemma 2.4 (see [10], p. 172). *Let G be an bounded open set in $\mathbf{R}^n$ of class $I(c, d_0)$. Then for every $\beta \in (\beta_0(G), 0)$ we have*

$$\int_G d_x^\beta dm(x) < \infty.$$

Lemma 2.5 (see [9], Lemma 2.11). *Let $c \in (0, 1/2]$, $0 < d_0 \leq \infty$, and let G be an open set in $\mathbf{R}^n$, $G \neq \mathbf{R}^n$, such that $(\mathbf{R}^n \setminus \partial G) \in I(c, d_0)$. Let $\beta \in (\beta_0(G), 0)$ and $s > n$. Then there exists a constant M such that for every $y \in \mathbf{R}^n \setminus \overline{G}$ with $d(y) < d_0$ we have*

$$\int_G |x - y|^{-s} d_x^\beta dm(x) < M d_y^{n+\beta-s}.$$

In particular, this lemma extends Lemma 2.4 to a wider class of domains. It is obvious that if $\partial G \in C^1$, then in Lemmas 2.4 – 2.5 we can replace $\beta_0(G)$ by -1.

For $\beta \geq 0$, Lemma 2.5 takes the following form.

Lemma 2.6. *Let G be an open set in $\mathbf{R}^n$, $G \neq \mathbf{R}^n$, let $\beta \geq 0$, and let $s > n + \beta$. Then there exists a constant M such that for every $y \in \mathbf{R}^n \setminus \overline{G}$ we have*

$$\int_G |x - y|^{-s} d_x^\beta dm(x) \leq M d_y^{n+\beta-s}.$$

Proof. For $x \in G$ and $y \in \mathbf{R}^n \setminus \overline{G}$ we have $d_x < |x - y|$. Therefore,

$$\int_G |x - y|^{-s} d_x^\beta dm(x) \leq \int_G |x - y|^{\beta-s} dm(x) \leq \int_{|x-y|\geq d(y)} |x - y|^{\beta-s} dm(x)$$
$$\leq \omega_{n-1} \int_{d(y)}^{+\infty} t^{n+\beta-s-1} dt \leq \omega_{n-1}(s - n - \beta)^{-1} d_y^{n+\beta-s},$$

where ω_{n-1} stands for the $(n-1)$-dimensional measure on ∂B. □

We need some further notation. Let $G \in E(c, d_0)$, and let $\mathcal{B}_a(\bar{x})$ be an a-covering of G. We set $\bar{x}' = \{x_k \in \bar{x} : d(x_k) < d_0\}$. By Definition 2.3, for every $\xi \in \bar{x}'$ there exists $z(\xi) \in \partial G$ and $\eta(\xi) \in \mathbf{R}^n \setminus \overline{G}$ such that $|\xi - z(\xi)| = d_\xi$ and $B(\eta(\xi), cd_\xi) \subset (\mathbf{R}^n \setminus G) \bigcap B(z(\xi), d_\xi)$. Put $\bar{y} = \{\eta(\xi) : \xi \in \bar{x}'\}$.

We next prove a technical lemma about covering by balls that lie outside G and are induced by balls of any a-covering of G. This lemma is similar to Lemma 2.1 in [9].

Lemma 2.7. *Let $G \in E(c, d_0)$, $c \in (0, 1/2]$, $0 < d_0 \leq \infty$, let $a \in (0, 1)$, and let $\mathcal{B}_a(\bar{x})$ be an a-covering of G. There exists a positive number N_1 such that every $y \in \mathbf{R}^n \setminus \overline{G}$ belongs to at most N_1 balls of the sequence $\{B(\eta, ad_\eta), \eta \in \bar{y}\}$. We can choose $N_1 \leq 21^n[ac(1 - a)]^{-n} + 1$.*

Proof. Fixing $y \in \bigcup_{\eta\in\bar{y}} B(\eta, ad_\eta)$, we choose and fix $\eta \in \bar{y}$, $\xi \in \bar{x}$, and $z \in \partial G$ such that $y \in B(\eta, ad_\eta)$, $|z - \xi| = d_\xi$, and $B(\eta, cd_\xi) \subset (\mathbf{R}^n \setminus G) \bigcap B(z, d_\xi)$. Then

$(1-a)d_\eta \le d_y \le (1+a)d_\eta$ and $cd_\xi \le d_\eta \le d_\xi$. Therefore, $d_\xi \le d_\eta/c \le d_y/[c(1-a)]$ and $d_y \le (1+a)d_\eta \le (1+a)d_\xi \le 2d_\xi$. Next, denoting $b = 1/[c(1-a)]$, we obtain

$$|y-\xi| \le |y-\eta| + |\eta - z| + |z-\xi| \le ad_\eta + d_\xi + d_\xi$$
$$\le ad_y/(1-a) + 2bd_y \le 5bd_y/2.$$

Hence,

$$B(\xi, ad_\xi) \subset B(y, 5bd_y/2 + ad_\xi) \subset B(y, 7bd_y/2).$$

Now we recall (see Lemma 2.1) that the balls $B(\xi, ad_\xi/3)$ and $B(\xi', ad(\xi')/3)$ are disjoint for every $\xi' \in \bar{x}$, $\xi' \neq \xi$. Moreover, $B(\xi, ad_y/6) \subset B(\xi, ad_\xi/3)$, and we arrive at the desired bound

$$N_1 \le m(B(y, 7bd_y/2)/m(B(\xi, ad_y/6)) + 1 = 21^n[ac(1-a)]^{-n} + 1.$$

□

The next lemma gives several equivalent conditions (see [9], Lemma 4.2).

Let $\mathcal{B}_a(\bar{x})$ be an a-covering of G, $\bar{x} = \{x_k\}$. For $\beta \in \mathbf{R}$ and a positive Borel measure μ we put $\mu(\bar{x}, a) = \{\mu(B(x_k, ad(x_k)))d(x_k)^\beta : x_k \in \bar{x}, k = 1, 2, \dots\}$, and for $x \in G$ we put $\mu(x, a) = \mu(B(x, ad_x))d_x^{\beta-(n/s)}$.

Lemma 2.8. *Let G be an open set in $\mathbf{R}^n$, $G \neq \mathbf{R}^n$, let $\beta \in \mathbf{R}$, and let $s \ge 1$. The following conditions are equivalent:*

(1) $\mu(\bar{x}, 1/2) \in l^s$;
(2) *there exists a number $a \in (0,1)$ and an a-covering $\mathcal{B}_a(\bar{\xi})$ of G such that $\mu(\bar{\xi}, a) \in l^s$;*
(3) *for every $a \in (0,1)$ and every a-covering $\mathcal{B}_a(\bar{\xi})$ of G we have $\mu(\bar{\xi}, a) \in l^s$;*
(4) *there exists $a \in (0,1)$ such that $\mu(x, a) \in L^s(G)$;*
(5) *for every $a \in (0,1)$ we have $\mu(x, a) \in L^s(G)$.*

Definition 2.9 (see [9]). *Let $A \ge 1$, let G be an open set in $\mathbf{R}^n$, and let $u : G \to [-\infty, +\infty)$ be an upper semicontinuous function. We say that $u \in S(A, G)$ if for every $x \in G$ and every $B(x, r) \subset G$, $r > 0$, we have*

$$u(x) \le A\left(\int_{B(x,r)} u(y)\, dm(y)/m(B(x,r))\right).$$

Various examples of functions of class $S(A, G)$ can be found in the paper [9].

3. Main lemmas

Lemma 3.1. *Let $c \in (0, 1/2]$, $0 < d_0 \le +\infty$, let $G \neq \mathbf{R}^n$ be an open set in $\mathbf{R}^n$ of class $E(c, d_0)$, and let $\beta \in (\beta_0, +\infty)$, where $\beta_0 = \beta_0(G)$ is as in Definition 2.3. Let $\alpha > 0$, $\gamma \ge 0$, $a \in (0,1)$, let $\mathcal{B}_a(\bar{x})$, $\bar{x} = \{x_k\}$, be an a-covering of G, and let $\bar{y} = \{y_k\}$ be as in Lemma 2.7. Then there exist constants $C_1 = C_1(\alpha)$ and $C_2 = C_2(\gamma)$ such that*

(i) *if* $\alpha > -\beta$, *then for every* $y \in \mathbf{R}^n \setminus G$ *(with* $d_y < d_0$*) we have*

$$\sum_{x_j \in \bar{x}} d(x_j)^{n+\beta} |x_j - y|^{-n-\beta-\alpha} \leq C_1 d_y^{-\alpha};$$

(ii) *for every* $x \in G$,

$$\sum_{y_k \in \bar{y}} d(y_k)^{n+\gamma} |x - y_k|^{-n-\gamma-\alpha} \leq C_2 d_x^{-\alpha}.$$

Proof. (i) We need the following estimate:

$$d(x_j)^{n+\beta} |y - x_j|^{-n-\beta-\alpha} \leq C \int_{B(x_j, ad(x_j))} |y - x|^{-n-\beta-\alpha} d_x^\beta dm(x). \tag{3.1}$$

To prove (3.1), observe that for every $x \in B(x_j, ad(x_j))$ we have $|y-x| \leq 2|y-x_j|$ and $(1-a)d(x_j) \leq d_x \leq 2d(x_j)$.

Thus, we obtain $d(x_j)^\beta \leq b^{|\beta|} d_x^\beta$ with $b = \max\{2, (1-a)^{-1}\}$. Hence, inequality (3.1) is fulfilled with $C = 2^{n+\beta+\alpha} b^{|\beta|} a^{-n} m(B(1))^{-1}$. By Lemma 2.1, every $x \in G$ belongs to at most N balls of the sequence $\{B(x_j, ad(x_j))\}$. Therefore, summing the inequalities (3.1), we obtain

$$\sum_{x_j \in \bar{x}} d(x_j)^{n+\beta} |x_j - y|^{-n-\beta-\alpha} \leq CN \int_G |y - x|^{-n-\beta-\alpha} d_x^\beta dm(x) \leq CNM d_y^{-\alpha}.$$

Here at the last step we used Lemma 2.5 ($\beta < 0$) and Lemma 2.6 ($\beta \geq 0$).

(ii) In this case it suffices to use Lemma 2.6 as above. □

Lemma 3.2. *Let* G, β_0, $\bar{x}$, *and* $\bar{y}$ *be as in Lemma* 3.1. *Let* $r > 0$. *For every* $s > r$, $x_j \in \bar{x}$, *and* $y_k \in \bar{y}$ *we denote* $A_{kj} = A_{kj}(s) = d(y_k)^{s-r} d(x_j)^r |x_j - y_k|^{-s}$. *There exist positive constants* C_3 *and* C_4 *such that*

(i) *for every* $t > (n + \beta_0)/r$ *with* $s > \max\{n/t, r\}$ *we have*

$$\sup_{y_k \in \bar{y}} \sum_{x_j \in \bar{x}} A_{kj}^t \leq C_3;$$

(ii) *for every* $t > 0$ *with* $s \geq n/t + r$ *we have*

$$\sup_{x_j \in \bar{x}} \sum_{y_k \in \bar{y}} A_{kj}^t \leq C_4.$$

We can choose $C_3 = C_1(t(s-r))$ *and* $C_4 = C_2(tr)$, *with* $C_1(\cdot)$ *and* $C_2(\cdot)$ *as in Lemma* 3.1.

Proof. To prove the lemma, we use Lemma 3.1. (i) Let $\alpha = t(s-r) > 0$ and $\beta = tr - n$. Then $\beta > \beta_0$ and $\alpha > -\beta$ because $t > (n+\beta_0)/r$ and $s > n/t$. Now we note that, by the definition of $\bar{y}$, we have $d(y_k) < d_0$ for every $y_k \in \bar{y}$, and we can apply Lemma 3.1(i).

In case (ii) we put $\alpha = tr > 0$ and $\gamma = t(s-r) - n$. Since $s \geq n/t + r$, we have obtained the conditions of Lemma 3.1(ii). □

Lemma 3.3. *Under the hypotheses of Lemma 3.2, let $p > 1$ and $r > (n+\beta_0)/p$. Then there exists a positive number s_0 and a positive constant C_5 such that for every $s > s_0$ and every sequence $\{\gamma_k\} \in l^p$, $\gamma_k \geq 0$, we have*

$$\sum_{x_j\in\bar{x}}(\sum_{y_k\in\bar{y}} A_{kj}\gamma_k)^p \leq C_5 \sum_k \gamma_k^p.$$

Proof. Let $q = p/(p-1)$. By the Hölder inequality, for every $\alpha \in (0,1)$ we have

$$\sum_{x_j\in\bar{x}}(\sum_{y_k\in\bar{y}} A_{kj}\gamma_k)^p \leq \sum_{x_j\in\bar{x}}(\sum_{y_k\in\bar{y}} A_{kj}^{(1-\alpha)q})^{p/q} \sum_{y_k\in\bar{y}} A_{kj}^{\alpha p}\gamma_k^p = I.$$

Let $\alpha \in ((n+\beta_0)/(pr), 1)$; we put $s_0 = \max\{n/[(1-\alpha)q] + r, n/(\alpha p)\}$. Then, by Lemma 3.3, for every $s > s_0$ we obtain $I \leq C_4^{p/q} C_3 \sum_k \gamma_k^p$. □

4. The proof of the main theorem

4.1. Proof of Theorem 1.1

Proof. For $p \leq q$ or $q < p \leq 2$ the implications (1) ⇒ (2) and (1) ⇒ (3), as well as (3) ⇒ (1) and (2) ⇒ (1), were proved in the paper [9]. So, we only must prove the implications (2) ⇒ (1) and (3) ⇒ (1) in the case where $p > 2$, $0 < q < p \leq +\infty$. Moreover, it suffices to consider a neighborhood of ∂G, that is the set of $x \in G$ with $d_x < d_0$ (see [9], p.193).

We prove the implication (2) ⇒ (1). First, let $p < +\infty$, and let $\mathcal{B}_a(\bar{x})$ be an a-covering of G; $a \in (0,1)$ will be chosen below. Let $x_k \in \bar{x}' = \{x_j \in \bar{x} : d(x_j) < d_0\}$, and let $\bar{y} = \{y_k\} \subset \mathbf{R}^n \setminus \overline{G}$ be as in Lemma 2.7. After relabeling if necessary, we may assume that $\bar{x}' = \{x_1, x_2, \dots\}$ and $\bar{y} = \{y_1, y_2, \dots\}$. For a fixed positive integer s, we put $l_k = (x_k - y_k)/|x_k - y_k|$ and

$$f_k(x) = (\partial/\partial l_k)^s\, |x - y_k|^{2-n}, \quad k = 1, 2, \dots.$$

Here the functions f_k are harmonic in $\mathbf{R}^n \setminus \{y_k\} \supset G$, and ($t = n + s - 2$)

$$f_k(x) = C_s^{(n/2)-1}(\cos\theta(x))|x - y_k|^{-t},$$

where $C_m^r(\cdot)$ stands for the Gegenbauer polynomial, and $\theta(x)$ is the angle between $x - y_k$ and $x_k - y_k$. We can and shall fix $\theta_0 \in (0, \pi/6]$, $A_0 > 0$, and $A_1 > 0$ such that $C_s^{n/2-1}\cos(\theta(x)) \geq A_0$ if $0 \leq |\theta(x)| \leq \theta_0$, and $|f_k(x)| \leq A_1|x - y_k|^{-t}$ for $x \in \mathbf{R}^n \setminus \{y_k\}$. By the choice of y_k we have $(1+c)d(x_k) \leq |x_k - y_k| \leq 2d(x_k)$; therefore, for every $x \in B(x_k, ad(x_k)) = B_k$ with $a = (1+c)\sin\theta_0 \leq 3/4$ we obtain

$$|f_k(x)| \geq A_0|x - y_k|^{-t} \geq A_0 3^{-t} d(x_k)^{-t},$$

whence, with some constant A_2,

$$\int_G |f_k|^q d\mu \geq \int_{B_k} |f_k|^q d\mu \geq A_2 \mu(B_k) d(x_k)^{-qt}. \tag{4.1}$$

Let $a_k \geq 0$, $k = 1, 2, \ldots$, be numbers to be chosen later on. Using a Khintchine type inequality (see [4, Theorem 1.d.6(i)] and [6]), from condition (2) we deduce that

$$\left(\int_G (\sum_{k=1}^{K} |a_k f_k|^2)^{q/2} d\mu\right)^{1/q} \leq A_3 \left(\int_G (\sum_{k=1}^{K} |a_k f_k(x)|^2)^{p/2} d_x^\beta dm(x)\right)^{1/p} \tag{4.2}$$

for every integer K, where A_3 is a positive constant independent of K.

Let N be as in Lemma 2.1(3). Applying inequality (4.1) we get

$$N \int_G (\sum_{m=1}^{K} |a_m f_m|^2)^{q/2} d\mu \geq \sum_{k=1}^{K} \int_{B_k} a_k^q |f_k|^q d\mu \geq A_2 \sum_{k=1}^{K} a_k^q \mu(B_k) d_k^{-qt}, \tag{4.3}$$

where $d_k = d(x_k)$. On the other hand, from the inequalities $|x - y_k| \geq (1-a)|x_j - y_k| \geq |x_j - y_k|/4$ and $d_x^\beta \leq b^{|\beta|} d_j^\beta$ with $b = \max\{2, (1-a)^{-1}\} \leq 4$ for $x_j \in \bar{x}$ and $x \in B_j$, we obtain

$$\int_G (\sum_{k=1}^{K} |a_k f_k(x)|^2)^{p/2} d_x^\beta dm(x) \leq \int_G (\sum_{k=1}^{K} a_k |f_k(x)|)^p d_x^\beta dm(x) \tag{4.4}$$

$$\leq A_1^p \sum_{j=1}^{+\infty} \int_{B_j} (\sum_{k=1}^{K} a_k |x - y_k|^{-t})^p d_x^\beta dm(x) \leq A_1^p 4^{tp+|\beta|} \sum_{j=1}^{+\infty} (\sum_{k=1}^{K} a_k |x_j - y_k|^{-t})^p d_j^{n+\beta}.$$

Let $r = (n+\beta)/p$, $\gamma_k = a_k d_k^{r-t}$, and $A_{kj} = d_j^r d_k^{t-r} |x_j - y_k|^{-t}$. Then

$$\sum_{j=1}^{+\infty} (\sum_{k=1}^{K} a_k |x_j - y_k|^{-t})^p d_j^{n+\beta} = \sum_{j=1}^{+\infty} (\sum_{k=1}^{K} A_{kj} \gamma_k)^p.$$

Note that $t = n + s - 2$ and $p > 1$. By Lemma 3.3, for sufficiently large s we have

$$\sum_{j=1}^{+\infty} (\sum_{k=1}^{K} A_{kj} \gamma_k)^p \leq C_5 \sum_{k=1}^{K} \gamma_k^p = C_5 \sum_{k=1}^{K} a_k^p d_k^{n+\beta-pt}. \tag{4.5}$$

Now, choosing a_k such that

$$a_k^q \mu(B_k) d_k^{-qt} = a_k^p d_k^{n+\beta-pt},$$

from inequalities (4.2)–(4.5) we see that

$$\sum_{k=1}^{K} \left[\mu(B_k) d_k^{-(n+\beta)q/p}\right]^{p/(p-q)} \leq A_4$$

for every K, where A_4 is a positive constant independent of K. This means that $\{\mu(B_k) d_k^{-(n+\beta)q/p}\} \in l^{p/(p-q)}$, and, by Lemma 2.8 ((2) $\Longleftrightarrow$ (5)), we get $\mu(B(x)) d_x^{-n-(\beta q/p)} \in L^{p/(p-q)}(G)$.

If $p = +\infty$, $q < \infty$ and $\mu(G) < \infty$, then

$$\left(\int_G |f|^q d\mu\right)^{1/q} \leq \mu(G)^{1/q} \sup_G |f|.$$

On the other hand, for $f \equiv 1$ from (2) we obtain $\mu(G) < \infty$.

Since $S(1, G) \subset S(A, G)$ and $|f| \in S(1, G)$ for every harmonic function f, the implication (2) $\Rightarrow$ (1) is proved completely.

The proof of the implication (3) $\Rightarrow$ (1) is similar. Here in place of the harmonic functions $f_k(x)$ we must use the analytic functions $(z - \xi_k)^{-t}$ with $\xi_k \in \mathbf{C} \setminus \bar{G}$ and with sufficiently large integer t. □

4.2. Remarks

(1) If $\partial G \in C^1$, in Theorem 1.1 we can replace $\beta_0(G)$ by -1.

(2) D.H. Luecking [L] proved the implication (3) $\Rightarrow$ (1) in Theorem 1.1 for every $p > q > 0$ in the case where $\beta = 0$ and G is the unit disk. This proof makes use of a result of E. Amar about interpolation sequences. On the other hand, Theorem 1.1 can be applied to the weighted interpolation problem in the Bergman space for $0 < q < p$ or $p \leq q$, $G \in E(c, d_0)$, and $\beta > \beta_0(G)$.

(3) Some further information on the plane domains such that conditions (1) and (2) in Theorem 1.1 are equivalent can be found in the papers [3], [7]. For example, if a domain G is finitely connected, then (1) $\Longleftrightarrow$ (3) (as well as (1) $\Longleftrightarrow$ (2)) if and only if the boundary ∂G does not contain isolated points and $\mathbf{C} \setminus G$ is an unbounded set.

References

[1] L. Carleson, *Private communication.*

[2] V.N. Kudashov and V.L. Oleinik, *An embedding theorem of Bergman space into Lebesgue space,* University of Helsinki, Reports of the Department of Mathematics, Preprint **42** (1993), 1–9.

[3] V.N. Kudashov and V.L. Oleinik, *Theorems on the embedding of Bergman Spaces into Lebesgue spaces for domains with nonsmooth boundary*, Functional Analysis and its Applications, **30(1)** (1996), 52–54. (A translation of Funkts. An. i ego Prilozh., **30(1)** (1996), 67–70.)

[4] J. Lindenstrauss and L. Tzafriri, *Classical Banach spaces II. Function spaces*, 1979, Springer-Verlag, New-York.

[5] D. Luecking, *Multipliers of Bergman spaces into Lebesgue spaces*, Proc. Edinburgh Math. Soc. **29** (1986), 125–131.

[6] V.L. Oleinik, *An embedding theorem for some classes of analytic functions*, in: Problems of math. physics. Leningrad, **11** (1986), 164–167. (Russian)

[7] V.L. Oleinik, *Carleson measures and uniformly perfect sets*, Zap. Nautch. Sem. POMI Steklov, **255** (1998), 92–103. (Russian)

[8] V.L. Oleinik and B.S. Pavlov, *Embedding theorems for weighted classes of harmonic and analytic functions*, J.Soviet Math. **2** (1974), 135–142 (A translation of Zap. Nautch. Sem. LOMI Steklov, **22** (1971), 92–102.)

[9] V.L. Oleinik and M.Vuorinen, *Embedding theorems for Bergman spaces in quasiconformal balls*, Manuscripta Math., **72** (1991), 181–203.

[10] Yu.G. Reshetnyak, *Stability theorems in geometry and analysis* (Russian), Nauka, Sibirsk. Otdelenie, Novosibirsk, 1982.

Department of Physics,
St.-Petersburg State University,
Ul'yanovskaya 1, Petrodvorets,
198904 St.-Petersburg, Russia
E-mail address: Vladimir.Oleinik@pobox.spbu.ru

Operator Theory:
Advances and Applications, Vol. 113

On the Zeros of Tails of Power Series

I. V. Ostrovskii

Dedicated to the memory of S. A. Vinogradov

1. Introduction

Let

$$f(z) = \sum_{k=0}^{\infty} a_k z^k \tag{1.1}$$

be a power series with a positive radius of convergence. Let

$$s_n(z) = \sum_{k=0}^{n} a_k z^k, \qquad t_n(z) = \sum_{k=n+1}^{\infty} a_k z^k$$

be its nth section and nth tail, respectively.

By now, the distribution of the zeros of $s_n(z)$ has been studied in detail; see, e.g., [2, 4, 6, 7], where further references can be found. The distribution of the zeros of the tails $t_n(z)$ has been paid less attention. The behavior of the zeros of the tails of some *concrete* power series was considered in [1, 3, 8, 12–14]. Many important results related to the tails of *general* power series were obtained in [2] and, especially, in [11]; however, only the *moduli* of the zeros of tails were treated there.

Some facts related to the *arguments* of the zeros of tails of power series with *infinite* radius of covergence were obtained in [10]. These facts show that some restrictions on the arguments of the zeros imply a bound for the growth of the entire function (1.1). Our aim in this paper is to obtain a similar result for power series with a *finite* radius of convergence. The main result is as follows.

Theorem. *Let $f(z)$ be a power series* (1.1) *convergent in the unit disc* $\mathbf{D}$. *Suppose there exist two different tails, $t_m(z)$ and $t_n(z)$, such that all zeros of $t_m(z)t_n(z)$ lie on a finite system of radii of* $\mathbf{D}$. *Then*

$$\log M(r,f) = O\left(\frac{1}{(1-r)^2}\right), \quad r \to 1, \tag{1.2}$$

where $M(r,f) = \max\{|f(z)| : |z| = r\}$.

Supported in part by INTAS 96-0858 grant.

The following example shows that the bound (1.2) is the best possible in the sense of order:

$$f(z) = P(z) + z^{n+1} \cos \frac{z}{(1-z)^2}, \quad z \in \mathbf{D}, \quad n \in \mathbf{N},$$

where $P(z)$ is a polynomial of degree $< n+1$. In this example, the tails

$$t_n(z) = z^{n+1} \cos \frac{z}{(1-z)^2}, \quad t_{n+1}(z) = z^{n+1} \left(\cos \frac{z}{(1-z)^2} - 1 \right)$$

have only real zeros in $\mathbf{D}$.

2. Proof of the theorem

Consider the function

$$q(z) = \frac{t_n(z)}{s_m(z) - s_n(z)}, \quad z \in \mathbf{D}, \tag{2.1}$$

and observe that

$$q(z) - 1 = \frac{t_m(z)}{s_m(z) - s_n(z)}.$$

Since $s_m(z) - s_n(z)$ has finitely many zeros, the condition of the theorem implies that all roots of the three equations

$$q(z) = 0, \quad q(z) = \infty, \quad q(z) = 1 \tag{2.2}$$

lie on a finite system of radii of $\mathbf{D}$. Denoting these radii by

$$\{z : \arg z = \alpha_j,\ 0 \le |z| < 1\}, \quad 0 \le \alpha_1 < \alpha_2 < \cdots < \alpha_p < 2\pi,$$

we consider the sectors

$$S_j = \{z : \alpha_j < \arg z < \alpha_{j+1},\ 0 < |z| < 1\}, \quad j = 1, \ldots, p, \quad \alpha_{p+1} = \alpha_1 + 2\pi.$$

In order to investigate the behavior of $q(z)$ and $f(z)$ in S_j, we need the function

$$w(\zeta; \gamma) = \frac{(2\zeta^{\pi/\gamma} - i)(2 - i\zeta^{\pi/\gamma})}{(2\zeta^{\pi/\gamma} + i)(2 + i\zeta^{\pi/\gamma})}, \quad 0 < \gamma \le 2\pi,\ 0 < \arg \zeta < \gamma.$$

This function conformally maps the sector $\{\zeta : 0 < \arg \zeta < \gamma,\ 0 < |\zeta| < 1\}$ onto $\mathbf{D}$. A direct calculation shows that

$$\begin{aligned} 1 - |w(\rho e^{i\theta}; \gamma)| &> (1 - |w(\rho e^{i\theta}; \gamma)|^2)/2 \\ &= \frac{12\rho^{\pi/\gamma}(1 - \rho^{2\pi/\gamma}) \sin(\pi\theta/\gamma)}{(4 + \rho^{2\pi/\gamma} - 4\rho^{\pi/\gamma} \sin(\pi\theta/\gamma))(1 + 4\rho^{2\pi/\gamma} + 4\rho^{\pi/\gamma} \sin(\pi\theta/\gamma))} \\ &> (4/15)\rho^{\pi/\gamma}(1 - \rho^{2\pi/\gamma}) \sin(\pi\theta/\gamma) \\ &> C(1 - \rho) \sin(\pi\theta/\gamma), \quad 0 < \theta < \gamma,\ 1/2 \le \rho < 1, \end{aligned} \tag{2.3}$$

where $C > 0$ is a constant independent of ρ and θ.

The function

$$w(ze^{-i\alpha_j}; \alpha_{j+1} - \alpha_j), \quad j \in \{1, 2, \ldots, p\},$$

conformally maps S_j onto $\mathbf{D}$. Let $z_j(w)$ be the function performing the inverse conformal mapping. Then the function

$$q(z_j(w)), \quad w \in \mathbf{D},$$

is analytic in $\mathbf{D}$ and does not assume the values 0 and 1 there. By the Schottky theorem (see, e.g., [5, p. 60]), we have the following estimate:

$$\log^+ |q(z_j(w))| = O\left(\frac{1}{1 - |w|}\right), \quad |w| \to 1. \tag{2.4}$$

Since (2.1) implies $f(z) = s_n(z) + (s_m(z) - s_n(z))q(z)$, we see that

$$f(z_j(w)) = s_n(z_j(w)) + [s_m(z_j(w)) - s_n(z_j(w))]q(z_j(w)).$$

The functions $s_m(z_j(w))$ and $s_n(z_j(w))$ are bounded in $\mathbf{D}$; therefore, (2.4) implies that

$$\log^+ |f(z_j(w))| = O\left(\frac{1}{1 - |w|}\right), \quad |w| \to 1.$$

Substituting

$$w = w(ze^{-i\alpha_j}; \alpha_{j+1} - \alpha_j), \quad z \in S_j,$$

we get

$$\log^+ |f(z)| \leq \frac{C_1}{1 - |w(ze^{-i\alpha_j}; \alpha_{j+1} - \alpha_j)|}, \quad z \in S_j,$$

and, taking (2.3) into account,

$$\log^+ |f(re^{i\varphi})| \leq \frac{C_2}{(1 - r)\sin[\pi(\varphi - \alpha_j)/(\alpha_{j+1} - \alpha_j)]}, \quad \alpha_j < \varphi < \alpha_{j+1}, \ \frac{1}{2} \leq r < 1,$$

where the constants $C_1, C_2 > 0$ are independent of $z = re^{i\varphi}$. Evidently, the validity of the latter inequality for $j = 1, 2, \ldots, p$, implies that

$$\log^+ |f(re^{i\varphi})| \leq \frac{C}{(1 - r)\prod_{j=1}^{p} |\varphi - \alpha_j|}, \quad 0 \leq \varphi < 2\pi, \ 0 < r < 1, \tag{2.5}$$

where $C > 0$ is independent of $re^{i\varphi}$.

To deduce (1.2) from (2.5), it suffices to apply the following lemma with

$$h(\zeta) = f(\zeta e^{i\alpha_j}), \ j = 1, 2, \ldots, p; \quad \delta = (1/2)\min\{(\alpha_{j+1} - \alpha_j) : 1 \leq j \leq p\}.$$

Lemma. *Let $h(\zeta)$ be a function analytic in the sector*

$$S = \{\zeta : |\arg \zeta| \leq \delta \leq \pi, \ 0 \leq |\zeta| < 1\}$$

and satisfying the condition

$$\log^+ |h(\zeta)| \leq \frac{B_1}{(1 - |\zeta|)|\arg \zeta|}, \quad \zeta \in S, \tag{2.6}$$

where $B_1 > 0$ *is a constant independent of* ζ. *Then*

$$\log^+ |h(\zeta)| = O\left(\frac{1}{(1-|\zeta|)^2}\right), \quad \zeta \in S, \ |\zeta| \to 1. \tag{2.7}$$

This lemma is contained implicitly in the well-known Sjöberg–Levinson–Domar log log-theorem (see, e.g., [9, p. 376]). We present its proof here for the reader's convenience.

Without loss of generality, we can assume that δ is small, namely, $\sin\delta \le \sqrt{2}/4$, because it suffices to prove (2.7) in any sector with smaller δ.

We fix R, $1/2 \le R \le 1$, and consider the function

$$h_R(\zeta) = h(\zeta)\exp\left\{-\frac{B_2}{(R-\zeta)^2}\right\}, \quad \zeta \in S\setminus\{R\}, \quad h_R(R) := 0,$$

where $B_2 > 0$ is a constant to be chosen later. This function is analytic in $S\setminus\{R\}$. Moreover, since

$$|h_R(\zeta)| = |h(\zeta)|\exp\left\{-\frac{B_2\cos(2\arg(R-\zeta))}{|R-\zeta|^2}\right\}, \tag{2.8}$$

it is continuous on

$$S_R := S\bigcap\{\zeta : |\arg(R-\zeta)| \le \frac{\pi}{8}\},$$

for $R < 1$. We show that the constant B_2 can be chosen independent of R and so large that

$$\max\{|h_R(\zeta)| : \ \zeta \in S_R\} \le B_3, \quad \text{for} \quad R < 1, \tag{2.9}$$

where the constant $B_3 > 0$ is also independent of R.

Putting

$$L_R^{(1)} = \partial S_R\bigcap\{\zeta : |\arg\zeta| = \delta\}, \qquad L_R^{(2)} = \partial S_R\bigcap\{\zeta : |\arg(R-\zeta)| = \pi/8\},$$
$$M_R^{(k)} = \max\{|h_R(\zeta)| : \zeta \in L_R^{(k)}\}, \quad k = 1, 2,$$

we have

$$\max\{|h_R(\zeta)| : \zeta \in \partial S_R\} = \max\{M_R^{(1)}, M_R^{(2)}\}.$$

An elementary geometric calculation shows that for $\zeta \in L_R^{(1)}$ and $R < 1$ we have

$$|\zeta| \le \left(R\sin\frac{\pi}{8}\right)/\sin\left(\delta+\frac{\pi}{8}\right) < \left(\sin\frac{\pi}{8}\right)/\sin\left(\delta+\frac{\pi}{8}\right) < 1.$$

Therefore, $M_R^{(1)}$ is bounded by a constant independent of R.

Next, (2.6) and (2.8) imply the inequality

$$M_R^{(2)} \le \max\left\{\exp\left(\frac{B_1}{(1-|\zeta|)|\arg\zeta|} - \frac{B_2\cos(\pi/4)}{|R-\zeta|^2}\right) : \zeta \in L_R^{(2)}\right\}.$$

We set

$$R - \zeta = t\exp\left(\pm\frac{i\pi}{8}\right), \ t > 0, \quad \text{for} \quad \zeta \in L_R^{(2)}, \ \ \frac{1}{2} \le R < 1.$$

Elementary geometric calculations show that, for these ζ,

$$0 \le t \le (R\sin\delta)/\sin\frac{\pi}{8} < R\cos\frac{\pi}{8},$$

$$1-|\zeta| > (1-|\zeta|^2)/2 > (R^2-|\zeta|^2)/2 = t(2R\cos\frac{\pi}{8}-t)/2 > (tR\cos\frac{\pi}{8})/2 > \frac{1}{8}t,$$

and, moreover, there is a positive constant B_4 independent of R and B_4 such that

$$|\arg\zeta| \ge B_4 t.$$

Therefore,

$$M_R^{(2)} \le \max\left\{\exp\left(\frac{8B_1}{B_4t^2} - \frac{B_2\cos(\pi/4)}{t^2}\right) : 0<t<\infty\right\}.$$

We conclude that if $B_2 \ge 8B_1/(B_4\cos(\pi/4))$, then $M_R^{(2)} \le 1$. Thus, (2.9) is true.

Fix any $\zeta \in \operatorname{int} S_1$. Since

$$\operatorname{int} S_1 \subset \bigcup_{1/2<R<1} S_R,$$

letting $R \uparrow 1$ in (2.9), we get

$$|h_1(\zeta)| \le B_3, \quad \zeta \in \operatorname{int} S_1.$$

Hence, for $\zeta \in \operatorname{int} S_1$,

$$|h(\zeta)| = \left|h_1(\zeta)\exp\frac{B_2}{(1-\zeta)^2}\right| \le B_3\exp\frac{B_2}{(1-|\zeta|)^2},$$

i.e., (2.7) is true. For $\zeta \in S\setminus\operatorname{int} S_1$ we have $|\arg\zeta| \ge B_5(1-|\zeta|)$, so that the validity of (2.7) is an immediate consequence of (2.6).

3. Some remarks

1°. One of the results of [10] mentioned in the Introduction is the following.

Let $f(z)$ be a power series (1.1) *with an infinite radius of convergence. Suppose that there exists* (i) *a finite system of rays*

$$P = \bigcup_{j=1}^{p}\{z : \arg z = \alpha_j,\ 0 \le |z| < \infty\}, \quad 0 \le \alpha_1 < \alpha_2 < \cdots < \alpha_p < 2\pi, \tag{3.1}$$

and (ii) *three pairwise different tails $t_m(z), t_n(z), t_l(z)$, such that all but finitely many zeros of $t_m(z)t_n(z)t_l(z)$ lie on P. Then*

$$\log M(r,f) = O(r^{\pi/\gamma}\log r), \quad r\uparrow\infty, \tag{3.2}$$

where $\gamma = \min\{(\alpha_{j+1}-\alpha_j) : 1 \le j \le p\}$ $(\alpha_{p+1} = \alpha_1 + 2\pi)$.

This result can be refined by replacing three tails in (ii) by two tails, $t_m(z)$ and $t_n(z)$, say. For the proof, it suffices to define $q(z)$ by the equation

$$q(z) = \frac{t_n(z)}{s_m(z)-s_n(z)}, \quad z \in \mathbf{C}$$

(but not by the equation on the last line on p. 1259 of [10]). The function $q(z)$ defined in this way is meromorphic in $\mathbf{C}$, and all but finitely many roots of the equations (2.2) lie on the rays (3.1). The remaining part of the proof in [10] can be repeated with only trivial changes.

2°. As mentioned in [10], estimate (3.2) can be refined in the following way by using the Nevanlinna theory for an angle:

$$\log M(r,f) = O(r^{\pi/\gamma}), \quad r \uparrow \infty. \tag{3.3}$$

This remark remains valid if one replaces three tails by two tails.

3°. Estimate (1.2) does not depend on the system of radii mentioned in the theorem, meanwhile (3.2) and (3.3) depend on the system of rays (3.1). The example

$$f(z) = P(z) + z^{n+1} \cos z^{p/2}, \quad n, p \in \mathbf{N},$$

where $P(z)$ is a polynomial of degree $< n+1$, and all zeros of $t_n(z)$ and $t_{n+1}(z)$ lie on the rays (3.1) with $\alpha_j = 2(j-1)\pi/p$, $j = 1, 2, \ldots, p$, $\gamma = 2\pi/p$, shows that this dependence is essential and, moreover, (3.3) is sharp in the sense of order.

4°. The question as to whether (1.2) can be sharpened if there is a finite system of radii such that more than two different tails (e.g., all tails) have all but finitely many zeros situated on that system remains open. A result related to the case of an infinite radius of convergence was obtained in [10].

5°. If we assume that there exist two different tails, $t_m(z)$ and $t_n(z)$, that have finitely many zeros in $\mathbf{D}$, then (1.2) can be improved; namely,

$$\log M(r,f) = O\left(\frac{1}{1-r}\right), \quad r \to 1. \tag{3.4}$$

The following example (due to A. M. Vishnyakova) shows that the bound (3.4) is also sharp in the sense of order:

$$f(z) = P(z) + z^{n+1} \exp\frac{2iz}{1+z^2}, \quad z \in \mathbf{D}, \quad n \in \mathbf{N},$$

where $P(z)$ is a polynomial of degree $< n+1$. In this example, both $t_m(z)$ and $t_n(z)$ have only one multiple zero at $z = 0$ in $\mathbf{D}$.

For the proof of (3.4), we observe that the function $q(z)$ defined by (2.1) admits values 0 and 1 only finitely many times in $\mathbf{D}$ and apply the Schottky theorem in the annulus $\{z : \rho < |z| < 1\}$ where $q(z)$ does not admit 0 and 1 at all.

I thank A. I. Il'inskii and A. M. Vishnyakova for their careful reading a preliminary version of this paper and for valuable remarks.

References

[1] J. D. Buckholtz, *A characterization of the exponential series*, Amer. Math. Monthly, **73** (1966), 121–123.

[2] J. D. Buckholtz and J. K. Shaw, *Zeros of partial sums and remainders of power series*, Trans. Amer. Math. Soc., **166** (1972), 269–284.

[3] J. Dieudonné, *Sur les zéros des polynômes-sections de* e^x, Bull. Soc. Math. France, **70** (1935), 333–351.

[4] A. Edrei, E. B. Saff, and R. S. Varga, *Zeros of sections of power series*, Lecture Notes in Math., **1002** (1983), 1–115.

[5] W. H. J. Fuchs, *Topics in the theory of functions of one complex variable*, D. Van Nostrand Company, Princeton, N.J., **1967**.

[6] T. Ganelius, *Sequences of analytic functions and their zeros*, Arkiv för Matematik, **3** (1953), 1–50.

[7] T. Ganelius, *The zeros of partial sums of power series*, Duke Math. J., **30** (1963), 533–540.

[8] G. W. Hedstrom and J. Korevaar, *The zeros of partial sums of certain small entire functions*, Duke Math. J., **30** (1963), 519–532.

[9] P. Koosis, *The logarithmic integral, I*, Cambridge University Press, **1988**.

[10] I. V. Ostrovskii, *Les séries de puissances dont les restes ont seulement des zéros non-positifs*, C. R. Acad. Sci. Paris, Sér. I, **325** (1997), 1257–1262.

[11] M. Pommiez, *Sur les restes successifs des séries de Taylor*, Ann. Fac. Sci. Univ. Toulouse, **(4) 24** (1960), 77–165.

[12] G. Szegö, *Über eine Eigenschaft der Exponentialreihe*, Sitzungsberichte der Berliner Math. Gesselschaft, **23** (1924), 50–64.

[13] C. Y. Yildirim, *A sum over zeros of partial sums of* e^x, J. Ramanujan Math. Soc., **6** (1991), 51–66.

[14] C. Y. Yildirim, *On the tails of the exponential series*, Canad. Math. Bull., **37** (1994), 278–286.

Department of Mathematics,
Bilkent University,
06533 Bilkent, Ankara, Turkey,
and
B. Verkin Institute for Low Temperature Physics and Engineering,
310164 Kharkov, Ukraine
E-mail address: iossif@fen.bilkent.edu.tr, ostrovskii@ilt.kharkov.ua

Operator Theory:
Advances and Applications, Vol. 113

Regularity Conditions for Vectorial Stationary Processes

Vladimir Peller

Dedicated to the memory of S. A. Vinogradov

Abstract. We study regularity conditions for vectorial stationary processes of full rank. We reduce the general case to the case of minimal processes, and improve and simplify earlier results of [P3]. For processes satisfying "strong" regularity conditions we obtain definitive results. For completely regular and p-regular processes we obtain sufficient conditions.

1. Vectorial stationary processes

In this paper we study regularity conditions for vectorial stationary processes. We simplify the approach given in [P3] and improve the results of [P3].

By a *vectorial stationary process* (in the wide sense) with discrete time we mean a sequence

$$\{x(t)\}_{t\in\mathbb{Z}}, \quad x(t) = \begin{pmatrix} x_1(t) & x_2(t) & \cdots & x_n(t) \end{pmatrix}^{\mathrm{t}} \tag{1.1}$$

of columns of real random variables such that $\mathbb{E}|x_j(t)|^2 < \infty$ and the *covariance matrix* $Q_{s,t}$,

$$Q_{s,t} = \{\mathbb{E}x_{j_1}(s)x_{j_2}(t)\}_{1\le j_1,j_2\le n} \tag{1.2}$$

depends only on $s-t$ ($\mathbb{E}$ stands for mathematical expectation). The most important class of stationary processes consists of *Gaussian processes* since the covariance matrix of Gaussian variables determines their mutual distribution. We refer the reader to [R], [IR], [DM], and [PK] for general information about stationary processes.

With the process (1.1) we associate the subspaces $\mathcal{X}_0$ (past) and $\mathcal{X}^m$ (future starting at moment m) of the space L^2 with respect to the probability measure defined by

$$\mathcal{X}_0 \stackrel{\text{def}}{=} \operatorname{span}\{x_j(t):\ t<0,\ 1\le j\le n\},$$
$$\mathcal{X}^m \stackrel{\text{def}}{=} \operatorname{span}\{x_j(t):\ t\ge m,\ 1\le j\le n\},$$

where "span" means closed linear span.

The author is partially supported by NSF grant DMS 9304011.

It is well known that for a stationary process with covariance matrix given by (1.2) there exists a matrix-valued positive measure M on the unit circle $\mathbb{T}$ whose Fourier coefficients satisfy

$$Q_{s,t} = \hat{M}(s-t), \quad s,t \in \mathbb{Z}.$$

The measure M is called the *spectral measure* of the process (1.1).

The process (1.1) is called *regular* if $\bigcap_{m\geq 0} \mathcal{X}^m = \{\mathbb{O}\}$. In this case M is absolutely continuous with respect to Lebesgue measure (see [R]). The density W of M with respect to Lebesgue measure is called the *spectral density* of the process. It is also well known (see [R]) that $\operatorname{rank} W(\zeta)$ is constant for almost all $\zeta \in \mathbb{T}$, it is called the *rank of the process.* If the rank is equal to n, the process is said to be of *full rank.* The process is regular and has full rank if and only if $\log\det W \in L^1$ (see [R]).

In this paper we consider only processes of full rank. We refer the reader to [I2] where the study of certain regularity conditions for general processes is reduced to the case of processes of full rank.

Let W be the density of the process (1.1). We identify the space $\operatorname{span}\{x_j(t) : t \in \mathbb{Z},\ 1 \leq j \leq n\}$ with the space $L^2(W)$ of $\mathbb{C}^n$-valued functions in the following way:

$$x_j(t) \mapsto \begin{pmatrix} \mathbb{O} \\ \vdots \\ \mathbb{O} \\ z^t \\ \mathbb{O} \\ \vdots \\ \mathbb{O} \end{pmatrix}$$

(z^t stands in the jth position). It is easy to see that this is a unitary map. It transforms $\mathcal{X}_0$ onto $H^2_-(W)$ and $\mathcal{X}^m$ onto $z^m H^2(W)$, where

$$H^2(W) \stackrel{\text{def}}{=} \operatorname{span}\{z^t c : \ t \geq 0, c \in \mathbb{C}^n\},$$

$$H^2_-(W) \stackrel{\text{def}}{=} \operatorname{span}\{z^t c : \ t < 0, c \in \mathbb{C}^n\}.$$

We denote by $\mathcal{P}_0$ and $\mathcal{P}^m$ the orthogonal projections from $L^2(W)$ onto $H^2_-(W)$ and $z^m H^2(W)$.

It is well known (see [WM], [R]) that if W is the density of a regular process of full rank, then there exist outer matrix functions Ψ and $\Psi_\#$ in the Hardy class H^2 such that

$$W = \Psi^*\Psi = \Psi_\#\Psi_\#^*. \tag{1.3}$$

The matrix function Ψ is unique modulo a left unitary constant factor while the function $\Psi_\#$ is unique modulo a right unitary constant factor.

Recall that an $n \times n$ matrix function Ψ of class H^2 is called *outer* if

$$\operatorname{span}\{z^m \Psi d : \ m \in \mathbb{Z}_+, d \in \mathbb{C}^n\} = H^2(\mathbb{C}^n).$$

A matrix function Θ is called *inner* if $\Theta^*(\zeta)\Theta(\zeta)$ is the identity matrix almost everywhere on $\mathbb{T}$.

Throughout this paper we adopt the following convention. If X is a space of scalar functions, then for a vector (or matrix) function F we say $F \in X$ if all entries of F are in X.

With the factorizations (1.3) we associate the *phase function* $U \stackrel{\text{def}}{=} \Psi_{\#}^{*}\Psi^{-1}$. It follows easily from (1.3) that U is a unitary-valued function. The phase function U is uniquely determined by W modulo left and right unitary constant factors.

2. Regularity conditions

We are going to study different regularity conditions for stationary processes, i.e., conditions of weak dependence of the future on the past.

For $m \in \mathbb{Z}_+$ we define the numbers

$$\rho_m(W) \stackrel{\text{def}}{=} \|\mathcal{P}_0\mathcal{P}^m\|.$$

Clearly, $\rho_m(W)$ is the cosine of the angle between $H^2_-(W)$ and $z^m H^2(W)$.

Definition. *The process is called* completely regular *if*

$$\lim_{m\to\infty} \rho_m(W) = 0. \tag{2.1}$$

For Gaussian processes condition (2.1) is equivalent to the condition

$$\lim_{m\to\infty} \sup\{|P(A\cap B) - P(A)P(B)| = 0 : \ A \in \Sigma_0,\, B \in \Sigma^m\} = 0,$$

where Σ_0 is the σ-algebra of events measurable with respect to $\mathcal{X}_0$, Σ^m is the σ-algebra of events measurable with respect to $\mathcal{X}^m$, and P is the probability measure [KR] (see also [PK]).

If the process satisfies the following stronger regularity condition

$$\rho_m(W) \le \text{const}(1+m)^{-\alpha}, \quad \alpha > 0,$$

we call it *completely regular* of order α.

In [Y] an important notion of canonical correlations of two subspaces of a Gaussian space was introduced. If G_1 and G_2 are such subspaces, the sequence of *canonical correlations* $\lambda_k(G_1, G_2)$ can be defined as the sequence of the singular values $s_k(\mathcal{P}_{G_1}\mathcal{P}_{G2})$ of the product of the orthogonal projections onto G_1 and G_2.

Definition. *Let* $0 < p < \infty$. *A Gaussian process is called* p-regular *if the sequence* $\lambda_k(\mathcal{X}_0, \mathcal{X}^0)$ *belongs to the sequence space* ℓ^p.

In other words, a process is p-regular if $\mathcal{P}_0\mathcal{P}^0$ belongs to the Schatten–von Neumann class $\boldsymbol{S}_p$. Recall that a Gaussian process is 2-regular if and only if it is *absolutely regular*, i.e.,

$$\lim_{m\to\infty} \mathbb{E} \sup_{A\in\Sigma^m} \|\mathbb{E}_{\Sigma_0}(\chi_A - P(A))\|^2 = 0,$$

where $\mathbb{E}_{\Sigma_0}$ is conditional expectation with respect to Σ_0 [IR] (see also [PK]).

Note also that a Gaussian process is 2-regular if and only if it is informationally regular (see [GY] for the definition and the proof).

Finally, we say that a process is *p-regular of order* α if

$$\|\mathcal{P}_0\mathcal{P}^m\|_{\boldsymbol{S}_p} \le \text{const}(1+m)^{-\alpha}.$$

In the scalar case the completely regular processes were described in [HS] and [Sa]: the process is completely regular if and only if its spectral density w admits a representation $w = |q|^2 e^{\varphi}$, where q is a polynomial whose zeros are on $\mathbb{T}$ and φ is a real function in the space VMO of functions of vanishing mean oscillation.

The process is α-regular if and only if $w = |q|^2 e^{\varphi}$, where q is a polynomial whose zeros are on $\mathbb{T}$ and φ is a real function in the Hölder–Zygmund class Λ_α. This result was obtained in [I1].

The p-regular processes were described in [IR] for $p = 2$, in [P1] for $1 \le p < \infty$ and in [P2] for $0 < p < 1$. Namely, the process is p-regular if and only if $w = |q|^2 e^{\varphi}$, where q is a polynomial whose zeros are on $\mathbb{T}$ and φ is a real function in the Besov space $B_p^{1/p}$ (see also [PK]).

Note that for different regularity conditions the proofs were different and some of them very complicated. In [PK] a unified approach was found which allowed the authors to find a single method to prove the above results. That approach is based on Hankel operators. In addition to that it was shown in [PK] that the process is p-regular of order α if and only if $w = |q|^2 e^{\varphi}$, where q is a polynomial whose zeros are on $\mathbb{T}$ and φ is a real function in the Besov space $B_{p\infty}^{1/p+\alpha}$.

Let us return now to the case of vectorial processes.

It was shown in [P3] that

$$\|\mathcal{P}_0\mathcal{P}^m\| = \|H_{z^m U}\| \quad \text{and} \quad \|\mathcal{P}_0\mathcal{P}^m\|_{\boldsymbol{S}_p} = \|H_{z^m U}\|_{\boldsymbol{S}_p}, \quad 0 < p < \infty.$$

Recall that for a function $\Phi \in L^\infty(\mathbb{M}_{n,n})$ taking values in the space $\mathbb{M}_{n,n}$ of $n \times n$ matrices the *Hankel operator* $H_\Phi : H^2(\mathbb{C}^n) \to H^2_-(\mathbb{C}^n) = L^2(\mathbb{C}^n) \ominus H^2(\mathbb{C}^n)$ is defined by

$$H_\Phi F = \mathbb{P}_- \Phi F,$$

where $\mathbb{P}_-$ is the orthogonal projection of $L^2(\mathbb{C}^n)$ onto $H^2_-(\mathbb{C}^n)$.

The *Toeplitz operator* $T_\Phi : H^2(\mathbb{C}^n) \to H^2(\mathbb{C}^n)$ is defined by

$$T_\Phi F = \mathbb{P}_+ \Phi F,$$

where $\mathbb{P}_+$ is the orthogonal projection of $L^2(\mathbb{C}^n)$ onto $H^2(\mathbb{C}^n)$.

As in the scalar case (see [PK]) the above regularity conditions can be expressed in terms of $\mathbb{P}_- U$, where U is a phase function of the process. Namely, the process is completely regular if and only if $\mathbb{P}_- U \in VMO$; the process is completely regular of order α if and only if $\mathbb{P}_- U \in \Lambda_\alpha$; the process is p-regular if and only if $\mathbb{P}_- U \in B_p^{1/p}$ (this follows from the fact that $H_\Phi \in \boldsymbol{S}_p$ if and only if $\mathbb{P}_- \Phi \in B_p^{1/p}$ which was established in [P1] for $1 \le p < \infty$, and in [P2] and [Se] for $0 < p < 1$); and the process is p-regular of order α if and only if $\mathbb{P}_- U \in B_{p,\infty}^{1/p+\alpha}$.

Let us introduce a more general notion. For a space X of functions on $\mathbb{T}$ we say that a process is *X-regular* if $\mathbb{P}_- U \in X$.

Completely regular vectorial process of order α were described in [I2]: these are processes whose weights can be represented as $W = Q^*W_1Q$, where Q is a polynomial such that the zeros of $\det Q$ are on $\mathbb{T}$ and W_1 is a matrix function invertible in Λ_α. Note that the proof given in [I2] is very complicated.

In [P3] Hankel operators were used to study regularity conditions for vectorial processes. In particular, a simple proof was found in [P3] to characterize completely regular processes of order α. The study of completely regular processes was reduced in [P3] to the case of so-called minimal processes.

Recall that the process (1.1) is called *minimal* if for each $t \in \mathbb{Z}$ and each j, $1 \le j \le n$, the function $x_j(t)$ does not belong to the closed linear span of the remaining functions, i.e.,

$$x_j(t) \notin \operatorname{span}\{x_k(s) : \ s \neq t \text{ or } k \neq j\}.$$

In the scalar case the minimal processes were characterized by Kolmogorov [K] who proved that the process is minimal if and only if the spectral density w satisfies the condition $w^{-1} \in L^1$. It is easy to prove the same result for vectorial processes.

Theorem 2.1. *Let W be the spectral density of a vectorial stationary process. Then the process is minimal if and only if $W^{-1} \in L^1$.*

The proof of this result is similar to the proof in the scalar case and can be found in [R].

Suppose that W is the density of a completely regular stationary process with phase function U determined by factorizations (1.3). Consider the Toeplitz operator T_U on $H^2(\mathbb{C}^n)$. It was shown in [P3] that T_U has dense range in $H^2(\mathbb{C}^n)$. It was also proved in [P3] that the process with the density W is minimal if and only if $\operatorname{Ker} T_U = \{\mathbb{O}\}$ (see [P3]).

In the next section we essentially improve and simplify results of [P3] and reduce the study of X-regular processes to the case of minimal processes for a large class of function spaces X.

We consider in this paper two classes of function spaces X. The first class consists of so-called *$\mathcal{R}$-spaces.* We refer the reader to [PK] where a formal definition is given. Roughly speaking X is an $\mathcal{R}$-space if one can determine whether a function φ belongs to X in terms of the singular values of the Hankel operators H_φ and $H_{\bar\varphi}$. For us the most important examples of $\mathcal{R}$-spaces are VMO and $B_p^{1/p}$ ($\varphi \in VMO$ if and only if H_φ and $H_{\bar\varphi}$ are compact; $\varphi \in B_p^{1/p}$ if and only if H_φ and $H_{\bar\varphi}$ belong to the Schatten–von Neumann class $\boldsymbol{S}_p$).

The second class consists of spaces $X \subset C(\mathbb{T})$ satisfying the following axioms:

(A1) *If $f \in X$, then $\bar f \in X$ and $\mathbb{P}_+ f \in X$;*

(A2) *X is a Banach algebra with respect to pointwise multiplication;*

(A3) *for every $\varphi \in X$ the Hankel operator H_φ is a compact operator from X_+ to X_-;*

(A4) *if $f \in X$ and f does not vanish on $\mathbb{T}$, then $1/f \in X$.*

Here for a function space $X \subset L^2$ we use the notation

$$X_+ = X \cap H^2; \quad X_- = X \cap H^2_-.$$

If (X) satisfies (A1), (A2), (A4) and the trigonometric polynomials are dense in X, then X also satisfies (A3). However, (A3) is satisfied for many nonseparable Banach algebras X (see [P4] for details where a general sufficient condition for a space X to satisfy (A3) is given). In particular, the spaces Λ_α, $0 < \alpha < \infty$, $B_{p,\infty}^{1/p+\alpha}$, $1 \le p < \infty$, $0 < \alpha < \infty$, satisfy (A1)–(A4).

In [P3] it was shown that if U is a unitary-valued matrix function such that the Toeplitz operator $T_U : H^2(\mathbb{C}^n) \to H^2(\mathbb{C}^n)$ has dense range and $\mathbb{P}_- U \in X$ then $U \in X$ for any $\mathcal{R}$-space X. Moreover, it was shown in [P3] that if in addition to that $\operatorname{Ker} T_U = \{\mathbb{O}\}$, then the operators $H_U^* H_U$ and $H_{U^*}^* H_{U^*}$ are unitarily equivalent.

In [P4] the same was proved for any function space X satisfying (A1)–(A4). Strictly speaking in [P4] this was proved for spaces X in which the trigonometric polynomials are dense and for nonseparable Banach algebras satisfying certain natural axioms. However, recently R. Alexeev observed that the reasoning in [P4] works for spaces satisfying the above axioms (A1)–(A4).

Those results allow us to conclude the following.

Theorem 2.2. *Let X be an $\mathcal{R}$-space or X be a space satisfying* (A1)–(A4). *Suppose that U is a phase function of a vectorial stationary process. Then the process is X-regular if and only if $U \in X$.*

In Section 4 we see that for spaces X satisfying (A1)–(A4) the X-regular processes can be characterized in the following way which is similar to the scalar case. Namely, the process is X-regular if and only if its density W can be represented as $W = Q^* W_1 Q$, where Q is a polynomial matrix function such that the zeros of $\det Q$ are on $\mathbb{T}$ and $\log W_1 \in X$.

However, it turns out that for completely regular processes the situation is different. In [TV] it was shown that there exists a function matrix Φ in VMO which takes self-adjoint values and such that the process with density $\exp \Phi$ is not completely regular.

In [TV] the following characterization of minimal vectorial completely regular processes was obtained. Let W be the density of a minimal vectorial process. The process is completely regular if and only if

$$\lim_{|J| \to 0} \left\| \left(\frac{1}{|J|} \int_J W dm \right)^{1/2} \left(\frac{1}{|J|} \int_J W^{-1} dm \right)^{1/2} \right\| = 1,$$

where the limit is taken over subarcs of $\mathbb{T}$ and $|J| \stackrel{\text{def}}{=} \boldsymbol{m}(J)$ is the Lebesgue measure of J.

Note, however, that this characterization does not generalize to the case of p-regular processes. In Section 5 we obtain necessary conditions for a process to be completely regular or p-regular in terms of matrix entries of the density W.

3. Reduction to the case of minimal processes

In this section we reduce the study of stationary processes satisfying different regularity conditions to the case of minimal processes.

The following theorem reduces the study of an arbitrary completely regular processes to the case of minimal processes. It improves Theorem 3.5 of [P3] while the approach given here simpler than that of [P3].

Theorem 3.1. *Let $\mathcal{W}$ be the density of a completely regular stationary process with phase function U determined by* (1.3). *Let $N = \dim \operatorname{Ker} T_U$. Then there exists an analytic polynomial matrix function Q such that* $\deg \det Q = N$, *all zeros of* $\det Q$ *are on $\mathbb{T}$, and*

$$W = Q^* W_1 Q, \tag{3.1}$$

where W_1 is the density of a minimal completely regular process which has a phase function U_1 satisfying $U_1 = \Pi U$, where Π is a polynomial inner function of size $n \times n$.

Proof. Consider the factorizations (1.3). Let $U = \Psi_{\#}^* \Psi^{-1}$ be the phase function. Since $U \in VMO$ (see Section 2), it follows that the Toeplitz operator T_U on $H^2(\mathbb{C}^n)$ is Fredholm. Indeed, this follows from the identities

$$I - T_U^* T_U = H_U^* H_U, \quad I - T_U T_U^* = H_{U^*}^* H_{U^*}$$

and the fact that the Hankel operators H_U and H_{U^*} are compact (see [P3] and [PK] for details).

By Simonenko's theorem [Si] the function U admits a factorization

$$U = \Phi_2^* \Lambda \Phi_1^{-1},$$

where Φ_1 and Φ_2 are outer matrix functions in H^2 such that $\Phi_1^{-1}, \Phi_2^{-1} \in H^2$ and Λ is a diagonal matrix function of the form

$$\Lambda = \begin{pmatrix} z^{d_1} & \mathbb{O} & \cdots & \mathbb{O} \\ \mathbb{O} & z^{d_2} & \cdots & \mathbb{O} \\ \vdots & \vdots & \ddots & \vdots \\ \mathbb{O} & \mathbb{O} & \cdots & z^{d_n} \end{pmatrix}, \quad d_1, \cdots d_n \in \mathbb{Z}.$$

Moreover, $\operatorname{ind} T_U = -\sum_{j=1}^n d_j = N$. It is easy to see that $d_j \le 0$, $1 \le j \le n$. Indeed, since T_U has dense range in $H^2(\mathbb{C}^n)$ (see Section 2), it follows that $\operatorname{Ker} T_U^* = \{\mathbb{O}\}$, and if $d_j > 0$ for some j, then $\Phi_2^{-1} e_j \in \operatorname{Ker} T_U^*$ (here e_j is the jth standard basis vector of $\mathbb{C}^n$).

Consider the unitary-valued function $V = \bar{z} U$. Clearly, T_V is Fredholm and $\operatorname{ind} T_V = \dim \operatorname{Ker} T_V = N + n$. It is easy to see that $\operatorname{Ker} T_V$ coincides with the subspace

$$L = \left\{ \Phi_1 \begin{pmatrix} q_1 \\ q_2 \\ \vdots \\ q_n \end{pmatrix} : \; q_j \text{ is a polynomial in } H^2, \; \deg q_j \le -d_j \right\}.$$

Indeed, L is obviously a subset of $\operatorname{Ker} T_V$ and $\dim L = N + n = \dim \operatorname{Ker} T_V$.

Since $V = \bar{z}\Psi_{\#}^*\Psi^{-1}$, it follows that $\Psi e_j \in \operatorname{Ker} T_V$ for $1 \le j \le m$. Hence,

$$\Psi e_j = \Phi_1 \begin{pmatrix} q_{1j} \\ q_{2j} \\ \vdots \\ q_{nj} \end{pmatrix},$$

where the q_{rj} are polynomials in H^2 such that $\deg q_{rj} \le -d_j$. Clearly, $\Psi = \Phi_1 Q$, where $Q = \{q_{rj}\}_{1 \le r,j \le n}$. We have $W = \Psi^*\Psi = Q^*\Phi_1^*\Phi_1 Q$.

Let us show that $\deg \det Q = N$ and the zeros of $\det Q$ lie on $\mathbb{T}$. Since $\deg q_{rj} \le -d_j$, it follows that $\deg \det Q \le -\sum_{j=1}^n d_j = N$.

We have

$$U = \Psi_{\#}^*\Psi^{-1} = \Psi_{\#}^* Q^{-1}\Phi_1^{-1}, \quad \text{on the other hand,} \quad U = \Phi_2^*\Lambda\Phi_1^{-1}.$$

Consequently, $\Phi_2^*\Lambda = \Psi_{\#}^* Q^{-1}$, and so $\Psi_{\#}^* = \Phi_2^*\Lambda Q$. Put

$$Q_{\#} \stackrel{\text{def}}{=} Q^*\Lambda^* = \Psi_{\#}\Phi_2^{-1}.$$

Clearly, $Q_{\#}$ is an outer function in H^1. On the other hand, $Q_{\#}$ is a trigonometric polynomial. Since, $\Lambda Q = Q_{\#}^*$, it follows that

$$\bar{z}^N \det Q = \overline{\det Q_{\#}}. \tag{3.2}$$

Moreover, both $\det Q$ and $\det Q_{\#}$ are polynomial outer functions. If $\deg \det Q < N$, it follows from (3.2) that $\det Q_{\#}(0) = 0$ which contradicts the fact that $\det Q_{\#}$ is outer. Hence, $\deg \det Q = N$.

Since $\deg Q$ is an outer function, it cannot have zeros in $\mathbb{D}$. On the other hand, it is easy to see from (3.2) that if $\det Q$ has a zero outside $\operatorname{clos} \mathbb{D}$, then $\det Q_{\#}$ has a zero in $\mathbb{D}$ which is impossible since $\det Q_{\#}$ is an outer function. Therefore all zeros of $\det Q$ lie on $\mathbb{T}$.

Consider the weight $W_1 \stackrel{\text{def}}{=} \Phi_1^*\Phi_1$. Then $W = Q^*W_1Q$ and $W_1^{-1} \in L^1$. Since $U = \Phi_2^*\Lambda\Phi_1^{-1}$ and U is unitary-valued, we have

$$U^*U = (\Phi_1^*)^{-1}\Lambda^*\Phi_2\Phi_2^*\Lambda\Phi_1^{-1}, \quad \text{and so} \quad \Lambda^*\Phi_2\Phi_2^*\Lambda = \Phi_1^*\Phi_1 = W_1.$$

Hence,

$$(\Lambda^*\Phi_2)^*(\Lambda^*\Phi_2) = W_1.$$

The function $\Lambda^*\Phi_2$ belongs to H^2. However, it is not outer unless Λ is the constant identity matrix. Let us find an outer function with the same property. Consider the function $(\Lambda^*\Phi_2)^{\mathrm{t}} = \Phi_2^{\mathrm{t}}\Lambda^*$. Since a square matrix function is outer if and only if its determinant is outer (see [N], Lect. I, Sect. 6), it follows that the function Φ_2^{t} is also outer. Consequently, the subspace

$$\{z^j \Phi_2^{\mathrm{t}} \Lambda^* x : \ j \in \mathbb{Z}_+, x \in \mathbb{C}^n\}$$

has finite codimension in $H^2(\mathbb{C}^n)$. Since this subspace is obviously invariant under multiplication by z, it follows that the function $\Phi_2^{\mathrm{t}}\Lambda^*$ admits a representation of the form $\Phi_2^{\mathrm{t}}\Lambda^* = B\Xi$, where B is a finite Blaschke–Potapov product of finite degree (see [N]). Since $\det(\Phi_2^{\mathrm{t}}\Lambda^*) = \det(B\Xi)$ and both $\det \Phi_2$ and $\det \Xi$ are outer functions, it follows that $\det B = \mathrm{const} \det \Lambda^* = \mathrm{const}\, z^N$. It is easy to see that B is a polynomial inner $n \times n$ function. Therefore the function $\Phi_\# \stackrel{\mathrm{def}}{=} \Lambda^* \Phi_2 \Pi^* = \Xi^{\mathrm{t}}$ is outer, where $\Pi \stackrel{\mathrm{def}}{=} B^{\mathrm{t}}$. Clearly, Π is a polynomial inner function of size $n \times n$. It is also easy to see that

$$\Phi_\# \Phi_\#^* = W_1 = \Phi_1^* \Phi^1.$$

These factorizations determine the phase function

$$U_1 = \Phi_\#^* \Phi_1^{-1} = \Pi \Phi_2^* \Lambda \Phi_1^{-1} = \Pi U$$

which completes the proof. □

Remark. *Let us note the following interesting identities which can be verified very easily:*

$$Q^* Q = Q_\# Q_\#^*; \quad W = Q_\# \Phi_2 \Phi_2^* Q_\#^*.$$

Theorem 3.2. *Suppose that X is a space of functions on $\mathbb{T}$ satisfying* (A1)–(A4) *or X is a linear $\mathcal{R}$-space such that $X \subset VMO$. Let W be the density of a completely regular vector process which is represented in the form of* (3.1), *where W_1 and Q are as in the statement of Theorem 3.1. Then the initial process is X-regular if and only if the process with density W_1 is X-regular.*

Proof. By Theorem 2.2, the process is X-regular if and only if the phase function belongs to X. The result follows now immediately from Theorem 3.1. □

4. Spectral characterizations for X satisfying (A1)–(A4)

In this section we give characterize X-regular vectorial stationary processes in terms of spectral densities for spaces X satisfying (A1)–(A4) and consider several important cases.

Theorem 4.1. *Let X be a space of functions on $\mathbb{T}$ which satisfies* (A1)–(A4). *Suppose that W is the density of a completely regular vectorial process. The following statements are equivalent:*

(i) *the process is X-regular;*

(ii) *W admits a representation $Q^* W_1 Q$, where Q is an analytic polynomial matrix function such that the zeros of* $\det Q$ *are on $\mathbb{T}$ and W_1 is invertible in X;*

(iii) *W admits a representation $Q^* \exp(\Upsilon)Q$, where Q is an analytic polynomial matrix function such that the zeros of* $\det Q$ *are on* $\mathbb{T}$ *and* Υ *is a matrix function in* X *which takes self-adjoint values.*

Proof. By Theorem 3.1 it is sufficient to assume that the process is minimal, i.e., $W^{-1} \in L^1$. Consider the factorizations (1.3) and let $U = \Psi_{\#}^* \Psi^{-1}$ be the corresponding phase function.

Suppose first that the process is X-regular, i.e., $\mathbb{P}_- U \in X$. In [P3] it was shown that $\Psi \in X$ and $\Psi^{-1} \in X$ which implies that $W = \Psi^* \Psi \in X$ and W is invertible in X (in [P3] this was shown for spaces X in which the trigonometric polynomials are dense or spaces satisfying the axioms (B1)–(B3) (see [P3])); however, the same proof works for arbitrary spaces satisfying (A1)–(A4)).

Suppose that W is invertible in X. Then it is well known (see [CG], [P4]) that $\Psi, \Psi_{\#}$ are invertible in X. Hence, $U \in X$.

It is obvious that (ii) is equivalent to (iii). $\square$

In particular the classes Λ_α, $0 < \alpha < \infty$, B_1^1, and $B_{p,\infty}^s$, $1 \le p < \infty$, $1/p < s < \infty$, satisfy (A1)–(A4) which implies the following facts.

Corollary 4.2. *Let W be the density of a completely regular stationary process. Then the process is completely regular of order α if and only if $W = Q^* W_1 Q$, where Q is an analytic polynomial matrix function such that the zeros of* $\det Q$ *are on* $\mathbb{T}$ *and W_1 is invertible in Λ_α.*

Recall that this result was first obtained in [I2] by a different method.

Corollary 4.3. *Let W be the density of a completely regular stationary process. Then the process is* 1*-regular if and only if $W = Q^* W_1 Q$, where Q is an analytic polynomial matrix function such that the zeros of* $\det Q$ *are on* $\mathbb{T}$ *and W_1 is invertible in B_1^1.*

Corollary 4.4. *Let W be the density of a completely regular stationary process and let $1 \le p < \infty$. Then the process is p-regular of order α if and only if $W = Q^* W_1 Q$, where Q is an analytic polynomial matrix function such that the zeros of* $\det Q$ *are on* $\mathbb{T}$ *and W_1 is invertible in $B_{p\infty}^{1/p+\alpha}$.*

5. The case of completely regular and p-regular processes

In this section we find necessary conditions for vectorial processes to be completely regular or p-regular. Then we consider a class of processes for which these conditions are also sufficient. Recall that Theorem 3.1 reduces the general case to the case of minimal processes. As we have already mentioned in Section 2, unlike the scalar case complete regularity for minimal processes is not equivalent to the condition $\log W \in VMO$ (see [TV]).

Theorem 5.1. *Let $W = \{w_{jk}\}_{1\le j,k\le n}$ be the density of a minimal vectorial process. If the process is completely regular, then*

$$\log w_{jj} \in VMO, \quad 1 \le j \le n, \quad \text{and} \quad \frac{w_{jk}}{w_{jj}^{1/2} w_{kk}^{1/2}} \in VMO, \quad 1 \le j,k \le n. \qquad (5.1)$$

Theorem 5.2. *Let $W = \{w_{jk}\}_{1\le j,k\le n}$ be the density of a minimal vectorial process and let $0 < p < \infty$. If the process is p-regular, then*

$$\log w_{jj} \in B_p^{1/p}, \quad 1 \le j \le n, \quad \text{and} \quad \frac{w_{jk}}{w_{jj}^{1/2} w_{kk}^{1/2}} \in B_p^{1/p}, \quad 1 \le j,k \le n. \qquad (5.2)$$

Recall that $\boldsymbol{S}_1$-regular processes are described in Corollary 4.4.

However, conditions (5.1) and (5.2) are not sufficient. If W is the density of a minimal process, then the Toeplitz operator T_U has trivial kernel (see [P3]) and as we have already mentioned in Section 2, the operators $H_U^* H_U$ and $H_{U^*}^* H_{U^*}$ are unitarily equivalent (see [P3]). Consider now the following factorizations of W^{-1}:

$$W^{-1} = (\Psi_{\#}^{-1})^* \Psi_{\#}^{-1} = \Psi^{-1}(\Psi^{-1})^*.$$

It is easy to see that the phase function of the process with density W^{-1} which corresponds to these factorizations is U^*. Hence, the minimal process with density W is completely regular (p-regular) if and only if the process with density W^{-1} is completely regular (p-regular). This allows us to obtain the following necessary conditions.

Theorem 5.3. *Let W be the density of a minimal vectorial process. If the process is completely regular, then the entries of W^{-1} satisfy* (5.1).

Theorem 5.4. *Let W be the density of a minimal vectorial process and let $0 < p < \infty$. If the process is p-regular, then the entries of W^{-1} satisfy* (5.2).

Let us show that conditions (5.1) for the spectral density W are not sufficient for complete regularity. Let $n = 2$. Consider the spectral density

$$W = \begin{pmatrix} 1 & w \\ w & 1 \end{pmatrix},$$

where w is a nonnegative function such that $w(\zeta) < 1$ almost everywhere and $\log(1-w) \in L^1$ and $(1-w)^{-1} \in L^1$. Then W is the density of a minimal regular process. If $w \in VMO$, then W satisfies (5.1). Clearly,

$$W^{-1} = (1-w^2)^{-1} \begin{pmatrix} 1 & -w \\ -w & 1 \end{pmatrix},$$

and so by Theorem 5.3, if W is completely regular, then $\log(1-w) \in VMO$. It is easy to construct a nonnegative function w such that $w(\zeta) < 1$ almost everywhere, $\log(1-w) \in L^1$, and $(1-w)^{-1} \in L^1$ but $\log(1-w) \notin VMO$. For example, we can take a function w which in a small neighborhood of 1 is defined by $w(e^{i\theta}) = 1-\theta^{1/2}$, which is smooth outside that neighborhood and is separated away from 1 there. By Theorem 5.3 the process with density W is not completely regular which proves that conditions (5.1) are not sufficient.

Similarly one can prove that conditions (5.2) are not sufficient.

I do not know whether it is true that if the entries of both W and W^{-1} satisfy (5.1), then the processes with density W is completely regular. It is also unknown whether conditions (5.2) for entries of both W and W^{-1} imply that the process with density W is p-regular.

To prove Theorems 5.1 and 5.2, we need the following fact.

Theorem 5.5. *Let $X = VMO$ or $X = B_p^{1/p}$, $0 < p < \infty$. Suppose that u is a unimodular function in X and φ is a bounded function such that $\mathbb{P}_- u\varphi \in X$. Then $\mathbb{P}_-\varphi \in X$.*

Proof. Suppose that $X = VMO$. It is well known (see e.g., [PK]) that the Toeplitz operator T_u is Fredholm. Multiplying u by z^m if necessary, we can reduce the problem to the case when T_u is invertible. Let us define the operator $\tilde{T}_u$ on H^2_- by

$$\tilde{T}_u f = \mathbb{P}_- u f, \quad f \in H^2_-.$$

Exactly the same argument as for the Toeplitz operator T_u proves that $\tilde{T}_u$ is invertible on H^2_-. The condition $\mathbb{P}_- u\varphi \in X$ means that the Hankel operator $H_{u\varphi}$ is compact. We have to prove that H_φ is compact. We have

$$H_{u\varphi} f = \mathbb{P}_- u\varphi f = \mathbb{P}_- u \mathbb{P}_- \varphi f + \mathbb{P}_- u \mathbb{P}_+ \varphi f = \tilde{T}_u H_\varphi f + H_u T_\varphi f.$$

By the hypotheses H_u is compact, T_φ is bounded. Hence, $H_u T_\varphi$ is compact, and so $\tilde{T}_u H_\varphi$ is compact. Since $\tilde{T}_u$ is invertible, it follows that H_φ is compact.

In the case $X = B_p^{1/p}$ the proof is the same. The only difference is that instead of the compactness condition we have to consider the membership of $\boldsymbol{S}_p$. □

Note that the theorem easily generalizes to the case of an arbitrary linear $\mathcal{R}$-space. For $X = VMO$ another proof has been found by S. Semmes.

Proof of Theorem 5.1. Suppose that

$$x(t) = \begin{pmatrix} x_1(t) & x_2(t) & \cdots & x_n(t) \end{pmatrix}^{\mathrm{t}}, \quad k \in \mathbb{Z},$$

is minimal and completely regular. Then for each j, $1 \le j \le n$, the scalar process

$$x_j(t), \quad t \in \mathbb{Z},$$

is also completely regular and minimal. It is easy to see that the density of this process is w_{jj}. Hence, $\log w_{jj} \in VMO$.

Suppose now that $j \ne k$. Consider functions $F, G \in H^2(W)$ of the following form.

$$F = \{f_l\}_{1 \le l \le n}, \quad f_l = \mathbb{O} \text{ if } l \ne k, \quad f_k = f \in H^2(w_{kk}),$$

$$G = \{g_l\}_{1 \le l \le n}, \quad g_l = \mathbb{O} \text{ if } l \ne j, \quad g_j = g \in H^2_-(w_{jj}).$$

Then $\|F\|_{H^2(W)} = \|f\|_{H^2(w_{kk})}$ and $\|G\|_{H^2_-(W)} = \|g\|_{H^2_-(w_{jj})}$.

Let h_j, $1 \le j \le n$, be scalar outer functions such that $|h_j|^2 = w_{jj}$. We have

$$\begin{aligned}\rho_m(W) &\ge \sup\{|(z^m F, G)_{L^2(W)}| : \|f\|_{H^2(w_{kk})} \le 1, \|g\|_{H^2_-(w_{jj})} \le 1\} \\ &= \sup\{|\int_{\mathbb{T}} z^m w_{jk} f g dm| : \|f\|_{H^2(w_{kk})} \le 1, \|g\|_{H^2_-(w_{jj})} \le 1\} \\ &= \sup\{|\int_{\mathbb{T}} z^m w_{jk} h_k^{-1/2} \xi \overline{h_j^{-1/2}} \eta dm| : \|\xi\|_{H^2} \le 1, \|\eta\|_{H^2_-} \le 1\} \\ &= \left\| H_{z^m \frac{w_{jk}}{h_k^{1/2} \overline{h_j^{1/2}}}} \right\| = \left\| H_{z^m u \frac{w_{jk}}{w_{kk}^{1/2} w_{jj}^{1/2}}} \right\|, \quad \text{where} \quad u \stackrel{\text{def}}{=} \frac{w_{kk}^{1/2} w_{jj}^{1/2}}{h_k^{1/2} \overline{h_j^{1/2}}}.\end{aligned}$$

Since $\log w_{jj} \in VMO$ and $\log w_{kk} \in VMO$, it follows that u is a unimodular function in VMO. Put

$$\varphi \stackrel{\text{def}}{=} \frac{w_{jk}}{w_{kk}^{1/2} w_{jj}^{1/2}}.$$

Since the matrix function

$$\begin{pmatrix} w_{jj} & w_{jk} \\ w_{jk} & w_{kk} \end{pmatrix}$$

is positive definite, it follows that $\varphi \in L^\infty$.

The condition $\rho_n(W) \to 0$ implies now that

$$\lim_{n \to \infty} \|H_{z^m u \varphi}\| = 0$$

which means that $\mathbb{P}_-(u\varphi) \in VMO$. By Theorem 5.5, $\varphi \in VMO$ which completes the proof. ☐

Proof of Theorem 5.2. The proof is similar to that of Theorem 5.1. First, for each j the process

$$x(t) = \begin{pmatrix} x_1(t) & x_2(t) & \cdots & x_n(t) \end{pmatrix}^{\mathrm{t}}, \quad k \in \mathbb{Z},$$

is minimal and p-regular. This implies that $\log w_{jj} \in B_p^{1/p}$ for $1 \le j \le n$.

Let now $j \ne k$. Take the operator $\mathcal{P}_0 \mathcal{P}^0$ on $L^2(W)$ and consider the operator given by the following Hermitian form:

$$(\mathcal{P}_0 \mathcal{P}^0 F, G),$$

where as in the proof of Theorem 5.1

$$F = \{f_l\}_{1 \le l \le n}, \quad f_l = \mathbb{O} \text{ if } l \ne k, \quad f_k = f \in H^2(w_{kk}),$$

$$G = \{g_l\}_{1 \le l \le n}, \quad g_l = \mathbb{O} \text{ if } l \ne j, \quad g_j = g \in H^2_-(w_{jj}).$$

This operator must be in $\boldsymbol{S}_p$ and the reasoning similar to that given in the proof of Theorem 5.1 shows that the Hankel operator $H_{u\varphi}$ must be in $\boldsymbol{S}_p$, where u and φ are defined in the same way as in the proof of Theorem 5.1. This means that $\mathbb{P}_-(u\varphi) \in B_p^{1/p}$ (see [P1], [P2]), and by Theorem 5.5, this implies that $u \in B_p^{1/p}$. ☐

Consider now the class of weights W for which the coordinate projections

$$F \mapsto f_j, \quad f_j(\zeta) = ((F(\zeta), e_j)_{\mathbb{C}^n})e_j,$$

are bounded in $L^2(W)$. It is easy to see that this condition is equivalent to the condition that

$$\sup_{\zeta \in \mathbb{T}} \|W^{1/2}(\zeta)e_j\|_{\mathbb{C}^n} \|W^{-1/2}(\zeta)e_j\|_{\mathbb{C}^n} < \infty, \quad 1 \le j \le n. \tag{5.3}$$

Theorem 5.6. *Suppose that W is the density of a minimal vectorial process which satisfies* (5.3). *If*

$$\log w_{jj} \in VMO, \quad 1 \le j \le n, \quad \text{and} \quad \frac{w_{jk}}{w_{jj}^{1/2} w_{kk}^{1/2}} \in VMO, \quad 1 \le j, k \le n,$$

then the process is completely regular.

Theorem 5.7. *Suppose that W is the density of a minimal vectorial process which satisfies* (5.3) *and let $0 < p < \infty$. If*

$$\log w_{jj} \in B_p^{1/p}, \quad 1 \le j \le n, \quad \text{and} \quad \frac{w_{jk}}{w_{jj}^{1/2} w_{kk}^{1/2}} \in B_p^{1/p}, \quad 1 \le j, k \le n,$$

then the process is p-regular.

It is easy to see that condition (5.3) allows one to reduce the estimation of $\|(\mathcal{P}_0\mathcal{P}^m F, G)\|_{L^2(W)}$ to the case when F and G have only one nonzero coordinate. Similarly, one can estimate $\|\mathcal{P}_0\mathcal{P}^m\|_{\boldsymbol{S}_p}$.

References

[CG] K. Clancey and I. Gohberg, *Factorization of matrix functions and singular integral operators*, Oper. Theory Adv. Appl., **3**, Birkhäuser Verlag, Basel–Boston–Stuttgart, **1981**.

[D] R. G. Douglas, *Banach algebra techniques in the theory of Toeplitz operators*, CBMS Amer. Math. Soc., **15**, Providence, RI, **1973**.

[DM] H. Dym and H. P. McKean, *Gaussian processes, function theory, and the inverse spectral problem*, Academic Press, New York–London, **1976**.

[GY] I. M. Gelfand and A. M. Yaglom, *Calculation of the amount of information about a random function contained in another such function*, Amer. Math. Soc. Translat., II Ser., **12** (1959), 199–246.

[HS] H. Helson and D. Sarason, *Past and future*, Math. Scand., **21** (1967), 5–16.

[I1] I. A. Ibragimov, *On the spectrum of stationary Gaussian sequences satisfying the strong mixing condition. I, Necessary conditions*, Teor, Veroyatnost. i Primen., **15** (1970), 24–37; English transl., Theory Probab. Appl., **15** (1970), 23–36.

[I2] I. A. Ibragimov, *Completely regular multidimensional stationary processes with discrete time*, Trudy Mat. Inst. Steklov., **111** (1970), 224–251; English transl., Proc. Steklov Inst. Math., **111** (1970), 269–301.

[IR] I. A. Ibragimov and Yu. A. Rozanov, *Gaussian stochastic processes*, Nauka, Moscow, **1970**. (Russian)

[IS] I. A. Ibragimov and V. A. Solev, *On a condition for the regularity of a Gaussian stationary sequence*, Zap. Nauchn. Semin. Leningrad. Otdel. Mat. Inst. Steklov. (LOMI), **12** (1969), 113-125; English transl., Sem. Math. V. A. Steklov Math. Inst. Leningrad, **12** (1971), 54–60.

[K] A. N. Kolmogorov, *Interpolation und Extrapolation von stationnaren zufalligen Folgen*, Izv. Akad. Nauk SSSR, ser. Mat., **5** (1941), 3–14.

[KR] A. N. Kolmogorov and Yu. A. Rozanov, *On the strong mixing for stationary Gaussian processes*, Teor. Veroyatnost. i Primen., **5** (1960), 222–227; English transl., Theor. Probability Appl., **5** (1960), 204–208.

[LS] G. S. Litvinchuk and I. M.Spitkovski, *Factorization of measurable matrix functions*, Oper. Theory Adv. Appl., **25**, Birkhäuser Verlag, Basel-Boston, MA, **1987**.

[N] N. K. Nikol'skii, *Treatise on the shift operator. Spectral function theory*, Springer-Verlag, Berlin–Heidelberg–New York–Tokyo, **1986**.

[P1] V. V. Peller, *Hankel operators of class S_p, and applications (rational approximation, Gaussian processes, the majorant problem for operators)*, Mat. Sbornik, **113** (1980), 538–581; English Transl., Math. USSR-Sb., **41** (1982), 443–479.

[P2] V. V. Peller, *A description of Hankel operators of class $\mathfrak{S}_p$ for $p > 0$, investigation of the rate of rational approximation and other applications*, Mat. Sb., **122** (1983), 481–510; English transl., Math. USSR-Sb., **50** (1985), 465–494.

[P3] V. V. Peller, *Hankel operators and multivariate stationary processes*, Proc. Symp. Pure Math., **51** (1990), 357–371.

[P4] V. V. Peller, *Factorization and approximation problems for matrix functions*, Amer. J. Math., **1998**.

[PK] V. V. Peller and S. V. Khrushchev, *Hankel operators, best approximation and stationary Gaussian processes*, Russian Math. Surveys, **37** (1982), 53–124.

[R] Yu. A. Rozanov, *Stationary stochastic processes*, Fizmatgiz, Moscow, **1963**. (Russian)

[Sa] D. Sarason, *An addendum to "Past and Future"*, Math. Scand., **30** (1972), 62–64.

[Se] S. Semmes, *Trace ideal criteria for Hankel operators and applications to Besov spaces*, Integral Equations Operator Theory, **7** (1984), 241–281.

[Si] I. B. Simonenko, *Some general problems of the theory of the Riemann boundary-value problem*, Izv. Akad. Nauk SSSR, Ser. Mat., **32** (1968), 1138–1146. (Russian)

[TV] S. R. Treil and A. L. Volberg, *Completely regular multivariate stationary processes and the Muckenhoupt condition*, preprint, **1997**.

[WM] N. Wiener and P. Masani, *The prediction theory of multivariate stochastic processes, I. The regularity conditions*, Acta Math., **98** (1957), 111–150.

[Y] A. M. Yaglom, *Stationary Gaussian processes satisfying the strong mixing condition and best predictable functionals*, Proc. Int. Research Seminar of the Statistical Laboratory, Univ. of Calif., Berkeley, **1963**, 241–252; Springer-Verlag, New York.

Department of Mathematics,
Kansas State University,
Manhattan, Kansas 66506, USA
E-mail address: `peller@math.ksu.edu`

Operator Theory:
Advances and Applications, Vol. 113

The Friedrichs Operator of a Planar Domain

Mihai Putinar and Harold S. Shapiro

Dedicated to the memory of S. A. Vinogradov

Abstract. Properties of the Friedrichs operator of a planar domain, originally intended for applications to planar elasticity, are related to the geometry and function theory of the domain. We show that the spectrum of the Friedrichs operator does not, in general, determine the domain, and we provide the additional necessary complete invariants. An analysis of the Friedrichs operator of generalized quadrature domains is carried out at the abstract level and on several examples. The study of Toeplitz operators on the Bergman space of the domain comes naturally into focus.

1. Introduction

K. Friedrichs [7] introduced implicitly a real-linear operator (below denoted F_Ω) on the Bergman space $AL^2(\Omega)$ of a plane domain, more precisely he studied the symmetric (not Hermitean) bilinear form

$$(f,g) \mapsto \operatorname{Re} \int_\Omega fg \, dA,$$

which defines, when represented against the L^2-scalar product, the operator F_Ω that we think it is justified to call "the Friedrichs operator" associated to the domain Ω. Above dA stands for the planar Lebesgue measure. The square S_Ω of F_Ω is a complex-linear, Hermitean and non-negative operator on $AL^2(\Omega)$.

These operators are interesting in several respects. First of all, the spectral properties of S_Ω reflect fine-structure properties of the geometric relations in the Hilbert space $L^2(\Omega)$ between its two subspaces consisting respectively of holomorphic and conjugate holomorphic functions. To be more specific, S_Ω is the "gap operator", in the terminology of [11], between these two subspaces. For example, the extreme situation where these spaces are mutually orthogonal (as happens for the so-called null quadrature domains, [19], e.g. the complement of an ellipse) corresponds to $F_\Omega = 0$.

The first author was partially supported by the National Science Foundation Grant DMS-9800666.

Secondly, these operators allow concise formulations of the boundary value problems for the biharmonic operator that are of importance in the theory of planar elasticity; in particular, in domains for which the operator F_Ω has finite dimensional range (identified in [24] as the class of quadrature domains), the exact solutions of the said boundary value problems are obtained by solving finite systems of linear equations. These aspects seem to have been the main motivation for Friedrichs' work. In elaborating his program, he proved a number of interesting results about these operators, concerning compactness, essential spectrum etc. in their dependence on the underlying domain.

The aim of this paper is to extend Friedrichs' pioneering work and take up some new aspects motivated by function-theoretic and functional-analytic considerations. We hope in future work to deal with the applications to the biharmonic equation. That there is a very intimate connection can be seen in a striking way from one of our results: F_Ω is injective if and only if the weakly elliptic equation

$$\overline{\partial}^2 u = 0$$

has no non-trivial solution in Ω vanishing on its boundary (note that the symbol of this operator, whose kernel comprises the so-called biholomorphic functions, divides that of the bi-Laplacian).

Another interesting aspect of the Friedrichs operator, which is here taken up in detail for the first time, is its close relation to the so-called (generalized) quadrature domains.

The present paper is organized as follows. Section 2 contains preliminaries and a few simple observations which relate the Friedrichs operator to Toeplitz-Bergman operators and to quadrature domains. Section 3 states Friedrichs' main theorem and its relevance for planar elasticity and an interesting inequality in function theory. In Section 4 we analyze some spectral properties of S_Ω; for instance we show that S_Ω cannot be Fredholm and that the elements in the kernel of S_Ω are remarkable analytic functions. Finally, in the same section we show that there are domains, not affinely equivalent, whose Friedrichs operators have the same spectrum. Section 5 deals with the relation between F_Ω and the generalized quadratures supported by thin subsets of Ω. A restriction operator between the Bergman space and the L^2-space of such a quadrature identity is instrumental in computing or at least having qualitative information about the eigenvalues of S_Ω and the corresponding eigenfunctions. This restriction operator has appeared in [21] and [23] in a functional-theoretic clarification of earlier work by S. Bergman. Section 6 contains some examples of domains on which the Friedrichs operator, or at least its spectrum and eigenfunctions, is within reach.

This paper is intended to be continued by another one, where different aspects related to the Friedrichs operator (already alluded to in the text below) will be developed.

2. Preliminaries

In this section we explain the notation, give some definitions and state a few basic observations concerning the defined objects.

2.1. Notation

Let Ω be a domain of the complex plane $\mathbf{C} = \mathbf{R} \oplus i\mathbf{R}$. In general we allow Ω to be unbounded, but we require that its boundary $\partial\Omega$ contains infinitely many points, so that Ω carries non-trivial, square integrable analytic functions. Such assumptions will depend and result from the context. The planar Lebesgue measure will be denoted by dA and the corresponding Lebesgue spaces will be $L^p(\Omega) = L^p(\Omega, dA), 1 \le p \le \infty$. For every $p, 1 \le p \le \infty$, we denote by $AL^p(\Omega)$ the space of complex analytic functions in Ω which belong to $L^p(\Omega)$. In addition, we put:

$$AL_0^2(\Omega) = \{f \in AL^2(\Omega); \int_\Omega f dA = 0\}.$$

Of course $AL^\infty(\Omega) = H^\infty(\Omega)$. In general, $H^p(\Omega)$, $1 \le p \le \infty$, will be the Hardy spaces of a domain Ω with piecewise smooth boundary. The space of analytic functions in an open set Ω will be denoted by $O(\Omega)$; for a closed set K, we denote by $O(K)$ the algebra of germs of analytic functions defined in neighbourhoods of K. The complex conjugation operator:

$$C : L^p(\Omega) \longrightarrow L^p(\Omega), \quad C(f) = \overline{f}, \;\; f \in L^p(\Omega),$$

is an isometry of $L^p(\Omega)$. Thus $CAL^p(\Omega)$ is the space of anti-analytic functions which are p-summable in Ω. We put $\|\phi\|_{p,\Omega}$ for the norm of the space $L^p(\Omega)$.

For a measure or distribution u, its closed support in the sense of distributions will be denoted by $\mathrm{supp}(u)$.

The **Bergman space** of the domain Ω is the space $AL^2(\Omega)$. It is a Hilbert space with reproducing kernel $K(z,w) = k_w(z)$, known as the **Bergman kernel** of the domain Ω. The orthogonal projection of $L^2(\Omega)$ onto $AL^2(\Omega)$ will be denoted by P_Ω, or in short, P (the **Bergman projection** of Ω). The scalar product and the norm in $L^2(\Omega)$ will be denoted by $\langle *, * \rangle$, respectively $\| * \|$. Therefore, for any element $f \in AL^2(\Omega)$ and $w \in \Omega$ one has $\langle f, k_w \rangle = f(w)$. The function identically equal to 1 will be denoted by $\mathbf{1}$.

Let $\phi \in L^\infty(\Omega)$ and let

$$T_\phi : AL^2(\Omega) \longrightarrow AL^2(\Omega), \quad T_\phi(f) = P(\phi f), \;\; f \in AL^2(\Omega),$$

be the **Bergman-Toeplitz operator** with symbol ϕ. It is a linear bounded operator of norm $\|T_\phi\| \le \|\phi\|_{\infty,\Omega}$. The operator $B = T_z$ corresponds to the multiplication on the Bergman space with the complex variable $z \in \Omega$. We call B the **Bergman shift** of Ω.

By convention we assume without explicit mention in the paper, that whenever the constant function $\mathbf{1}$ comes into discussion, the area of Ω is finite, so that $\mathbf{1} \in AL^2(\Omega)$. Also, we speak of the Bergman shift B only when the domain Ω is bounded, so that this operator satisfies $\|B\| < \infty$.

Finally, for an abstract linear bounded operator T acting on a Hilbert space H we denote by $\sigma(T)$ the spectrum of T and by $\sigma_{\text{ess}}(T)$ the essential spectrum of T. The identity operator on H is denoted by I.

2.2. Definitions

The **Friedrichs operator** of the domain Ω is the anti-linear operator $F = PC : AL^2(\Omega) \longrightarrow AL^2(\Omega)$. Thus, for a function $f \in AL^2(\Omega)$ we have:

$$F(f)(z) = \int_\Omega K(z,w)\overline{f(w)}dA(w). \tag{1}$$

Obviously, $\|F(f)\| \le \|f\|$, and if the domain Ω is bounded, $F(\mathbf{1}) = \mathbf{1}$. For a pair of functions $f, g \in AL^2(\Omega)$ we note the identities:

$$\langle f, Fg\rangle = \langle g, Ff\rangle = \int_\Omega (fg)dA. \tag{2}$$

The square $S = F^2$ of the Friedrichs operator is $\mathbf{C}$-linear and:

$$\langle f, Sf\rangle = \langle Ff, Ff\rangle.$$

Therefore S is a self-adjoint operator on the Bergman space, satisfying $0 \le S \le I$.

Actually, the operator S is the "gap operator" between the closed subspaces $AL^2(\Omega)$, $CAL^2(\Omega)$ of $L^2(\Omega)$, see for more details [11], p. 197. Indeed, denoting by $Q = CPC$ the orthogonal projection of $L^2(\Omega)$ onto $CAL^2(\Omega)$, one finds that:

$$S = PCPC|_{AL^2(\Omega)} = PQ|_{AL^2(\Omega)}.$$

Next we introduce a class of domains on which the Friedrichs operator is simpler (finite rank) and more accessible. A domain $\Omega \subset \mathbf{C}$ is called after Aharonov and Shapiro ([2]) a **quadrature domain** if there exists a distribution u of finite support in Ω, with the property that:

$$\int_\Omega f dA = u(f), \qquad f \in AL^2(\Omega).$$

The dimension of the space $L^2(\Omega) \ominus Ker(u)$ is the **order** $o(\Omega)$ of Ω. A quadrature domain of order 1 is necessarily a disk, [2]. By writing

$$u(f) = \sum_{i=1}^n \sum_{j=0}^{m_i} c_{ij} f^{(j)}(a_i), \qquad f \in AL^2(\Omega), \tag{3}$$

with $a_i \in \Omega, 1 \le i \le n$, and $c_{im_i} \ne 0$, we find that:

$$\mathbf{1} = \sum_{i=1}^n \sum_{j=0}^{m_i} \overline{c_{ij}} \frac{\partial^j k_{a_i}}{\partial \overline{a_i}^j}. \tag{4}$$

We call a_i the **nodes** of the quadrature identity (3), and c_{ij} its **weights**.

Quadrature domains are very rigid; for instance their boundary is an irreducible real algebraic curve with very special singularity points. The class of simply connected quadrature domains coincides with the conformal images of the unit disk by rational maps. For further details concerning quadrature domains see [24] and the references cited there.

2.3. First observations

Several simple facts concerning the Friedrichs operator are already within reach. The notations are those introduced before.

Lemma 2.1. *Assume the domain Ω is bounded and let $f \in L^2(\Omega)$ be a solution of the equation $Sf = f$. Then f is a constant function.*

Proof. Indeed, from $Sf = f$ one obtains $\|Ff\| = \|f\|$, hence $Cf = PCf$. Thus Cf is both an anti-analytic and an analytic function, so f is necessarily a constant. □

Lemma 2.2. *Let $\phi \in H^\infty(\Omega)$. Then:*

$$FT_\phi = T_\phi{}^* F. \tag{5}$$

Proof. A pair of functions $f, g \in AL^2(\Omega)$ satisfies:

$$\begin{aligned}\langle f, T_\phi{}^* Fg\rangle &= \langle \phi f, PCg\rangle = \langle \phi f, Cg\rangle \\ &= \langle f, C(\phi g)\rangle = \langle f, PC(\phi g)\rangle = \langle f, F(T_\phi g)\rangle.\end{aligned}$$

□

Since $F\mathbf{1} = \mathbf{1}$, the next result follows.

Corollary 2.3. *For every function $\phi \in H^\infty(\Omega)$ one has:*

$$F\phi = T_\phi{}^* \mathbf{1}. \tag{6}$$

In particular, if $AL^\infty(\Omega)$ is dense in the Bergman space, then F is the unique anti-linear operator on $L^2(\Omega)$, satisfying relation (6).

As an application we derive the following known theorem, see [24] Theorem 8.4.

Theorem 2.4 (H. S. Shapiro). *Let Ω be a domain with boundary consisting of finitely many continua. Then Ω is a quadrature domain if and only if* $\operatorname{rank}(F) < \infty$.

In that case $o(\Omega) = \operatorname{rank}(F)$.

Proof. The regularity of $\partial\Omega$ in the statement implies that $AL^\infty(\Omega)$ is dense in $AL^2(\Omega)$ and moreover, there are no bounded point evaluations for $AL^2(\Omega)$ lying on $\partial\Omega$, see for details [25].

Let Ω be a quadrature domain, and assume for clarity that its nodes a_i are simple. Let $\phi \in H^\infty(\Omega)$. According to the identity $T_\phi^* k_{a_i} = \overline{\phi(a_i)}$ and formula (4), $F\phi$ belongs to the linear span of $k_{a_i}, 1 \le i \le o(\Omega)$. The case of multiple nodes is a simple generalization of this observation.

Conversely, if $\operatorname{rank}(F)$ is finite and N is large enough, then the vectors $(T_z^*)^k\mathbf{1}$, $1 \le k \le N$, are linearly dependent. Consequently there exists a polynomial $p(z)$ of minimal degree, satisfying $T_p^*\mathbf{1} = 0$. Since there are no bounded point evaluations in $\partial\Omega$, the zeroes a_i of p lie in Ω. Then formula (4) follows. □

Actually, using more recent investigations concerning finite codimensional invariant subspaces of the Bergman space, cf. [3], we can complement the latter theorem by the following observation.

Theorem 2.5. *Let Ω be a bounded domain, no boundary component of which is a point. The following are equivalent:*
a) $\mathrm{im}(S)$ *has finite dimension k;*
b) $\ker(S)$ *has finite codimension k;*
c) $\ker(F)$ *has finite codimension k;*
d) Ω *is a quadrature domain of order k.*

Proof. The equivalence between a) and b) follows from the fact that $S^* = S$. We also know that $\ker(F) = \ker(S)$ and that d) $\Rightarrow$ c). It remains to prove that c) implies d).

Assume that $\ker(F)$ has finite codimension k. Since $\ker(F)$ is invariant under multiplication by $H^\infty(\Omega)$, the Axler-Bourdon theorem [3] shows that there exists a polynomial $p(z)$, having k zeroes in Ω, with the property that $\ker(F) = \{pf;\ f \in AL^2(\Omega)\}$. Thus, all elements g in $AL^2(\Omega)$ vanishing with the corresponding multiplicity on the zeroes a_i of p, satisfy $\int_\Omega g dA = 0$. Therefore the vector $\mathbf{1}$ is a linear combination of the evaluation functions k_{a_i} and possibly some of their derivatives. This implies that Ω is a quadrature domain. □

In particular, $\mathrm{rank}(F) = 1$ if and only if Ω is a disk.

3. Planar elasticity and the Friedrichs inequality

In this section we briefly recall and comment on a part of the results due to Friedrichs [7]. A bounded planar domain Ω will be regarded as a thin, deformable plate. We adopt the hypothesis that both strain and stress tensors of Ω depend only on the two coordinates in the plane of Ω. Two fundamental questions in elasticity theory ask to find the (equilibrum) distribution of stress through Ω, in the case when the stress, respectively the strain, is known on $\partial\Omega$. A long time ago it was realized (it seems by J. C. Maxwell, [14]) that these questions can be interpreted as boundary value problems for the bi-Laplacian, hence the natural connection with complex functions in Ω. Along these lines, the two problem were satisfactorily solved at the beginning of this century by Kolosov and Muskhelishvili, see [14] for details. Their method, which is considered the standard one, consists in transforming the problems into singular integral equations (with Cauchy type kernels) and then to use the Fredholm alternative for describing the set of solutions.

Less known is the approach proposed by Friedrichs [7]. By using variational methods for the energy of the plate, he has reduced the two questions to integral equations with operators involving only the area measure on Ω. This seems to be the first occurrence and application of the Bergman space to mathematical physics. We sketch below Friedrichs' equations and a way of representing their solutions. For complete details the reader can consult [16] besides Friedrichs' paper.

Specifically, the two basic problems of planar elasticity correspond to finding elements $u, v \in AL^2(\Omega)$ such that:

$$(I + F)u = f, \tag{7}$$

respectively

$$(\kappa - F)v = g, \tag{8}$$

where $f, g \in L^2(\Omega)$ are given and $\kappa > 1$ is a constant depending on the material. The next result is the key to analyzing these equations.

Theorem 3.1 (Friedrichs). *Assume that the bounded domain Ω has piece-wise smooth boundary, with finitely many corners of interior angles α_k, $0 < \alpha_k \le 2\pi$, $1 \le k \le n$. Put $\alpha_0 = \pi$ for the generic smooth point of $\partial\Omega$.*

Then $\sigma_{\text{ess}}(S) \supset \{|\frac{\sin \alpha_k}{\alpha_k}|^2; 0 \le k \le n\}$.

Let us remark that, only external cusps are excluded in the statement from the singularities of the boundary. As we shall see later in Section 6, the inclusion in Friedrichs' theorem may be strict.

To solve equation (8) it is enough to remark that

$$\kappa^2 - S = \kappa^2 - F^2 = (\kappa - F)(\kappa + F)$$

is an invertible operator ($S \le I$). Thus $v = (\kappa - F)^{-1} g = (\kappa^2 - S)^{-1}(\kappa + F)g$ is the unique solution of (8). In order to solve equation (7) we either decompose f into $f = \langle f, \mathbf{1}\rangle\mathbf{1} + (f - \langle f, \mathbf{1}\rangle\mathbf{1})$ and correspondingly split (7) along $\mathbf{C1}$ and $\mathbf{1}^\perp$, or apply $(I - F)$ to (7). Either way, according to Lemma 2.1, the equation on the orthogonal complement of $\mathbf{1}$ has a unique solution. The equation along $\mathbf{1}$ reduces to $(I + F)\sigma = \lambda$, with complex numbers σ, λ. This is possible if and only if $\lambda \in \mathbf{R}$, and in that case $\operatorname{Re}\sigma = \lambda/2$. All in all, equation (7) is solvable if and only if $\langle f, \mathbf{1}\rangle \in \mathbf{R}$. In that case the solution is:

$$u = \langle f, \mathbf{1}\rangle/2 + iy + (I + F)^{-1}(f - \langle f, \mathbf{1}\rangle\mathbf{1}),$$

with an arbitrary parameter $y \in \mathbf{R}$.

As a function theoretic application of Theorem 3.1, Friedrichs has established the following inequality, which actually gives the title of his paper.

Corollary 3.2. *Let Ω be a domain satisfying the conditions in Theorem* 3.1. *Then there exists a positive constant $c = c(\Omega) < 1$, with the property:*

$$|\int_\Omega f^2 dA| \le c \int_\Omega |f|^2 dA, \qquad f \in AL_0^2(\Omega). \tag{9}$$

We call (9) the **Friedrichs inequality** of the domain Ω. A different proof and a generalization to $L^1(\Omega)$ of Friedrichs' inequality was given in [20]. Another generalization of Friedrichs' inequality to $AL^p(\Omega), 1 < p < \infty$, where Ω is a domain satisfying the interior cone condition at each boundary point, appears in [22]. The main idea there is to use the theory of strongly elliptic systems of differential operators in order to establish that the algebraic sum $AL^p(\Omega) + CAL^p(\Omega)$ is closed in $L^p(\Omega), 1 < p < \infty$, whenever Ω satisfies the interior cone condition. As a matter of fact Friedrichs' theorem above and Corollary 3.2 hold for such domains.

Proof of Corollary 3.2. Since the self-adjoint operator $I - S$ is Fredholm and $\mathbf{1}$ spans its kernel, we obtain:

$$c^2 = \sup\{\langle Sf, f\rangle | \langle f, \mathbf{1}\rangle = 0, \|f\| \leq 1\} < 1.$$

Therefore:

$$|\int_\Omega f^2 dA| = |\langle f, Ff\rangle| \leq \|f\|\|Ff\| \leq \|f\|\langle Sf, f\rangle^{1/2} \leq c\|f\|^2. \qquad \square$$

Actually the above argument can be reversed and it yields

Proposition 3.3. *A bounded planar domain Ω satisfies Friedrichs' inequality if and only if* $\sup\{\lambda; \lambda \in \sigma_{\text{ess}}(S)\} < 1$.

Proof. Let $\gamma = \sup\{\lambda | \lambda \in \sigma_{\text{ess}}(S)\}$ and assume $\gamma < 1$. We know that 1 is the largest eigenvalue of S, of multiplicity one, cf. Lemma 2.1. Hence the interval $(\gamma, 1)$ contains only finitely many elements in $\sigma(S)$, all lying in the point spectrum of S. Let c^2 be the largest element, if the latter set is non-empty, or let $c^2 \in (\gamma, 1)$ be arbitrary, if the set is empty. Then

$$\langle Sf, f\rangle \leq c^2\|f\|^2, \qquad f \in AL_0^2(\Omega).$$

Thus, Friedrichs' inequality holds for Ω.

Conversely, if Friedrichs' inequality is satisfied, then obviously $\gamma < 1$. $\square$

4. Spectral theory of the Friedrichs operator

In this section we investigate the kernel and the range of the Friedrichs operator, its essential spectrum and other spectral invariants. We will prove that, in general, the spectrum of F does not determine the underlying domain, even up to affine motions. Simple arguments show that translations and homotheties of Ω in the complex plane produce unitarily equivalent Friedrichs operators.

4.1. Fredholm theory

Below, an $\mathbf{R}$-linear operator T is called **Fredholm** if both $\ker(T)$ and $\text{coker}(T)$ are finite dimensional (the latter condition implies that $\text{im}(T)$ is closed). Most of the Fredholm theory carries to $\mathbf{R}$-linear operators, cf. [11].

Theorem 4.1. *Let Ω be a planar domain and let F be its Friedrichs operator. Then* $\ker(F)$ *is 0- or ∞-dimensional and F is not Fredholm. In particular,* $0 \in \sigma_{\text{ess}}(S)$.

Proof. According to relation (5), $\ker(F)$ is $H^\infty(\Omega)$-invariant. Hence its dimension is either zero, or infinity.

Assume by contradiction that F is Fredholm, that is $\ker(F) = 0$ and $\text{coker}(F)$ has finite dimension. Pick distinct points $w_j \in \Omega, 1 \leq j \leq N$. If N is big enough, then there exists a non-trivial linear combination $v = \sum_{j=1}^N \alpha_j k_{w_j}$ belonging to $\text{im}(F)$, the range of F. Let $p(z)$ be a non-trivial polynomial which vanishes at the points $w_j, 1 \leq j \leq N$.

It is easy to verify that the evaluation vector k_w satisfies $T_\phi{}^* k_w = \overline{\phi(w)} k_w$ for every function $\phi \in H^\infty(\Omega)$. Thus $T_p^* v = 0$. But relation (5) implies $F(pg) = T_p^* v = 0$, where $Fg = v$, a contradiction to the fact that F is injective.

Since $S = F^2$, the operator S cannot be Fredholm. □

For a quadrature domain, the Friedrichs operator has infinite dimensional kernel and finite dimensional range. There are many examples of domains for which the Friedrichs operator is injective. For instance the ellipse or an annulus, see Section 6. The preceding proof has the following corollary which may be of independent interest.

Corollary 4.2. *For every domain Ω possessing injective Friedrichs operator, no linear combination of the evaluation functions k_w, $w \in \Omega$, lies in* $\operatorname{im}(F)$.

Next, keeping in mind the example of quadrature domains, we focus on the behaviour of all functions belonging to $\ker(F)$. We say that the domain Ω admits a **generalized quadrature** supported by the closed set $K \subset \overline{\Omega}$, if there exists a complex measure μ on K, so that the total variation $|\mu|$ of μ does not charge $\partial\Omega$, and with the property that:

$$\int_\Omega f dA = \int_K f d\mu, \qquad f \in O(\overline{\Omega}).$$

In that case $\partial\Omega \setminus K$ is a real analytic (possibly singular) curve.

Instead of the test space $O(\overline{\Omega})$ above one can take $AL^1(\Omega)$, by assuming in this case that

$$\int_K \frac{|d\mu|(z)}{dist(z, \partial\Omega)} < \infty.$$

If the domain Ω allows a generalized quadrature in the above sense, then $\partial\Omega \setminus K$ is a real analytic curve possibly with inward pointing cusps, cf. [24] Chapter 5.

Again an example is the ellipse, in which case K is the segment joining the foci.

Proposition 4.3. *Let Ω be a domain with the property that $O(\overline{\Omega})$ is dense in $AL^2(\Omega)$. Suppose that Ω satisfies a generalized quadrature formula supported by a compact set K, so that ∂K does not separate the plane.*

Then either $\ker(F) = 0$, *or the set K can be chosen discrete and contained in the zero set of a non-trivial function in $AL^2(\Omega)$. In that case,*

$$\ker(F) = \{f \in AL^2(\Omega); f|_K = 0\}.$$

Proof. According to Mergelyan's theorem, the functions $g \in AL^2(\Omega)$ are dense in $C(K)$, the algebra of continuous functions on K. Let $f \in \ker(F)$. Then

$$0 = \langle f, Fg\rangle = \int_\Omega (fg) dA = \int_K (fg) d\mu.$$

Consequently $f d\mu = 0$ and we can choose K to be contained in the zero set of f. With this choice, any function $h \in AL^2(\Omega)$ vanishing on K is necessarily annihilated by the Friedrichs operator. □

The argument in the preceding proof can be extended to classes of sets K whose boundary may separate the plane; for instance by approximating the continuous functions on K with rational functions with poles off K. We do not expand here these details.

We do not know other domains than quadrature domains for which the Friedrichs operator has closed range.

4.2. The kernel of the Friedrichs operator

This part of the paper deals with functions which are annihilated by the Friedrichs operator. The following example shows that, in the absence of a nice generalized quadrature formula, Proposition 4.3 may fail.

Theorem 4.4. *There exists a Jordan domain with C^∞ boundary, with the property that its Friedrichs operator annihilates a nowhere vanishing function.*

In other terms we will construct a domain Ω and a function $g \in AL^2(\Omega)$, with the property that $g(z) \neq 0, z \in \Omega$, and

$$\int_\Omega fg dA = 0, \qquad f \in AL^2(\Omega). \tag{10}$$

The proof is based on the following result, of independent interest. Below we denote by $H^2(\mathbf{D})$ the Hardy space of the unit disk.

Proposition 4.5. *There exists a singular inner function J in the unit disk, with the property that $H^2(\mathbf{D}) \ominus JH^2(\Omega)$ contains a univalent function $u \in H^2(\mathbf{D}) \cap C^\infty(\partial\mathbf{D})$.*

Proof of Theorem 4.4. Assume the existence of the univalent function u in Proposition 4.5. A power series expansion argument shows that, for any pair $f, g \in O(\overline{\mathbf{D}})$, relation:

$$\langle f', g\rangle_{2,\mathbf{D}} = \langle f, zg\rangle_{2,\partial\mathbf{D}} \tag{11}$$

holds, where the right hand side is the scalar product in $H^2(\mathbf{D})$. Therefore,

$$\langle u', Jf\rangle_{2,\mathbf{D}} = \langle u, zJf\rangle_{2,\partial\mathbf{D}}.$$

By changing variables $w = u(z)$, and letting $z = v(w)$, we have a biholomorphic map $u : \mathbf{D} \longrightarrow \Omega$, with inverse v defined on the bounded domain Ω. According to the previous computations, for all $f \in O(\overline{\mathbf{D}})$, we obtain:

$$\begin{aligned} 0 = \langle Jf, u'\rangle_{2,\mathbf{D}} &= \int_{\mathbf{D}} \overline{u'(z)} J(z) f(z) dA(z) \\ &= \int_\Omega \overline{u'(v(w))} J(v(w)) f(v(w)) v'(w) \overline{v'(w)} dA(w) \\ &= \int_\Omega J(v(w)) f(v(w)) v'(w) dA(w). \end{aligned}$$

Thus, denoting $g(w) = J(v(w))v'(w)$, we see that:

$$\int_{\Omega} |g|^2 dA \leq \int_{\Omega} |v'(w)|^2 dA(w) = |\mathbf{D}| = \pi,$$

so $g \in AL^2(\Omega)$ and g does not vanish on Ω. Since the functions $f(v(w)), f \in O(\overline{\mathbf{D}})$ are dense in $AL^2(\Omega)$, relation (10) follows. □

Proof of Proposition 4.5. Let I be a singular inner function with associated measure $d\mu$, singular with respect to dt:

$$I(z) = \exp[\int_{-\pi}^{\pi} \frac{z + e^{it}}{z - e^{it}} d\mu(t)].$$

If the support of $d\mu$ is small enough, for instance a point, there exists an outer function O, vanishing on that set, so that $\overline{O}I \in C^r(\partial\mathbf{D})$, for a prescribed $r > 1$, including the possibility $r = \infty$.

Let $f(z) = \overline{zO(z)}I(z)$ and denote by $\Pi : L^2(\partial\mathbf{D}) \longrightarrow H^2(\partial\mathbf{D})$ be the Szegö projection. Then Πf is a non-trivial element of $(IH^2(\partial\mathbf{D}))^{\perp}$. Indeed, for all $b \in H^{\infty}(\mathbf{D})$ the following identities, in the scalar product of H^2:

$$\langle \Pi f, Ib \rangle = \langle \Pi(\overline{zO}I), Ib \rangle = \langle \overline{zO}I, Ib \rangle = \langle I, zOIb \rangle = \langle 1, zOb \rangle = 0,$$

hold.

Moreover, $\Pi f \neq 0$, because otherwise, $0 = \langle \overline{zO}I, b \rangle = \langle I, zOb \rangle$. But O is outer, so $OH^{\infty}(\partial\mathbf{D})$ is dense in the Hardy space. This would imply $0 = I'(0) = \langle I, z \rangle$, a contradiction.

Next we will produce from f a univalent function u as in the statement. If $u(z) = z + \sum_{j=2}^{\infty} c_j z^j$, we use

$$\sum_{j=2}^{\infty} j|c_j| < 1 \tag{12}$$

as an injectivity criterion for u, see [26] p. 212, Problem 8.

Since all backward shifts preserve $(IH^2)^{\perp}$, we can assume that:

$$f(z) = b_0 + z + \sum_{j=2}^{\infty} b_j z^j.$$

We will use an averaging process, in order to annihilate a part of the Taylor coefficients of f. Pick a positive integer k, let $\omega = e^{2\pi i/k}$ and write $h(z) = (f(z) - b_0)/z$. Then:

$$\frac{h(z) + h(\omega z) + \ldots + h(\omega^{k-1} z)}{k} = \sum_{j=0}^{\infty} b_{kj+1} z^{kj}.$$

Thus,

$$\frac{f(z) - b_0}{z} + \frac{f(\omega z) - b_0}{\omega z} + \ldots + \frac{f(\omega^{k-1} z) - b_0}{\omega^{k-1} z} = k \sum_{j=0}^{\infty} b_{kj+1} z^{kj},$$

or,

$$u(z) = \frac{f(z) + \overline{\omega} f(\omega z) + \ldots + \overline{\omega^{k-a}} f(\omega^{k-a} z)}{k} = \sum_{j=0}^{\infty} b_{kj+1} z^{kj+1}.$$

Now the function u is no more orthogonal to IH^2. Instead, it is orthogonal to JH^2, where

$$J(z) = I(z)I(\omega z) \ldots I(\omega^{k-1} z).$$

By picking k large enough we can assure that condition (12) is met. Moreover, u inherits from f the regularity to the boundary. □

The above proof allows to construct a univalent function u, orthogonal to JH^2, so that $u \in C^\infty(\overline{\mathbf{D}})$. Moreover, u can be a small perturbation of the identity map, so the domain $\Omega = u(\mathbf{D})$ can be constructed to be arbitrarily close to the unit disk. A morphological feature of the function $g \in \ker(F_\Omega)$ is that it tends to zero rather fast along at least one path in Ω leading to a boundary point.

The next result shows that the phenomenon described in Theorem 4.4 cannot hold for domains with smooth real analytic boundary.

Theorem 4.6. *Let Ω be a domain with generalized quadrature formula given by a measure with compact support in Ω. If* $\ker F$ *contains a non-trivial function f, then f must have a zero in Ω.*

Proof. Let E be a compact subset of Ω containing the support of a generalized quadrature positive measure μ. Suppose that $F(f) = 0$ and that f does not vanish in Ω. By Runge's theorem there exists a sequence R_n of rational functions without poles in $\overline{\Omega}$ converging uniformly to $1/f$ on E. Thus $R_n f$ converges uniformly to $\mathbf{1}$ on E, so:

$$\lim_{n\to\infty} \int_\Omega R_n f dA = \lim_{n\to\infty} \int_E R_n f d\mu = \int_E d\mu = Area(\Omega).$$

But all the integrals $\int_\Omega R_n f dA$ are zero by assumption, a contradiction. □

Since all domains with smooth real analytic boundary possess such generalized quadrature formulas, see for instance [9], the next corollary follows.

Corollary 4.7. *If Ω is a bounded domain with smooth real analytic boundary, then any non-trivial function in* $\ker(F)$ *vanishes somewhere in Ω.*

Next we relate the kernel of the Friedrichs operator to biholomorphic functions. Recall that a **biholomorphic function** u is a solution of the equation $\overline{\partial}^2 u = 0$. Such functions are always representable as $u(z) = f(z) + \overline{z} g(z)$, with f, g analytic. The maximum principle does not hold for this class of functions, for instance there are domains Ω for with the property that there exist biholomorphic functions in Ω, which are identically zero on the boundary, see [6].

Theorem 4.8. *Let Ω be a bounded domain with smooth boundary. Then there exists a biholomorphic function u in Ω, continuous on $\overline{\Omega}$ and such that $u|_{\partial\Omega} = 0$ if and only if the kernel of the Friedrichs operator is non-trivial.*

Proof. Assume that $f \in \ker(F)$. According to Havin's lemma [10] there exists a function $u \in W_0^{(1,2)}(\Omega)$ with the properties: $\overline{\partial} u = f$ and $\overline{\partial}^2 u = 0$, in Ω.

Conversely, assume that $v \in W^{(1,2)}(\Omega)$ satisfies $\overline{\partial}^2 v = 0$ and $v|_{\partial\Omega} = 0$. Then, again by Havin's lemma, $f = \overline{\partial} u \in AL^2(\Omega)$ and by Stokes theorem $\int_\Omega fg dA = 0$, $g \in AL^2(\Omega)$. Thus $Ff = 0$. □

4.3. The inverse spectral problem

A natural question is whether the spectrum of the Friedrichs operator characterizes the domain, up to the group of affine transformations of $\mathbf{C}$. Here by affine transformation we mean a map of the form $z \to az + b$ with $a, b \in \mathbf{C}$, $a \neq 0$. The next result settles in the negative this question.

Proposition 4.9. *There exists a continuous family of quadrature domains of order three, with the same Friedrichs operator (up to unitary equivalence) and such that no two domains in the family are related by an affine transformation of* $\mathbf{C}$.

Proof. We base our proof on a count of parameters. Let

$$f_{c,d}(z) = \frac{c_1}{z - d_1} + \frac{c_2}{z - d_2} + \ldots + \frac{c_n}{z - d_n},$$

be a rational function with poles outside the closed unit disk and such that it is univalent on $\mathbf{D}$. We can choose the n-tuples $c = (c_1, \ldots, c_n), d = (d_1, \ldots, d_n)$ with independent entries, lying in open balls $c \in B_r(c^0) \subset \mathbf{C}^n$ and $d \in B_r(d^0) \subset \mathbf{C}^n$. Thus the family of quadrature domains:

$$\Omega_{c,d} = f_{c,d}(\mathbf{D}),$$

depends on $4n$ real parameters. Two such domains are affinely equivalent if and only if there is a Moebius transform M of the unit disk onto itself and complex numbers a, b, $(a \neq 0)$, such that:

$$f_{c,d}(z) = a f_{c',d'}(M(z)) + b.$$

Therefore, the orbits of this group action on $f_{c,d}$ have dimension at most 7.

In conclusion there is a family depending on $4n - 7$ parameters, of mutually non-equivalent domains. On the other hand, the spectrum of the square of the Friedrichs operator (which has rank exactly equal to n) contains n eigenvalues (counting multiplicities), among which the highest is always 1. This leaves at most $n-1$ independent eigenvalues. The statement follows as soon as $4n - 7 - (n-1) = 3n - 6 > 0$. The first value is $n = 3$. □

The borderline case $n = 2$ will be illustrated in Section 6, where we will show that $\sigma(F) = \{1, \lambda\}$ parametrizes (through λ) a continuous family of quadrature domains of order 2.

A variety of variants of the inverse spectral problem can now be foreseen. We discuss below a situation in which the Friedrichs operator and its square (modulo unitary equivalence) determine the quadrature domain (modulo a finite set).

Let Ω be a bounded quadrature domain of order n and let $p(z)$ be the monic polynomial of degree n vanishing at the nodes of Ω. We will denote by

$$a_{kl} = \int_\Omega z^k \overline{z}^l dA(z),$$

the moments of Ω. It is proved in [17] that, knowing $p(z)$ and the partial matrix of moments $(a_{kl})_{0\leq k\leq l\leq n-1}$, the domain Ω can be reconstructed. More specifically, the irreducible polynomial $P(z,\overline{z})$ which defines $\Omega \equiv \{z \in \mathbf{C}; P(z,\overline{z}) < 0\}$ is precisely:

$$P(z,\overline{z}) = \|p(z)\|^2 \exp[\frac{-1}{\pi}\sum_{k,l=0}^{n-1}\frac{a_{kl}}{z^{k+1}\overline{z}^{l+1}}] - \text{(terms in } 1/z \textit{ or } 1/\overline{z}). \tag{13}$$

Above, "$\equiv$" means equal up to a finite set.

Let $B = T_z$ be the Bergman shift on Ω. Since Ω is a quadrature domain, the space

$$\operatorname{im}(F) = \bigvee_{k=0}^{n-1} B^{*k}\mathbf{1} = \bigvee_{k=0}^{\infty} B^{*k}\mathbf{1}$$

has dimension $n = o(\Omega)$. The operator $S = F^2$ has the same range and it is of rank n, whence $S|_{\operatorname{im} F}$ is injective. For a pair of elements $f, g \in \operatorname{im}(F)$ we remark the identity:

$$\langle (S|_{\operatorname{im} F})^{-1} F f, F g\rangle = \langle g, f\rangle. \tag{14}$$

Indeed, denoting $Ff = F^2 h$, we have

$$\langle (S|_{\operatorname{im} F})^{-1} F^2 h, F g\rangle = \langle h, Fg\rangle = \langle g, Fh\rangle = \langle g, f\rangle.$$

Theorem 4.10. *Two quadrature domains Ω, Ω' with Friedrichs operators F, S, respectively F', S', coincide up to a finite set if and only if there exists a unitary operator $U : \operatorname{im}(F) \longrightarrow \operatorname{im}(F')$ with the properties:*

$$\begin{cases} U(S|_{\operatorname{im} F}) = (S'|_{\operatorname{im} F'})U \\ UFq = F'q, \quad q \in \mathbf{C}[z], \quad \deg(q) \leq \max(o(\Omega), o(\Omega')). \end{cases} \tag{15}$$

Proof. Assume that the unitary U in the statement exists. The relation $UF = F'$ is symmetric. So, if a polynomial q of degree less or equal than $o(\Omega)$ satisfies $Fq = 0$, then $F'q = 0$, and conversely. Therefore $o(\Omega) = o(\Omega')$ and the two quadrature domains have the same nodes, counting also their multiplicities. Let $p(z)$ be the monic polynomial of degree $o(\Omega)$ vanishing at these nodes.

According to relations (14), for all $k, l < o(\Omega)$, we obtain:

$$\begin{aligned} \langle z^k, z^l\rangle_{2,\Omega'} &= \langle (S'|_{\operatorname{im} F'})^{-1} F'(z^l), F'(z^k)\rangle_{2,\Omega'} \\ &= \langle U(S|_{\operatorname{im} F})^{-1} F(z^l), UF(z^k)\rangle_{2,\Omega'} \\ &= \langle (S|_{\operatorname{im} F})^{-1} F(z^l), F(z^k)\rangle_{2,\Omega} = \langle z^k, z^l\rangle_{2,\Omega}. \end{aligned}$$

Then we apply (13) and the proof is complete. □

Corollary 4.11. *Two quadrature domains Ω, Ω' with the same nodes coincide if and only if relation* (15) *holds for all polynomials q of degree* $\deg(q) < o(\Omega)$.

In other terms, the above criterion says that a quadrature domain is determined up to a finite set by the pair of matrices $(S|_R, B^*|_R)$, both acting on the finite dimensional space $R = \mathrm{im}(F)$. The quadrature identity can easily be obtained in these terms:

$$\int_\Omega f dA = \langle \mathbf{1}, f^*(B^*|_R)\mathbf{1}\rangle, \qquad f \in AL^2(\Omega),$$

where $f^*(z) = \overline{f(\overline{z})}$. Indeed, for $f \in H^\infty(\Omega)$ we have:

$$\langle \mathbf{1}, f^*(B^*|_R)\mathbf{1}\rangle = \langle \mathbf{1}, T_f^*\mathbf{1}\rangle = \langle T_f\mathbf{1}, \mathbf{1}\rangle = \langle f, \mathbf{1}\rangle.$$

Note that the quadrature nodes a_j coincide, including multiplicities, with the spectrum of $(B^*|_R)^*$. Moreover, the correlation matrix formed by the eigenvectors k_{a_j} of $B^*|_R$ is a part $(K = K(a_j, a_k))$ of the Bergman kernel. Another similar, and related, set of matricial invariants for a quadrature domains appears in [17].

5. Generalized quadratures and the restriction operator

We continue the analysis begun in Proposition 4.3, of the relation between the Friedrichs operator and generalized quadratures. By exploiting the explicit integral representations for the Friedrichs operator and the restriction operator to a quadrature space (to be defined below), we describe a possible way of computing the eigenvalues, and sometimes even the eigenfunctions, of S. The basic data needed for such computations are the Bergman kernel of the domain and the (generalized) quadrature formula. The last section of this paper contains several examples obtained from this principle.

First an observation of independent interest which might be useful in later proofs.

Lemma 5.1. *Let Ω be a domain with a generalized quadrature given by the compactly supported complex measure $d\mu$. Then $S = F^2$ is trace class and:*

$$Tr(S) = \int_{\Omega\times\Omega} K(z,w)^2 d\mu(z)d\mu(w). \tag{16}$$

Proof. The nuclearity of S is well known (see also Theorem 5.7 below). For more general results (although only in the case of simply connected domains) see [13]. By iterating formula (1) we obtain for every pair of functions $f, g \in AL^2(\Omega)$:

$$\begin{aligned}\langle Sf, g\rangle &= \int_{\Omega\times\Omega} K(z,w)f(z)\overline{g(w)}dA(z)dA(w)\\ &= \int_{\Omega\times\Omega} K(z,w)f(z)\overline{g(w)}d\mu(z)d\mu(w).\end{aligned}$$

Let $u_j(z), 1 \le j < \infty$, be a complete system of orthonormalized eigenfunctions of S. Then

$$\begin{aligned} Tr(S) &= \sum_{j=1}^{\infty} \langle Su_j, u_j \rangle \\ &= \int_{\Omega\times\Omega} K(z,w)[\sum_{j=1}^{\infty} u_j(z)\overline{u_j(w)}]d\mu(z)d\mu(w) \\ &= \int_{\Omega\times\Omega} K(z,w)^2 d\mu(z)d\mu(w). \end{aligned}$$

□

Recall that every bounded domain with smooth real analytic boundary fulfills the condition in Lemma 5.1. Also, Lemma 5.1 remains valid for some domains allowing quadrature with measures which are not necessarily carried by a compact subset; this time the convergence of the integral (16) must be assumed. Formula (16) remains valid, with the same proof, for distributions instead of quadrature measures (for instance in the case of quadrature domains).

In order to give an application of Lemma 5.1, we consider a quadrature domain Ω with simple nodes:

$$\int_{\Omega} f dA = \sum_{j=1}^{n} c_j f(a_j), \qquad f \in AL^2(\Omega). \tag{17}$$

Let us denote by K the positive definite matrix $[K(a_i,a_j)]_{i,j=1}^n$ and by $Q = diag(c_1, \dots, c_n)$ the positive matrix given by the quadrature weights.

Then, for $f \in AL^2(\Omega)$, we obtain:

$$(Ff)(z) = \int_{\Omega} K(z,w)\overline{f(w)}dA(w) = \sum_j K(z,a_j)\overline{c_j f(a_j)}.$$

The space $\mathrm{im}(F)$ is generated by the linearly independent vectors $k_{a_j}, 1 \le j \le n$. With respect to this basis F is represented, as an antilinear operator, by the matrix $K^t Q^*$. This yields the following formula.

Lemma 5.2. *Let Ω be a quadrature domain with quadrature* (17). *Then*

$$\det(S) = |c_1 c_2 \dots c_n|^2 \det(K)^2. \tag{18}$$

Concerning the range of the Friedrichs operator, we note the following generalization of Corollary 4.2.

Proposition 5.3. *Let Ω be a domain with* $\ker(F) = 0$. *Let $A \subset \Omega$ be the zero set of a non-trivial $H^\infty(\Omega)$-function and let H_A be the closed linear subspace of $AL^2(\Omega)$ spanned by k_w, $w \in A$.*

Then $\mathrm{im}(F) \cap H_A = 0$.

Proof. The proof repeats that of Theorem 4.1. Namely if A is the zero set of $\phi \in H^\infty(\Omega)$, then $0 = T_\phi^* Fg = F(T_\phi)g$ for a non-trivial vector $Fg \in H_A \cap \mathrm{im}(F)$. This implies, via the uniqueness principle, that $Fg = 0$. □

One consequence is in order at this point. Assume that F is injective and compact, and let $f_j, 0 \le j < \infty$, be an orthonormal system of eigenfunctions for S:

$$\int_\Omega f_i f_j dA = \lambda_i \int_\Omega f_i \overline{f_j} dA, \qquad i, j \ge 0. \tag{19}$$

We can arrange the eigenvalues $\lambda_j > 0$ in decreasing order ($\lambda_0 = 1, f_0 = const$). In other terms:

$$F(f_j) = \lambda_j (f_j), \quad S(f_j) = \lambda_j^2 f_j, \qquad j \ge 0.$$

Fix a point $a \in \Omega$. Then, by expanding the evaluation function k_a in a series in f_j, and applying Proposition 5.3, or even Corollary 4.2, we see that the series cannot terminate. Thus we have proved the next corollary.

Corollary 5.4. *Let Ω be a domain with compact, injective Friedrichs operator. Then for every point of Ω there are infinitely many Friedrichs eigenfunctions non-vanishing at that point.*

The spaces H_A, with A a subset of Ω, appear in the reconstruction of an analytic function from its values on A; more specifically they appear in the extremal problem of finding the function of least L^2-norm when its values on A are prescribed, cf. [23]. We elaborate below such an aspect.

Let Ω be a (not necessarily bounded) domain with Bergman kernel $K(z, w)$. The Friedrichs operator F is given by the integral formula (1), while its square S acts as:

$$Sf(z) = \int_\Omega L(z, w) f(w) dA(w), \qquad f \in AL^2(\Omega), \tag{20}$$

where $L(z, w) = \int_\Omega K(z, \zeta) K(w, \zeta) dA(\zeta)$. Note an unusual fact, namely that the kernel $L(z, w)$ is symmetric in z, w, and not Hermitean-symmetric as would necessarily be the case for the kernel representing a self-adjoint operator on the whole space $L^2(\Omega)$.

Consider a positive measure μ on Ω, whose support E may intersect $\partial\Omega$, but such that $\mu(\partial\Omega) = 0$. We suppose that $E \cap \Omega$ is a set of uniqueness for $AL^2(\Omega)$, and also that there exists a positive constant $C = C(\Omega, \mu)$ with the property:

$$\int_E |f|^2 d\mu \le C \int_\Omega |f|^2 dA, \qquad f \in AL^2(\Omega). \tag{21}$$

In this case the restriction operator $R(f) = f|_E$ is well defined and bounded as a linear operator $R : AL^2(\Omega) \longrightarrow L^2(E, \mu)$. By our assumption, R is injective, and we will also assume that R has dense range. Its adjoint has the explicit representation:

$$R^* \psi(z) = \int_E K(z, w) \psi(w) d\mu(w), \qquad z \in \Omega, \ \psi \in L^2(E, \mu). \tag{22}$$

We study below an important particular case of this setting. Let us denote by $HL^1(\Omega), SL^1(\Omega)$ the spaces of integrable harmonic, respectively subharmonic, functions in Ω. Suppose that Ω is symmetric with respect to the real axis and admits a quadrature identity of the type:

$$\int_\Omega u dA = \int u d\mu, \qquad u \in HL^1(\Omega), \tag{23}$$

where μ is a positive measure on $\mathbf{R} \cap \Omega$. Due to the symmetry assumption and a theorem of Sakai, [19] §9, we have in these conditions the much stronger estimate:

$$\int s d\mu \leq \int_\Omega s dA, \qquad s \in SL^1(\Omega). \tag{24}$$

In particular relation (21) is satisfied with the constant $C = 1$.

Let $f^*(z) = \overline{f(\overline{z})}$, $z \in \Omega$ be the involution induced by the symmetry with respect to the axis $\mathbf{R}$.

Theorem 5.5. *Let Ω be a symmetric domain with respect to the real axis, which satisfies a generalized quadrature* (23) *and let*

$$R : AL^2(\Omega) \longrightarrow L^2(\mathbf{R}, \mu)$$

be the corresponding restriction operator. Then $S = (R^*R)^2$ and $S(f^*) = (Sf)^*$, $f \in AL^2(\Omega)$.

Proof. By the symmetry assumption and the transformation law of the Bergman kernel under this symmetry, we have $K(x, y) = K(y, x) \in \mathbf{R}$ for $x, y \in \mathbf{R} \cap \Omega$.

For a function $f \in AL^2(\Omega)$ we have:

$$R^*Rf(z) = \int K(z, x)f(x)d\mu(x),$$

so

$$\begin{aligned}(R^*R)^2 f(z) &= \int K(z, y)[\int K(y, x)f(x)d\mu(x)]d\mu(y)\\ &= \int f(x)[\int K(z, y)K(y, x)d\mu(y)]d\mu(x)\\ &= \int f(x)[\int K(z, y)K(x, y)d\mu(y)]d\mu(x).\end{aligned}$$

By using twice the quadrature identity we obtain:

$$\begin{aligned}(R^*R)^2 f(z) &= \int f(x)[\int K(z, w)K(x, w)dA(w)]d\mu(x)\\ &= \int f(x)L(z, x)d\mu(x) = \int f(w)L(z, w)dA(w) = Sf(z).\end{aligned}$$

From the same computations we infer $S(f^*) = (Sf)^*$. □

The commutativity between S and the involution "$*$" in Theorem 5.5 implies that all spectral subspaces of S are invariant under $*$, therefore they are generated by invariant functions: $f^* = f$. In particular, it will be enough in such situations to solve the eigenfunction equation $Sf = \lambda f$ for f satisfying $f^* = f$, or equivalently, $f(x) \in \mathbf{R}$, $x \in \mathbf{R} \cap \Omega$.

Corollary 5.6. *In the conditions of Theorem* 5.5, *the operator* $\sqrt{S}$ *is unitarily equivalent to the integral operator* $RR^* : H \longrightarrow H$:

$$(RR^*f)(x) = \int_{\mathbf{R}\cap\Omega} K(x,y)f(y)d\mu(y), \qquad f \in H,$$

where H *is the closure of the range of* R *in* $L^2(\mathbf{R},\mu)$.

Proof. For the proof is suffices to use relation (22) and recall that the operators R^*R and $RR^*|_H$ are unitarily equivalent, for a proof see for instance [22]. □

In most applications $H = L^2(\mathbf{R},\mu)$.

The next result was stated without proof in [24] (see the remarks following [24], Theorem 8.5).

Theorem 5.7. *Let* Ω *be a bounded domain which admits a generalized quadrature formula with the associated measure compactly supported by* Ω. *Then* S *is compact and its sequence of eigenvalues* λ_n *(arranged in decreasing order) satisfies*:

$$\limsup_{n\to\infty} \lambda_n^{1/n} < 1. \tag{25}$$

Proof. Assume that the measure μ is supported by the compact set $E \subset \Omega$ and it satisfies:

$$\int_\Omega f dA = \int_E f d\mu, \qquad f \in AL^2(\Omega).$$

Then it is known that $\partial\Omega$ consists of finitely many analytic arcs. Since $Ff(z) = \int_\Omega K(z,w)\overline{f(w)}dA(w) = \overline{\int_E K(w,z)f(w)d\mu(w)}$, there exists a positive constant $C(= \max_{z\in E} K(z,z))$ such that:

$$\|Ff\| \leq C \max_{z\in E} |f(z)|, \qquad f \in AL^2(\Omega).$$

Consequently,

$$\|Sf\| = \|F(Ff)\| \leq C\|Ff\|_{\infty,E} \leq C^2\|Ff\|_{2,\Omega} \leq C^3\|f\|_{\infty,E}.$$

Thus, there exists a positive constant B with the property:

$$\|Sf\|_{2,\Omega} \leq B\|f\|_{\infty,E}, \qquad f \in AL^2(\Omega). \tag{26}$$

This proves the compactness, even nuclearity, of S.

In order to estimate the decay of the eigenvalues of S we use the Fischer-Courant minimax criterion: $\lambda_0 = \max\{\langle Sf, f\rangle;\ \|f\| = 1\} = 1$, and for $n \geq 1$:

$$\lambda_n = \min_{X_n} \max\{\langle Sf, f\rangle,\ f \in X_n,\ \|f\| = 1\}, \tag{27}$$

where X_n ranges over all subspaces of $AL^2(\Omega)$ having codimension n.

Let now G be an open subset of Ω such that $E \subset G$ and $\overline{G} \subset \Omega$. Then we can find a function $\phi \in H^\infty(\Omega)$ such that:

$$m_1 = \sup_{z\in E} |\phi(z)| < \inf_{z\in\Omega\setminus\overline{G}} |\phi(z)| = m_2. \tag{28}$$

For example, if ϕ is a nonconstant function in $H^\infty(\Omega)$ satisfying $\lim_{z\to\zeta} |\phi(z)| = 1$ for all $\zeta \in \partial\Omega$, then one can define G so that relation (28) holds.

From (28) it is clear that the set of zeroes of ϕ is non-empty and finite; let r denote their number, counting multiplicities.

We shall first derive an upper bound for λ_n when n is divisible by r, say $n = kr$. We will need the following lemma, whose proof is an application of the maximum principle, and left to the reader.

Lemma 5.8. *If K is a compact subset of Ω, then there is a positive constant $C = C(\Omega, K)$ such that:*

$$\int_\Omega |g|^2 dA \leq C \int_{\Omega\setminus K} |g|^2 dA, \qquad g \in AL^2(\Omega).$$

Let Y_n denote the subspace of $AL^2(\Omega)$ consisting of all functions $f = \phi^k g$, with $g \in AL^2(\Omega)$ where ϕ satisfies (28). This space has codimension $n = kr$, so by (27):

$$\lambda_n \leq \max\{\langle Sf, f\rangle; f \in Y_n, \ \|f\| = 1\},$$

and, using (26):

$$\lambda_n \leq B \sup\{\|f\|_{\infty,E}; \ f \in Y_n, \ \|f\| = 1\}.$$

Let $f = \phi^k g$ be an element of Y_n of norm 1. According to (28) we have:

$$1 = \int_\Omega |f|^2 dA \geq \int_{\Omega\setminus\overline{G}} |f|^2 dA \geq m_2^{2k} \int_{\Omega\setminus\overline{G}} |g|^2 dA.$$

Hence

$$\int_{\Omega\setminus\overline{G}} |g|^2 dA \leq (m_2)^{-2k},$$

so by the lemma, $\|g\|^2 \leq C(m_2)^{-2k}$.

From here on, we let C_j denote constants that depend only on the geometric configuration Ω, E, G. Then

$$|g(z)|^2 \leq C_1(m_2)^{-2k}, \qquad z \in E,$$

so

$$\|f\|_{\infty,E} \leq C_2(m_1)^k(m_2)^{-k} = C_2\rho^k = C_2\rho_0^n,$$

where $\rho = m_1/m_2 < 1$ and $\rho_0 = \rho^{1/r}$.

For an arbitrary n, let k be the greatest integer less than or equal to n/r. Then

$$\lambda_n \leq \lambda_{kr} \leq C_2\rho_0^{kr} \leq C_2\rho_0^{n-r} = (C_2\rho_0^{-r})\rho_0^n.$$

In conclusion,

$$\limsup_{n\to\infty} \lambda_n^{1/n} \leq \rho_0$$

and the theorem is proved. □

6. Examples

This section contains examples of domains on which the Friedrichs operator is relatively simple or at least its spectrum is computable. Due to space constraints, part of the routine computations are omitted.

6.1. Quadrature domains of order two

For such domains the spectrum of the Friedrichs operator is computable in terms of the Bergman kernel. For instance, let Ω be a bounded quadrature domain of order two, with a double point at $z = 0$:

$$\int_{\Omega} f dA = c_0 f(0) + c_1 f'(0), \qquad f \in AL^2(\Omega).$$

Then **1** and $u = \frac{\partial K(z,w)}{\overline{w}}|_{w=0}$ generate $\mathrm{im}(F)$. Since these vectors are orthogonal, the second eigenvalue λ of F is given by the formula $Fu = \lambda u$. But

$$\begin{aligned} Fu(z) &= \int_{\Omega} K(z,w)\overline{u(w)} dA(w) \\ &= \overline{c_0} K(z,0)\overline{u(0)} + \overline{c_1}\frac{\partial K(z,w)}{\overline{w}}|_{w=0}\overline{u(0)} + \overline{c_1} K(z,0)\overline{u'(0)} = \overline{c_1 \overline{u(0)}} u(z). \end{aligned}$$

Thus

$$\lambda = \overline{c_1} \frac{\partial K(0,w)}{\overline{w}}|_{w=0}.$$

By choosing $\Omega_a = \{z^2 + az; |z| < 1\}$, with the real parameter $a \geq 2$, a short computation using the change of variable formula for the Bergman kernel yields:

$$\lambda(\Omega_a) = \frac{4}{a^4}.$$

Thus, in the family of quadrature domains $(\Omega_a)_{a\geq 2}$, the spectrum of their Friedrichs operator distinguishes each pair of them.

Similar computations can be carried out in the case of two distinct nodes, say $\pm a$. The eigenvectors are then **1** and $k_a - k_{-a}$.

6.2. An annulus

Let $A_t = \{z;\ e^{-t/2} < |z| < e^{t/2}\}$ be an annulus with a fixed parameter $t > 0$. It is easy to verify that, for conveniently chosen constants $a_n > 0$, the system of functions $f_0 = \mathbf{1}$, $f_{\pm n} = z^n \pm a_n z^{-n}$, $n > 0$, is a doubly orthogonal basis of $AL^2(A_\rho)$:

$$\sigma_k \langle f_k, f_l \rangle = \langle f_k, F f_l \rangle = \int_{A_\rho} f_k f_l dA, \quad k, l \in \mathbf{Z}.$$

The eigenvalues of $\sqrt{S}$ will then be $\lambda_k = |\sigma_k|$, $k \in \mathbf{Z}$.

We indicate below how to compute these numbers. First, the orthogonality condition:

$$0 = \langle f_n, f_{-n} \rangle = 2\pi \int_1^{\rho} (|z|^{2n} - a_n^2 |z|^{-2n}) \rho d\rho,$$

yields:

$$a_n^2 = \begin{cases} \frac{\sinh 2t}{2t}, \; n = 1, \\ \\ \frac{\sinh(n+1)t}{n+1} \frac{n-1}{\sinh(n-1)t}, \; n > 1. \end{cases}$$

Consequently we find the eigenvalues of $\sqrt{S}$:

$$\lambda_n = \lambda_{-n} = \frac{a_n \sinh t}{\frac{\sinh(n+1)t}{n+1}}, \qquad n > 0.$$

In order to unify the formulas, we put by convention:

$$t = \frac{\sinh nt}{n}|_{n=0}.$$

Therefore, for all values $n \in \mathbf{Z}$ one obtains:

$$\lambda_n = \sqrt{\frac{(n+1)\sinh t}{\sinh(n+1)t}} \sqrt{\frac{(n-1)\sinh t}{\sinh(n-1)t}}. \tag{29}$$

Note that $\lambda_0 = 1$ and $\lambda_n = \lambda_{-n}$ decays exponentially at infinity.

Thus, in the case of the annulus, the spectrum of S determines the domain, up to homotheties and translations, in the family of all planar annuli.

The above computations can be reobtained in a more invariant way, starting with the generalized quadrature formula satisfied by the annulus:

$$\int_{A_t} f dA = \sinh t \int_{-\pi}^{\pi} f(e^{i\theta}) d\theta, \qquad f \in AL^2(A_t).$$

For details see [19] or [24]. The Bergman kernel of A_t can easily be obtained from orthonormalizing the system $z^n, n \in \mathbf{Z}$:

$$K_{A_t}(z, w) = \frac{1}{2\pi} \sum_{n=-\infty}^{\infty} \frac{n+1}{\sinh(n+1)t} (z\overline{w})^n,$$

see for instance [15], Chapter VII, Section 5.

Thus, according to Theorem 5.5, an eigenfunction f of the operator S satisfies the equation:

$$(\sinh t)^2 \int_{-\pi}^{\pi} \int_{-\pi}^{\pi} K(e^{iu}, e^{i\theta}) K(e^{iv}, e^{i\theta}) d\theta f(e^{iv}) dv = \lambda f(e^{iu}).$$

In particular one finds that the latter eigenvalue problem has the convolution form:

$$\int_{-\pi}^{\pi} G(u - v) f(e^{iv}) dv = \lambda f(e^{iu}).$$

More specifically, the evaluation of the integral in θ above yields:

$$G(u - v) = \frac{1}{2\pi} \sum_{n=-\infty}^{\infty} \frac{(n+1)\sinh t}{\sinh(n+1)t} \frac{(n-1)\sinh t}{\sinh(n-1)t} e^{in(u-v)}.$$

In conclusion, by Fourier transform, the eigenvalues of S are exactly the Fourier coefficients of the function G, including multiplicities. Thus formulas (29) are reproved.

6.3. The ellipse

The ellipse fulfills the conditions in Theorem 5.5. We show below how to identify the eigenfunctions of the Friedrichs operator by a more conceptual method than the original computations of Friedrichs [7].

Let $\mathbf{C}_n[z]$ be the space of polynomials of degree less than or equal to n.

Proposition 6.1. *The Friedrichs operator of an ellipse satisfies*

$$F\mathbf{C}_n[z] \subset \mathbf{C}_n[z], \qquad n \geq 0.$$

Proof. Let E be an ellipse and let p be a polynomial of degree n. Let $g = Fp$, so that $\overline{p} = g + G$, where $G \in L^2(E) \ominus AL^2(E)$. According to Havin's Lemma, [10]:

$$p = \overline{g} + \partial v / \partial \overline{z},$$

where the function v belongs to the Sobolev space $W_0^{1,2}(E)$. Let h be a holomorphic primitive of g, so that, integrating the last identity we obtain:

$$\overline{z}p = \overline{h} + v + H,$$

where $H \in AL^2(E) \cap C(\overline{E})$.

Therefore, $\overline{z}p(z) = \overline{h(z)} + H(z)$ for $z \in \partial E$. Since the solution of the Dirichlet problem on E with real polynomial data on ∂E is a real polynomial of the same total degree, or less (cf. [12]), it follows that $\partial^{n+2}(\overline{h} + H) = 0 = \overline{\partial}^{n+2}(\overline{h} + H)$, in E. Thus h is a polynomial of degree at most equal to $n+1$, and consequently g is a polynomial of degree at most n. □

Let E be an ellipse with foci at ± 1. Then E is a generalized quadrature domain with measure $d\mu = c(1-x^2)^{1/2}dx$, supported by the interval $[-1, 1]$, see [5]. According to Proposition 4.3, in this case $\ker(F) = 0$. Thus $F\mathbf{C}_n[z] = \mathbf{C}_n[z]$ for all $n \geq 0$. The latter increasing chain of invariant subspaces diagonalizes F, or equivalently S, and this yields the following result.

Corollary 6.2. *The complete system of eigenfunctions of the Friedrichs operator of an ellipse with foci at ± 1 is provided by the Chebyshev polynomials of the second kind.*

In view of Theorem 5.5, the same eigenfunctions are shared by the modulus of the restriction operator R. Let us mention without giving here the details, that in this case the Bergman shift is represented by a non-selfadjoint Jacobi type matrix with only two non-zero diagonals (adjacent to the main diagonal, which is zero).

As for the computation of the corresponding eigenvalues of F we outline Friedrichs' computation, [7]. Let $E(t)$ be the ellipse centered at $z = 0$, with semi-axes $\cosh(t), \sinh(t)$, respectively. Let $U_n(\cos z) = \frac{\sin(n+1)z}{\sin z}$, $n \geq 0$, be the Chebyshev polynomials of the second kind, which we already know that are orthogonal on $E(t)$ and they are the eigenvectors of the corresponding Friedrichs operator, or better its square S. Thus the eigenvalues of $\sqrt{S}$ are:

$$\lambda_n(E(t)) = \frac{\int_{E(t)} U_n^2 dA}{\int_{E(t)} |U_n|^2 dA}, \qquad n \geq 0. \tag{30}$$

The numerator can be computed from the quadrature formula for the ellipse, cf. [24] formula (3.10):

$$\int_{E(t)} U_n^2 dA = \pi \cosh(t) \sinh(t) = (\pi/2) \sinh(2t).$$

Since the map $w = \cos(z)$ transforms conformally the rectangle $R(t) = \{z = x + iy;\ 0 < x < 2\pi,\ 0 < y < t\}$ onto $E(t)$ minus a segment of its major axis, the denominator in the above formula can be computed by a change of variables:

$$\begin{aligned}\int_{E(t)} |U_n(w)|^2 dA(w) &= \int_{R(t)} |U_n(\cos z)|^2 |\sin z|^2 dA(z) \\ &= [\pi/(2n+2)] \sinh(2(n+1)t).\end{aligned}$$

Thus, returning to equation (24) we find:

$$\lambda_n(E(t)) = \frac{(n+1)\sinh 2t}{\sinh 2(n+1)t}, \qquad n \geq 0.$$

Note that the first eigenvalue is $\lambda_0(E(t)) = 1$ and that $\lambda_n(E(t))$ decays exponentially when n tends to infinity. The limiting case $t \to 0$ produces after rescaling a strip, which will be the subject of the next example. In that case the operator S has only continuous spectrum, equal to the interval $[0, 1]$, as this limiting process suggests.

6.4. A strip

Let c be a positive constant. The strip $\Omega = \{z = x + iy;\ |y| < c\}$ fulfills the conditions in Theorem 5.5. Indeed, it is known that:

$$\int_\Omega u dA = 2c \int_{-\infty}^{\infty} u(x, 0) dx, \qquad u \in HL^1(\Omega). \tag{31}$$

We outline the proof, which is needed again for the next example (the wedge). More details can be found in [20] (and the reference [Shapiro 1987] there). By approximation, it is enough to prove this quadrature identity for u fast decreasing at infinity and we can also assume that $u = \operatorname{Re} f$ with f analytic in Ω. By Cauchy's theorem, the integral $\int_{-\infty}^{\infty} f(x + iy) dx$ is independent of y, $|y| < c$. So,

$$\int_\Omega f dA = \int_{-c}^{c} \int_{-\infty}^{\infty} f(x + iy) dx dy = 2c \int_{-\infty}^{\infty} f(x) dx.$$

In order to use Theorem 5.5, we need the Bergman kernel of Ω. Without loss of generality we can work next with the strip $|y| < \pi/2$. Let $g : \Omega \longrightarrow \mathbf{D}$ be the conformal map which satisfies $g(0) = 0, g'(0) > 0$:

$$g(z) = \frac{e^z - 1}{e^z + 1}, \quad z \in \Omega.$$

The kernel is therefore:

$$K(z, w) = \frac{1}{\pi} \frac{g'(z)\overline{g'(w)}}{(1 - g(z)\overline{g(w)})^2}.$$

Note in addition that the Bergman kernel enjoys the invariance property with respect to translations in the real direction: $K(z, w) = K(z + a, w + a)$ $a \in \mathbf{R}$. Therefore, for real entries x, ξ we obtain $K(x, \xi) = k(x - \xi)$, where k is a real function of a real variable. To find it, we take $\xi = 0$ and use the above formula:

$$k(x) = \frac{1}{\pi} g'(x)g'(0) = \frac{1}{\pi} \frac{e^x}{(e^x + 1)^2} = \frac{1}{4\pi} \frac{1}{\cosh^2(x/2)}, \quad x \in \mathbf{R}.$$

According to Corollary 5.6 the operator $\sqrt{S}$ is unitarily equivalent to the integral operator:

$$(RR^* f)(x) = \frac{1}{4} \int_{-\infty}^{\infty} \frac{1}{\cosh^2((x - t)/2)} f(t)dt, \qquad f \in L^2(\mathbf{R}). \tag{32}$$

By taking the Fourier transform and using the known integral:

$$\int_{-\infty}^{\infty} \frac{e^{ix\xi}}{\cosh^2(x/2)} dx = \frac{4\pi\xi}{\sinh \pi\xi},$$

the operator (32) is unitarily equivalent to the multiplication operator:

$$\hat{f}(\xi) \mapsto \frac{\pi\xi}{\sinh \pi\xi} \hat{f}(\xi),$$

still in $L^2(\mathbf{R})$, by Plancherel theorem.

The multiplier $m(\xi) = \frac{\pi\xi}{\sinh \pi\xi}$ obviously satisfies the inequalities $0 < m(\xi) \leq 1$, $\xi \in \mathbf{R}$. In conclusion, using standard spectral theory for self-adjoint operators, cf. [18] Chapter VII, we have proved the following result.

Proposition 6.3. *The modulus $\sqrt{S}$ of the Friedrichs operator of a strip is unitarily equivalent to the operator $M_t \oplus M_t$, where M_t is the multiplication with the variable, acting on $L^2([0, 1], dt)$.*

6.5. A wedge

Let $W_\alpha = \{z \in \mathbf{C}; |\arg z| < \alpha/2\}$ be a wedge of angle $0 < \alpha < 2\pi$. Denote $\beta = \pi/\alpha$. By using the conformal map:

$$\phi : W_\alpha \longrightarrow \mathbf{D}, \qquad \phi(z) = \frac{z^\beta - 1}{z^\beta + 1}, \quad z \in W_\alpha,$$

one finds the Bergman kernel of W_α:

$$K(z, w) = \frac{1}{\pi} \frac{\phi'(z)\overline{\phi'(w)}}{(1 - \phi(z)\phi(w))^2} = \frac{\beta}{\alpha} \frac{z^{\beta-1}\overline{w}^{\beta-1}}{(z^\beta + \overline{w}^\beta)^2}.$$

Equivalently, by using the automorphism group $z \to tz, t > 0$ of the wedge, one can start with the invariance formula $K(z,t) = \frac{1}{t^2}K(\frac{z}{t},1),\ z \in W_\alpha, t > 0$ and reach by analytic continuation the same formula for K.

For a function $f \in AL^2(W_\alpha)$ rapidly decreasing at infinity, the transformation of the area integral in polar coordinates yields:

$$\int_{W_\alpha} f dA = \int_{-\alpha/2}^{\alpha/2} d\theta \int_0^\infty f(re^{i\theta}) r dr = \int_{-\alpha/2}^{\alpha/2} e^{-2i\theta} d\theta \int_{l_\theta} f(z) z dz,$$

where l_θ is the ray of angle θ. Since the function f decreases fast enough at infinity, Cauchy's theorem shows that the line integral along l_θ does not depend on θ. In this way one obtains the generalized quadrature formula satisfied by the wedge (see for more details [20]):

$$\int_{W_\alpha} f dA = \sin\alpha \int_0^\infty f(x) x dx, \qquad f \in AL^2(W_\alpha). \tag{33}$$

We mention that it is known that the above relation remains valid for $f \in AL^1(W_\alpha)$, see [24] and the references cited there.

According to Corollary 5.6, the spectrum of $\sqrt{S}$ is identical to the spectrum of the operator:

$$(RR^* f)(x) = \frac{\beta \sin\alpha}{\alpha} \int_0^\infty \frac{(xt)^{\beta-1}}{(x^\beta + t^\beta)^2} t f(t) dt. \tag{34}$$

The change of variables $x = e^u$, $t = e^v$ and the transformation $F(u) = f(e^u)e^u$ maps $L^2([0,\infty), xdx)$ unitarily onto $L^2(\mathbf{R}, dx)$. The integral operator in (34) becomes:

$$\int_{-\infty}^\infty \frac{e^{(\beta-1)(u+v)}}{(e^{\beta u} + e^{\beta v})^2} f(e^u) e^{2u} du = e^{-v} \int_{-\infty}^\infty \frac{e^{\beta(u+v)}}{(e^{\beta u} + e^{\beta v})^2} F(u) du.$$

Thus the integral operator (34) is unitarily equivalent to:

$$F(u) \mapsto \frac{\beta \sin\alpha}{4\alpha} \int_{-\infty}^\infty \frac{F(u)}{\cosh^2 \beta\frac{u-v}{2}} du.$$

By passing to Fourier transforms, as in the previous example, we infer that $\sqrt{S}$ is unitarily equivalent to the multiplication operator:

$$\hat{F}(\xi) \mapsto \hat{F}(\xi) \frac{\sin\alpha}{\alpha} \frac{\pi\xi/\beta}{\sinh(\pi\xi/\beta)}.$$

Therefore, the following proposition holds.

Proposition 6.4. *The modulus $\sqrt{S}$ of the Friedrichs operator of the wedge W_α is unitarily equivalent to the operator*

$$\frac{\sin\alpha}{\alpha}(M_t \oplus M_t) : L^2([0,1], dt) \oplus L^2([0,1], dt) \longrightarrow L^2([0,1], dt) \oplus L^2([0,1], dt).$$

In conclusion, we stress the unitary equivalence:

$$(\frac{\sin\alpha}{\alpha})^2 S_\Omega \equiv S_{W_\alpha},$$

between the squares of the Friedrichs operators of a strip Ω and a wedge W_α (of angle α).

Final comments

Below we list a few subjects which will be expanded in the next article.

Various generalizations of the Friedrichs operator come now naturally into discussion. For instance the Friedrichs operator for the Hardy space of a domain with nice boundary appears as a "defect" operator in the corresponding Herglotz-Riesz formula; or a natural notion of Friedrichs operators with symbols is very much related to Hankel operators on the Bergman space.

As mentioned in the introduction, the connection between the Friedrichs operator and the boundary value problems for the bi-Laplacian or $\overline{\partial}^2$ is only touched in the present paper.

Most likely, the list of computable spectra of S_Ω's can be enlarged for domains with simple Bergman kernels and "mother bodies", (the latter in the terminology of [8]).

Part of the results and proofs above make sense in several complex variables. We leave the interested reader the easy task of finding these precise statements in $\mathbf{C}^n$. The specific, new features of the Friedrichs operator in several complex variables will be investigated elsewhere.

References

[1] Yu. A. Amenzade, *Theory of elasticity*, Mir, Moscow, 1979.

[2] D. Aharonov and H. S. Shapiro, *Domains on which analytic functions satisfy quadrature identities*, J. d'Analyse Math. **30** (1976), 39–73.

[3] S. Axler and P. Bourdon, *Finite codimensional invariant subspaces of Bergman spaces*, Trans. Amer. Math. Soc. **306** (1988), 805–817.

[4] St. Bergman, *The kernel function and conformal mapping*, Second (rev.) ed., Amer. Math. Soc. Math. Surveys **5**, Providence, R.I., 1970.

[5] Ph. Davis, *Interpolation and Approximation*, Dover Publ., New York, 1975.

[6] J. Détraz, *Problème de Dirichlet pour le système* $\partial^2 f/\partial\overline{z_i}\partial\overline{z_j} = 0$, Arkiv för Mat., **26**(1988), 173–184.

[7] K. Friedrichs, *On certain inequalities for analytic functions and for functions of two variables*, Trans. Amer. Math. Soc. **41**(1937), 321–364.

[8] B. Gustafsson, *On mother bodies of convex polyhedra*, SIAM J. Math. Analysis **29**(1998), 1106–1117.

[9] B. Gustafsson, M. Sakai, and H. S. Shapiro, *On domains in which harmonic functions satisfy generalized mean value properties*, Potential Analysis **7**(1997), 467–484.

[10] V. Havin, *Approximation in the mean by analytic functions*, Dokl. Akad. Nauk SSSR **178**(1968), 1025–1028.

[11] T. Kato, *Perturbation theory for linear operators*, Springer Verlag, Berlin, 1995.

[12] D. Khavinson and H. S. Shapiro, *Dirichlet's problem when the data is an entire function*, Bull. London Math. Soc. **24**(1992), 456–468.

[13] P. Lin and R. Rochberg, *On the Friedrichs operator*, Proc. Amer. Math. Soc. **123**(1995), 3335–3342.

[14] N. I. Muskhelishvili, *Some basic problems of the mathematical theory of elasticity*, Noordhoff, Groningen, 1963.

[15] Z. Nehari, *Conformal mapping*, Dover, New York, 1975.

[16] S. Norman, *The Friedrichs operator*, research report TRITA-MAT-1987-16, Royal Inst. of Technology, 1987.

[17] M. Putinar, *Extremal solutions of the two-dimensional L-problem of moments*, J. Funct. Analysis **136** (1996), 331–364.

[18] M. Reed and B. Simon, *Methods of modern mathematical physics, Vol. 1*, Academic Press, New York, 1972.

[19] M. Sakai, *Quadrature Domains*, Lect. Notes Math. **934**, Springer-Verlag, Berlin, 1982.

[20] H. S. Shapiro, *Some inequalities for analytic functions integrable over a plane domain*, in proc. of conf. "Approximation and Function Spaces" Gdansk 1979, North Holland, 1981, pp. 645–666.

[21] H. S. Shapiro, *Stefan Bergman's theory of doubly-orthogonal functions. An operator-theoretic approach*, Proc. Royal Irish Acad., Section A **79:A-6**(1979), 49–58.

[22] H. S. Shapiro, *On some Fourier and distribution-theoretic methods in approximation theory*, in *Approximation Theory. III*, Proc. Conf. held in Austin, Texas, 1980 (W. Cheney et al., eds.), Academic Press, San Diego, 1980, pp. 87–124.

[23] H. S. Shapiro, *Reconstructing a function from its values on a subset of its domain—a Hilbert space approach*, J. Approx. Th. **46**(1986), 385–402.

[24] H. S. Shapiro, *The Schwarz function and its generalization to higher dimensions*, Wiley-Interscience, New York, 1992.

[25] A. L. Shields, *Weighted shift operators and analytic function theory*, Math. Surveys Vol. **13** (1974), Amer. Math. Soc., Providence, R.I., pp. 49–128.

[26] E. C. Titchmarsh, *The theory of functions*, Oxford Univ. Press, London, 1952.

Department of Mathematics, University of California, Santa Barbara, CA 93106, U.S.A.
E-mail address: `mputinar@math.ucsb.edu`

Mathematics Institute, Royal Institute of Technology, Stockholm S 100 44, Sweden.
E-mail address: `shapiro@math.kth.se`

Operator Theory:
Advances and Applications, Vol. 113

Parametrical Representations of Some Classes of Holomorphic Functions in the Disk

F. A. Shamoyan and E. N. Shubabko

Dedicated to the memory of S. A. Vinogradov

Let $\mathbf{D} = \{z : |z| < 1\}$ be the unit disk on the complex plane $\mathbf{C}$, let $\mathbf{\Gamma}$ be its boundary, and let $H(\mathbf{D})$ be set of all holomorphic functions in $\mathbf{D}$. For $\alpha > 0$, we denote by N_α the following class of functions:

$$N_\alpha = \{f \in H(\mathbf{D}) : T(r, f) \leq \frac{c_f}{(1-r)^\alpha};\ 0 \leq r \leq 1\},$$

where $T(r, f)$ is the Nevanlinna characteristic of f. It is obvious that if $\alpha = 0$, then $N_0 = N$ (N is the standard Nevanlinna class). The following classical result of R. Nevanlinna (see [1]) is well known: $f \in N$ if and only if

$$f(z) = c_\lambda z^\lambda B(z, z_k) \exp \int_{-\pi}^{\pi} \frac{d\mu(\theta)}{(1 - e^{-i\theta} z)}, \quad z \in \mathbf{D},$$

where c_λ is a complex number, λ is a non-negative integer, $B(z, z_k)$ is a Blaschke product, $\{z_k\}_{k=1}^{+\infty}$ is any sequence of points in $\mathbf{D}$ satisfying

$$\sum_{k=1}^{+\infty} (1 - |z_k|) < +\infty,$$

$\mu(\theta)$ is any real measure on $[-\pi, \pi]$.

It is natural to ask about similar representations for other values $\alpha > 0$. In this paper we show that they do exist.

We need the following notation. Let $\{z_k\}_{k=1}^{+\infty}$ be any sequence of points in $\mathbf{D}$ satisfying

$$\sum_{k=1}^{+\infty} (1 - |z_k|)^{\beta+2} < +\infty, \tag{1}$$

where $\beta > -1$. In [2] it was established that under condition (1) the infinite product

$$\Pi_\beta(z, z_k) = \prod_{k=1}^{+\infty} \left(1 - \frac{z}{z_k}\right) \exp\left\{-\frac{\beta}{\pi} \int_D \frac{(1 - |\zeta|^2)^\beta \ln |1 - \frac{\zeta}{z_k}|}{(1 - \bar{\zeta} z)^{\beta+2}}\right\} dm_2(\zeta) \tag{2}$$

converges uniformly inside $\mathbf{D}$.

We denote by $B_\gamma^{1,\infty}$ $(\gamma > 0)$ the standard Besov class on the unit circle $\mathbf{\Gamma}$.

Theorem 1. *If $\alpha > 0$, $\beta > \alpha - 1$, then the following statements are equivalent:*
1) $f \in N_\alpha$,
2) *f admits a representation*

$$f(z) = c_\lambda z^\lambda \Pi_\beta(z, z_k) \exp \int_{-\pi}^{\pi} \frac{\psi(e^{i\theta})}{(1 - e^{-i\theta}z)^{\beta+2}} d\theta, \quad z \in \mathbf{D}, \tag{3}$$

where c_λ is a complex number, λ is a non-negative integer, $\Pi_\beta(z, z_k)$ is a product of the form (2), $\{z_k\}_{k=1}^{+\infty}$ is any sequence of points in $\mathbf{D}$ satisfying

$$n(r) = \{\text{card} \;\; z_k : |z_k| < r\} \leq \frac{C}{(1 - r)^{\alpha+1}}, \tag{4}$$

$\psi(e^{i\theta})$ is any real function belonging to $B^{1,\infty}_{\beta-\alpha+1}$.

For $\beta = \alpha$ we have:

Theorem 2. *If $\alpha > 0$, then the following statements are equivalent:*
1) $f \in N_\alpha$,
2) *f admits a representation*

$$f(z) = c_\lambda z^\lambda \Pi_\alpha(z, z_k) \exp \int_{-\pi}^{\pi} \frac{\psi(e^{i\theta})}{(1 - e^{-i\theta}z)^{\alpha+2}} d\theta, \quad z \in \mathbf{D},$$

where c_λ is a complex number, λ is a non-negative integer, $\Pi_\alpha(z, z_k)$ is the product (2), $\{z_k\}_{k=1}^{+\infty}$ is any sequence of points in $\mathbf{D}$ satisfying $n(r) \leq \frac{C}{(1-r)^{\alpha+1}}$, $\psi(e^{i\theta})$ is any real function belonging to $B^{1,\infty}_1$.

Remark 1. *Theorem 1 implies that condition (4) completely characterizes the zero sets of the class N_α.*

Remark 2. *It should be noted that the method of the proof of Theorem 1 allows us to obtain similar representations for classes of functions holomorphic in $\mathbf{D}$ for which $T(r, f) \leq C_f \varphi(\frac{1}{1-r})$, where φ is a function varying regularly on the interval $(0, +\infty)$ (see [3]).*

Remark 3. *An analog of Theorem 1 for the classes of functions holomorphic in $\mathbf{D}$ and satisfying*

$$\int_0^1 \omega(1 - r) T^p(r, f) dr < +\infty \quad (0 < p < +\infty),$$

where ω is a function varying regularly on $(0, 1)$ (see [3]) was obtained by the first author in [4]. (For $p = 1$ see [5], [6] and [7].)

In the proof of the theorem we use the following auxiliary statements.

Lemma 1 (see [5]). *If $\beta > -1$ and*

$$\sum_{k=1}^{+\infty} (1 - |z_k|)^{\beta+2} < +\infty,$$

then for the product $\Pi_\beta(z,z_k)$ *we have*

$$\ln|\Pi_\beta(z,z_k)| \le \text{const}\sum_{k=0}^{+\infty}\left(\frac{1-|z_k|^2}{|1-z\bar{z}_k|}\right)^{\beta+2}.$$

Lemma 2 (see [8]). *The Besov class* $B_\gamma^{1,\infty}$ $(\gamma>0)$ *coincides with the class of functions* f *the harmonic continuation of which inside* $\mathbf{D}$, *i.e.,*

$$u(re^{i\varphi}) = \frac{1}{2\pi}\int_{-\pi}^{\pi}\frac{(1-r^2)f(e^{i\theta})}{1-2r\cos(\theta-\varphi)+r^2}d\theta,$$

satisfies the estimate

$$\int_{-\pi}^{\pi}\left|\frac{\partial^k u(re^{i\varphi})}{\partial r^k}\right|d\varphi \le \frac{C}{(1-r)^{k-\gamma}},$$

where k *is a positive integer and* $k>\gamma$.

The following statement can be deduced from the results of N. K. Nikol′skiĭ (in [9]), where certain estimates were obtained in the uniform metric for sufficiently weights. We will offer a simple proof in case of the L_1 metric.

Lemma 3. *If* $u(z)$ *is a function harmonic in* $\mathbf{D}$ *and* v *is the harmonic conjugate of* u, $v(0)=0$, *then the inequality*

$$\int_{-\pi}^{\pi}|u(re^{i\varphi})|d\varphi \le \frac{A}{(1-r)^\alpha},\quad \alpha>0,\quad 0\le r<1,$$

implies

$$\int_{-\pi}^{\pi}|v(re^{i\varphi})|d\varphi \le \frac{C(A)}{(1-r)^\alpha}.$$

Proof. Let $\beta>\alpha+1$, and let $f(z)=u(z)+iv(z)$. Representing $f(z)$ in terms of $u(z)$ (see [2]), we obtain

$$f(z) = \frac{\beta+1}{\pi}\int_0^1\int_{-\pi}^{\pi}\frac{(1-\rho^2)^\beta u(re^{i\theta})}{(1-re^{-i\theta}z)^{\beta+2}}\rho d\rho d\theta;$$

therefore,

$$\int_{-\pi}^{\pi}|f(re^{i\varphi})|d\varphi \le \frac{\beta+1}{\pi}\int_{-\pi}^{\pi}\left(\int_0^1\int_{-\pi}^{\pi}\frac{(1-\rho^2)^\beta|u(re^{i\theta}|}{|1-re^{-i\theta}z|^{\beta+2}}\rho d\rho d\theta\right)d\varphi.$$

From this we deduce the estimate

$$\begin{aligned}\int_{-\pi}^{\pi}|f(re^{i\varphi})|d\varphi &\le \frac{C(\beta+1)}{\pi}\int_0^1\frac{(1-\rho^2)^\beta}{(1-\rho r)^{\beta+1}}\frac{d\rho}{(1-\rho)^\alpha}\\ &\le 2^\beta\frac{C(\beta+1)}{\pi}\left(\int_0^r\frac{d\rho}{(1-\rho)^{\alpha+1}}+\int_r^1\frac{d\rho}{(1-r\rho)^{\alpha+1}}\right)\\ &\le \frac{C_1(\beta+1)}{\alpha}\frac{1}{(1-r)^\alpha}.\end{aligned}$$

The lemma is proved. □

Corollary. *Let $f(z) = u(z) + iv(z)$. Assume that $f \in H^1$, $v(0) = 0$, and $u(e^{i\varphi}) \in B^{1,\infty}_\gamma$ $(\gamma > 0)$. Then $f(e^{i\varphi})$ belongs to $B^{1,\infty}_\gamma$.*

The proof of this statement immediately follows from Lemma 3 and the representation of $B^{1,\infty}_\gamma$ in terms of derivatives (see [10]).

Proof of the theorem. First, we prove the implication 1) $\Longrightarrow$ 2). If $f \in N_\alpha$, then, clearly,

$$N(R) = \int_o^R \frac{n(t)}{t} dt \le \frac{c}{(1-R)^\alpha}.$$

Choosing $R = r + \frac{1-r}{2}$, where $0 < r < 1$, we obtain

$$\int_o^{r+\frac{1-r}{2}} \frac{n(t)}{t} dt \le \frac{c_1}{(1-r)^\alpha}, \qquad \int_r^{r+\frac{1-r}{2}} \frac{n(t)}{t} dt \le \frac{c_1}{(1-r)^\alpha},$$

$$n(r)\frac{1-r}{2} \le \frac{c_1}{(1-r)^\alpha},$$

i.e.,

$$n(r) \le \frac{C}{(1-r)^{\alpha+1}} \tag{5}$$

We claim that (5) implies the convergence of the series $\sum_{k=1}^{+\infty}(1-|z_k|)^{\beta+2}$. Indeed,

$$\sum_{|z_k|<R} (1-|z_k|)^{\beta+2} = \int_o^R (1-t)^{\beta+2} dn(t).$$

Integrating by parts and using (5), we obtain

$$\sum_{|z_k|<R} (1-|z_k|)^{\beta+2} \le (\beta+2) \int_o^R \frac{(1-t)^{\beta+1}}{(1-t)^{\alpha+1}} dt + o(1) \le \text{const} < +\infty.$$

The last-written integral is bounded because $\beta > \alpha - 1$.

From the convergence of this series we deduce that the infinite product $\Pi_\beta(z, z_k)$ constructed by the zeros $\{z_k\}_{k=1}^{+\infty}$ of the function $f(z)$ uniformly converges inside $\mathbf{D}$ (see [2]). We claim that (5) implies the relation $\Pi_\beta(z, z_k) \in N_\alpha$. By Lemma 1, we have

$$\ln|\Pi_\beta(z, z_k)| \le c_1 \sum_{k=0}^{+\infty} \left(\frac{1-|z_k|^2}{|1-z\bar{z}_k|}\right)^{\beta+2}.$$

Hence

$$\int_{-\pi}^{\pi} \ln^+|\Pi_\beta(re^{i\varphi}, z_k)| d\varphi \le c_1 \sum_{k=1}^{+\infty} (1-|z_k|^2)^{\beta+2} \int_{-\pi}^{\pi} \frac{d\varphi}{|1-re^{i\varphi}\bar{z}_k|^{\beta+2}}$$

$$\le c_2 \sum_{k=1}^{+\infty} \frac{(1-|z_k|^2)^{\beta+2}}{(1-r|z_k|)^{\beta+1}} = c_2 \int_0^1 \frac{(1-t^2)^{\beta+2}}{(1-rt)^{\beta+1}} dn(t).$$

Integrating by parts and again using (5), we obtain

$$I = \int_0^1 \frac{(1-t^2)^{\beta+2}}{(1-rt)^{\beta+1}} dn(t)$$
$$= \int_0^1 \left(2t(\beta+2) \left(\frac{1-t^2}{1-rt} \right)^{\beta+1} - r(\beta+1) \left(\frac{1-t^2}{1-rt} \right)^{\beta+2} \right) dn(t)$$
$$\leq c \int_0^1 n(t) \left(\frac{1-t^2}{1-rt} \right)^{\beta+1} dt.$$

Yet another application of (5) yields

$$I \leq c \int_0^1 \left(\frac{1-t^2}{1-rt} \right)^{\beta+1} \frac{dt}{(1-t)^\alpha} \leq c_1 \int_0^1 \frac{(1-t)^{\beta-\alpha}}{(1-rt)^{\beta+1}} dt.$$

We split the latter integral in the sum of two integrals $\int_0^r + \int_r^1 = I_1 + I_2$.

$$I_1 = \int_0^r \frac{(1-t)^{\beta-\alpha}}{(1-rt)^{\beta+1}} dt \leq \int_0^r \frac{dt}{(1-t)^{\alpha+1}} = \frac{1}{\alpha(1-r)^\alpha} - \frac{1}{\alpha}.$$
$$I_2 = \int_r^1 \frac{(1-t)^{\beta-\alpha}}{(1-rt)^{\beta+1}} dt \leq \frac{1}{(1-r)^{\beta+1}} \int_r^1 (1-t)^{\beta-\alpha} dt = \frac{1}{(1+\beta-\alpha)} \frac{1}{\alpha(1-r)^\alpha}.$$

Thus, $I \leq \frac{C}{(1-r)^\alpha}$, i.e., $\Pi_\beta(z, z_k) \in N_\alpha$.

Now, we prove the representation (3) for f. Without loss of generality we assume that $f(0) = 1$. We denote $g(z) = \frac{f(z)}{\Pi_\beta(z,z_k)}$. Using elementary properties of the Nevanlinna characteristic, we see that $T(r, g) < T(r, f) + T(r, \frac{1}{\Pi_\beta})$. Applying the Jensen formula $T(r, \frac{1}{\Pi_\beta}) = T(r, \Pi_\beta) - \ln|\Pi_\beta(0, z_k)|$ (see [1]), finally we get $T(r, g) \leq \frac{C}{(1-r)^\alpha}$, that is, $g(z) \in N_\alpha$.

We show that the function $h(z) = \ln g(z)$, $z \in \mathbf{D}$ (where the principal branch of the logarithm is meant) satisfies

$$\int_{-\pi}^{\pi} |h(re^{i\varphi})| d\varphi \leq \frac{\text{const}}{(1-r)^\alpha} \quad (0 < r < 1). \tag{6}$$

Indeed,

$$\int_{-\pi}^{\pi} \ln^+ |g(re^{i\varphi})| d\varphi \leq \frac{c_1}{(1-r)^\alpha}, \quad r \in (0,1).$$

Since

$$\frac{1}{2\pi} \int_{-\pi}^{\pi} \ln |g(re^{i\varphi})| d\varphi = \ln |g(0)|$$

and

$$\frac{1}{2\pi} \int_{-\pi}^{\pi} |\ln |g(re^{i\varphi})|| d\varphi = \frac{1}{\pi} \int_{-\pi}^{\pi} \ln^+ |g(re^{i\varphi})| d\varphi - \ln |g(0)|,$$

we see that

$$\int_{-\pi}^{\pi} |\ln |g(re^{i\varphi})|| d\varphi \leq \frac{c_2}{(1-r)^\alpha},$$

i.e., the function $h(z)$ satisfies the estimate

$$\int_{-\pi}^{\pi} |Re\ h(re^{i\varphi})| d\varphi \leq \frac{c_2}{(1-r)^{\alpha}}.$$

By Lemma 3, $h(z)$ satisfies (6).

We put

$$\psi(z) = \int_0^1 (1-t)^{\beta} h(tz) dt \quad (z \in \mathbf{D}). \tag{7}$$

Obviously, $\psi(z) \in H(\mathbf{D})$. We prove that $\psi(e^{i\varphi}) \in B^{1,\infty}_{\beta-\alpha+1}$ $(\beta > \alpha - 1)$. By the Hardy–Littlewood theorem, it suffices to obtain the estimate

$$\int_{-\pi}^{\pi} |\psi^{(k)}(re^{i\varphi})| d\varphi \leq \frac{\text{const}}{(1-r)^{k-(\beta-\alpha+1)}} \quad (0 < r < 1), \tag{8}$$

where k is a natural number such that $k > \beta - \alpha + 1$ (see [11]). k-fold differentiation of (7) yields

$$\psi^{(k)}(z) = \int_0^1 (1-t)^{\beta} t^k h^{(k)}(tz) dt.$$

Therefore

$$\int_{-\pi}^{\pi} |\psi^{(k)}(re^{i\varphi})| d\varphi \leq \int_0^1 (1-t)^{\beta} \int_{-\pi}^{\pi} |h^{(k)}(tre^{i\varphi})| dt d\varphi.$$

By (6), we obtain

$$\int_{-\pi}^{\pi} |\psi^{(k)}(re^{i\varphi})| d\varphi \leq c_3 \int_0^1 \frac{(1-t)^{\beta}}{(1-rt)^{\alpha+k}} dt \tag{9}$$

(we have used the estimate

$$\int_{-\pi}^{\pi} |h^{(k)}(\rho e^{i\varphi})| d\varphi \leq \frac{c(k)}{(1-\rho)^k} \int_{-\pi}^{\pi} |h(Re^{i\varphi})| d\varphi, \quad 0 < \rho < R < 1,$$

see [11]). But from (9) it is easy to deduce the inequality

$$\int_{-\pi}^{\pi} |\psi^{(k)}(re^{i\varphi})| d\varphi \leq \frac{c_4}{(1-r)^{k-(\beta-\alpha+1)}}.$$

Thus, (8) is established. Since ψ belongs to the Hardy class H_1, we obtain

$$\psi(z) = \frac{1}{2\pi} \int_{-\pi}^{\pi} \frac{\psi(e^{i\theta}) e^{i\theta}}{e^{i\theta} - z} d\theta = \frac{1}{2\pi} \int_{-\pi}^{\pi} \frac{\psi(e^{i\theta})}{1 - e^{-i\theta} z} d\theta.$$

Using (7), finally we obtain

$$h(z) = D^{\beta+1} \psi(z) = \frac{1}{2\pi} \int_{-\pi}^{\pi} \frac{\psi(e^{i\theta})}{(1 - e^{-i\theta} z)^{\beta+2}} d\theta,$$

where $D^{\beta+1}$ is the operator of fractional differentiation, that is, if $\psi(z) = \sum_{k=0}^{\infty} a_k z^k$, then

$$D^{\beta+1}\psi(z) = \sum_{k=0}^{\infty} \frac{\Gamma(\beta+2+k)}{\Gamma(\beta+2)\Gamma(k+1)} a_k z^k$$

(see [12]). Thus, if $f \in N_\alpha$, then f admits a representation in the form (3). Since $\psi \in H_1$ and

$$\frac{1}{2\pi}\int_{-\pi}^{\pi} \frac{\overline{\psi(e^{i\theta})}}{(1-e^{-i\theta}z)^{\beta+2}} d\theta = \frac{1}{2\pi}\int_{-\pi}^{\pi} \overline{\psi}(e^{i\theta}) d\theta = c_0(\psi),$$

we have

$$\frac{1}{2\pi}\int_{-\pi}^{\pi} \frac{\psi(e^{i\theta})}{(1-e^{-i\theta}z)^{\beta+2}} d\theta = \frac{1}{2\pi}\int_{-\pi}^{\pi} \frac{\psi(e^{i\theta}) + \overline{\psi(e^{i\theta})}}{(1-e^{-i\theta}z)^{\beta+2}} d\theta - c_0(\psi),$$

therefore, we may assume that in (3) the function ψ is real.

Now, we prove the implication 2) $\Longrightarrow$ 1). Suppose $f(z)$ is represented as in (3). Using the first part of the theorem, we can state that the product $\Pi_\beta(z, z_k)$ is in N_α. We must prove that the function

$$g(z) = \exp \frac{1}{2\pi}\int_{-\pi}^{\pi} \frac{\psi(e^{i\theta})}{(1-e^{-i\theta}z)^{\beta+2}} d\theta,$$

where $\psi(e^{i\varphi}) \in B^{1,\infty}_{\beta-\alpha+1}$, belongs to N_α. Without loss of generality, we may assume that $\beta < \alpha + 1$. By the corollary to Lemma 3, we see that $\psi = \psi_1 + \psi_2$, where $\psi_1 \in H^1$, $\psi_2 \perp H^1$. Therefore, the function

$$h(z) = \frac{1}{2\pi}\int_{-\pi}^{\pi} \frac{\psi(e^{it})}{(1-e^{-it}z)^{\beta+2}} dt$$

can be represented as follows:

$$\begin{aligned} h(z) &= \frac{1}{2\pi}\int_{-\pi}^{\pi} \frac{\psi(e^{i(\theta+t)}) + \psi(e^{i(\theta-t)}) - 2\psi(e^{i\theta})}{(1-e^{-it}z)^{\beta+2}} dt + 2\psi(e^{i\theta}) - \psi(0) \\ &= \frac{1}{2\pi}\int_{-\pi}^{\pi} \frac{\Delta_t^2\psi(e^{i\theta})}{(1-e^{-it}z)^{\beta+2}} dt + 2\psi(e^{i\theta}) - \psi(0), \end{aligned}$$

where $\Delta_t^2\psi(e^{i\theta}) = \psi(e^{i(\theta+t)}) + \psi(e^{i(\theta-t)}) - 2\psi(e^{i\theta})$. From this we have

$$\begin{aligned} \frac{1}{2\pi}\int_{-\pi}^{\pi} |h(re^{i\theta})| d\theta \le{}& \frac{1}{4\pi^2}\int_{-\pi}^{\pi}\int_{-\pi}^{\pi} \frac{|\Delta_t^2\psi(e^{i\theta})|}{|1-e^{-it}r|^{\beta+2}} dt d\theta \\ &+ \frac{1}{\pi}\int_{-\pi}^{\pi} |\psi(e^{i\theta})| d\theta + \frac{1}{2\pi}\int_{-\pi}^{\pi} |\psi(0)| d\theta. \end{aligned}$$

Since $|\psi(0)| \leq \frac{1}{2\pi}\int_{-\pi}^{\pi}|\psi(e^{i\theta})|d\theta$, finally we obtain

$$\frac{1}{2\pi}\int_{-\pi}^{\pi}|h(re^{i\theta})|d\theta \leq c\int_{-\pi}^{\pi}\int_{-\pi}^{\pi}\frac{|\Delta_t^2\psi(e^{i\theta})|}{|1-e^{-it}r|^{\beta+2}}dtd\theta$$
$$\leq c_1\int_{-\pi}^{\pi}\frac{|t|^{\beta-\alpha+1}}{|1-e^{-it}r|^{\beta+2}}dt \leq \frac{c_3}{(1-r)^{\alpha}}.$$

The theorem is proved. □

References

[1] R. Nevanlinna, *Eindeutige Analytischefunktionen*, Springer, Berlin, 1936.

[2] M. M. Dzhrbashyan, *On the representation problem of analytic functions*, Soobshch. Inst. Matem. i Mekh. Akad. Nauk Arm. SSR, **2** (1948), 3–40. (Russian)

[3] E. Seneta, *Regulary varying functions*, Nauka, Moscow, 1985. (Russian translation)

[4] F. A. Shamoyan, *Zero sets and factorization of certain weighted classes of functions holomorphic in the disk*, Siberian Mathematical Journal (to appear). (Russian)

[5] F. A. Shamoyan, *M. M. Dzhrbashyan's factorization theorem and characterization of zeros of functions analytic in the disk with a majorant of bounded growth*, Izv. Akad. Nauk Arm. SSR, Matematika, **13** (1978), no. 5-6, 405–422. (Russian)

[6] A. E. Dzhrbashyan and F. A. Shamoyan, *Topics in the Theory of A^p_α Spaces*, Teubner-Texte zur Math., b. 105, Leipzig, 1988.

[7] F. A. Shamoyan, *On the parametric representation of Nevanlinna-Dzhrbashyan classes*, Dokl. Akad. Nauk Arm. SSR, **90** (1990), no. 3, 99–103. (Russian)

[8] H. Triebel, *Theory of Function Space*, Birkhäuser Verlag, Basel – Boston – Stuttgart, 1983.

[9] N. K. Nikol′skiĭ, *Selected problems in weighted approximations and spectral analysis*, Proc. Steklov Inst. Math., **120** (1974).

[10] E. M. Stein, *Singular Integrals and the differentiability properties of functions*, Princeton Univers. Press, Princeton, New Jersey, 1970.

[11] P. L. Duren. *Theory of H^p Spaces*, Academic Press, New York, 1970.

[12] M. M. Dzhrbashyan, *Integral transform and representation of functions in the complex domain*, Nauka, Moscow, 1966. (Russian)

Bryansk State Ped. University
Bejinskaya, 14
241036 Bryansk
Russia
E-mail address: sham@bgpi.bitmcnit.bryansk.su

Operator Theory:
Advances and Applications, Vol. 113

Double Power Series and Reproducing Kernels

Serguei Shimorin

Dedicated to the memory of S. A. Vinogradov

Abstract. We consider some properties of the reproducing kernels for Hilbert spaces of functions analytic in the unit disk; these preperties are related to double power series expansions. It turns out that some structure properties of the matrices of coefficients of these expanisons are determined by certain quadratic inequalities involving the operator of multiplication by an independent variable. As examples, we consider Dirichlet type spaces and weighted Bergman spaces.

1. Introduction

Banach and Hilbert spaces of analytic functions were one of the main objects of S. A. Vinogradov's research interests. Many-hour talks with him had great impact on the author's interest in problems in this field of mathematics, forming the author's understanding of this theory. The present paper is devoted to a specific important aspect of the theory, namely, to the reproducing kernels of Hilbert spaces of analytic functions.

Reproducing kernels for function spaces are known to be a very useful tool for solving different problems pertaining to function and operator theory, to partial differential equations, and to other topics. However, no explicit formulas for them are known except for a fairly limited class of special Hilbert spaces. Therefore, the estimates and structure properties of reproducing kernels obtained without referring to an explicit formula are very important. The present paper deals with a special situation, namely, with Hilbert spaces of functions analytic in the unit disk on the complex plane. We examine some structure properties of the reproducing kernels for such spaces related to their double power series representation, and reformulate these properties in terms of quadratic inequalities involving the operator of multiplication by an independent variable.

If X is a Hilbert space of functions analytic in the unit disk $\mathbb{D} = \{z \in \mathbb{C} : |z| < 1\}$, then the *reproducing kernel* K_X of X is a function of two variables z, λ such that $K_X(\cdot, \lambda) \in X$, and $f(\lambda) = (f, K_X(\cdot, \lambda))_X$ for any $\lambda \in \mathbb{D}$ and any $f \in X$. The well-known relation $K_X(\lambda, z) = \overline{K_X(z, \lambda)}$ implies that the function $K_X(z, \lambda)$

Supported in part by the Russian Foundation for Basic Research (grant no. 96-01-00693).

is analytic in z and antianalytic in λ in the bidisk $\mathbb{D}\times\mathbb{D}$. Therefore, it admits a double power series representation

$$K_X(z,\lambda)=\sum_{n,k\geq 0}\hat{K}_X(n,k)z^n\bar{\lambda}^k. \tag{1}$$

Suppose, in addition, that the space X is invariant with respect to the *shift operator* S of multiplication by z, $(Sf)(z)=zf(z)$. It turns out that certain properties of the operator S imply structure properties of the matrix $\hat{K}_X$, which, in turn, result in some nontrivial relations for the kernel function K_X.

With any infinite matrix $\hat{A}=(\hat{A}(n,k))_{n,k\geq 0}$ with slowly growing entries we can associate a function $A(z,\lambda)$ of two variables defined on $\mathbb{D}\times\mathbb{D}$ as the sum of a double power series,

$$A(z,\lambda)=\sum_{n,k\geq 0}\hat{A}(n,k)z^n\bar{\lambda}^k. \tag{2}$$

Viewed as a function of two variables, such a function A is positive definite (i.e., it is the reproducing kernel for some Hilbert function space) if and only if the matrix $\hat{A}$ is positive definite. Indeed, the positive definiteness of A can be derived from the positive definiteness of $\hat{A}$ by straightforward computation; in the reverse direction we may use the formula

$$\hat{A}(n,k)=\frac{1}{r^{n+k}}\int_{\mathbb{T}}\int_{\mathbb{T}}A(r\zeta,r\eta)\bar{\zeta}^n\eta^k\,dm_1(\zeta)\,dm_1(\eta).$$

Here $\mathbb{T}=\partial\mathbb{D}$ is the unit circle, and dm_1 is the normalized arc measure on $\mathbb{T}$.

We introduce two operations, σ and σ^* (the forward and backward matrix shifts), that act on the infinite matrices by the rule

$$(\sigma\hat{A})(n,k):=\begin{cases}0 & \text{if } n=0 \text{ or } k=0,\\ \hat{A}(n-1,k-1) & \text{if } n\geq 1 \text{ and } k\geq 1,\end{cases}$$

and

$$(\sigma^*\hat{A})(n,k):=\hat{A}(n+1,k+1).$$

The same letters σ and σ^* will denote the corresponding operations on the functions represented by the double power series (2):

$$(\sigma A)(z,\lambda):=\sum_{n,k\geq 0}(\sigma\hat{A})(n,k)z^n\bar{\lambda}^k,$$

$$(\sigma^* A)(z,\lambda):=\sum_{n,k\geq 0}(\sigma^*\hat{A})(n,k)z^n\bar{\lambda}^k,$$

Obviously, the operations σ and σ^* preserve positive definiteness and satisfy the simple relations

$$(\sigma A)(z,\lambda)=\bar{\lambda}zA(z,\lambda)\quad\text{and}\quad\sigma^*\sigma=I.$$

We are interested in the monotonicity and convexity type properties of the matrices $\hat{K}_X$ in the decomposition (1), i.e., in the positive definiteness and the negative definiteness of the matrices $(I-\sigma^*)\hat{K}_X$, and $(I-\sigma^*)^2\hat{K}_X$. (In the simplest case where the monomials $(z^n)_{n\geq 0}$ form an orthogonal basis of X, the matrix $\hat{K}_X$ is a diagonal one with the entries $\hat{K}_X(n,n) = \|z^n\|_X^{-2}$, and the positive definiteness of $(I-\sigma^*)\hat{K}_X$ and $(I-\sigma^*)^2\hat{K}_X$ corresponds to the monotonicity and convexity of the diagonal sequence $(\hat{K}_X(n,n))_{n\geq 0}$.) Our study is based on the interpretation of the functions $\sigma^{*n}K_X$ as reproducing kernels for the space X supplied with some equivalent norm, and on an abstract comparision principle for sums of reproducing kernels.

Also, we consider the positive definiteness of the functions

$$((I-\sigma)^n K_X)(z,\lambda) = (1-\bar{\lambda}z)^n K_X(z,\lambda).$$

For two functions $K(\cdot,\cdot)$, and $K'(\cdot,\cdot)$ defined on $E \times E$, where E is some abstract set, we shall write $K' \ll K$ if the difference $K(\cdot,\cdot) - K'(\cdot,\cdot)$ is positive definite as a function of two variables.

2. Main theorems

We recall the following two facts from the general theory of reproducing kernels (see Aronszajn [2]).

Proposition 2.1. *Suppose that $H_1,\dots,H_n$ are Hilbert spaces of functions defined on the same set E, and that K_i are the reproducing kernels for H_i, $i=1,\dots,n$. Then the sum $\sum_{i=1}^n K_i$ is the reproducing kernel for the Hilbert space $H = H_1 + \cdots + H_n$ consisting of the functions $f = \sum_{i=1}^n f_i$ with $f_i \in H_i$, supplied with the norm*

$$\|f\|^2 := \inf\left(\sum_{i=1}^n \|f_i\|^2\right),$$

where the infimum is taken over all the decompositions $f = \sum_{i=1}^n f_i$ with $f_i \in H_i$.

Remark. Aronszajn proved this fact only for $n=2$. The general case can easily be obtained by induction.

Proposition 2.2. *Suppose that H and H' are Hilbert spaces of functions defined on the same set E; let K,K' be the reproducing kernels for H, H', respectively. Then the relation $K' \ll K$ is equivalent to the inclusion $H' \subset H$ with $\|g\|_H \leq \|g\|_{H'}$ for any $g \in H'$.*

Combining these two facts, we immediately arrive at the following abstract comparision principle for sums of reproducing kernels (we also use the fact that, after multiplying the norm of some Hilbert function space by a positive constant c, the corresponding new reproducing kernel can be obtained by dividing the initial one by c^2).

Proposition 2.3. *Suppose that $H_1, \dots, H_n$, and $H_1', \dots, H_m'$ are Hilbert spaces of functions defined on the same set E; let K_i, K_j' be the reproducing kernels for H_i, H_j', respectively, and let $c_1, \dots, c_n$ and $c_1', \dots, c_m'$ be some positive constants. Then the relation*

$$\sum_{j=1}^{m} c_j' K_j' \ll \sum_{i=1}^{n} c_i K_i$$

is fulfilled if and only if for any $g_1, \dots, g_m$ with $g_j \in H_j'$ there exist $f_1, \dots, f_n$ with $f_i \in H_i$ such that

$$\sum_{j=1}^{m} g_j = \sum_{i=1}^{n} f_i$$

and

$$\sum_{i=1}^{n} c_i^{-1} \|f_i\|_{H_i}^2 \le \sum_{j=1}^{m} {c_j'}^{-1} \|g_j\|_{H_j'}^2.$$

In the most interesting particular cases where $m = 1$ or $n = 1$, we obtain the following statements.

1) We have $\sum_{j=1}^{m} c_j' K_j' \ll K$ if and only if $H_1' + H_2' + \cdots + H_m' \subset H$ and

$$\left\| \sum_{j=1}^{m} g_j \right\|_H^2 \le \sum_{j=1}^{m} {c_j'}^{-1} \|g_j\|_{H_j'}^2.$$

2) We have $K' \ll \sum_{i=1}^{n} c_i K_i$ if and only if $H' \subset H_1 + H_2 + \cdots + H_n$ and any $g \in H'$ can be written as $g = \sum_{i=1}^{n} f_i$ with $f_i \in H_i$ in such a way that

$$\sum_{i=1}^{n} c_i^{-1} \|f_i\|_{H_i}^2 \le \|g\|_{H'}^2.$$

Now we turn to the kernel function K_X. It is easy to check that, for $n \ge 0$, the function $(\sigma^n K_X)(z, \lambda) = (\bar{\lambda} z)^n K_X(z, \lambda)$ is the reproducing kernel for the space $S^n X$ supplied with the range-norm

$$\|f\|_{S^n X} := \|g\|_X,$$

where $f \in S^n X$, $f = S^n g$. The interpretation of the function $\sigma^{*n} K_X$ as a reproducing kernel is slightly more complicated. Suppose that our space X satisfies the following additional (not too restrictive) assumptions:

(A) X contains the function $\mathbf{1}$ (and, with it, all polynomials);

(B) SX is a closed subspace in X, i.e., S is left-invertible in X.

Under these assumptions, the following statement is true.

Proposition 2.4. *For $n \ge 1$, the function $\sigma^{*n} K_X$ is the reproducing kernel for the space X supplied with the norm $\|f\|_{X,n} := \|P_{E_n} S^n f\|_X$, where E_n is the orthogonal complement in X to the family of monomials $\mathbf{1}, z, \dots, z^{n-1}$.*

Proof. The function $K_X(z,\bar\lambda)$ is uniformly bounded on the compact subsets of $\mathbb{D}\times\mathbb{D}$, because this function is analytic in the bidisk. Hence, the evaluation functionals $f\to f(\lambda)$, whose norms are $(K_X(\lambda,\lambda))^{1/2}$, are uniformly bounded on the compact subsets of $\mathbb{D}$. This implies that the functionals $f\to f^{(n)}(\lambda)$ are bounded in X for $\lambda\in\mathbb{D}$ $(n\ge 0)$. Therefore, the functions $D_X^{(n)}(z):=\left.\frac{\partial^n}{\partial\bar\lambda^n}K_X(z,\lambda)\right|_{\lambda=0}$ belong to X and represent the functionals of differentiation at the origin,

$$(f,D_X^{(n)})_X=f^{(n)}(0).$$

For the function K_X we can write the following decomposition:

$$K_X(z,\lambda)=(\sigma^n\sigma^{*n}K_X)(z,\lambda)+\sum_{l=0}^{n-1}\frac{1}{l!}D_X^{(l)}(z)\bar\lambda^l+\sum_{p=0}^{n-1}\left(\sum_{l=n}^{\infty}\hat K_X(p,l)\bar\lambda^l\right)z^p.$$

This yields

$$(P_{E_n}S^n\sigma^{*n})K_X(z,\lambda)=\frac{1}{\bar\lambda^n}P_{E_n}\left(K_X(z,\lambda)-\sum_{l=0}^{n-1}\frac{1}{l!}D_X^{(l)}(z)\bar\lambda^l\right)$$

(the operators S^n and P_{E_n} act with respect to the variable z). Since the function $K_X(\cdot,\lambda)-\sum_{l=0}^{n-1}\frac{1}{l!}D_X^{(l)}(\cdot)\bar\lambda^l$ is orthogonal to all monomials $\mathbf{1},z,\dots,z^{n-1}$, we have

$$(P_{E_n}S^n\sigma^{*n})K_X(z,\lambda)=\frac{1}{\bar\lambda^n}\left(K_X(z,\lambda)-\sum_{l=0}^{n-1}\frac{1}{l!}D_X^{(l)}(z)\bar\lambda^l\right).$$

Further, the norm $\|\cdot\|_{X,n}$ is equivalent to the original norm in X, because S^nX is a closed subspace in X by our assumption (B), and the angle between the linear span $\mathcal{L}(\mathbf{1},\dots,z^{n-1})$ and S^nX is positive. It remains to verify the reproducing property of the function $\sigma^{*n}K_X$ with respect to the norm $\|\cdot\|_{X,n}$. For any $f\in X$ we have

$$\begin{aligned}(P_{E_n}S^nf,P_{E_n}S^n\sigma^{*n}K_X(\cdot,\lambda)_X&=(S^nf,P_{E_n}S^n\sigma^{*n}K_X(\cdot,\lambda)_X\\&=\frac{1}{\lambda^n}\left(S^nf,K_X(\cdot,\lambda)-\sum_{l=0}^{n-1}\frac{1}{l!}D_X^{(l)}(\cdot)\bar\lambda^l\right)_X=f(\lambda),\end{aligned}$$

which completes the proof. □

Now we can combine the above interpretation of the functions σ^nK_X and $\sigma^{*n}K_X$ as reproducing kernels with the comparision principle given by Proposition 2.3.

Theorem 2.5. *Let n be a positive integer. We put $n_1=[n/2]$ and $m_1=[(n-1)/2]$. The function $((I-\sigma)^nK_X)(z,\lambda)=(1-\bar\lambda z)^nK_X(z,\lambda)$ is positive definite if and* *only if for any* $g_0, \ldots, g_{m_1} \in X$ *there exist* $f_0, \ldots, f_{n_1} \in X$ *such that*

$$\sum_{j=0}^{m_1} S^{2j+1} g_j = \sum_{i=0}^{n_1} S^{2i} f_i$$

and

$$\sum_{i=0}^{n_1} \binom{n}{2i}^{-1} \|f_i\|_X^2 \leq \sum_{j=0}^{m_1} \binom{n}{2j+1}^{-1} \|g_j\|_X^2.$$

Theorem 2.6. *Suppose that* X *satisfies the above assumptions* (A) *and* (B), *and let* n *be a positive integer. We put* $n_1 = [n/2]$ *and* $m_1 = [(n-1)/2]$.

(A) *The function* $(I - \sigma^*)^n K_X$ *is positive definite if and only if for any* $g_0, \ldots, g_{m_1} \in X$ *there exist* $f_0, \ldots, f_{n_1} \in X$ *such that*

$$\sum_{j=0}^{m_1} g_j = \sum_{i=0}^{n_1} f_i$$

and

$$\sum_{i=0}^{n_1} \binom{n}{2i}^{-1} \|f_i\|_{X,2i}^2 \leq \sum_{j=0}^{m_1} \binom{n}{2j+1}^{-1} \|g_j\|_{X,2j+1}^2.$$

(B) *The function* $(I - \sigma^*)^n K_X$ *is negative definite if and only if for any* $f_0, \ldots, f_{n_1} \in X$ *there exist* $g_0, \ldots, g_{m_1} \in X$ *such that*

$$\sum_{i=0}^{n_1} f_i = \sum_{j=0}^{m_1} g_j$$

and

$$\sum_{j=0}^{m_1} \binom{n}{2j+1}^{-1} \|g_j\|_{X,2j+1}^2 \leq \sum_{i=0}^{n_1} \binom{n}{2i}^{-1} \|f_i\|_{X,2i}^2.$$

Here $\|\cdot\|_{X,k}$ *are the norms defined in Proposition* 2.4.

Proof. To prove Theorem 2.5, we write the relation $(I - \sigma)^n K_X \gg 0$ in the form

$$\sum_{l=0}^{n_1} \binom{n}{2l} \sigma^{2l} K_X \gg \sum_{j=0}^{m_1} \binom{n}{2j+1} \sigma^{2j+1} K_X.$$

Now, interpreting the functions $\sigma^k K_X$ as reproducing kernels of the spaces $S^k X$ with the range-norm, and applying Proposition 2.3, we obtain the claim. Combining similar arguments with Proposition 2.4, we prove Theorem 2.6. □

3. Particular cases and examples

Now we turn to particular cases of Theorems 2.5 and 2.6, and to examples.

In the simplest case where $n = 1$, Theorem 2.5 says that the function

$$(1 - \bar{\lambda}z)K_X(z, \lambda)$$

is positive definite if and only if S is a contraction. This is a well-known fact; see, e.g., [5, Chapter 8]. In this case we also have $(I - \sigma^*)K_X \ll 0$, since $(I - \sigma^*) = -\sigma^*(I - \sigma)$. Therefore, the contractive operator S corresponds to the matrix $\hat{K}_X$ of increasing type.

For $n = 1$, Theorem 2.6A asserts that the decreasing type property $(I - \sigma^*)K_X \gg 0$ is equivalent to the inequality $\|f\|_X \leq \|P_{E_1}Sf\|_X$ for any $f \in X$. In particular, the operator S must be norm increasing, i.e., it must possess a contractive left inverse operator. If the space X is such that $K_X(\cdot, 0) = \mathbf{1}$, then $P_{E_1}Sf = Sf$ for any $f \in X$, and in this case the decreasing type property of the matrix $\hat{K}_X$ is equivalent to the expansion property $\|Sf\|_X \geq \|f\|_X$ of S.

The next interesting particular case of Theorem 2.6A is that with $n = 2$. We see that the convexity type property $(I - 2\sigma^* + \sigma^{*2})\hat{K}_X \gg 0$ of the matrix $\hat{K}_X$ is equivalent to the following property: any $f \in X$ can be represented as $f = f_0 + f_1$ with $\|f_0\|_X^2 + \|P_{E_2}S^2 f_1\|_X^2 \leq \frac{1}{2}\|P_{E_1}Sf\|_X^2$. There is a simple condition ensuring such a representation. Suppose that $K_X(\cdot, 0) = \mathbf{1}$, and that S possesses the concavity property $2\|Sf\|_X^2 \geq \|f\|_X^2 + \|S^2 f\|_X^2$ for any $f \in X$. Then for any $f \in X$ we can take $f_0 = f_1 = f/2$, which yields

$$\|f_0\|_X^2 + \|P_{E_2}S^2 f_1\|_X^2 \leq \frac{1}{4}\left(\|f\|_X^2 + \|S^2 f\|_X^2\right) \leq \frac{1}{2}\|Sf\|_X^2 = \frac{1}{2}\|P_{E_1}Sf\|_X^2.$$

Also, it is easily seen that the above concavity property of S implies that $\|Sf\| \geq \|f\|$ for any f; thus, in this case, the matrix $\hat{K}_X$ is of decreasing type.

The reproducing kernels K_X such that $K_X(\cdot, 0) = \mathbf{1}$ and the matrices $\hat{K}_X$ are of decreasing and convex type possess the following important property.

Proposition 3.1 (see [6]). *Suppose that the reproducing kernel K_X satisfies the conditions*

(i) $K_X(\cdot, 0) = \mathbf{1}$;
(ii) $(I - \sigma^*)\hat{K}_X \gg 0$;
(iii) $(I - \sigma^*)^2\hat{K}_X \gg 0$.

Then for any $z, \lambda \in \mathbb{D}$ we have

$$\operatorname{Re} K_X(z, \lambda) \geq \frac{1}{2}.$$

For the diagonal matrices $\hat{K}_X$, Proposition 3.1 corresponds to the well-known fact that the cosine trigonometric series

$$\frac{a_0}{2} + \sum_{n=1}^{\infty} a_n \cos nx$$

with nonnegative, decreasing, and convex sequence of coefficients $(a_n)_{n\geq 0}$ is the Fourier series of a nonnegative Borel measure.

Together with the above discussion, Proposition 3.1 immediately implies the following theorem.

Theorem 3.2. *Suppose that*
(i) $K_X(\cdot,0)=\mathbf{1}$;
(ii) $2\|Sf\|_X^2 \geq \|f\|_X^2 + \|S^2f\|_X^2$ *for any* $f\in X$.
Then for any $z,\lambda\in\mathbb{D}$ *we have*

$$\operatorname{Re} K_X(z,\lambda)\geq \frac{1}{2}. \tag{3}$$

As examples of spaces satisfying the conditions of Theorem 3.2 we mention the Dirichlet type spaces $D(\mu)$ introduced by Richter [4] and Aleman [1]. If μ is a nonnegative finite Borel measure on $\overline{\mathbb{D}}$, then the space $D(\mu)$ consists of all functions f in the Hardy space $H^2(\mathbb{D})$ for which

$$\|f\|^2_{D(\mu)} := \|f\|^2_{H^2} + \int_{\overline{\mathbb{D}}}\int_{\mathbb{T}} \left|\frac{f(\zeta)-f(u)}{\zeta-u}\right|^2 dm_1(\zeta)\,d\mu(u) < +\infty.$$

If $d\mu = dm_1$, the space $D=D(dm_1)$ is the classical Dirichlet space with the norm

$$\|f\|_D^2 = \sum_{n\geq 0}(n+1)|\hat{f}(n)|^2.$$

Any Dirichlet type space $D(\mu)$ satisfies the assumptions of Theorem 3.2. The identity $K_{D(\mu)}(\cdot,0)=\mathbf{1}$ follows immediately from the definition of the norm in $D(\mu)$. The concavity inequality (ii) is a consequence of the identity

$$\|Sf\|^2_{D(\mu)} = \|f\|^2_{D(\mu)} + \int_{\overline{\mathbb{D}}} |f(u)|^2\,d\mu(u),$$

which can be proved directly (see [1, Chapter 4]). Therefore, for any space $D(\mu)$ we have $\operatorname{Re} K_{D(\mu)}(z,\lambda)\geq \frac{1}{2}$. This property of the reproducing kernels is closely related to the contractive multiplier property of the $D(\mu)$-inner functions; see [6].

Finally, we discuss the case where $n=2$ in the context of Theorem 2.6B. We see that the negative definiteness of the matrix $(I-\sigma^*)^2\hat{K}_X$ (the concavity type property) is equivalent to the inequality $\frac{1}{2}\|P_{E_1}(Sf_0+Sf_1)\|_X^2 \leq \|f_0\|_X^2 + \|P_{E_2}S^2f_1\|_X^2$ for any $f_0,f_1\in X$. To obtain a simple condition ensuring this inequality, it is convenient to impose a stronger requirement on K_X, namely, $(\sigma-2I+\sigma^*)K_X \ll 0$. (The identity $\sigma^*(\sigma-2I+\sigma^*)=(I-\sigma^*)^2$ shows that this requirement implies the concavity property). By Proposition 2.3, this requirement is equivalent to the inequality $\frac{1}{2}\|Sf_0+f_1\|_X^2\leq \|f_0\|_X^2+\|P_{E_1}Sf_1\|_X^2$. In the particular case where $K_X(\cdot,0)=\mathbf{1}$, it takes a simpler form

$$\frac{1}{2}\|Sf_0+f_1\|_X^2 \leq \|f_0\|_X^2 + \|Sf_1\|_X^2. \tag{4}$$

Theorem 3.3. *Suppose that $K_X(\cdot,0) = \mathbf{1}$ and that inequality (4) is fulfilled for any $f_0, f_1 \in X$. Then the matrix $\hat{K}_X$ possesses the concavity type property $(I-\sigma^*)^2\hat{K}_X \ll 0$. Moreover, the kernel K_X admits the representation*

$$K_X(z,\lambda) = \frac{1-\bar{\lambda}zL_X(z,\lambda)}{(1-\bar{\lambda}z)^2}, \tag{5}$$

where L_X is some positive definite function.

Proof. From the above discussion it follows that $(\sigma - 2I + \sigma^*)K_X \ll 0$, whence $(I-\sigma^*)^2\hat{K}_X \ll 0$. To prove (5), we observe that the condition $K_X(\cdot,0) = \mathbf{1}$ implies that

$$(\sigma^* K_X)(z,\lambda) = \frac{K_X(z,\lambda)-1}{\bar{\lambda}z}.$$

Hence, (5) is equivalent to $L_X = (-\sigma + 2I - \sigma^*)K_X$, and we see that L_X is positive definite. □

As examples of Hilbert spaces for which the operator S satisfies (4), we consider the weighted Bergman spaces with logarithmically subharmonic weights. If w is a weight function in the unit disk $\mathbb{D}$, then the weighted Bergman space $L^2_a(\mathbb{D},w)$ is the space of all functions analytic in $\mathbb{D}$ and square area integrable with the weight w; this space is endowed with the natural norm

$$\|f\|^2_w = \int_{\mathbb{D}} |f(z)|^2\, w(z)\, dm_2(z),$$

where dm_2 stands for the normalized area measure.

In [3] it was proved that inequality (4) is fulfilled in the weighted Bergman space $L^2_a(\mathbb{D},w)$ provided that the weight function w is logarithmically subharmonic in $\mathbb{D}$. In the particular case where w is of the form $w(z) = \sum_l |g_l(z)|^2$ where the g_l are analytic functions, this inequality follows from the identity

$$2\|F_0\|^2 + 2\|SF_1\|^2 - \|SF_0 + F_1\|^2 = \sum_{n\geq 1} \frac{n}{(n+1)(n+2)} \left| \frac{n+2}{n}\hat{F}_0(n-1) - \hat{F}_1(n)\right|^2,$$

which can be verified by direct calculations involving the Taylor coefficients. Here $\|\cdot\|$ denotes the norm in the unweighted Bergman space $L^2_a(\mathbb{D},\mathbf{1})$.

For the space $L^2_a(\mathbb{D},w)$, the condition $K_X(\cdot,0) = \mathbf{1}$ means that the weight function w satisfies the reproducing property

$$\int_{\mathbb{D}} h(z)w(z)\, dm_2(z) = h(0) \tag{6}$$

for any harmonic function h that is the real part of a function in $L^2_a(\mathbb{D},w)$, in particular, for any harmonic polynomial. The weighted Bergman spaces with logarithmically subharmonic weights satisfying the reproducing property (6) provide an important example of spaces X for which the reproducing kernel K_X admits representation in the form (5). In [3], this representation was used for proving the inequality

$$2\,\mathrm{Re}\,K_X(z,\lambda) - 1 \le -\frac{1}{|z-\lambda|^2}\left(\frac{1}{w(z)} + \frac{1}{w(\lambda)}\right), \quad z, \lambda \in \mathbb{T},\ z \neq \lambda;$$

this was an essential point in the proof of the positivity of the Green function for the weighted biharmonic operator $\Delta w^{-1}\Delta$.

References

[1] A. Aleman, *The multiplication operator on Hilbert spaces of analytic functions,* Habilitationsschrift,(1993), Hagen.

[2] N. Aronszajn, *Theory of reproducing kernels,* Trans.A.M.S., **68** (1950), 337–404.

[3] H. Hedenmalm, S. Jakobsson, and S. Shimorin, *An Hadamard maximum principle for biharmonic operators with applications to the Bergman spaces,* Preprint, (1998).

[4] S. Richter, *A representation theorem for cyclic analytic two-isometries,* Trans. A.M.S., **328** (1991), 325–349.

[5] S. Saitoh, *Theory of reproducing kernels and its applications,* Pitman Research Notes in Mathematics, **189** 1988, Longman Scientific & Technical, Harlow.

[6] S. Shimorin, *Reproducing kernels and extremal functions in Dirichlet type spaces,* Zapiski Nauch. Sem. POMI, **225** (1998), 198–220 (Russian).

Centre for Mathematical Sciences,
Lund University,
P.O.Box 118
S-221 00 Lund, Sweden
E-mail address: shimorin@maths.lth.se

Operator Theory:
Advances and Applications, Vol. 113

Outer Functions in Yet Another Class of Analytic Functions

N. A. Shirokov

Dedicated to the memory of S. A. Vinogradov

Abstract. Let ω be a positive continuous function on $(0, \pi]$ satisfying the condition

$$c_1 \left(\frac{x}{y}\right)^\alpha \leq \frac{\omega(y)}{\omega(x)} \leq c_2 \left(\frac{y}{x}\right)^\alpha, \quad 0 < x < y \leq \pi,$$

with some $0 < \alpha < 1$. For a natural number n, let $\Lambda^{n-1} Z_\omega$ denote the class of functions f analytic in the unit disc $\mathbf{D}$ and continuous in $\overline{\mathbf{D}}$ such that the derivatives $f', \dots, f^{(n-1)}$ possess the same property and, moreover,

$$\left| f^{(n-1)}(e^{i(\theta+h)}) - 2f^{(n-1)}(e^{i\theta}) + f^{(n-1)}(e^{i(\theta-h)}) \right| \leq c_f h\, \omega(h),$$
$$0 < h \leq \pi, \quad 0 \leq \theta < \pi.$$

The outer functions (in the sense of the Nevanlinna inner-outer factorization) belonging to $\Lambda^{n-1} Z_\omega$ are completely described.

1. Introduction

Let C_A be the disc-algebra, i.e., the algebra of all functions analytic in the unit disc $\mathbf{D}$ and continuous in its closure $\overline{\mathbf{D}}$, and let $X \subset C_A$ be a subspace of C_A endowed with some Banach norm. Each function $f \in X$ can be factorized in the sense of Nevanlinna in the product $f = I \cdot F$ of the so-called inner factor I and outer function F (the definitions are well known). It has turned out that the character of the norm in X is intimately related to the properties of outer functions belonging to X and that the outer factor in the Nevanlinna factorization of a function $f \in X$ is strongly influenced by the inner factor (see [1]–[3]; an alternative approach to this topic may be found in [4]–[6]). It is interesting that, in fact, certain versions of one and the same condition arise in [1]–[3]; they change in details only, in accordance with the choice of X. In the present paper we consider yet another class X for which the description of outer functions found in [1]–[3] persists.

2. Notation and statement

Let ω be a positive function continuous on $(0,\infty)$ such that

$$c_1\left(\frac{x}{y}\right)^\alpha \le \frac{\omega(y)}{\omega(x)} \le c_2\left(\frac{y}{x}\right)^\alpha, \quad 0<x\le y\le\infty; \quad \omega(x)=\omega(\pi) \ \text{ for } x>\pi; \quad (2.1)$$

here $0\le\alpha.<1$ and c_1, c_2 are some constants independent of x, y. Next, Z_ω denotes the class of all complex-valued functions on the unit circle $\mathbf{T}$ such that

$$|\Delta^2 f(e^{i\theta},h)|\le c_f h\omega(h), \qquad 0<h\le\pi,$$

where $\Delta^2 f(e^{i\theta},h)=f(e^{i(\theta+h)})-2f(e^{i\theta})+f(e^{i(\theta-h)})$; if $\omega(h)\equiv 1$, we get the usual Zygmund class. Further, we put

$$C^o Z_\omega = Z_\omega, \qquad C^{n-1}Z_\omega=\{f: D^{n-1}f\in Z_\omega\}, \qquad n\ge 2,$$

where $D=-ie^{-i\theta}\frac{d}{d\theta}$. If C^s denotes the usual Hölder class of order s on $\mathbf{T}$, then the class $C^{n-1}Z_\omega$ may "fluctuate" between $C^{n-\alpha}$ and $C^{n+\alpha}$, "passing" through the "level" $C^{n-1}Z=\{f: D^{n-1}f\in Z\}$. Finally, $\Lambda^{n-1}Z_\omega\subset C^{n-1}Z_\omega$ denotes the class of all functions analytic in $\mathbf{D}$ with boundary values (which are at least $C^{1-\alpha}$-smooth) belonging to $C^{n-1}Z_\omega$.

For $z\in\mathbf{D}$ we denote $\gamma(z)=\{\zeta\in\mathbf{T}; |\zeta-z|\le 2(1-|z|)\}$, and for a function $f\in C(\mathbf{T})$ we put

$$M_f(z)=\max_{\zeta\in\gamma(z)}|f(z)|.$$

The length of an arc $I\subset\mathbf{T}$ is denoted by $|I|$. The relation $a\asymp b$ $(a,b>0)$ means that $c_1 a\le b\le c_2 a$ with some constants c_1, c_2 independent on the (important) parameters involved. Constants arising in estimates will be enumerated independently within any particular proof, but in important cases we will recall the origin of some constants in the notation. Now, we are able now to formulate the main results of the paper.

Theorem 2.1. *If $f\in\Lambda^{n-1}Z_\omega$ $(n\ge 1)$, f is not identically zero, $|\Delta^2 f(e^{i\theta},h)|\le h\omega(h)$, then there exists a constant $c=c(f,n,\omega)$ such that for any $z\in\mathbf{D}$ satisfying*

$$M_f(z)\ge(1-|z|)^n\omega(1-|z|) \qquad (2.2)$$

we have

$$\int_{\mathbf{T}}\left|\log\left|\frac{f(\zeta)}{M_f(z)}\right|\right|\frac{1-|z|^2}{|\zeta-z|^2}|d\zeta|\le c. \qquad (2.3)$$

Theorem 2.2. *Suppose f is a complex-valued function defined on $\mathbf{T}$, f is not identically zero, and $|\Delta^2 D^{n-1}f(e^{i\theta},h)|\le h\omega(h)$, $\int_{\mathbf{T}}\log|f(\zeta)||d\zeta|>-\infty$. Denote by F an outer function in $\mathbf{D}$ introduced by the condition $|F||_{\mathbf{T}}=|f||_{\mathbf{T}}$. Suppose that there exists a constant c independent of $z\in\mathbf{D}$ such that (2.3) is fulfilled for any $z\in\mathbf{D}$ satisfying (2.2). Then $F\in\Lambda^{n-1}Z_\omega$.*

Remark 2.3. *We emphasize that, actually, Theorems 1 and 2 give a complete description of the outer functions in the space $\Lambda^{n-1}Z_\omega$. In other subspaces of C_A, such as the Hölder classes, the classes $H^p_n=\{f: f^{(n)}\in H^p\}$, or the analytic Besov* *classes, the descriptions of outer functions given in* [1]–[3] *also involves* (2.3), *which must be true for* z *satisfying a condition (similar to* (2.2)) *that depends upon the class under consideration.*

3. Some technical facts about $C^{n-1}Z_\omega$ and $\Lambda^{n-1}Z_\omega$

Here we list some properties of functions belonging to $C^{n-1}Z_\omega$ or $\Lambda^{n-1}Z_\omega$, under the assumption that ω satisfies (2.1). The proof of these statements are quite similar to the proofs in the case of $C^{n-1}Z$ (see [1], Ch. 2 §3), so, we omit them.

Lemma 3.1. *Suppose* $f \in C^{n-1}Z_\omega$, $n \geq 2$, *and* $|\Delta^2 f^{n-1}(e^{i\theta}, h)| \leq dh\omega(h)$, $0 < h \leq \pi$. *If* $f(\zeta_0) = 0$, $\zeta_0 \in \mathbf{T}$, *we put* $F(\zeta) = f(\zeta)/(\zeta - \zeta_0)$. *Then* $F \in C^{n-2}Z_\omega$ *and*

$$|\Delta^2 F^{n-1}(e^{i\theta}, h)| \leq cdh\omega(h), \qquad 0 < h \leq \pi.$$

For the future use we need the following function $\Omega(t,v)$, $0 < t < v < \infty$:

$$\Omega(t,v) = \int_t^v \frac{\omega(s)}{s} ds. \tag{3.0}$$

For $v \geq 2t$, we have the following elementary estimates:

$$\Omega(t,v) \geq c_1\omega(t), \quad \Omega(t,v) \geq c_2\omega(v), \quad \frac{\Omega(t,v)}{\omega(v)} \leq c_3 \left(\frac{v}{t}\right)^\alpha, \quad \frac{\Omega(t,v)}{\omega(t)} \leq c_3 \left(\frac{v}{t}\right)^\alpha; \tag{3.1}$$

next, for $ct \geq v \geq 2t$ we have $\Omega(t,v) \asymp \omega(v)$.

Lemma 3.2.

a) *Suppose* $\zeta_0, \ldots, \zeta_n \in \mathbf{T}$, $|\zeta_j - \zeta_{j+1}| \asymp |\zeta_k - \zeta_{k+1}|$, $0 \leq j,k \leq n-1$, $E = \cup_{\nu=0}^n \{\zeta_\nu\}$, ζ_ν *lies on the smaller arc with ends at* ζ_0 *and* ζ_n. *If* $f \in C^{n-1}Z_\omega$, $|\Delta^2 f^{n-1}(e^{i\theta}, h)| \leq \lambda h\omega(h)$, $0 < h \leq \pi$, *and* $f(\zeta_\nu) = 0$, $\nu = 0, \ldots, n$, *then we have the following estimates for the derivatives of* f *of order up to* $n-1$:

$$|f^{(\nu)}(\zeta)| \leq c_1\lambda \operatorname{dist}(\zeta, E)|\zeta_n - \zeta_0|^{n-\nu-1}\Omega(\operatorname{dist}(\zeta,E), 2(|\zeta_n - \zeta_0| + \operatorname{dist}(\zeta,E)) \tag{3.2}$$

provided that $\operatorname{dist}(\zeta,E)) \leq |\zeta_n - \zeta_0|$, $\nu = 0, \ldots, n-1$;

$$|f^{(\nu)}(\zeta)| \leq c_1\lambda \operatorname{dist}(\zeta,E)\Omega(|\zeta_n - \zeta_0|, 2(|\zeta_n - \zeta_0| + \operatorname{dist}(\zeta,E)) \tag{3.3}$$

provided that $\operatorname{dist}(\zeta,E)) > |\zeta_n - \zeta_0|$, $\nu = 0, \ldots, n-1$.

b) *Let* $\zeta_0, \ldots, \zeta_n$ *be a collection of (not necessarily distinct) points of* $\mathbf{T}$ *that lie on the smaller arc joining* ζ_0 *and* ζ_n, *let* $E = \cup_{\nu=0}^n \{\zeta_\nu\}$, *and let a function* f *be as in part* a). *Then for* ζ *lying outside the arc* (ζ_0, ζ_n) *we have*

$$|f(\zeta)| \leq c_3\lambda \left(\operatorname{dist}(\zeta,E)) + |\zeta_n - \zeta_0|\right)^n \Omega(|\zeta_n - \zeta_0|, 2(|\zeta_n - \zeta_0| + \operatorname{dist}(\zeta,E)), \tag{3.4}$$

and for ζ *lying inside this arc we have*

$$|f(\zeta)| \leq c_3\lambda|\zeta_n - \zeta_0|^n\omega(|\zeta_n - \zeta_0|). \tag{3.5}$$

Lemma 3.3.

a) *Suppose that* $I_0 \subset I \subset \mathbf{T}$ *are arcs,* $|I| \le A|I_0|$, $f \in C^{n-1}Z_\omega$, $|\Delta^2 f^{n-1}(e^{i\theta}, h)| \le \lambda h\omega(h)$, *and* $\max_{\zeta\in I_0}|f(\zeta)| \ge \alpha|I_0|^n\omega(|I_0|)$. *Then*

$$\max_{\zeta\in I_0}|f(\zeta)| \le \max_{\zeta\in I}|f(\zeta)| \le c_1 \max_{\zeta\in I_0}|f(\zeta)|, \qquad c_1 = c_1(n, A, \alpha, \lambda).$$

b) *There exists* $\sigma_0 = \sigma_0(c_2, \lambda, n, \omega)$ *such that if* $0 < \sigma \le \sigma_0$ *and* $|I_0|/|I| = \sigma/2$, $\max_{\zeta\in I}|f(\zeta)| \ge c_2\frac{1}{\sigma}|I_0|^n\Omega(|I_0|, |I|)$, *then*

$$\max_{\zeta\in I_0}|f(\zeta)| \ge |I_0|^n\omega(|I_0|).$$

Lemma 3.4. *If* $f \in \Lambda^{n-1}Z_\omega$, $n \ge 1$, $|\Delta^2 f(e^{i\theta}, h)| \le \lambda h\omega(h)$, $\zeta_0 \in \mathbf{T}$, $\zeta_\nu = (1-\nu h)\zeta_0$, $\nu = 1, \ldots, n+1$, *then*

$$\left|\sum_{\nu=0}^{n+1}(-1)^\nu C_{n+1}^\nu f(\zeta_\nu)\right| \le c_1\ \lambda h^n\omega(h), \qquad c_1 = c_1(n, \lambda, \omega).$$

4. Proof of Theorem 2.1

We begin with a simple but important observation concerning the integral in (2.3).

Lemma 4.1. *Suppose* $f \in C^{n-1}Z_\omega$, $|\Delta^2 f^{(n-1)}(e^{i\theta}, h)| \le h\omega(h)$, $\int_{\mathbf{T}} \log|f(\zeta)||d\zeta| > -\infty$. *If* (2.3) *is true whenever* $z \in \mathbf{D}$ *and* $M_f(z) \ge A(1-|z|)^n\omega(1-|z|)$ *with some fixed* $A > 1$, *then* (2.3) *is also true (with a different constant* c*) if* $z \in \mathbf{D}$ *satisfies condition* (2.2).

Proof. We take any $z \in \mathbf{D}$ satisfying

$$M_f(z) \ge (1-|z|)^n\omega(1-|z|)$$

and choose $z_0 \in \gamma(z)$ such that $|f(z_0)| = M_f(z)$, $\sigma = A^{-\frac{1}{n-\alpha}}$, $z_1 = 1-\sigma(1-|z|)z_0$. Lemma 3.3 implies $M_f(z_1) \asymp M_f(z)$; using elementary estimates, we get

$$\begin{aligned}\int_{\mathbf{T}}\left|\log\left|\frac{f(\zeta)}{M_f(z)}\right|\right|\frac{1-|z|^2}{|\zeta-z|^2}|d\zeta| &\le c_1\int_{\mathbf{T}}\left|\log\left|\frac{f(\zeta)}{M_f(z)}\right|\right|\frac{1-|z_1|^2}{|\zeta-z_1|^2}|d\zeta| \\ &\le c_1\int_{\mathbf{T}}\left|\log\left|\frac{f(\zeta)}{M_f(z_1)}\right|\right|\frac{1-|z_1|^2}{|\zeta-z_1|^2}|d\zeta| + c_2. \qquad (4.1)\end{aligned}$$

But by (2.1) we have

$$\begin{aligned}M_f(z_1) \ge M_f(z) \ge (1-|z|)^n\omega(1-|z|) &= \frac{(1-|z_1|)^n}{\sigma^n}\omega\left(\frac{1-|z_1|}{\sigma}\right) \\ &\ge \frac{(1-|z_1|)^n}{\sigma^n}\sigma^\alpha\omega(1-|z_1|) = A(1-|z_1|)\omega(1-|z_1|);\end{aligned}$$

so, the assumption of the lemma combined with (4.1) implies

$$c_1\int_{\mathbf{T}}\left|\log\left|\frac{f(\zeta)}{M_f(z_1)}\right|\right|\frac{1-|z_1|^2}{|\zeta-z_1|^2}|d\zeta| + c_2 \le c_3. \qquad \square$$

Remark 4.2. *The proof of Lemma* 4.1 *also shows that if* (2.2) *implies* (2.3), *then the condition* $M_f(z) \geq \beta(1-|z|)^n\omega(1-|z|)$ (*where* $0<\beta<1$) *implies* (2.3) *with a different constant* $\tilde{c} = \tilde{c}(\beta)$.

Next, we need the following elementary assertions.

Lemma 4.3.
a) *If* P *is a nonzero polynomial, then*

$$\int_{\mathbf{T}} \left|\log\left|\frac{P(\zeta)}{M_P(z)}\right|\right| \frac{1-|z|^2}{|\zeta-z|^2}|d\zeta| \leq c_1, \qquad c_1 = c_1(\deg P).$$

b) *If* $1>|z|\geq 1/2,\ z_0 = z/|z|,\ a,\ b,\ s>0$, *then*

$$\frac{1}{2\pi}\int_{\mathbf{T}} \log(a+b|\zeta-z_0|^s)\frac{1-|z|^2}{|\zeta-z|^2}|d\zeta| \leq \log(c_2(a+b|z-z_0|^s)),$$

where $c_2 = c_2(s)$ *is independent of* a, b, *and* z.

For a point $z\in\mathbf{D}$, we introduce the region

$$U(z) = \{\zeta\in\mathbf{D} : \frac{1-|z|}{2} \leq 1-|\zeta| \leq (n+1)(1-|z|), |\arg\frac{\zeta}{z}| \leq \pi(1-|z|)\}.$$

Now, we start the proof of Theorem 2.1.

Proof. Take $A = 2\pi(c_1(\text{Lemma 3.2}) + c_2(\text{Lemma 3.2}) + c_1(\text{Lemma 3.4}))$ and suppose that $z\in\mathbf{D}$ and $M_f(z) > A(1-|z|)^n\omega(1-|z|)$. We choose $n+1$ equidistributed points ζ_ν, $0\leq\nu\leq n$, that lie on the arc $\gamma(z)$ joining ζ_0 and ζ_n. The polynomial P is the interpolating polynomial of degree not exceeding n such that $P(\zeta_\nu) = f(z_\nu)$, $0\leq\nu\leq n$. Lemma 3.2 and the choice of A yield

$$|P(\zeta)| \leq |f(\zeta)| + |f(\zeta)-P(\zeta)| \leq$$
$$M_f(z) + c_1(\text{Lemma 3.2})|\zeta_0-\zeta_n|^n\omega(|\zeta_0-\zeta_n|) \leq 2M_f(z), \qquad \zeta\in\gamma(z).$$

Hence, putting $\rho = 1-|z|$ we get

$$|P(\zeta)| \leq 2M_f(z)c_1\left(\frac{|\zeta-z|+\rho}{\rho}\right)^n, \qquad \zeta\in\mathbf{T}\setminus\gamma(z), \qquad c_1 = c_1(n). \tag{4.2}$$

Another application of Lemma 3.2, together with (3.1), gives

$$\begin{aligned}|f(\zeta)-P(\zeta)| &\leq c_2(\text{Lemma 3.2})\,(|\zeta-z|+|\zeta_n-\zeta_0|)^n\,\Omega(|\zeta_n-\zeta_0|, 2(|\zeta-z|+|\zeta_n-\zeta_0|))\\ &\leq c_2(\text{Lemma 3.2})\,(|\zeta-z|+|\zeta_n-\zeta|)^n\,\omega(|\zeta_n-\zeta_0|)c_4(3.1)\\ &\times\left(\frac{2(|\zeta-z|+|\zeta_n-\zeta_0|)}{|\zeta_n-\zeta_0|}\right)^\alpha \leq c_3\left(\frac{|\zeta-z|+\rho}{\rho}\right)^{n+\alpha} M_f(z).\end{aligned} \tag{4.3}$$

Now, (4.2) and (4.3) imply

$$|f(\zeta)| \leq |P(\zeta)| + |f(\zeta)-P(\zeta)| \leq c_4\left(\frac{|\zeta-z|+\rho}{\rho}\right)^{n+\alpha} M_f(z), \qquad \zeta\in\mathbf{T}. \tag{4.4}$$

If we employ the usual notation $\log^+$, $\log^-$ ($\log = \log^+ - \log^-$), then from (4.4) and Lemma 4.3 we deduce

$$\frac{1}{2\pi}\int_{\mathbf{T}} \log^+ \left|\frac{f(\zeta)}{M_f(z)}\right| \frac{1-|z_*|^2}{|\zeta - z_*|^2}|d\zeta| \le \frac{1}{2\pi}\int_{\mathbf{T}} \log^+ \frac{|f(\zeta)| + M_f(z)}{M_f(z)} \frac{1-|z_*|^2}{|\zeta - z_*|^2}|d\zeta|$$

$$= \frac{1}{2\pi}\int_{\mathbf{T}} \log \frac{|f(\zeta)| + M_f(z)}{M_f(z)} \frac{1-|z_*|^2}{|\zeta - z_*|^2}|d\zeta|$$

$$\le \frac{1}{2\pi}\int_{\mathbf{T}} \log\left(c_4\left(\frac{|\zeta - z| + \rho}{\rho}\right)^{n+\alpha} + 1\right) \frac{1-|z_*|^2}{|\zeta - z_*|^2}|d\zeta| \le \log c_5, \qquad z_* \in U(z). \tag{4.5}$$

We put

$$B = \frac{1}{2\pi}\int_{\mathbf{T}} \log^- \left|\frac{f(\zeta)}{M_f(z)}\right| \frac{1-|z_*|^2}{|\zeta - z_*|^2}|d\zeta|.$$

Since

$$\frac{1-|z_*|^2}{|\zeta - z_*|^2} \ge c_6 \frac{1-|z|^2}{|\zeta - z|^2}, \qquad \zeta \in \mathbf{T}, \qquad z_* \in U(z), \qquad c_6 = c_6(n),$$

from (4.5) we obtain

$$|f(z_*)| \le \exp\left(\frac{1}{2\pi}\int_{\mathbf{T}} \log|f(\zeta)| \frac{1-|z_*|^2}{|\zeta - z_*|^2}|d\zeta|\right)$$

$$= M_f(z)\exp\left(\frac{1}{2\pi}\int_{\mathbf{T}} \log\left|\frac{f(\zeta)}{M_f(z)}\right| \frac{1-|z_*|^2}{|\zeta - z_*|^2}|d\zeta|\right)$$

$$= M_f(z)\exp\left(\frac{1}{2\pi}\int_{\mathbf{T}} \log^+\left|\frac{f(\zeta)}{M_f(z)}\right| \frac{1-|z_*|^2}{|\zeta - z_*|^2}|d\zeta|\right)$$

$$\times \exp\left(-\frac{1}{2\pi}\int_{\mathbf{T}} \log^-\left|\frac{f(\zeta)}{M_f(z)}\right| \frac{1-|z_*|^2}{|\zeta - z_*|^2}|d\zeta|\right) \le M_f(z) c_5((4.5)) e^{-c_6 B}. \tag{4.6}$$

Choose now $z_0 \in \gamma(z)$ such that $|f(z_0)| = M_f(z)$ and put $z_\nu = (1 - \nu\rho)z_0$, $\nu = 1, \ldots, n+1$, $\rho = 1 - |z_0|$. Then Lemma 3.4 yields

$$\left|\sum_{\nu=0}^{n+1} (-1)^\nu C_{n+1}^\nu f(\zeta_\nu)\right| \le c_1(\text{Lemma 3.4})\rho^n \omega(\rho), \qquad c_1 = c_1(n, \lambda, \omega). \tag{4.7}$$

Finally, (4.6), (4.7), and the relations $z_\nu \in U(z)$, $1 \le \nu \le n+1$, imply

$$M_f(z) = |f(z_0)| \le \left|\sum_{\nu=0}^{n+1} (-1)^\nu C_{n+1}^\nu f(\zeta_\nu)\right| + \sum_{\nu=1}^{n+1} C_{n+1}^\nu |f(\zeta_\nu)|$$

$$\le c_1(\text{Lemma 3.4})\rho^n\omega(\rho) + \sum_{\nu=1}^{n+1} C_{n+1}^\nu M_f(z) c_5((4.5)) e^{-c_6 B}$$

$$\le c_1(\text{Lemma 3.4})\rho^n\omega(\rho) + 2^{n+1} M_f(z) c_5((4.5)) e^{-c_6 B}. \tag{4.8}$$

Supposing that

$$2^{n+1} c_5((4.5)) e^{-c_6 B} < \frac{1}{2},$$

from (4.8) we obtain

$$M_f(z) < c_1(\text{Lemma 3.4})\rho^n\omega(\rho) + \frac{1}{2}M_f(z), \quad M_f(z) < 2c_1(\text{Lemma 3.4})\rho^n\omega(\rho),$$

which contradicts the choice of A. Hence, we have

$$c_5((4.5))e^{-c_6 B} \geq \frac{1}{2}$$

which means that $B \leq c_7$; recalling (4.5), we get

$$\int_{\mathbf{T}} \log\left|\frac{f(\zeta)}{M_f(z)}\right| \frac{1-|z_*|^2}{|\zeta - z_*|^2}|d\zeta| = \int_{\mathbf{T}} \log^+\left|\frac{f(\zeta)}{M_f(z)}\right| \frac{1-|z_*|^2}{|\zeta - z_*|^2}|d\zeta|$$
$$+ \int_{\mathbf{T}} \log^-\left|\frac{f(\zeta)}{M_f(z)}\right| \frac{1-|z_*|^2}{|\zeta - z_*|^2}|d\zeta| \leq c_8.$$

Applying Lemma 4.1, we finish the proof of Theorem 2.1. □

5. The first crucial step in the proof of Theorem 2.2

In what follows we denote by f a function satisfying the conditions of Theorem 2. The symbol Ω will denote the function defined in (3.0). Next, z will stand for a point of $\mathbf{D}$ satisfying $|z| > 0.9$; we put $\rho = 1 - |z|$; further, z_ν^* $(0 \leq \nu \leq n)$ will be some points equidistributed on the arc $\gamma(z)$ joining z_0^* and z_n^*. By P we denote an interpolating polynomial, $\deg P \leq n$, such that $P(\zeta_\nu^*) = f(\zeta_\nu^*)$, $0 \leq \nu \leq n$. In Sections 5–8 we will use the following technical notation:

$$q_n' = 1 - \frac{1}{4(n+1)^2}, \qquad q_n'' = 1 + \frac{1}{4(n+1)^2};$$

$D(z,r)$ is the disc $\{\zeta \in \mathbf{C} : |\zeta - z| \leq r\}$; $D(z,r_1,r_2) = \{\zeta \in \mathbf{C} : r_1 \leq |\zeta - z| \leq r_2\}$;

$$\sigma_P(z,V) = \frac{V^n\Omega(\rho, 2V)}{\max_{\zeta\in D(z,V)}|P(\zeta)|}. \tag{5.0}$$

Lemma 5.1. *Retaining the above notation, we take $v \geq \rho$, $4/3 \leq V/v \leq 5/2$. Put $L_0 = D(z, q_n''v, q_n'V)$, $L = D(z,v,V)$ and suppose that $P(z)$ has no zeros in $L \setminus L_0$ and has $m \geq 1$ zeros $\lambda_1, \ldots, \lambda_m$ in L_0. Define*

$$r(\lambda) = \prod_{\nu=1}^{m}(\lambda - \lambda_\nu), \tag{5.1}$$

and let

$$Q(\lambda) \text{ be an outer function in } \mathbf{D} \text{ such that } |Q||_{\mathbf{T}} = |r||_{\mathbf{T}}. \tag{5.1a}$$

Next, put $I = \mathbf{T} \cap L$, and, finally,

$$\phi(\lambda) = Q(\lambda)\exp\left(\frac{1}{2\pi}\int_I \log\left|\frac{f(\zeta)}{P(\zeta)}\right| \frac{\zeta+\lambda}{\zeta-\lambda}|d\zeta|\right). \tag{5.2}$$

Then there exists $\sigma_1 = \sigma_1(n,\omega) > 0$ such that if

$$\sigma = \sigma_P(z,V) \leq \sigma_1, \tag{5.3}$$

then

$$\left|\phi^{(\nu)}(\lambda)|_{\lambda=z}\right| \le c_1 V^{m-\nu}, \nu \ge 0, \quad c_1 = c_1(\nu, n, f, \omega), \tag{5.4}$$

$$\left|\phi^{(\nu)}(\lambda)|_{\lambda=z}\right| \le c_2 \sigma V^{m-\nu}, \nu \ge m+1, \quad c_2 = c_2(\nu, n, f, \omega). \tag{5.5}$$

Remark 5.2. *The most difficult case of Lemma* 5.1 *is that of small* V; *so, we shall present the arguments under the assumption* $V < 0.1$. *The case of* $V \ge 0.1$ *is similar but much simpler; if* $I = \varnothing$, *then* $\phi(\lambda) = Q(\lambda)$, $Q(\lambda)$ *is a polynomial,* $\deg Q = m$, *and it follows that* $\phi^{(\nu)} = Q^{(\nu)} \equiv 0$, $\nu \ge m+1$; *thus, obviously,* (5.5) *is valid also in this case.*

A slightly different form of the previous assertion is valid for a disc instead of an annulus.

Lemma 5.3. *Let* $2\rho \le V \le 3\rho$, $L_0 = D(z, q_n' V)$, $L = D(z, V)$, $I = \mathbf{T} \cap L$. *Suppose that* P *has no zeros in* $L \setminus L_0$, *and let* $\lambda_1, \dots, \lambda_m$ $(m \ge 1)$ *be all zeros of* P *in* L_0. *If* r, Q, ϕ *are defined us in* (5.1), (4.1a), (5.2), *then there exists* $\sigma_1 = \sigma_1(n, \omega) > 0$ *such that* (4.3) *implies* (5.4) *and* (5.5).

The proofs of Lemmas 5.1 and 5.3 are similar; so, we give the details only for Lemma 5.1.

Proof. First, we verify (5.4)

Applying Lemma 3.2, we find $\sigma_3 = \sigma_3(n, \omega)$ such that the condition $\sigma = \sigma_P(z, V) \le \sigma_3$ implies

$$\max_{D(z,V)} |P| \asymp \max_{D(z,V)\cap\mathbf{T}} |P| \asymp \max_{D(z,V)\cap\mathbf{T}} |f| \ge (2V)^n \omega(2V). \tag{5.6}$$

Hence, for $\tilde{\zeta} = (1-2V)z/|z|$ we get the inequality $M_f(\tilde{\zeta}) \ge (1-|\tilde{\zeta}|)^n \omega(1-|\tilde{\zeta}|)$, which, combined with (2.3), (5.6), and Lemma 4.3, yields

$$\int_{\mathbf{T}} \left|\log\left|\frac{f(t)}{P(t)}\right|\right| \frac{1-|\tilde{\zeta}|^2}{|t-\tilde{\zeta}|^2}|dt| \le \int_{\mathbf{T}} \left|\log\left|\frac{f(t)}{M_f(\tilde{\zeta})}\right|\right| \frac{1-|\tilde{\zeta}|^2}{|t-\tilde{\zeta}|^2}|dt|$$

$$+ \int_{\mathbf{T}} \left|\log\left|\frac{P(t)}{M_P(\tilde{\zeta})}\right|\right| 1-|\tilde{\zeta}|^2|1-\tilde{\zeta}||dt| + \left|\log\frac{M_P(\tilde{\zeta})}{M_f(\tilde{\zeta})}\right| \le c_3 + c_4 + c_5 = c_6. \tag{5.7}$$

We observe that for $t \in I$ we have $|t-\tilde{\zeta}| \asymp |t-z| \asymp V$; thus, (5.7) implies

$$\int_I \left|\log\left|\frac{f(t)}{P(t)}\right|\right| \frac{1-|\tilde{\zeta}|^2}{|t-\tilde{\zeta}|^2}|dt| \le \int_{\mathbf{T}} \left|\log\left|\frac{f(t)}{P(t)}\right|\right| \frac{1-|\tilde{\zeta}|^2}{|t-\tilde{\zeta}|^2}|dt| \le c_6,$$

$$\int_I \left|\log\left|\frac{f(t)}{P(t)}\right|\right| |dt| \le c_7 V,$$

$$\int_I \left|\log\left|\frac{f(t)}{P(t)}\right|\right| \frac{|dt|}{|z-t|^\nu} \le c_8 V^{1-\nu}, \quad \nu \ge 0, \quad c_8 = c_8(n, \nu, \omega). \tag{5.8}$$

Differentiating ϕ and using (5.8), we easily obtain (5.4) q.e.d.

Now, we verify (5.5)

Case a): suppose that $m < n$. As before, we put $\tilde{\zeta} = (1-2V)z/|z|$; let $R(\lambda) = \frac{P(\lambda)}{r(\lambda)}$, $H = f - P$, $R_0 = |R(z)|$. Lemma 3.2 implies

$$|H^{(\nu)}(\lambda)| \le c_1 V^{n-\nu}\Omega(\rho, 2V), \quad \lambda \in \gamma(\tilde{\zeta}), \quad 0 \le \nu \le n-1. \tag{5.9}$$

Since $\max_{D(z,V)} |P| \asymp R_0 V^m$, we can rewrite (5.8) in the form

$$|H^{(\nu)}(\lambda)| \le c_1 \frac{V^n \Omega(\rho, 2V)}{\max_{D(z,V)} |P|} R_0 V^{m-n} = \sigma R_0 V^{m-n}, \quad 0 \le \nu \le n-1.$$

Since $R(\lambda) \neq 0$, $\lambda \in L$, we find

$$\left| \left(\frac{H}{R}\right)^{(\nu)} (\lambda) \right| \le c_3 \sigma V^{m-\nu}, \quad \lambda \in \gamma(\tilde{\zeta}), \quad 0 \le \nu \le n-1. \tag{5.10}$$

Put $h = H/R$, then $f/P = 1 + H/P = 1 + r/h$,

$$\phi(\lambda) = Q(\lambda) \exp\left(\frac{1}{2\pi}\int_I \log\left|1 + \frac{h(t)}{r(t)}\right| \frac{t+\lambda}{t-\lambda}|dt|\right)$$

and, thus, inequalities (5.10) for the functions $h^{(\nu)}$ provide possibilities to apply Lemma 1.9 of [1, Ch. 2, §1] to ϕ. That lemma shows that there exists $\sigma_3 > 0$, $\sigma_3 = \sigma_3(n) > 0$ such that if $\sigma \le \sigma_3$, then estimates (5.4) are true. We emphasize that the current assumption $m < n$ is decisive for the applicability of the lemma mentioned above.

Case b): suppose that $m = n$. This case is more difficult. We remind the reader that we assume that $V < 0.1$. We begin with some preliminary observations. Fix a point x_0^* $\partial D(z, V) \cap \mathbf{T}$ and choose arbitrary (perhaps coinciding) points x_ν^*, $1 \le \nu \le n$, lying on the arc $\mathbf{T} \cap D(z, q'_{n+1}V)$. Put $X^* = (x_1^*, \dots, x_n^*)$ and let P_{X^*} be the interpolating polynomial, $\deg P_{X^*} \le n$, such that $P_{X^*}(x_\nu^*) = f(x_\nu^*)$, $0 \le \nu \le n$. Then Lemma 2 gives the estimate

$$|f(\lambda) - P_{X^*}(\lambda)| \le c_4 V^n \Omega(\rho, 2V), \quad \lambda \in D(z, V) \cap \mathbf{T}, \tag{5.11}$$

with a constant c_4 independent of V, ρ, λ, and X^*. Using (5.9) (with $\nu = 0$) and (5.11), we get

$$|P(\lambda) - P_{X^*}(\lambda)| \le c_5 V^n \Omega(\rho, 2V), \quad \lambda \in D(z, V) \cap \mathbf{T}, \quad c_5 = c_1 + c_4. \tag{5.12}$$

Since $m = n$ and the annuli $D(z, q'_n V, V)$ and $D(z, v, q''_n v)$ do not contain any zero of P, we see that all n zeros of P lie in $D(z, q''_n v, q'_n V)$; together with (5.12), this implies

$$\left|\frac{P(\lambda) - P_{X^*}(\lambda)}{P(\lambda)}\right| \le c_6 \sigma, \quad \lambda \in D(z, q''_{n+1} v, q'_{n+1} V) \tag{5.13}$$

(we notice that in (5.13) we use q''_{n+1}, q'_{n+1} instead of q''_n, q'_n). Thus, (5.13) shows that if $\sigma \le \sigma_n$, $c_6 \sigma_4 = 1/2$, then, by the argument principle, the polynomial P_{X^*} has n zeros lying in the annulus $D(z, q''_{n+1} v, q'_{n+1} V)$ (precisely as for P). We are going to replace the polynomial P by P_{X^*} in (5.2) with a suitable choice of X^*.

Let $\zeta_1(X^*)$, $\dots$,$\zeta_n(X^*)$ be all zeros of P_{X^*}. We enumerate them by the following rule: take two circular arcs $\gamma_+(z, V)$ and $\gamma_-(z, V)$ that pass through the origin and are tangent the circle $\partial D(z, q'_{n+1}V)$ at the points $\partial D(z, q'_{n+1}V) \cap \mathbf{T}$; let $O(z,V)$ be another point of intersection of $\gamma_+(z,V)$ and $\gamma_-(z,V)$; next, for each $\zeta \in D(z, q'_{n+1}V)$ we introduce the circle $\Gamma(\zeta, z, V)$ passing through the origin, $O(z,V)$, and ζ; the latter circle intersects the arc $\partial D(z, q'_{n+1}V) \cap \mathbf{T}$ at a unique point, which we call $\xi(\zeta)$. Finally, applying this construction to the points $\zeta_\nu(X^*)$ in the role of ζ, we reenumerate them in such a way that

$$\arg \xi(\zeta_\nu(X^*)) \le \arg \xi(\zeta_{\nu+1}(X^*)), \qquad 1 \le \nu \le n-1. \tag{5.14}$$

We put $\xi(X^*) = (\xi(\zeta_1(X^*)), \dots, \xi(\zeta_n(X^*))$. Then the vector $\xi(X^*)$ depends continuously on X^*. We have already noticed that for $\sigma \le \sigma_4$ all n zeros of P_{X^*} lie in $D(z, q'_{n+1}V)$; taking (5.14) into account, we get continuous mappings

$$X^*(x_1^*, \dots, x_n^*) \to (\zeta_1(X^*), \dots, \zeta_n(X^*)) \to (\xi(\zeta_1(X^*)), \dots, \xi(\zeta_n(X^*))),$$

i.e., we have a continuous mapping of

$$\underbrace{\tau \times \cdots \times \tau}_{n \text{ times}} \text{ into } \underbrace{\tau \times \cdots \times \tau}_{n \text{ times}}, \quad \tau = \mathbf{T} \cap D(z, q'_{n+1}V).$$

The Brauer theorem guarantees the existence of a fixed point of this mapping, i.e., there exist points $x_1^+, \dots, x_n^+ \in \tau$ such that

$$\xi(\zeta_\nu(X^+)) = x_\nu^+, \ 1 \le \nu \le n, \ X^+ = (x_1^+, \dots, x_n^+). \tag{5.15}$$

We put $P_+ = P_{X^+}$ and denote by a, a_+ the senior coefficients of P and P_+, $r_+ = P_+/a_+$. Let Q_+ be an outer function in $\mathbf{D}$ such that $|Q_+||_{\mathbf{T}} = |r_+||\mathbf{T}$. Now we can rewrite the function $\phi(\lambda)$ in the following form:

$$\phi(\lambda) = \left[Q_+(\lambda) \exp \frac{1}{2\pi} \int_I \log \left|\frac{f(t)}{P_+(t)}\right| \frac{\lambda + t}{t - \lambda} |dt|\right]$$
$$\times \left[\frac{Q(\lambda)}{Q_+(\lambda)} \exp \frac{1}{2\pi} \int_I \log \left|\frac{P_+(t)}{P(t)}\right| \frac{\lambda + t}{t - \lambda} |dt|\right] \overset{\text{def}}{=} \phi_+(\lambda) q(\lambda). \tag{5.16}$$

Passing to the limit as $\lambda \to \infty$ in (5.13) and recalling that P has no zeros outside $D(z, q'_{n+1}V))$, we obtain

$$\left|\frac{a_+}{a} - 1\right| = \left|\frac{a_+ - a}{a}\right| \le c_6 \sigma. \tag{5.17}$$

Now, we take the constant $\tilde{\sigma}_5 \le 1/2$ occurring in Lemma 1.8 in [1, Ch. 2, §1]; then, by (5.17), for $c_6\sigma \le \tilde{\sigma}_5$ we can apply that lemma, which will lead to the following estimates valid for $\sigma \le \sigma_6$, $\sigma_6 = \tilde{\sigma}_5/c_6$:

$$|q(z)| \le c_7, \ c_7 = c_7(n), \tag{5.18}$$

$$|q^{(\nu)}(z)| \le c_8 \sigma V^{-n}, \ \nu \ge 1, \ c_8 = c_8(n, \nu). \tag{5.19}$$

Now, we analyze the function ϕ_+. First we modify the expression for $\sigma = \sigma_P(z, V)$. We have

$$\max_{D(z,V)} |P| \asymp |a| V^n;$$

so,

$$\sigma_P(z,V) = \frac{V^n \Omega(\rho, 2V)}{\max_{D(z,V)} |P|} \asymp \frac{\Omega(\rho, 2V)}{|a|}, \tag{5.20}$$

but $|a| \asymp |a_+|$ by (5.18), and

$$\sigma_P(z,V) \asymp \frac{\Omega(\rho, 2V)}{|a|} \geq c_9 \frac{\omega(V)}{|a_+|}, \quad c_9 = c_9(n, \alpha). \tag{5.21}$$

Put $h_+ = f - P_+$ and

$$g(x) = \frac{h_+(x)}{(x - x_1^+) \cdots (x - x_{n-1}^+)}, \ n \geq 2, \ g(x) = h_+(x), \ n = 1.$$

Lemma 3.1 implies the relations $g \in Z_\omega$ and $\Delta^2 g(x, \delta) \leq c_{10} \delta \omega(\delta)$. Moreover, we have $g(x_0^+) = 0$, $g(x_n^+) = 0$ for any $n \geq 1$. Applying Lemma 3.2, we obtain

$$|g(z)| \leq c_{11} \ |x - x_n^+| \Omega(|x - x_n^+|, 2V). \tag{5.22}$$

Now $f = P_+ + h_+$ and

$$\frac{f(x)}{P_+(x)} = 1 + \frac{(x - x_1^+) \cdot (x - x_{n-1}^+)}{a_+(x - \zeta_1^+) \cdot (x - \zeta_{n-1}^+)} \frac{g(x)}{x - \zeta_n^+},$$

where $\zeta_\nu^+ = \zeta_\nu(X^+)$. Condition (5.15) yields

$$\left| \frac{x - x_\nu^+}{x - \zeta_\nu^+} \right| \leq c, \ x \in \mathbf{T}, \ 1 \leq \nu \leq n,$$

and, combined with (5.20), (5.21), and (5.22), this implies the estimate

$$\begin{aligned} \left| \frac{1}{a_+} \frac{(x - x_1^+) \cdot (x - x_{n-1}^+)}{(x - \zeta_1^+) \cdot (x - \zeta_{n-1}^+)} \frac{g(x)}{(x - \zeta_n^+)} \right| &\leq c_{12} \frac{1}{|a_+|} \left| \frac{x - x_n^+}{x - \zeta_n^+} \right| \Omega(|x - x_n^+|, 2V) \\ &\leq c_{13}' \sigma \frac{\Omega(|x - x_n^+|, 2V)}{\omega(V)}, \ x \in \mathbf{T} \cap D(z, V). \end{aligned} \tag{5.23}$$

Observing that

$$\max_{\mathbf{T} \cap D(z,V)} |P| \geq c_{13} \max_{D(z,V)} |P| = c_{13} \frac{1}{\sigma} V^n \Omega(\rho, 2V)$$

and applying Lemma 3 for $\sigma \leq \sigma_7$, we get

$$\max_{\mathbf{T} \cap D(z,V)} |f| \geq c_{13} \frac{1}{2\sigma} V^n \Omega(\rho, 2V) \leq c_{14} \ \frac{1}{\sigma} \ V^n \ \omega(V). \tag{5.24}$$

But for $x \in \mathbf{T} \cap D(z,V)$ and $|x - x_n^+| \geq V\sigma/4$ the last term in (5.23) admits the estimate

$$(5.23) \leq c_{13} \sigma \frac{\Omega(\frac{\sigma}{4} V, 2V)}{\omega(V)} \leq c_{15} \sigma \sigma^{-\alpha} \leq \frac{1}{2}$$

for $\sigma \leq \sigma_8$, which, finally, leads to the inequality

$$\left| \log \left| \frac{f(x)}{P_+(x)} \right| \right| \leq c_{16} \sigma \frac{\Omega(|x - x_n^+|, 2V)}{\omega(V)}, \ x \in \mathbf{T} \cap D(z,V), |x - x_n^+| \geq \frac{\sigma}{4} V. \tag{5.25}$$

Now, for $x \in \mathbf{T} \cap D(z, V)$ such that $|x - x_n^+| < \frac{\sigma}{4} V$ we use (5.24) and Lemma 3.3 to see that for $\sigma \le \sigma_8$ and for $\tilde{\tilde{\zeta}} = (1 - \sigma V) x_n^+$ we have

$$M_f(\tilde{\tilde{\zeta}}) \ge (\sigma V)^n \omega(\sigma V). \tag{5.26}$$

Condition (5.26) means that condition (2.2) of Theorem 2.2 is fulfilled for the point $\tilde{\tilde{\zeta}}$, implying inequality (2.3) for that point, i.e.,

$$\int_{\mathbf{T}} \left| \log \left| \frac{f(t)}{M_f(\tilde{\tilde{\zeta}})} \right| \right| \frac{1 - |t|^2}{|t - \tilde{\tilde{\zeta}}|^2} |dt| \le c.$$

From this we obtain

$$\int_{\gamma(\tilde{\tilde{\zeta}})} \left| \log \left| \frac{f(t)}{M_f(\tilde{\tilde{\zeta}})} \right| \right| |dt| \le c_{17}(1 - |\tilde{\tilde{\zeta}}|) = c_{17} \sigma V. \tag{5.27}$$

Using (5.24) and again Lemmas 3.3 and 4.3, we get

$$M_f(\tilde{\tilde{\zeta}}) \asymp \max_{D(x_n^+, \sigma V) \cap \mathbf{T}} |f| \asymp \max_{D(x_n^+, \frac{\sigma}{4} V, \sigma V) \cap \mathbf{T}} |f| \asymp \max_{D(x_n^+, \frac{\sigma}{4} V, \sigma V) \cap \mathbf{T}} |P_+| \asymp M_{P_+}(\tilde{\tilde{\zeta}}),$$

which allows us to reshape (5.27):

$$\int_{\gamma(\tilde{\tilde{\zeta}})} \left| \log \left| \frac{f(t)}{P_+(t)} \right| \right| |dt| \le \int_{\gamma(\tilde{\tilde{\zeta}})} \left| \log \left| \frac{f(t)}{M_f(\tilde{\tilde{\zeta}})} \right| \right| |dt|$$

$$+ \int_{\gamma(\tilde{\tilde{\zeta}})} \left| \log \left| \frac{M_f(\tilde{\tilde{\zeta}})}{M_{P_+}(\tilde{\tilde{\zeta}})} \right| \right| |dt| + \int_{\gamma(\tilde{\tilde{\zeta}})} \left| \log \left| \frac{M_{P_+}(\tilde{\tilde{\zeta}})}{P_+(t)} \right| \right| |dt| \le c_{18}\, \sigma\, V. \tag{5.28}$$

Combining (5.25) with (5.28) and recalling (3.1), we find:

$$\int_I \left| \log \left| \frac{f(t)}{P_+(t)} \right| \right| |dt| = \int_{I \setminus \gamma(\tilde{\tilde{\zeta}})} \left| \log \left| \frac{f(t)}{P_+(t)} \right| \right| |dt| + \int_{I \cap \gamma(\tilde{\tilde{\zeta}})} \left| \log \left| \frac{f(t)}{P_+(t)} \right| \right| |dt|$$

$$\le c_{16} \int_{I \setminus \gamma(\tilde{\tilde{\zeta}})} \frac{\sigma \Omega(|t - x_n^+|, 2V)}{\omega(V)} |dt| + c_{18} \sigma V$$

$$\le c_{19} \sigma \int_{\frac{\sigma}{4} V}^{2V} \frac{\Omega(t, 2V)}{\omega(V)} |dt| + c_{18} \sigma V \le c_{20} \sigma V. \tag{5.29}$$

From the relations $I = \mathbf{T} \cap D(z, v, V)$, $v \asymp V$, $|t - z| \asymp V$, $t \in I$, and (5.29) we deduce that

$$\int_I \left| \log \left| \frac{f(t)}{P_+(t)} \right| \right| \frac{|dt|}{|t - z|^\nu} \le c_{21} \sigma V^{1-\nu}, \ \nu \ge 0, \ c_{21} = c_{21}(\nu). \tag{5.30}$$

Finally, after ν-fold differentiation of the function ϕ_+ (where $\nu \ge n + 1$), we get a sum each term of which involves a derivative of some order of the exponential factor (because $\deg Q_+ = n$). Using (5.30), (5.16), (5.18), (5.19), we finish the proof of Lemma 5.1. □

6. The second crucial step in the proof of Theorem 2.2

Our next step consists in the following result.

Lemma 6.1. *Let z, f, P be the same as in Lemma 5.1. Suppose that P has no zeros in $D(z, q_n' v, q_n'' V)$, where $v \geq 1 - |z|$, $V/v \geq 4/3$, and has $m \leq n$ zeros in $D(z, v)$. Put $I = D(z, v, V) \cap \mathbf{T}$,*

$$\phi(\lambda) = \exp\left(\frac{1}{2\pi}\int_I \log\left|\frac{f(t)}{P(t)}\right|\left|\frac{t+\lambda}{t-\lambda}\right||dt|\right), \tag{6.0}$$

and let $\sigma_P(z, V)$ be defined by (5.0) (we recall that $\rho = 1 - |z|$). Then there exists $\tilde{\sigma} = \tilde{\sigma}(n) > 0$ such that if $\sigma_P(z, V) \leq \tilde{\sigma}$, then

$$|\phi^{(\nu)}(z)| \leq c_1 v^{-\nu}, \nu \geq 0; \tag{6.1}$$

$$|\phi^{(\nu)}(z)| \leq c_2(\sigma_{\dots}(\cdot, v)v^{-\nu} + \sigma_{\dots}(\cdot, V)V^{-\nu}),\ \nu \geq 1,\ \nu \neq n-m; \tag{6.2}$$

$$|\phi^{(n-m)}(z)| \leq c_3 \sigma_{\dots}(\cdot, V) V^{-(n-m)} \log\frac{V}{v}\ ,\ n - m \geq 1. \tag{6.3}$$

Lemma 6.2. *Suppose P has no zeros in $D(z, q_n'' V)$, $V \geq 2\rho$. If $I = \mathbf{T} \cap D(z, V)$, $v = \rho$, and $\phi(\lambda)$ is defined by (6.0), then there exists a constant $\tilde{\sigma}$ such that the inequality $\sigma_P(z, V) \leq \tilde{\sigma}$ implies (6.1) and (6.2).*

We prove only the more difficult Lemma 6.1.

Proof. Denote by $\lambda_1, \dots, \lambda_m$ all zeros of P lying in $D(z, v)$ and put $\psi = \log\phi$,

$$Q(\lambda) = \frac{P(\lambda)}{\prod_{\nu=1}^m(\lambda - \lambda_\nu)},\ Q = |Q(z)|.$$

Lemma 3.3 yields the estimate

$$|f(x) - P(x)| \leq c_1|x - z|^n \Omega(\rho, 2|x - z|),\ x \in I;$$

thus, it is easy to check that there exists $\tilde{\sigma}_1 \geq 0$, $\tilde{\sigma}_1 = \tilde{\sigma}_1(n)$, such that if $\sigma = \sigma_P(z, V) \leq \tilde{\sigma}_1$, then

$$|\phi^{(r)}(\zeta)|_{\zeta=z}| \leq c_2 \int_I \frac{|x-z|^n \Omega(\rho, 2|x-z|)|dx|}{Q|x-z|^m|x-z|^{(r+1)}} \leq \frac{c_3}{Q}\int_v^V t^{n-m-r-1}\Omega(\rho, 2t)dt$$

$$\leq \frac{c_4}{Q}\begin{cases} V^{n-m-r}\Omega(\rho, 2V) & \text{if } n-m-r>0,\\ \log\frac{V}{v}\Omega(\rho, 2V) & \text{if } n-m-r=0,\\ v^{n-m-r}\Omega(\rho, 2v) & \text{if } n-m-r>0. \end{cases} \tag{6.4}$$

Now, (6.1) is a consequence of the Cauchy formula for derivatives, so we must check only (6.2) and (6.3). By the elementary formula

$$\phi^{(\nu)} = \sum_{r_j>0, r_1+\cdots+r_l=\nu} C_\nu^{r_1,\dots,r_l}\phi\psi^{(r_1)}\cdot\psi^{(r_l)},$$

it suffices to estimate separately each term $t_{r_1,\dots,r_l} \stackrel{\text{def}}{=} |\psi^{(r_1)}\cdot\psi^{(r_l)}|$. We consider several possibilities.

Case 1. For $j = 1, \dots, l$, either $r_j < n - m$ for all j or $r_j > n - m$ for all j.

Then (6.4) implies

$$t_{r_1,\dots,r_l} \le c_5 1 Q^l w^{(n-m)l-\nu} \left(\Omega(\rho,2w)\right)^l$$

$$= c_5 w^n \Omega(\rho,2w) Q w^{m+\nu} \left(w^n \Omega(\rho,2w) Q w^m\right)^{l-1} \le c_6 \left(\frac{\sigma_P(z,v)}{v^\nu} + \frac{\sigma_P(z,V)}{V^\nu}\right), \tag{6.5}$$

where $w = V$ if $r_j < n-m$ and $w = v$ if $r_j > n-m$.

Case 2. $r_1,\dots,r_k < n-m$, $r_{k+1} = \dots + r_l = n-m$.

Let

$$\nu_0 = (l-k)(n-m), \ \nu_1 = \sum_{j=1}^{k} r_j.$$

We note that $n > m$ in this case. Further, we get

$$t_{r_1,\dots,r_l} \le c_7 \frac{1}{Q^l} V^{(n-m)k-\nu_1} \left(\Omega(\rho,2V)\right)^l \left(\log \frac{V}{v}\right)^{l-k}$$

$$= \frac{v^{n-m-\nu}}{Q} \frac{\Omega(\rho,2v)}{Q^{l-1}} \left(\frac{v}{V}\right)^{\nu_1} V^{k(n-m)} \left(\Omega(\rho,2V)\right)^l \left(\log \frac{V}{v}\right)^{l-k} \frac{v^{(n-m)(l-k-1)}}{\Omega(\rho,2v)}. \tag{6.6}$$

If $l-k=1$, then $\nu_1 \ge 1$. It is easy to check that the assumption (2.1) implies the condition

$$\frac{\Omega(\rho,2V)}{\Omega(\rho,2v)} \le c_8 \left(\frac{V}{v}\right)^\alpha,$$

so we get

$$\frac{1}{Q^{-1}l} \left(\frac{v}{V}\right)^{\nu_1} V^{(n-m)(l-1)} \left(\Omega(\rho,2V)\right)^l \frac{\log \frac{V}{v}}{\Omega(\rho,2v)}$$

$$= \left(\frac{V^{n-m}\Omega(\rho,2V)}{Q}\right)^{l-1} \left(\frac{v}{V}\right)^{\nu_1} \log \frac{V}{v} \frac{\Omega(\rho,2V)}{\Omega(\rho,2v)}$$

$$\le c_7 \left[\left(\frac{v}{V}\right)^{1-\alpha} \log \frac{V}{v}\right] \cdot \left[\left[\frac{v}{V}\right)^\alpha \cdot \frac{\Omega(\rho,2V)}{\Omega(\rho,2v)}\right] \le c_8 \tag{6.7}$$

(in (6.7) we have used also the equivalence $\sigma_P(z,V) \asymp Q^{-1}V^{n-m}\Omega(\rho,2V)$). Now, (6.6), (6.7) and our assumptions yield

$$t_{r_1,\dots,r_l} \le c_8 \sigma_P(z,v) v^{-\nu}. \tag{6.8}$$

But if $l-k>1$, then

$$\frac{1}{Q^{l-1}} \left(\frac{v}{V}\right)^{\nu_1} V^{(n-m)k} \left(\Omega(\rho,2V)\right)^l \left(\log \frac{V}{v}\right)^{l-k} \frac{v^{(n-m)(l-k-1)}}{\Omega(\rho,2v)}$$

$$= \frac{1}{Q^{l-1}} V^{(n-m)k} \left(\Omega(\rho,2V)\right)^l \left[\left(\log \frac{V}{v}\right)^{l-k} \cdot \frac{v^{(n-m)(l-k-1)}}{\Omega(\rho,2v)}\right]$$

$$\le c_9 \frac{1}{Q^{l-1}} V^{(n-m)k} \left(\Omega(\rho,2V)\right)^l \frac{V^{(n-m)(l-k-1)}}{\Omega(\rho,2V)} = c_9 \left(\frac{V^{(n-m)}\Omega(\rho,2V)}{Q}\right)^{l-1} \le c_{10},$$

and again

$$t_{r_1,\dots,r_l} \le c_{11}\sigma_P(z,v)v^{-\nu}. \tag{6.9}$$

Case 3. $l = 1$, $r = n - m$.

It suffices to apply (6.4).

Case 4. There exist k, L such that $0 < k + L < l$ and $r_1, \dots, r_k < n - m$; $r_{k+1}, \dots, r_{k+L} = n - m$; $r_{k+L+1}, \dots, r_l > n - m$.

Since $r_j \ge 1$ for all j, we have $n > m$. Put

$$\nu_0 = (n-m)L, \ \nu_1 = \sum_{j=1}^{k} r_j, \ \nu_2 = \sum_{j=k+L+1}^{l} r_j.$$

Then $\nu_2 > 0$, $\nu_0 + \nu_1 + \nu_2 = \nu$. Inequalities (6.4) yield

$$t_{r_1,\dots,r_l} \le c_{12}\left(\frac{1}{Q^l}V^{(n-m)k-\nu_1}\left(\Omega(\rho,2V)\right)^k\right)\left(\Omega(\rho,2V)\right)^L$$

$$\times\left(\log\frac{V}{v}\right)^L v^{(n-m)(l-k-L)-\nu_2}\left(\Omega(\rho,2V)\right)^{l-k-L} = c_{12}\left(\frac{1}{Q^l}v^{n-m-\nu}\Omega(\rho,2V)\right)$$

$$\times\left[V^{(n-m)k-\nu_1}\left(\Omega(\rho,2V)\right)^{k+L}\left(V^{n-m}\Omega(\rho,2V)\right)^{l-k-L-1}v^{\nu_1+\nu_0}\left(\log\frac{V}{v}\right)^L\right]$$

$$\le c_{13}\left(\frac{1}{Q}v^{n-m-\nu}\left(\Omega(\rho,2v)\right)\right)\frac{1}{Q^{l-1}}$$

$$\times\left[V^{(n-m)k-\nu_1}\left(\Omega(\rho,2V)\right)^{k+L}V^{\nu_1+\nu_0}\cdot V^{n-m}\left(\Omega(\rho,2V)\right)^{l-k-L-1}\right]$$

$$\le c_{14}\sigma_P(z,v)\cdot v^{-\nu}\left(\frac{V^{n-m}\Omega(\rho,2V)}{Q}\right)^{l-1} \le c_{15}\sigma_P(z,v)v^{\nu}. \tag{6.10}$$

Estimates (6.5)–(6.10) finish the proof of Lemma 8. □

7. The third crucial step in the proof of Theorem 2.2

Suppose we are under conditions of Theorem 2.2. It is standard that $F \in \Lambda^{n-1}Z_\omega$ if and only if

$$|F^{(n+1)}(z)| \le c\,|\omega(1-|z|)(1-|z|)^{-1}, \ z \in \mathbf{D}. \tag{7.1}$$

Also, it is clear that it suffices to verify (7.1) only for $|z| > 0.9$. We choose a big number A independent on z (the exact value of A will be specified in (7.7), (7.8)). Consider two possible situations.

Situation 1. We have

$$M_f(z) \le A(1-|z|)^n\omega(1-|z|). \tag{7.2}$$

Put $\rho = 1 - |z|$, and let P be an interpolating polynomial for f, $\deg P \le n$, constructed for $n+1$ equidistributed points on the arc $\gamma(z)$ including two endpoints. Lemma 3.3 yields

$$|f(t) - P(t)| \le c_2\rho^n\omega(\rho), \ t \in \gamma(z).$$

By (6.2), we get

$$M_P(z) \le c_3\rho^n\omega(\rho), c_3 = c_2 + A,$$

and so

$$|P(\tau)| \le c_4\left(\frac{|\tau - z| + \rho}{\rho}\right)^n M_P(z) \le c_5\left(\frac{|\tau - z| + \rho}{\rho}\right)^n \rho^n\omega(\rho),\ \tau \in \mathbf{T}. \quad (7.3)$$

Applying (7.3), Lemma 3.3, and (3.1), we obtain

$$\begin{aligned}|f(\tau)| &\le |f(\tau) - P(\tau)| + |P(\tau)|\\ &\le c_6|\tau - z|^n\Omega(\rho, 2|t - z|) + c_5\left(\frac{|\tau - z| + \rho}{\rho}\right)^n \rho^n\omega(\rho)\\ &\le \omega(\rho)\left(c_7|\tau - z|^n\left(\frac{|\tau - z| + \rho}{\rho}\right)^\alpha + c_5\left(\frac{|\tau - z| + \rho}{\rho}\right)^n \rho^n\right)\\ &\le c_8\left(\frac{|\tau - z| + \rho}{\rho}\right)^n (|\tau - z| + \rho)^n\omega(\rho),\ \tau \in \mathbf{T}\setminus\gamma(z). \quad (7.4)\end{aligned}$$

By (7.2), estimate (7.4) is also true for $\tau \in \gamma(z)$. Now, applying Lemma 4.3 and (7.4) for $|\lambda - z| = \rho/2$, we get

$$\begin{aligned}|F(\lambda)| &= \exp\left(\frac{1}{2\pi}\int_{\mathbf{T}} \log|f(\tau)|\frac{1 - |\lambda|^2}{|\tau - \lambda|^2}|d\tau|\right)\\ &\le \exp\left(\frac{1}{2\pi}\int_{\mathbf{T}} \log\left[c_8\left(\frac{|\tau - z| + \rho}{\rho}\right)^n (|\tau - z| + \rho)^n\omega(\rho)\right]\frac{1 - |\lambda|^2}{|\tau - \lambda|^2}|d\tau|\right)\\ &\le c_9(|\lambda - z| + \rho)^n\omega(\rho) \le c_{10}\rho^n\omega(\rho). \quad (7.5)\end{aligned}$$

Now, by the Cauchy inequalities and (7.5) we obtain

$$|F^{(n+1)}(\lambda)| \le (n+1)!2^{n+1}\rho^{-n-1}\max_{|\lambda - z| = \rho/2}|F(z)| \le c_{11}\rho^{-1}\omega(\rho), \quad (7.6)$$

as required (see (7.1)).

Situation 2. We have $M_f(z) \ge A\rho^n\omega(\rho)$.

Let a polynomial P be as in Section 1. First of all we want to determine A. Let $\tilde{A}$ be the minimal number satisfying the conditions

$$\frac{1}{\tilde{A}} \le \sigma_1(\text{Lemma 5.1}), \frac{1}{\tilde{A}} \le \sigma(\text{Lemma 6.1}), \quad \tilde{A} \ge (2c_1 + 1)c_2 \quad (7.7)$$

where $c_1 = c_1(\text{Lemma 3.3})$, $c_2 = c_2(n)$ is the constant in the inequality

$$\max_{D(z,\tau)}|P| \le c_2 \max_{D(z,\tau)\cap\mathbf{T}}|P|,\ \tau \ge 2\rho, \tau \le 2.$$

Let $\tilde{c} = \tilde{c}(n, \omega, \alpha)$ be a number such that

$$((6\rho)^n + \rho^n)\Omega(\rho, 12\rho) \le \tilde{c}\rho^n\omega(\rho).$$

Finally, let

$$A = \tilde{c}\tilde{A}(2c_1 + 2). \quad (7.8)$$

We observe that if $M_f(z) \geq A\rho^n\omega(\rho)$, then Lemma 3.3 yields

$$M_P(z) \geq M_f(z) - c_1\rho^n\omega(\rho) \geq \frac{1}{2}M_f(z) \geq \tilde{c}\tilde{A}\rho^n\omega(\rho)$$

and

$$\max_{D(z,6\rho)} |P| \geq M_P(z) \geq \tilde{c}\tilde{A}\rho^n\omega(\rho) \geq \tilde{A}((6\rho)^n + \rho^n)\Omega(\rho, 12\rho). \tag{7.9}$$

Now, we introduce an important number s:

$$s = \max\{\delta : \max_{D(z,\tau)} |P| \geq \tilde{A}(\tau^n + \rho^n)\Omega(\rho, 2\tau) \text{ for any } \tau \text{ with } \rho \leq \tau \leq \delta\}. \tag{7.10}$$

The possibility $s = +\infty$ is excluded because $\Omega(\rho, 2\tau) \to \infty$ as $\tau \to \infty$, by (2.1a). Since we are in situation 2, by (7.9) we see that $s \geq 6\rho$. This is a technical inequality, which will permit us to make some geometric constructions. Let Ξ be the zero set of P. Using the Dirichlet principle, we can construct several concentric circles centered at z, which determine a disc L_0, annuli L_j, $1 \leq j \leq N$, and the complement L_{N+1} to a disc, such that (we recall that $q_n' = 1 - \frac{1}{4(n+1)^2}$, $q_n'' = 1 + \frac{1}{4(n+1)^2}$):

$$N \leq 2n+1 \tag{7.11}$$

$$\left.\begin{array}{c} L_0 = D(z, V_0),\ L_j = D(z, V_{j-1}, V_j),\ 1 \leq j \leq N, \\ L_{N+1} = \mathbf{C} \setminus D(z, V_N), \\ \text{and } V_0 \geq \frac{4}{3}\rho,\ V_j < V_{j+1},\ 0 \leq j \leq N-1, \end{array}\right\} \tag{7.12}$$

$$\left.\begin{array}{c} \text{if } L_0 \cap \Xi = \varnothing, \text{ then } D(z, q_n''V_0) \cap \Xi = \varnothing; \\ \text{if } L_j \cap \Xi = \varnothing, \text{ then } D(z, q_n'V_{j-1}, q_n''V_0) \cap \Xi = \varnothing,\ 1 \leq j \leq N, \\ \text{and } \frac{V_j}{V_{j-1}} \geq \frac{4}{3}; \\ \text{if } L_{N+1} \cap \Xi = \varnothing, \text{ then } (\mathbf{C} \setminus D(z, q_n'V_N)) \cap \Xi = \varnothing; \end{array}\right\} \tag{7.13}$$

$$\left.\begin{array}{c} \text{if } L_0 \cap \Xi \neq \varnothing, \text{ then } L(z, q_n'V_0, V_0) \cap \Xi = \varnothing \text{ and } V_0 \leq 2\rho; \\ \text{if } L_j \cap \Xi \neq \varnothing,\ 1 \leq j \leq N, \text{ then } \left(L_j \setminus L(z, q_n''V_{j-1}, q_n'V_j)\right) \cap \Xi = \varnothing, \\ \text{and } \frac{4}{3} \leq \frac{V_j}{V_{j-1}} \leq \frac{5}{2}; \\ \text{if } L_{N+1} \cap \Xi \neq \varnothing, \text{ then } L(z, V_N, q_n''V_N) \cap \Xi = \varnothing. \end{array}\right\} \tag{7.14}$$

Having ensured (7.11)–(7.14) (the construction is elementary), we introduce a new number s_0 in terms of s and V_j. Namely, we put:

$$\left.\begin{array}{l} (\alpha)\left\{\begin{array}{c} \text{if there are some } V_k \text{ such that } \frac{s}{3} \leq V_k \leq s, \\ \text{then } s_0 = \max\{V_k : \frac{s}{3} \leq V_k \leq s\}, \end{array}\right. \\ (\beta)\left\{\begin{array}{c} \text{if there are no } V_k \text{ such that } \frac{s}{3} \leq V_k \leq s, \\ \text{then } s_0 = s/2 \end{array}\right. \end{array}\right\}. \tag{7.15}$$

Finally, we modify the sets L_j in the following way: if $V_N \leq s/3$, we put $\tilde{L}_{N+1} = L(z, V_N, s_0)$, $\tilde{L}_{N+2} = \mathbf{C} \setminus D(z, s_0)$ and define $K = N+1$; if $V_N \geq s/3$ and the number s_0 is chosen in accordance with (7.15α), then $s_0 = V_{k_0}$, $k_0 \leq N$, and we put $\tilde{L}_{k_0+1} = \mathbf{C} \setminus D(z, V_{k_0}) = \mathbf{C} \setminus D(z, s_0)$, $K = k_0$; if $V_N > s$ and the number s_0 is chosen in accordance with (7.15β), then we take k_0 with $V_{k_0} < s/3$, $V_{k_0+1} > s$ and put $\tilde{L}_{k_0+1} = D(z, V_{k_0}, s_0)$, $\tilde{L}_{k_0+2} = \mathbf{C} \setminus D(z, s_0)$, $K = k_0 + 1$. Having done this, we restore the notation L_j for the modified sets. Then the index will satisfy the restriction $j \le K+1$, all properties (7.11)–(7.15) will persist and, moreover, we shall have

$$K \le N+1 \text{ and } V_K = s_0. \tag{7.16}$$

Let m_j, $0 \le j \le K+1$, be the number of zeros of P lying in L_j, $m_j \ge 0$. If $m_j > 0$, let $\zeta_{j1}, \dots, \zeta_{jm_j}$ be all these zeros. We put

$$P_j(\zeta) = \prod_{l=1}^{m_j} (\zeta - \zeta_{jl}),\ 0 \le j \le K. \tag{7.17}$$

If $m_j = 0$, we put

$$P_j(\zeta) = 1, \quad 0 \le j \le K. \tag{7.18}$$

Next, we define

$$P_{K+1} = \frac{P}{\prod_{j=0}^{K} P_j}. \tag{7.19}$$

Let Q_j, $0 \le j \le K+1$, be an outer function in $\mathbf{D}$ such that $|Q_j||_{\mathbf{T}} = |P_j||_{\mathbf{T}}$, let $I_j = \mathbf{T} \cap L_j$ $(0 \le j \le K+1)$, and let

$$\phi_j(\zeta) = Q_j(\zeta) \exp\left(\frac{1}{2\pi}\int_{I_j} \log\left|\frac{f(t)}{P(t)}\right| \frac{t+\zeta}{t-\zeta}|dt|\right), 0 \le j \le K+1. \tag{7.20}$$

We note that $\cup_{j=0}^{K+1} I_j = \mathbf{T}$, $\prod_{j=0}^{K+1} Q_j = Q$, where Q be an outer function in $\mathbf{D}$ such that $|Q||_{\mathbf{T}} = |P||_{\mathbf{T}}$. Now we write down a decisive identity involving the functions ϕ_j. Namely, we have

$$\begin{aligned} F(\zeta) &= \exp\left(\frac{1}{2\pi}\int_{\mathbf{T}} \log|f(t)| \frac{t+\zeta}{t-\zeta}|dt|\right) \\ &= Q(\zeta)\exp\left(\frac{1}{2\pi}\int_{\mathbf{T}} \log\left|\frac{f(t)}{P(t)}\right| \frac{t+\zeta}{t-\zeta}|dt|\right) \\ &= \prod_{j=0}^{K+1}\left[Q_j(\zeta)\exp\left(\frac{1}{2\pi}\int_{I_j} \log\left|\frac{f(t)}{P(t)}\right| \frac{t+\zeta}{t-\zeta}|dt|\right)\right] = \prod_{j=0}^{K+1} \phi_j(\zeta),\ \zeta \in \mathbf{D}. \end{aligned} \tag{7.21}$$

8. The fourth crucial step in the proof of Theorem 2.2

The construction of the sets L_j and the functions Q_j, ϕ_j given in (7.11)–(7.21) begins with fixing a point $z \in \mathbf{D}$, $|z| > 0.9$. Now, we differentiate (7.21) in ζ and put $\zeta = z$ in the resulting formula. Differentiation yields

$$F^{(n+1)}(\zeta) = \sum_{\substack{\nu_j \ge 0 \\ \nu_0 + \cdot + \nu_{K+1} = n+1}} \frac{(n+1)!}{\nu_0! \cdots \nu_{K+1}!} \prod_{j=0}^{K+1} \phi_j^{(\nu_j)}(\zeta). \tag{8.1}$$

We estimate each summand in (8.1) separately. The functions $\phi_j^{(\nu)}(\zeta)|_{\zeta=z}$, $0 \le j \le K$, were treated in Lemmas 5.1 and 6.1. It remains to handle the functions $\phi_{K+1}^{(\nu)}(\zeta)|_{\zeta=z}$. We need the following crucial lemma, in which inequality (2.3) plays a central role.

Lemma 8.1. *Let the symbols f, P mean the same as in the beginning of the analysis of situation* 1 *in Section* 6, *and let the number s_0 be defined by* (7.15) *If*

$$\psi(\zeta) = \frac{1}{2\pi}\int_{\mathbf{T}\setminus D(z,s_0)} \log\left|\frac{f(t)}{P(t)}\right| \frac{t+\zeta}{t-\zeta}|dt|, \quad h(\zeta) = \exp\psi(\zeta),$$

then

$$|h^{(\nu)}(\zeta)| \le c_1 s_0^{-\nu}, \quad \nu \ge 0, \quad c_1 = c_1(n,\nu,\omega). \tag{8.2}$$

Proof. We suppose that $\mathbf{T}\setminus D(z,s_0) \ne \varnothing$. Then (7.7) and (7.15) yield

$$\begin{aligned}\max_{D(z,s_0)\cap\mathbf{T}} |P| &\ge \frac{1}{c_2(7.7)} \max_{D(z,s_0)} |P| \\ &\ge \frac{1}{c_2(7.7)}(2c_1(\text{Lemma } 3.3)\cdot c_2(7.7)\cdot s_0^n\cdot \Omega(\rho,2s_0)) \\ &= 2c_1(\text{Lemma } 3.3)\cdot s_0^n\cdot\Omega(\rho,2s_0).\end{aligned} \tag{8.3}$$

Applying Lemma 3.3, from (8.3) we deduce

$$\begin{aligned}\max_{D(z,s_0)\cap\mathbf{T}} |f| \ge \max_{D(z,s_0)\cap\mathbf{T}} |P| - \max_{D(z,s_0)\cap\mathbf{T}} |f-P| \ge \\ 2c_1(\text{Lemma } 3.3)\cdot s_0^n\cdot\Omega(\rho,2s_0) - c_1(\text{Lemma } 3.3)\cdot s_0^n\cdot\Omega(\rho,2s_0) = \\ c_1(\text{Lemma } 3.3)\cdot s_0^n\cdot\Omega(\rho,2s_0) \ge c_3 s_0^n\omega(s_0).\end{aligned} \tag{8.4}$$

We choose σ from the condition $\gamma(\tilde z) = D(z,s_0)\cap\mathbf{T}$, $\tilde z = (1-\sigma)\frac{z}{|z|}$. Then $\sigma \asymp s_0$ and (8.4) means that

$$M_f(\tilde z) \ge c_6(1-|\tilde z|)^n\omega(1-|\tilde z|). \tag{8.5}$$

The remark after Lemma 4.1 and formula (8.5) yield

$$\int_{\mathbf{T}} \left|\log\left|\frac{f(t)}{M_f(\tilde z)}\right|\right| \frac{1-|\tilde z|^2}{|t-\tilde z|^2}|dt| \le c_7. \tag{8.6}$$

Observe that (8.3) and Lemma 3 imply

$$\max_{D(z,s_0)\cap\mathbf{T}} |f-P| \le 1/2 \max_{D(z,s_0)\cap\mathbf{T}} |P|,$$

hence

$$M_f(\tilde z) = \max_{D(z,s_0)\cap\mathbf{T}} |f| \asymp \max_{D(z,s_0)\cap\mathbf{T}} |P| = M_P(\tilde z) \tag{8.7}$$

Using Lemma 4.3, (8.6), and (8.7) we get

$$\begin{aligned}\int_{\mathbf{T}} \left|\log\left|\frac{f(t)}{P(t)}\right|\right| \frac{1-|\tilde z|^2}{|t-\tilde z|^2}|dt| \le \int_{\mathbf{T}} \left|\log\left|\frac{f(t)}{M_f(\tilde z)}\right|\right| \frac{1-|\tilde z|^2}{|t-\tilde z|^2}|dt| \\ + \int_{\mathbf{T}} \left|\log\left|\frac{M_f(\tilde z)}{M_P(\tilde z)}\right|\right| \frac{1-|\tilde z|^2}{|t-\tilde z|^2}|dt| + \int_{\mathbf{T}} \left|\log\left|\frac{P(t)}{M_P(\tilde z)}\right|\right| \frac{1-|\tilde z|^2}{|t-\tilde z|^2}|dt| \le c_8.\end{aligned} \tag{8.8}$$

Since $|t-\tilde{z}| \asymp |t-z|$ for $\mathbf{T}\setminus D(z,s_0)$, (8.8) gives

$$\int_{\mathbf{T}\setminus D(z,s_0)} \left|\log\left|\frac{f(t)}{P(t)}\right|\right| \frac{1}{|t-\tilde{z}|^2}|dt| \le \frac{c_9}{s_0}. \tag{8.9}$$

By (8.9), we obtain

$$|\Re\psi(z)| = \frac{1}{2\pi}\int_{\mathbf{T}\setminus D(z,s_0)} \left|\log\left|\frac{f(t)}{P(t)}\right|\right| \frac{1-|z|^2}{|t-z|^2}|dt| \le \frac{c_9}{2\pi}\frac{2\rho}{s_0} \le c_{10}, \tag{8.10}$$

$$|\psi^{(\nu)}(\zeta)|_{\zeta=z}| \le \frac{\nu!}{\pi}\int_{\mathbf{T}\setminus D(z,s_0)} \left|\log\left|\frac{f(t)}{P(t)}\right|\right| \frac{|dt|}{|t-z|^{\nu+1}}$$

$$\le \frac{\nu!}{\pi s_0^{\nu-1}}\int_{\mathbf{T}\setminus D(z,s_0)} \left|\log\left|\frac{f(t)}{P(t)}\right|\right| \frac{|dt|}{|t-z|^{2}} \le \frac{c_{11}}{s_0^{\nu}}. \tag{8.11}$$

Now we differentiate the function $h=\exp\psi$ in the variable ζ and get

$$h^{(\nu)} = \sum C^{(\nu)}_{r_1,\dots,r_j}\psi^{(r_1)}\cdots\psi^{(r_j)}\cdot h \tag{8.12}$$

with some numerical coefficients $C^{(\nu)}_{r_1,\dots,r_j}$, where $r_1+\dots+r_j=\nu$, $r_l>0$. Combining (8.10)–(8.12), we obtain (8.2). □

Corollary 8.2. *For $\nu=1,2,\dots$ we have*

$$|\psi^{(\nu)}_{K+1}(\zeta)|_{\zeta=z}| \le c_{12}|\psi_{K+1}(z)|\cdot V_K^{-\nu},\quad c_{12}=c_{12}(\nu,n). \tag{8.13}$$

Proof. We have $\phi_{K+1}=Q_{K+1}\cdot h$, where Q_{K+1} is a polynomial that does not vanish in the disc $D^*(z,q_n'V_K)=D^*(z,q_n's_0)$ symmetric to $D(z,q_n's_0)$ with respect to the unit circle $\mathbf{T}$. Consequently,

$$\left.\begin{aligned} &|Q_{K+1}(\zeta)| \asymp |Q_{K+1}(z)|,\quad \zeta\in D(z,s_0/2),\\ &|Q^{(\nu)}_{K+1}(\zeta)|_{\zeta=z}| \le c_{13}|Q_{K+1}(z)|s_0^{-\nu},\quad \nu\ge 1. \end{aligned}\right\} \tag{8.14}$$

Differentiating the function ϕ_{K+1} and combining (8.2), (8.10), and (8.14), we get (8.13). □

9. The end of the proof of Theorem 2.2 in situation 2

We use identity (8.1). Put

$$t_{\nu_0,\dots,\nu_{K+1}} = \prod_{j=0}^{K+1} |\phi_j^{(\nu_j)}(\zeta_j)|\Big|_{\zeta=z}. \tag{9.1}$$

Since we are in situation 2, we may apply Lemmas 5.1, 5.3, 6.1, 6.2, and Corollary 8.2. We recall that the numbers $\nu_0,\dots,\nu_{K+1}$ occurring in (9.1) satisfy $\nu_0+\dots+\nu_{K+1}=n+1$. Since $m_0+\dots+m_{K+1}=\deg P\le n$, there exists j, $0\le j\le K+1$, such that $\nu_j>m_j$. Let j_0 be the smallest index with this property. We consider several possible cases.

Case 1. $j_0=K+1$.

Then $\nu_{K+1} > 0$, $\mathbf{T} \setminus D(z, s_0) \neq \varnothing$ (otherwise $t_{\nu_0,\ldots,\nu_{K+1}} = 0$), $\nu_j = 0$ if $m_j = 0$, $j \leq K$. Putting $Q = |\phi_{K+1}(z)|$ and applying the lemmas mentioned above, we get

$$t_{\nu_0,\ldots,\nu_{K+1}} \leq c_1 \prod_{j=0}^{K} V_j^{m_j-\nu_j} \cdot Q V_K^{-\nu_{K+1}}$$

$$\leq c_1 Q \prod_{j=0}^{K} V_K^{m_j-\nu_j} \cdot V_K^{-\nu_{K+1}} = c_1 Q V_K^{n_*} \cdot V_K^{-n-1}, \tag{9.2}$$

where $n_* = \sum_{j=0}^{K} m_j$. Notice that

$$\max_{D(z,V_K)} |P| \asymp Q V_k^{n_*} \tag{9.3}$$

and that, by the choice (7.15) of $s_0 = V_K$, we have

$$\frac{V_K^n \Omega(\rho, 2V_K)}{\max_{D(z,V_K)} |P|} \asymp \frac{1}{\tilde{A}} \asymp 1. \tag{9.4}$$

Combining (9.2)–(9.4), we get

$$c_1 \frac{Q V_K^{n_*}}{V_K^{n+1}} \leq c_2 \max_{D(z,V_K)} \frac{|P|}{V_K^{n+1}} \asymp \frac{V_K^n \Omega(\rho, 2V_K)}{V_K^{n+1}} = \frac{\Omega(\rho, 2V_K)}{V_K} \leq c_3 \frac{\omega(\rho)}{\rho} \tag{9.5}$$

because by (3.1) we have

$$\frac{\Omega(\rho, 2V_K)}{\omega(\rho)} \leq c \left(\frac{V_K}{\rho} \right)^{\alpha}, \; 0 < \alpha < 1.$$

Case 2. $j_0 \leq K, \; m_{j_0} > 0$.

This means that $\nu_j = 0$ for all $j < j_0$ such that $m_j = 0$, and $\nu_j \leq m_j$ for $j < j_0$, $m_j > 0$. Put $\sigma_{j_0} = \sigma_P(\rho, V_{j_0})$, where $\sigma_P(\rho, V_{j_0})$ is defined in (5.0). As before, let $Q = |\phi_{K+1}(z)|$. Then, by Lemmas 5.1–8.1 we obtain

$$t_{\nu_0,\ldots,\nu_{K+1}} \leq c_4 \prod_{j=0}^{j_0-1} V_j^{m_j-\nu_j} \cdot \sigma_{j_0} \cdot V_{j_0}^{m_{j_0}-\nu_{j_0}} \cdot \prod_{j=j_0+1}^{K} V_{j-1}^{m_j-\nu_j} \cdot Q V_K^{-\nu_{K+1}}$$

$$\leq c_4 \prod_{j=0}^{j_0-1} V_{j_0}^{m_j-\nu_j} \cdot \sigma_{j_0} \cdot V_{j_0}^{m_{j_0}-\nu_{j_0}} \cdot \prod_{j=j_0+1}^{K} V_j^{m_j} \cdot \prod_{j=j_0+1}^{K+1} V_{j_0}^{-\nu_j} \cdot Q$$

$$\asymp \sigma_{j_0} \max_{D(z,V_{j_0})} |P| \cdot V_{j_0}^{-n-1} = \frac{\Omega(\rho, 2V_{j_0})}{V_{j_0}} \leq c_3 \frac{\omega(\rho)}{\rho} \tag{9.6}$$

Case 3. $j_0 \leq K$, $m_{j_0} = 0$.

Then $m_j \geq \nu_j$ for $0 \leq j < j_0$. Let $V_{-1} = \rho$, $Q = |\phi(_{K+1}(z)|$,

$$\tilde{\sigma}_j = \frac{V_l^n \Omega(\rho, 2V_j) \log \frac{2V_j}{\rho}}{\max_{D(z,V_j)} |P|}.$$

Lemma 6.1 yields

$$|\phi_{j_0}^{(\nu_{j_0})}(\zeta)|_{\zeta=z}| \le c_6 \left(\tilde{\sigma}_{j_0-1} V_{j_0-1}^{-\nu_{j_0}} + \tilde{\sigma}_{j_0} V_{j_0}^{-\nu_{j_0}}\right),$$

hence

$$t_{\nu_0,\ldots,\nu K+1} \le c_7 \sum_{\delta=0}^{1} \prod_{j=0}^{j_0-1} V_j^{m_j-\nu_j} \cdot \tilde{\sigma}_{j_0-\delta} \cdot V_{j_0-\delta}^{-\nu_{j_0}} \cdot \prod_{j=j_0+1}^{K} V_j^{m_j-\nu_j} \cdot Q V_K^{-\nu_{K+1}}$$

$$\le c_7 \sum_{\delta=0}^{1} \prod_{j=0}^{j_0-1} V_{j_0}^{m_j-\nu_j} \cdot \tilde{\sigma}_{j_0-\delta} \cdot \prod_{j=j_0+1}^{K} V_j^{m_j} \cdot \prod_{j=j_0}^{K+1} V_{j_0-\delta}^{-\nu_j} \cdot Q$$

$$\asymp \sum_{\delta=0}^{1} \tilde{\sigma}_{j_0-\delta} \max_{D(z,V_{j_0-\delta})} \frac{|P|}{V_{j_0-\delta}^{n+1}} = \sum_{\delta=0}^{1} \frac{\Omega(\rho, 2V_{j_0-\delta}) \log \frac{2V_{j_0-\delta}}{\rho}}{V_{j_0-\delta}} \le c_9 \frac{\omega(\rho)}{\rho}. \tag{9.7}$$

Identity (8.1) and estimates (9.5)–(9.7) complete the analysis of situation 2.2 in the proof of Theorem 2.2. This yields estimate (7.1) and, thus, finishes the proof of the theorem.

References

[1] N. A. Shirokov, *Analytic functions smooth up to the boundary,* Lecture Notes in Math., **1312** (1988).

[2] N. A. Shirokov, *Outer functions from analytic Besov's classes,* Zapiski nauchn. sem. POMI, **217** (1994), 172–217,(Russian).

[3] N. A. Shirokov, *Inner functions in analytic Besov's classes,* Algebra i analiz, **8** (1996), 193–221, (Russian).

[4] K. M. Dyakonov, *Equivalent norms on Lipschitz-type space of holomorphic functions,* Acta Mathematica, **178** (1997), 143–167.

[5] K. M. Dyakonov, *The moduli of holomorphic functions in Lipschitz spaces,* Michigan Math. J., **44** (1997), 139–147.

[6] K. M. Dyakonov, *Besov spaces and outer functions,* Michigan Math. J., **45** (1998), 143–157.

Department of Mathematics,
SPb Electrotechnical University,
Prof. Popova str. 5
Saint-Petersburg, 197376, Russia
E-mail address: kam@kam.sbp.su

Operator Theory:
Advances and Applications, Vol. 113

Estimates for the Approximation Numbers of the Weighted Riemann-Liouville Operator in the Spaces L_p

Michael Solomyak

To the memory of S. A. Vinogradov, friend and colleague

1. Introduction

We study the operator

$$T = T_{\nu,v} : f(x) \mapsto \frac{v(x)}{x^\nu} \int_0^x (x-y)^{\nu-1} f(y)dy \tag{1}$$

in the spaces $L_p(\mathbb{R}_+)$, $1 < p < \infty$. In (1), ν is a given real positive number, and v is a function, which is assumed to belong to the space $L_{p,\mathrm{loc}}(\mathbb{R}_+)$. We shall show that, to a large extent, the properties of the operator (1) can be described in terms of the number sequence $\boldsymbol{\eta}_p(v) = \{\eta_{p,k}(v)\}$, $k \in \mathbb{Z}$:

$$\eta_{p,k}(v) = \left(\int_{\Delta_k} \frac{|v(x)|^p}{x} dx\right)^{\frac{1}{p}}, \qquad \Delta_k = (2^k, 2^{k+1}).$$

We start with boundedness and compactness criteria.

Theorem 1. *Let* $1 < p < \infty$ *and* $p\nu > 1$. *Then the operator* (1) *is bounded in* $L_p(\mathbb{R}_+)$ *if and only if* $\boldsymbol{\eta}_p(v) \in l_\infty$, *and it is compact if and only if* $\eta_{p,k}(v) \to 0$ *as* $|k| \to \infty$. *There is a constant* $C = C(\nu, p) > 0$ *such that*

$$C^{-1}\|\boldsymbol{\eta}_p(v)\|_\infty \le \|T_{\nu,v}\| \le C\|\boldsymbol{\eta}_p(v)\|_\infty. \tag{2}$$

This result is not new; for $\nu > 1$ it was proved (in an equivalent form) by Stepanov [14] and by Martín-Reyes and Sawyer [9]; see also the survey paper [15]. Actually, in these papers a more complicated case of the operators with two weights was considered. For arbitrary $\nu > \frac{1}{p}$ the result (and its extension to the operators acting from L_p to L_q with $p \ne q$) was recently proved by Prokhorov [12]. I am grateful to V.D.Stepanov for information on the subject. In Section 2 we present our proof of Theorem 1. We do this because some objects introduced in the course of the proof are useful also for the proof of Theorem 2.

1991 *Mathematics Subject Classification.* Primary 47G10, 47B06.

Our main goal is to estimate the approximation numbers of the operator (1). Recall that for a bounded operator $S : X \to Y$, where X and Y are Banach spaces, its approximation numbers are defined as

$$a_j(S) = \inf_{\operatorname{rank} S' < j} \|S' - S\|, \qquad j \in \mathbb{N}.$$

See the books [3, 7, 11] for a detailed account of their properties.

Given a countable family $\boldsymbol{h} = \{h_j\}_{j \in J}$ of real numbers, it is convenient to describe its behaviour in terms of the counting function $n(\varepsilon, \boldsymbol{h}) = \#\{j \in J : |h_j| > \varepsilon\}$. In what follows $n(\varepsilon, S)$ stands for the counting function of the sequence $\{a_j(S)\}$.

We estimate $n(\varepsilon, T_{\nu,v})$ only for $\nu \geq 1$.

Theorem 2. *Let $1 < p < \infty$ and $\nu \geq 1$. Then*

$$\sup_{\varepsilon > 0} \varepsilon^{\frac{1}{\nu}} n(\varepsilon, T_{\nu,v}) \leq C \sum_{k \in \mathbb{Z}} \eta_{p,k}^{\frac{1}{\nu}}(v), \qquad C = C(\nu, p), \tag{3}$$

and, for any $q > \frac{1}{\nu}$,

$$\sup_{\varepsilon > 0} \varepsilon^{q} n(\varepsilon, T_{\nu,v}) \leq C \sup_{\varepsilon > 0} \varepsilon^{q} n(\varepsilon, \boldsymbol{\eta}_p(v)), \qquad C = C(\nu, p, q). \tag{4}$$

Remarks.

1) Estimate (4) is informative only if the series on the right-hand side of (3) diverges.
2) The assumption $\nu \geq 1$ in Theorem 2 is probably excessive. We expect that the results extend to all $\nu > \frac{1}{p}$, as in Theorem 1. However, the approach based on the "Schur test" (Proposition 7) requires this assumption.
3) For $n(\varepsilon, T_{\nu,v})$, the techniques developed in the paper gives only upper estimates. By analogy with the case $p = 2$ (see [10]), we expect that these estimates are sharp in order.

If X and Y are Hilbert spaces, then the approximation numbers of an operator $S : X \to Y$ coincide with its singular numbers. For integral operators acting in L_2-spaces, many results on the behaviour of their singular numbers are known; see, e.g., [7] and, especially, the survey paper [2]. For the operator (1) this behaviour was studied in detail in [10].

Much less is known for $p \neq 2$. In the paper [4] the case where $\nu = 1$ was analyzed for $1 < p < \infty$. The operator

$$T : f(x) \mapsto v(x) \int_0^x u(x) f(x) dx \tag{5}$$

with two weights was studied, and some two-sided estimates for $a_j(T)$ were found. In principle, the upper estimates obtained in [4] should be regarded as a more subtle result compared to the quasinorm estimates given by Theorem 2 (for $\nu = 1$ and $u \equiv 1$, when both results apply). However, the functionals involved in the estimates in [4] are hardly computable in all but the simplest cases, so that the results are not easy to use. The problem was revisited in [5], where the method became more transparent. In particular, some results of [10] were extended there to the operator (5) viewed as acting in L_p with $p \neq 2$. In the recent paper [6] the borderline cases $p = 1, \infty$ were studied. It is unclear at the moment whether the approach of [4,5] and [6] applies to the operator (1) with $\nu \neq 1$.

For the proof of Theorem 2 we apply piecewise-polynomial approximations. The approach to piecewise-polynomial approximations we make use of, was suggested in [1] and turned out to be quite efficient in many spectral problems involving weight functions. A systematic account of this approach in application to estimates for the singular numbers of integral operators acting in weighted L_2-spaces was presented in [2]. Here we show that this approach works equally well for estimating the approximation numbers of integral operators in L_p, $p \neq 2$.

Our notations are rather standard. We write $\|\cdot\|_p$ for the norm in L_p. Different constant factors in estimates are usually denoted by C or c, without indices. Here is our way to denote the repeated constants: C_{13} is the constant that appears for the first time in estimate (13).

2. Proof of Theorem 1

The following decomposition of the operator (1) is useful when proving both Theorems 1 and 2. Let χ_j stand for the characteristic function of the interval $I_j = (2^j, 2^{j+1})$. Consider the family of operators

$$T_{k,l} = \chi_l T \chi_k, \qquad k, l \in \mathbb{Z}. \tag{6}$$

It is clear that $T_{k,l} = 0$ for $l < k$. Also, we denote

$$T_j = \sum_{l \in \mathbb{Z}} T_{l-j,l}, \qquad j = 0, 1, \dots$$

The study of the operators T_j, due to their block-diagonal structure, reduces to the study of their blocks $T_{k,l}$. In its turn, at least formally,

$$T = \sum_{j=0}^{\infty} T_j. \tag{7}$$

Boundedness: upper estimate. Given a function $f \in L_p(\mathbb{R}_+)$, we denote $f_k = \chi_k f$ and estimate $T_{k,l} f = \chi_l T f_k$. We start with $l \leq k+1$. If $x \in I_k \cup I_{k+1}$, then

$$\begin{aligned}
|(Tf_k)(x)| &\leq \frac{|v(x)|}{x^\nu} \int_{2^k}^{x} (x-y)^{\nu-1} |f_k(y)| dy \\
&\leq \frac{|v(x)|}{x^\nu} \left(\int_{2^k}^{x} (x-y)^{(\nu-1)p'} dy \right)^{\frac{1}{p'}} \|f_k\|_p \\
&= c \frac{|v(x)|}{x^\nu} (x - 2^k)^{\nu - \frac{1}{p}} \|f_k\|_p \leq c x^{-\frac{1}{p}} |v(x)| \|f_k\|_p, \qquad c = c(\nu, p).
\end{aligned}$$

It follows that

$$\|T_{k,k}f_k\|_p \le c\,\eta_{p,k}\|f_k\|_p, \qquad \|T_{k,k+1}f_k\|_p \le c\,\eta_{p,k+1}\|f_k\|_p. \tag{8}$$

The estimates for $l > k+1$ are easier to obtain because then $x \in I_l$ and $y \in I_k$ imply $x/2 \le x - y \le x$. Thus, with $c = \max(1, 2^{1-\nu})$, for $x \in I_l$ we get

$$|(T_{k,l}f_k)(x)| \le c\frac{|v(x)|}{x}\int_{I_k}|f(y)|dy \le c\frac{|v(x)|}{x^{\frac{1}{p}}}2^{-\frac{l-k}{p'}}\|f_k\|_p.$$

Therefore,

$$\|T_{k,l}f_k\|_p^p \le c^p\eta_{p,l}^p(v)2^{-(l-k)(p-1)}\|f_k\|_p^p, \qquad l > k+1.$$

The latter inequality and (8) can be written in a unified form:

$$\|T_{k,l}\| \le c\eta_{p,l}(v)2^{-\frac{l-k}{p'}}, \qquad l \ge k. \tag{9}$$

In view of the block-diagonal structure of the operators T_j, from (9) we obtain

$$\|T_j\| \le \sup_{l\in\mathbb{Z}}\|T_{l-j,l}\| \le c2^{-\frac{j}{p'}}\|\boldsymbol{\eta}_p(v)\|_\infty, \qquad j = 0, 1, \dots. \tag{10}$$

If $p > 1$, the terms on the right-hand sides constitute a convergent series. So, the series in (7) converges in norm and we get the upper estimate in (2).

Boundedness: lower estimate. Apply T to the function χ_l. Then, for $x \in I_{l+1}$,

$$|(T\chi_l)(x)| \ge \frac{|v(x)|}{x^\nu}\int_{2^l}^{2^{l+1}}(x-y)^{\nu-1}dy \ge c|v(x)| \ge c2^{\frac{l+1}{p}}x^{-\frac{1}{p}}|v(x)|. \tag{11}$$

Therefore,

$$\|T\chi_l\|_p \ge c'\eta_{p,l+1}(v)\|\chi_l\|_p. \tag{12}$$

The lower estimate in (2) follows immediately.

Compactness: sufficiency. Assuming first that $v(x) = 0$ outside a finite interval $[a, b] \subset \mathbb{R}_+$, we set $u_0(x) = x^{-\nu}|v(x)|$; then $u_0 \in L_p(a, b)$. Evidently, T can be viewed as an operator in $L_p(0, b)$. The Hölder inequality yields $|Tf(x)| \le C(b)\|f\|_p u_0(x)$. In the terminology of [8, Section 6], this means that the operator T is u_0-*bounded.* This implies its compactness ([8, Th. 6.3]).

For general v, the assumption $\eta_{p,k}(v) \to 0$ and inequality (11) show that $T_{\nu,v}$ can be represented as the sum of a convergent (in norm) series of compact operators; therefore, it is compact itself.

Compactness: necessity. Let T be compact. The sequence $\varphi_l = 2^{-\frac{l}{p}}\chi_l$, $l \in \mathbb{Z}$ converges to 0 weakly in L_p, whence $T\varphi_l \to 0$ in norm. Now, inequality (12) implies that $\eta_{p,l}(v) \to 0$ as $l \to \infty$.

3. Theorem 2: Reduction to Main Lemma

We shall deduce Theorem 2 from the following statement.

Lemma 3 (Main Lemma). *Let $T_{k,l}$ be the operators* (6). *Then for any $k, l \in \mathbb{Z}$ we have*

$$n(\varepsilon, T_{k,l}) \le C_{13} 2^{-\frac{l-k}{p'}} \eta_{p,l}^{\frac{1}{\nu}}(v) \varepsilon^{-\frac{1}{\nu}}, \qquad C_{13} = C(\nu, p). \tag{13}$$

We postpone the proof until the end of the paper. Now we show how to obtain Theorem 2 from this lemma.

We recall the Weyl inequality for the approximation numbers:

$$a_{m+n-1}(A+B) \le a_m(A) + a_n(B) \tag{14}$$

(see, e.g., [3, Prop. I.2.2]. Translated into the language of counting functions and extended to the sums of an arbitrary number of operators, this inequality takes the form

$$n(\varepsilon, \sum_j A_j) \le \sum_j n(\varepsilon\theta_j, A_j), \qquad \theta_j \ge 0,\ \sum_j \theta_j = 1. \tag{15}$$

Proof of (3). Due to the block-diagonal structure of the operators T_j, we have

$$n(\varepsilon, T_j) \le \sum_{l \in \mathbb{Z}} n(\varepsilon, T_{l-j,l}). \tag{16}$$

Suppose that $\boldsymbol{\eta}_p(v) \in l_{\frac{1}{\nu}}$ and denote $M = \sum_{l\in\mathbb{Z}} \eta_{p,l}^{\frac{1}{\nu}}(v)$. Then (16) and (13) imply

$$n(\varepsilon, T_j) \le C_{13} M 2^{-\frac{j}{p'}} \varepsilon^{-\frac{1}{\nu}}, \qquad j = 0, 1, \dots$$

In accordance with to (15), this yields

$$n(\varepsilon, T_{\nu,v}) \le C_{13} M \sum_{j=0}^{\infty} 2^{-\frac{j}{p'}} (\varepsilon\theta_j)^{-\frac{1}{\nu}}.$$

Here we choose $\theta_j = c2^{-\kappa j}$, where c is the normalizing factor. If $\kappa < \nu/p'$, the series converges, and we arrive at (3).

Proof of (4). We need the following elementary fact.

Lemma 4. *Let $E_k = \{e_{k,n}\}_{n\in\mathbb{N}}$, $k \in K$, be a countable family of sequences. Suppose that*

$$n(\varepsilon, E_k) \le A_k \varepsilon^{-\alpha} \tag{17}$$

and

$$n(\varepsilon, \{A_k\}) \le B\varepsilon^{-\beta} \tag{18}$$

with some $\beta > 1$, so that, in general, the series $\sum_k A_k$ diverges. Then for the "double sequence" $E = \{e_{k,n}\}_{k\in K, n\in\mathbb{N}}$ we have

$$n(\varepsilon, E) \le \beta(\beta-1)^{-1} B \varepsilon^{-\alpha\beta}. \tag{19}$$

Proof. From (17) it follows that $n(\varepsilon, E_k) = 0$ if $A_k < \varepsilon^\alpha$. Therefore,

$$\begin{aligned} n(\varepsilon, E) &= \sum_k n(\varepsilon, E_k) \le \varepsilon^{-\alpha} \sum_{k: A_k \ge \varepsilon^\alpha} A_k = -\varepsilon^{-\alpha} \int_{\varepsilon^\alpha}^\infty \tau dn(\tau, \{A_k\}) \\ &= \varepsilon^{-\alpha} \Big(\int_{\varepsilon^\alpha}^\infty n(\tau, \{A_k\}) d\tau + \varepsilon^\alpha n(\varepsilon^\alpha, \{A_k\}) \Big) \\ &\le B\varepsilon^{-\alpha} \Big(\int_{\varepsilon^\alpha}^\infty \tau^{-\beta} d\tau + \varepsilon^{\alpha - \alpha\beta} \Big), \end{aligned}$$

which implies (19). □

Now, suppose that $\boldsymbol{\eta}_p(v) \in l_{q,\infty}$ for some $q > \nu^{-1}$, that is

$$n(\varepsilon, \boldsymbol{\eta}_p(v)) \le D\varepsilon^{-q}. \tag{20}$$

Fixing j, we apply Lemma 4 to the sequences $E_k = \{a_n(T_{k,k+j})\}$, $k \in \mathbb{Z}$. By (13), inequality (17) is satisfied with $\alpha = \nu^{-1}$ and $A_k = C_{13} 2^{-\frac{j}{p'}} \eta_{p,k+j}^{\frac{1}{\nu}}$. Together with (20), this yields inequality (18) with $\beta = q\nu$ and $B = DC_{13}^{q\nu} 2^{-\frac{jq\nu}{p'}}$. Now, from (16) and (19) we obtain

$$n(\varepsilon, T_j) \le C 2^{-\frac{jq\nu}{p'}} \varepsilon^{-q}.$$

Passage to the operator T, based upon (15), is the same as in the proof of (3), and we skip it.

4. Piecewise-polynomial approximations

The rest of the paper is devoted to the proof of Main Lemma. We deduce it from some results on weighted approximation of functions in the Sobolev spaces $W^{l,1}$. The results we need were obtained, in a more general setting, in [1]. In this Section we reproduce some of this material. For the reader's convenience, the proofs are presented.

Let $\mathcal{P}_r$ stand for the linear space of all polynomials on $\mathbb{R}$ of degree at most r, $r = 0, 1, \dots$. For an arbitrary finite interval Δ, the space $\mathcal{P}_r$ can be regarded as a subspace of $L_1(\Delta)$. Choose a linear projection $P_r : L_1(0,1) \to \mathcal{P}_r$. We recall that any such projection can be represented as

$$(P_r u)(t) = \sum_{j=0}^r t^j \int_0^1 u(s) m_j(s) ds, \tag{21}$$

where the functions $m_j \in L_\infty(0,1)$ constitute a family biorthogonal to the family $\{t^j\}$, $0 \le j \le r$. By $P_{r,\Delta}$ we denote the "transfer" of the projection P_r to the interval $\Delta \in \mathbb{R}$:

$$(P_{r,\Delta} u) \circ \varphi_\Delta = P_r (u \circ \varphi_\Delta), \qquad \varphi_\Delta(x) = a + (b-a)x.$$

Next, suppose that we are given a partition Ξ of an interval $I \subset \mathbb{R}$ into the union of a finite number of its subintervals: $\Xi = \{\Delta_1, \dots, \Delta_n\}$; here we do not distinguish between closed and open intervals. The mapping $P_{r,\Xi} : u \mapsto f$, where $f \restriction \Delta_j = P_{r,\Delta_j} u$, is a projection of the space $L_1(I)$ onto its subspace consisting of the piecewise-polynomial functions (of degree $\le r$). The operators $P_{r,\Xi}$ will serve us as the instrument of approximation in the weighted L_1-spaces. Note that $\operatorname{rank} P_{r,\Delta} = r+1$ and $\operatorname{rank} P_{r,\Xi} = (r+1)n$.

In what follows $|I|$ stands for the length of an interval $I \subset \mathbb{R}$.

Lemma 5. *Let $I \subset \mathbb{R}$ be a finite interval, let $0 \le V \in L_1(I)$, and let $l \ge 1$. Then for any $r \ge l-1$ and any $n \in \mathbb{N}$ there is a partition $\Xi = \Xi(V, r, n)$ of I into n subintervals such that*

$$\int_I |u - P_{r,\Xi} u| V dx \le C_{22} |I|^{l-1} n^{-l} \int_I |u^{(l)}| dx \int_I V dx, \qquad u \in W^{l,1}(I). \tag{22}$$

For the proof we need the following lemma; it is a slightly facilitated version of Theorem 2.2 in [1].

Lemma 6. *Let a finite interval I and a number $\alpha \ge 0$ be given, and let $V \ge 0$ be a function of class $L_1(I)$. Then for an arbitrary $n \ge 1$ there exists a partition $\Xi = \Xi(V, \alpha, n)$ of I into n subintervals $\Delta_1, \dots, \Delta_n$ such that*

$$\max_{k=1,\dots,n} \left(|\Delta_k|^\alpha \int_{\Delta_k} V dx \right) \le |I|^\alpha n^{-(\alpha+1)} \int_I V dx. \tag{23}$$

Proof. First, we note that inequalities (23) for different intervals I can be obtained one from another by scaling; thus, it suffices to prove (23) for $I = (0,1)$. Next, we may assume that V is normalized, i.e., $\int_0^1 V dx = 1$.

For $n = 1$ the result of the lemma is obvious. Assuming it for some $n \ge 1$, we take a point $A \in (0,1)$ (its choice will be specified later on). By our hypothesis, there is a partition of $(0, A)$ into n subintervals such that (23) is satisfied. Adding the interval $(A, 1)$, we obtain a partition of $(0,1)$ into $n+1$ subintervals for which

$$\max_{k=1,\dots,n+1} |\Delta_k|^\alpha \int_{\Delta_k} V dx \le \max\left(A^\alpha n^{-(\alpha+1)} \int_0^A V dx;\ (1-A)^\alpha \int_A^1 V dx \right).$$

Denote $X = X(A) = \int_0^A V dx$. Using continuity arguments, we see that it is possible to choose A in such a way that

$$A^\alpha n^{-(\alpha+1)} X = (1-A)^\alpha \int_A^1 V dx = (1-A)^\alpha (1-X).$$

This gives $X^{-1} = 1 + A^\alpha (1-A)^{-\alpha} n^{-(1+\alpha)}$, or

$$\left(A^\alpha n^{-(1+\alpha)} X\right)^{-1} = n^{1+\alpha} A^{-\alpha} + (1-A)^{-\alpha} =: \Psi(A).$$

The minimal value of $\Psi(A)$ for $0 < A < 1$ is $(n+1)^{1+\alpha}$, which implies that for the point A chosen we have

$$A^\alpha n^{-(\alpha+1)} \int_0^A V dx = (1-A)^\alpha \int_A^1 V dx = \Psi^{-1}(A \le (n+1)^{-(1+\alpha)}. \qquad \square$$

Proof of Lemma 5. For any partition Ξ we have

$$\int_I |u - P_{r,\Xi}u|V\,dx = \sum_{k=1}^{n} \int_{\Delta_k} |u - P_{r,\Delta_k}u|V\,dx. \tag{24}$$

For any interval $\Delta \in \mathbb{R}$, the following inequality is satisfied:

$$\|u - P_{r,\Delta}u\|_{L_\infty(\Delta)} \le C_{25}|\Delta|^{l-1} \int_\Delta |u^{(l)}|\,dx, \qquad u \in W^{l,1}(\Delta), \tag{25}$$

where the constant factor C_{25} does not depend on Δ. To see this, consider first $\Delta = (0,1)$. For $r \ge l-1$, the expression $\int_0^1(|u^{(l)}| + |P_r u|)dx$ defines a norm equivalent to the standard norm of the space $W^{l,1}(0,1)$ given by $\sum_{j=0}^{l} \int_0^1 |u^{(j)}|dx$; this is a particular case of the theorem on equivalent norms in the Sobolev spaces $W^{l,p}$ (see, e.g., [13, Sec. 114]). Taking the embedding $W^{l,1} \subset L_\infty$ into account, we obtain $\|u\|_{L_\infty(0,1)} \le c\int_0^1(|u^{(l)}| + |P_r u|)dx$. Replacing here u by $u - P_{r,\Delta_k}u$, we come to (25) for $\Delta = (0,1)$, with $C_{25} = c$. Scaling extends this to an arbitrary $\Delta \subset \mathbb{R}$.

From (24) and (25) it follows that

$$\begin{aligned}\int_I |u - P_{r,\Xi}u|V\,dx &\le C_{25}\sum_{k=1}^{n} |\Delta_k|^{l-1} \int_{\Delta_k} |u^{(l)}|\,dx \int_{\Delta_k} V\,dx \\ &\le C_{25}\int_I |u^{(l)}|\,dx \max_{k=1,\dots,n}\left(|\Delta_k|^{l-1}\int_{\Delta_k} V\,dx\right).\end{aligned} \tag{26}$$

Now the desired result follows from Lemma 4 (with $\alpha = l-1$). □

We conclude this section with a few remarks concerning the vector-valued version of the projections $P_{r,\Delta}$ and $P_{r,\Xi}$. Let X be a Banach space, and let $\mathcal{P}_r^X$ be the space of all polynomials of degree $\le r$ with coefficients in X. Formula (21) applies to the X-valued functions and defines a projection $P_r^X : L_1((0,1);X) \to \mathcal{P}_r^X$. Evidently, for any $t \in (0,1)$,

$$\|(P_r^X u)(t)\|_X \le C_{27}\int_0^1 \|u(t)\|_X\,dt, \qquad C_{27} = \sum_{j=0}^{k} \|m_j\|_\infty, \tag{27}$$

which yields

$$\|P_r^X u\|_{L_p((0,1),X)} \le C_{27}\|u\|_{L_p((0,1),X)}, \qquad 1 \le p \le \infty. \tag{28}$$

The projections P_{r,Δ_k}^X and $P_{r,\Xi}^X$ are introduced in the same way as for the scalar case. An inequality similar to (28) is valid for the operators P_{r,Δ_k}^X, with the same constant factor C_{27}; this follows from (28) by scaling. In its turn, this implies that for any interval $I \subset \mathbb{R}$ and its arbitrary partition Ξ we have

$$\|P_{r,\Xi}^X u\|_{L_p(I,X)} \le C_{29}\|u\|_{L_p(I,X)}, \qquad 1 \le p \le \infty. \tag{29}$$

5. Integral operators

Here we apply the results of Section 4 to estimates for the approximation numbers of integral operators with kernels in $W^{l,1}$.

Let I_0 and I_1 be two finite intervals, and let two weight functions $0 \le V_0 \in L_1(I_0)$ and $0 \le V_1 \in L_1(I_1)$ be given. Suppose also that $K(x,y)$ is a measurable function on $I_0 \times I_1$. The next statement is the classical "Schur test".

Proposition 7. *Suppose that*

$$A_0 = \sup_{x\in I_1} \int_{I_0} |K(x,y)|V_0(y)dy < \infty, \qquad A_1 = \sup_{y\in I_0} \int_{I_1} |K(x,y)|V_1(x)dx < \infty.$$

Then for any p, $1 \le p \le \infty$, the integral operator

$$\mathcal{K} : L_p(I_0, V_0) \to L_p(I_1, V_1), \qquad (\mathcal{K}f)(x) = \int_{I_0} K(x,y)f(y)V_0(y)dy \tag{30}$$

is bounded, and $\|\mathcal{K}\| \le A_0^{\frac{1}{p'}} A_1^{\frac{1}{p}}$.

Theorem 8. *Under the assumptions of Proposition 7, suppose that the (distributional) derivatives of the kernel K of some order m satisfy*

$$B_0 = B_0(K,m) = \sup_{x\in I_1} \int_{I_0} |K_y^{(m)}(x,y)|dy < \infty \tag{31}$$

and

$$B_1 = B_1(K,m) = \sup_{y\in I_0} \int_{I_1} |K_x^{(m)}(x,y)|dx < \infty. \tag{32}$$

Then for the approximation numbers $a_n(\mathcal{K})$ we have

$$a_n(\mathcal{K}) \le C\Big(|I_0|^{m-1}B_0 \int_{I_0} V_0 dy\Big)^{\frac{1}{p'}} \Big(|I_1|^{m-1}B_1 \int_{I_1} V_1 dx\Big)^{\frac{1}{p}} n^{-m}, \qquad n > 2m. \tag{33}$$

Proof. For $i = 0,1$ and a given $n \in \mathbb{N}$, let Ξ_i be the partition of the interval I_i, constructed in accordance with Lemma 4 for $r = m-1$ and the weight function V_i. We denote by P_0^y and P_1^x the corresponding operator P_{r,Ξ_i}, acting in the variable written in the upper index. Consider the kernel $K' = K'(x,y)$,

$$K' = P_0^y K + P_1^x K - P_0^y P_1^x K;$$

then $\operatorname{rank} K' \le 2mn$. We have

$$K' - K = (P_0^y - I)(P_1^x - I)K = (P_1^x - I)(P_0^y - I)K.$$

Inequality (22) and the assumption (31) yield

$$\sup_{x\in I_1} \int_{I_0} |(P_0^y - I)K(x,y)|V_0(y)dy \le C_{22}B_0|I_0|^{m-1}n^{-m}.$$

This means that the function $(P_0^y - I)K$ can be regarded as an element of the space $L_\infty(I_1, L_{1,V_0}(I_0))$. Now, from (29) (for $p = \infty$) it follows that

$$\sup_{x\in I_1} \int_{I_0} |K'(x,y) - K(x,y)|V_0(y)dy \le C'|I_0|^{m-1}B_0 n^{-m}, \qquad C' = 1 + C_{27}.$$

Interchanging the roles of x and y, we obtain a similar inequality for the integral in dx. We recall that $\operatorname{rank} K' \le 2mn$. Using the definition of approximation numbers and Proposition 7, we get

$$a_{2mn+1}(\mathcal{K}) \le \|K' - K\|_{L_p(I_0,V_0)\to L_p(I_1,V_1)}$$
$$\le C'\left(|I_0|^{m-1}B_0 \int_{I_0} V_0 dy\right)^{\frac{1}{p'}} \left(|I_1|^{m-1}B_1 \int_{I_1} V_1 dx\right)^{\frac{1}{p}} n^{-m}, \quad n \ge 1.$$

This implies (33) with $C = C'(2m)^m$. □

6. The kernel $(x-y)_+^{\nu-1}$

A direct application of Theorem 8 to this kernel does not give a sharp decreasing rate for its approximation numbers. Indeed, its smoothness properties cannot be described in terms of the Sobolev spaces $W^{l,1}$, provided the segments $\bar{I}_0$ and $\bar{I}_1$ have nonempty intersection. However, an elementary interpolation argument saves the situation.

Proposition 9. (i) *Let $\nu = l + \theta$, where $l = 0, 1, \dots$ and $0 < \theta < 1$. Then for any $h > 0$ there is a decomposition*

$$t_+^{\nu-1} = \varphi_h(t) + \psi_h(t) \tag{34}$$

such that

$$\int_{\mathbb{R}} |\varphi_h^{(l+1)}(t)|dt \le C_{35}h^{\theta-1}, \qquad \int_{\mathbb{R}} |\psi_h^{(l)}(t)|dt \le C'_{35}h^{\theta}. \tag{35}$$

(ii) *Let $l \in \mathbb{N}$. Then for any $h > 0$ there is a decomposition $t_+^l = \varphi_h(t) + \psi_h(t)$ such that $\int_{\mathbb{R}} |\varphi_h^{(l+1)}(t)|dt \le \tilde{C}h^{-1}$ and $\int_{\mathbb{R}} |\psi_h^{(l-1)}(t)|dt \le \tilde{C}'h$.*

This result should be viewed as well known; we only mention that for the proof of (i) one takes $\varphi_h(t) = \frac{1}{2h}\int_{-h}^{h}(t+s)_+^{\nu-1}ds$ (the Steklov mean of $t_+^{\nu-1}$), and, for the proof of (ii), $\varphi_h(t) = h^{-2}\int_{-h}^{h}(h-|s|)(t+s)_+^l ds$ (the Steklov mean of the second order). In both cases, naturally, $\psi_h(t) = t_+^{\nu-1} - \varphi_h(t)$. The estimates for the derivatives of φ_h and ψ_h are quite standard and we skip them.

Theorem 10. *Let $\nu \ge 1$, and let $\mathcal{K}_\nu$ be the integral operator* (30) *with the kernel*

$$K(x,y) = (x-y)_+^{\nu-1}. \tag{36}$$

Then

$$a_n(\mathcal{K}_\nu) \le C_{37}\left(|I_0|^{\nu-1}\int_{I_0} V_0 dy\right)^{\frac{1}{p'}}\left(|I_1|^{\nu-1}\int_{I_1} V_1 dx\right)^{\frac{1}{p}} n^{-\nu}$$
$$C_{37} = C(\nu, p), \qquad n > 4([\nu]+1). \tag{37}$$

Proof. The cases of noninteger ν and of integer ν (except for $\nu = 1$) can be treated in the same way. For definiteness, let $\nu \notin \mathbb{N}$. We make use of the decomposition (34), with a parameter h to be chosen later on. Then

$$(x-y)_+^{\nu-1} = K_h(x,y) + K_h'(x,y) := \varphi_h(x-y) + \psi_h(x-y).$$

From (35) it follows that, uniformly with respect to the intervals I_0 and I_1,

$$B_i(K_h, l+1) \le C_{35}h^{\theta-1}, \qquad B_i(K_h', l) \le C_{35}'h^{\theta}, \qquad i = 0, 1.$$

By Theorem 8,

$$a_n(\mathcal{K}_h) \le Ch^{\theta-1}\left(|I_0|^l\int_{I_0} V_0 dy\right)^{\frac{1}{p'}}\left(|I_1|^l\int_{I_1} V_1 dx\right)^{\frac{1}{p}} n^{-(l+1)}, \qquad n > 2(l+1),$$

and

$$a_n(\mathcal{K}_h') \le Ch^{\theta}\left(|I_0|^{l-1}\int_{I_0} V_0 dy\right)^{\frac{1}{p'}}\left(|I_1|^{l-1}\int_{I_1} V_1 dx\right)^{\frac{1}{p}} n^{-l}, \qquad n > 2l.$$

Using Weyl's inequality (14), from here we deduce that, at least for $n > 2l+2$,

$$\begin{aligned} a_{2n-1}(\mathcal{K}_\nu) &\le a_n(\mathcal{K}_h) + a_n(\mathcal{K}_h') \\ &\le C\left(\int_{I_0} V_0 dy\right)^{\frac{1}{p'}}\left(\int_{I_1} V_1 dx\right)^{\frac{1}{p}} \\ &\quad \times \left(h^{\theta-1}|I_0|^{\frac{l}{p'}}|I_1|^{\frac{l}{p}}n^{-(l+1)} + h^{\theta}|I_0|^{\frac{l-1}{p'}}|I_1|^{\frac{l-1}{p}}n^{-l}\right). \end{aligned}$$

Taking $h = |I_0|^{\frac{1}{p'}}|I_1|^{\frac{1}{p}}n^{-1}$ and changing the value of the constant factor, we get (37).

Proposition 9(ii) does not apply to $l = 0$ (that is, to $\nu = 1$). Consider this case separately. First, we derive an estimate which replaces Lemma 5 in this case.

Given a function $0 \le V \in L_1(I)$ and a number $n \in \mathbb{N}$, take a partition $\Xi = \{\Delta_1, \dots, \Delta_n\}$ of the interval I, such that

$$\int_{\Delta_1} V dx = \dots = \int_{\Delta_n} V dx = \frac{1}{n}\int_I V dx.$$

Let $u(x) = \chi_+(x-a)$, where χ_+ is the Heaviside function. For this u, the right-hand side of identity (24) (in which we take $r = 0$) reduces to the single term corresponding to the interval $\Delta_k \ni a$. Thus,

$$\int_I |u - P_{0,\Xi}u| V dx = \int_{\Delta_k} |u - P_{0,\Delta_k}u| V dx \le \int_{\Delta_k} V dx = n^{-1}\int_I V dx.$$

Leaning upon this estimate, for the proof of (37) with $\nu = 1$ we can repeat the argument in the proof of Theorem 8. □

7. Proof of Main Lemma

Setting $V_0(y) \equiv 1$ and $V_1(x) = \frac{|v(x)|^p}{x^{\nu p}}$, we observe that

$$\int_{\Delta_l} V_1(x)dx \le 2^{-l(p\nu-1)}\eta^p_{p,l}(v). \tag{38}$$

Consider the integral operators

$$\mathfrak{L}_{k,l} : L_p(\Delta_k) \to L_p(\Delta_l, V_1) : \qquad (\mathfrak{L}_{k,l}f)(x) = \int_{\Delta_k} (x-y)_+^{\nu-1} f(y)dy.$$

It is clear that the approximation numbers of the operators $\mathfrak{L}_{k,l}$ and $T_{k,l}$ coincide. Theorem 10 applies to $\mathfrak{L}_{k,l}$, and from (37) and (38) we see that, for $n \ge 4([\nu]+1)$,

$$a_n(T_{k,l}) \le C_{37} 2^{\frac{k\nu}{p'}} \left(2^{-l\nu(p-1)}\eta^p_{p,l}(v)\right)^{\frac{1}{p}} n^{-\nu} = C_{37}\eta_{p,l}(v)2^{-\frac{(l-k)\nu}{p'}} n^{-\nu}. \tag{39}$$

For $1 \le n < 4([\nu]+1)$ we use inequality (10):

$$a_n(T_{k,l}) \le \|T_{k,l}\| \le c\eta_{p,l}(v)2^{-\frac{(l-k)\nu}{p'}} \le c(4([\nu]+1))^\nu \eta_{p,l}(v) 2^{-\frac{(l-k)\nu}{p'}} n^{-\nu}.$$

It follows that, actually, (39) is fulfilled (possibly, with another constant factor) for all $n \in \mathbb{N}$. This is equivalent to (13).

Acknowledgement

The paper was prepared during my visit (supported by the EPSRC grant Gr/MO 8301) to the University of Sussex. I take this opportunity to express my gratitude to Professor D. E. Edmunds for his hospitality at this University and also for numerous discussions on the subject of the paper. I am also grateful to Professors W. D. Evans and V. D. Stepanov for their helpful remarks.

References

[1] M. Sh. Birman and M. Solomyak, *Piecewise polynomial approximations of functions of classes* W_p^α, Mat. Sb. (N.S.), **73(115)** (1967) no. 3, 331–355 (in Russian); English transl., Math. USSR Sb., **73 (115)** (1967), 295–317.

[2] M. Sh. Birman and M. Solomyak, *Estimates for the singular numbers of integral operators*, Uspehi Mat. Nauk, **32, 1(193)** (1977), 17–84 (in Russian); English transl., Russ. Math. Surveys, **32 (1)** (1977), 15–89.

[3] D. E. Edmunds and W. D. Evans, *Spectral theory and differential operators*, Oxford University Press, NY, **1987**.

[4] D. E. Edmunds, W. D. Evans, and D. J. Harris, *Approximation numbers of certain Volterra integral operator*, J. London Math. Soc. (2), **37 (3)** (1988), 71–489.

[5] D. E. Edmunds, W. D. Evans, and D. J. Harris, *Two-sided estimates of the approximation numbers of certain Volterra integral operators*, Studia Math., **124 (1)** (1997), 59–80.

[6] W. D. Evans, D. J. Harris, and J. Lang, *On the approximation numbers of Hardy-type operators in L_∞ and L_1*, Studia Math., **130 (2)** (1998), 171–192.

[7] H. König, *Eigenvalue distribution of compact operators*, Birkhäuser Verlag, Basel–Boston, **1986**.

[8] Krasnoselskii et al., *Integral operators in spaces of summable functions*, "Nauka", Moscow, **1966** (Russian); English transl., Noordhoff International Publishing, Leiden, **1976**.

[9] F. J. Martín-Reyes and E. Sawyer, *Weighted inequalities for Riemann-Liouville fractional integrals of order one and greater*, Proc. Amer. Math. Soc., **106** (1989).

[10] J. Newman and M. Solomyak, *Two-sided estimates on singular values for a class of integral operators on the semiaxis*, Integr. Equat. Oper. Th., **20** (1994), 335–349.

[11] A. Pietsch, *Operator ideals*, North-Holland, Amsterdam, **1980**.

[12] D. V. Prokhorov, *On the boundedness and compactness of a class of integral operators*, Research Report No. 98/33, Khabarovsk: Computer Center FEB RAS, **1998**.

[13] V. I. Smirnov, *A course of higher mathematics*, Gosudarstv. Izdat. Fiz.-Mat. Lit., Moscow, **1959** (Russian); English transl., Pergamon Press, Oxford, NY, **1964**.

[14] V. D. Stepanov, *Two-weight estimates for Riemann-Liouville integrals*, Report No. 39, Ceskoslovenska Akademie Věd, Matematický Ústav, Praha, **1988**.

[15] V. D. Stepanov, *Weighted norm inequalities for integral operators and related topics*, Nonlinear analysis, function spaces and applications (Prague, 1994), **5**, 139–175, Prometheus, Prague, **1994**.

Department of Theoretical Mathematics,
The Weizmann Institute of Science,
Rehovot 76100, Israel
E-mail address: `solom@weizmann.wisdom.ac.il`

Operator Theory:
Advances and Applications, Vol. 113

Special Transformations of Cauchy Type Integral Spaces

Genrikh Ts. Tumarkin

Dedicated to the memory of S. A. Vinogradov

Abstract. We consider the space $K_p(G)$, $p \geq 1$, of analytic functions $f(z)$ representable by Cauchy type integrals in a simply connected domain G with rectifiable boundary γ. We suppose that the density function ω is in $L_p(\gamma)$. Let $z = \phi(w)$ ($z = \phi^-(w)$) be a conformal mapping of the unit disk D onto G (or of the domain $|w| > 1$ onto the complement to $\overline{G}$, respectively). An important result of G. David (for $p > 1$) describes the class $\mathcal{K}_D$ of all domains for which the mapping $z = \phi(w)$ induces the linear transformation

$$f \mapsto f(\phi)[\phi']^{1/p}$$

of $K_p(G)$ to $K_p(D)$. Earlier (1960), the author introduced the class $\mathcal{K}$ of domains with a similar property for $p = 1$. Now we introduce more general classes of domains, that include $\mathcal{K}_D$ and $\mathcal{K}$, and investigate some properties of these classes with applications to approximation problems. Using the mapping ϕ^-, we construct a special isometric transformation of $K_p(G)$ onto $K_p(D)$ (for a general domain G) that allows us to get results on approximation by rational functions with preassigned poles.

1. Introduction

We consider the space $K_p(G)$, $p \geq 1$, of analytic functions $f(z)$ representable by Cauchy type integral (**CTI**) in a simply connected domain G with a rectifiable boundary γ:

$$f(z) = \int_\gamma \frac{\omega(\zeta)d\zeta}{\zeta - z}, \quad z \in G, \quad \omega(\zeta) \in L_p(\gamma), \quad p \geq 1. \tag{1}$$

We assume that the complement G^- to the closure of G contains ∞. The norm in the space $K_p(G)$ is defined in a natural way:

$$\|f\|^p_{K_p} = \inf \int_\gamma |\omega(\zeta)|^p |d\zeta|,$$

where the infimum is taken over all possible ω in (1). Let $z = \phi(w)$ (and $z = \phi^{-}(w), \phi^{-}(\infty) = \infty$) be a conformal mapping of the unit disk D onto G (or of the domain $|w| > 1$ onto G^{-}, respectively); $H_p(D)$ denotes the Hardy space.

An important 1984 result of G. David [1] describes the class $\mathcal{K}_D$ of all domains for which the mapping $z = \phi(w)$ induces a linear transformation of $K_p(G)$ into $H_p(D) = K_p(D)$ for $p > 1$. Earlier in 1960 (see [9]), the author introduced a class $\mathcal{K}$ of domains with a similar property for $p = 1$. Here we **define** these classes:

$$G \in \mathcal{K}_D \quad \text{if} \quad (f \in K_p(G) \Rightarrow f(\phi)\,[\phi']^{\frac{1}{p}} \in K_p(D)), \quad p > 1. \tag{2}$$

and

$$G \in \mathcal{K} \quad \text{if} \quad (\forall f \in K_1(G) \Rightarrow f(\phi)(\phi') \in K_1(D)). \tag{3}$$

The investigation of Cauchy type integrals has made a significant progress recently. However, many problems dealing with the boundary properties of **CTI** in general domains are still open. It seems to be useful to describe the classes of domains G for which there exist "good" transformations of spaces of **CTI** in G to spaces of **CTI** in the disk D. Such transformations are viewed as "good" if some properties of **CTI** in D are valid for the corresponding **CTI** in G.

Here, in Section 3, we investigate the properties of certain classes of domains, which include the classes $\mathcal{K}_D$ and $\mathcal{K}$, and in Section 4 we give some applications to approximation problems. Using the mapping $z = \phi^{-}(w)$, in Section 5 we introduce a special isometric transformation of $K_p(G)$ onto $K_p(D)$ for a general domain G that allows us to get results on approximation of **CTI** by rational functions with preassigned poles in $K_p(G)$.

In what follows, we use some known facts about the boundary behaviour of analytic functions; these facts can be found in [3], [6], and [7].

2. Known results on Cauchy type integrals

I. The angular boundary values are known to exist a.e. on γ for any $f(z) \in K_p(G)$, $p \geq 1$ (Calderon, 1977).

II. David (1984, [1]) completely described the class of γ for which the singular Cauchy operator is bounded in L_2: γ must be regular in the Ahlfors sense (A-regular), i.e.,

$$\text{measure}\,\{\gamma \cap \{|\zeta - \zeta_0| < \rho\}\} < C\rho, \quad \zeta_0 \in \gamma.$$

The case of the disk was thoroughly investigated.

The following result of M. Riesz [1927] is well known:

$$K_p(D) = H_p(D), \quad p > 1.$$

In 1932 (see [8]), V. I. Smirnov proved that

$$K_1(D) \subset H_\delta, \quad \delta < 1.$$

David's theorem may be formulated as follows:

$$K_p(G) = E_p(G) \Leftrightarrow \gamma \text{ is } A\text{-regular}, \quad p > 1.$$

Here $E_p(G)$ is the Smirnov class:

$$f(z) \in E_p(G) \Leftrightarrow \exists \gamma_n \to \gamma,\ \gamma_n \subset G,\ \int_{\gamma_n} |f(z)|^p |dz| \leq C(f).$$

It is known (V. Smirnov, 1932) that

$$f(z) \in E_p(G) \Leftrightarrow f[\phi(w)][\phi'(w)]^{\frac{1}{p}} \in H_p(D).$$

Once again, we can formulate David's theorem, using Riesz's theorem and Smirnov's results, as follows:

for all $p > 1$,

$$\{f(z) \in K_p(G) \Rightarrow f[\phi(w)][\phi'(w)]^{\frac{1}{p}} \in K_p(D)\} \Leftrightarrow \{\gamma \text{ is } A\text{-regular}\}.$$

Previously, in [9], [14], [15] (1960, 1963, 1971), the author introduced and investigated the class $\mathcal{K}$ of domains G with the property (3) or with equivalent property for the spaces $K_S(G)$ of Cauchy–Stieltjes type integrals: $G \in \mathcal{K}$ if

$$f(z) \in K_S(G) \Rightarrow f[\phi(w)]\phi'(w) \in K_S(D). \tag{4}$$

For the space E_1 of analytic functions $f(z)$ that can be expressed by Cauchy integrals (the density function coincides with the boundary values of $f(z)$), a similar property is well known for a general domain G with rectifiable boundary γ (Smirnov, 1932), namely,

the possibility of the representation

$$f(z) = \frac{1}{2\pi i}\int_\gamma \frac{f(\zeta)d\zeta}{\zeta - z}, \quad z \in G,$$

is equivalent to that of the representation

$$f[\phi(w)]\phi'(w) = \frac{1}{2\pi}\int_0^{2\pi} \frac{e^{i\theta} f[\phi(e^{i\theta})]\phi'(e^{i\theta})d\theta}{e^{i\theta} - w}. \tag{5}$$

On the other hand, not all $f(z) \in K_1(G)$ representable by Cauchy type integrals have the property similar to (5) for a general domain G with rectifiable boundary. This follows from the next necessary condition [Tumarkin, 1963] for G to belong to the class $\mathcal{K}$ defined by (3):

$$G \in \mathcal{K} \Rightarrow G \in S.$$

Here S is the Smirnov class of domains, i.e., $G \in S$ if $\ln|\phi'(w)|$ admits the Poisson integral representation in the domain $|w| < 1$.

The Smirnov class is rather large, but not all domains with rectifiable boundary belong to S. The first examples of "pathological" domains G that fail to belong to S were constructed in [5] (see also [7]). Another approach to the construction of domains G that do not belong to S was given in [2].

Note that one of **sufficient conditions** for G to belong to $\mathcal{K}$ is that the curve $\gamma : \zeta = \zeta(\tau)$ is of bounded rotation,

$$\mathrm{Var}_\gamma \arg \zeta'(\tau) < \infty,$$

or, more generally,

$$\mathrm{Var}_\gamma \arg(\zeta - \zeta_0) < A, \quad \zeta, \zeta_0 \in \gamma \tag{6}$$

(see [12–14]).

Any domain $G \in \mathcal{K}$ has the following characteristic property (Tumarkin, [9]).

The transformation induced by a conformal mapping $z = \phi(w), w = \psi(z)$ of the set of rational functions $\{R(z)\}$ with poles in D to the set of rational functions $\{r(z)\}$ with poles in G is bounded:

$$\{R(w) \leftrightarrow r(z)\} \Rightarrow \{\|r(z)\|_{C(\gamma)} \leq A_G \|R\|_{C[|w|=1]}\}. \tag{7}$$

Here $r(z)$ is the sum of the principal parts of the expansions of $R(\phi(z))$.

We note that, later, without mentioning Tumarkin's results, some authors (E. Dynkin, J. Andersson, etc. (see [4])) used a property similar to (7) to introduce the Faber class of domains (using the mapping $z = \phi^-(w)$.) Some properties established by Tumarkin (1965–1971, [12], [13], [14]) were proved once again (e.g., in [4], the sufficient condition (6) for $G \in \mathcal{K}$ or $G^- \in \mathcal{K}$ was attributed to Kővari (1972)).

3. The class $\mathcal{K}^{(0)}$ and its subclasses

Now we introduce a class of domains $\mathcal{K}^{(0)}$ containing as subclasses the class $\mathcal{K}$ defined by (3) and the David class $\mathcal{K}_D$ defined by (2).

Definition. $G \in \mathcal{K}^{(0)}$ *if for any* $f(z) \in K_\infty(G)$ *the function* $f[\phi(w)]\phi'(w)$ *can be represented in* D *by a Cauchy–Stieltjes type integral.*

I. **Necessary and sufficient condition** for G to belong to $\mathcal{K}^{(0)}$ (similar to the condition in (7) ensuring that G belongs to $\mathcal{K}$) is as follows: $G \in \mathcal{K}^{(0)}$ *if and only if*

$$(R(w) \leftrightarrow r(z)) \Rightarrow (\|r(z)\|_{L_1(\gamma)} \leq A_G \|R\|_{C[|w|=1]}).$$

Here $R(w)$ is a rational function with all poles lying inside D, and the corresponding rational function $r(z)$, the sum of the principal parts of $R(\psi(z))$, has all poles inside G. We denoted by $w = \psi(z)$ a conformal mapping G onto D.

II. Another **necessary and sufficient condition** for G to belong to $\mathcal{K}^{(0)}$ looks like this:

$$\{h^-_\infty(w) \in H_\infty\{|w| > 1\}, h^-_\infty(\infty) = 0\} \Rightarrow \{f^-(z) \in E_1(G^-)\}. \tag{8}$$

Here

$$f^-(z) = \int_\gamma \frac{h^-_\infty[\psi(\zeta)]d\zeta}{\zeta - z}, \quad z \in G^-.$$

(Note that $G \in \mathcal{K}$ if and only if $\sup |f^-(z)| < \infty$).

III. **A necessary condition** for $G \in \mathcal{K}^{(0)}$ is $G \in S$, i.e.,

$$\mathcal{K}^{(0)} \subset S.$$

We introduce **two subclasses** of $\mathcal{K}^{(0)}$, the classes $\tilde{\mathcal{K}}^{(0)}$ and $\hat{\mathcal{K}}^{(0)}$.

Definition.

$$G \in \tilde{\mathcal{K}}^{(0)} \iff (\{f(z) \in K_\infty(G)\} \Rightarrow \{f[\phi(w)]\phi'(w) \in H_1\}).$$

Definition.

$$G \in \hat{\mathcal{K}}^{(0)} \iff \int_\gamma \sup |r(\zeta)||d\zeta| < \infty,$$

where the supremum is taken over all rational r for which the corresponding R satisfies

$$\|R(w)\|_{C\{|w|=1\}} \leq 1.$$

The class $\tilde{\mathcal{K}}^{(0)}$ contains $\mathcal{K}_D$, and the class $\hat{\mathcal{K}}^{(0)}$ contains $\mathcal{K}$.

IV. **A sufficient condition** for $G \in \hat{\mathcal{K}}^{(0)}$:

$$\int_\gamma \operatorname{Var}_{\zeta\in\gamma} \arg(\zeta - \zeta_0)|d\zeta_0| < \infty. \tag{9}$$

V. If $G \in \mathcal{K}^{(0)}$, then

$$\{f(z) \in K_\infty(G)\} \Rightarrow \{f(z) \in N^+(G)\},$$

where N^+ is the Smirnov subclass of the Nevanlinna class N.

Using the mapping ϕ^- in place of ϕ, we extend all the definitions and properties of the above classes to the case of the domain G^-.

4. Applications to approximation of Cauchy type integrals

Let $G^- \in \hat{\mathcal{K}}^{(0)}$ (in particular, γ satisfies (9)). Let $F(w) \in C_A$ (C_A is the space of functions analytic in $|w| < 1$ and continuous in $|w| \leq 1$) and let $f(z)$ correspond to $F(w)$:

$$f(z) = \frac{1}{2\pi i}\int_\gamma \frac{F[\psi^-(\zeta)]d\zeta}{\zeta - z}, \quad z \in G. \tag{10}$$

Here $w = \psi^-(z)$ is a conformal mapping of G^- onto the domain $|w| > 1$. Let a pair of functions $f(z)$, $F(w)$ be as in (10), and let another pair $r(z)$, $R(w)$ consist of a rational function $R(w)$ with poles in $|w| > 1$ and the corresponding $r(z)$ (with poles in G^-); $r(z)$ is the sum of the principal parts and constants in the expansions of $R(\psi^-(z))$. Note that if $F(w) = R(w)$, then in (10) we have $f(z) = r(z)$.

For $z \in G$, we have the **inequality**

$$|f(z) - r(z)| \leq |Q(z)| \, \|F(e^{i\theta}) - R(e^{i\theta})\|_{C\{|w|=1\}}. \tag{11}$$

The function Q(z) may be expressed in terms of the Green function $g(z, z_0)$ of G:

$$\ln|Q(z)| = \frac{1}{2\pi}\int_\gamma \ln\nu(\zeta)\,\frac{\partial g}{\partial n}|d\zeta|, \quad \nu(\zeta) = \sup|r(\zeta)|,$$

where the supremum is taken over all rational r for which the corresponding R satisfies

$$\|R(w)\|_{C\{|w|=1\}} \leq 1.$$

From the definition of $\hat{\mathcal{K}}^{(0)}$ it follows that $Q(z) \in E_1(G)$. Using (11), for the γ-n-level curves of the Green function g we obtain

$$\int_{\gamma_n} |f(z) - r(z)|\,|dz| \le A\|F(e^{i\theta}) - R(e^{i\theta})\|_{C\{|w|=1\}},$$

$$\gamma_n \subset G, \quad \gamma_n \to \gamma, \quad A = \int_\gamma \nu(\zeta)|d\zeta|. \tag{11a}$$

Consequently,

$$\int_\gamma |f(\zeta) - r(\zeta)|\,|d\zeta| \le A\|F(e^{i\theta}) - R(e^{i\theta})\|_{C\{|w|=1\}}. \tag{11b}$$

Inequalities (11), (11a), and (11b) generalize the known results on rational approximation of analytic functions in the Faber domains (see [4]).

5. Isometric transformations of CTI spaces

To solve rational approximation problems for **CTI** (in the space $K_p(G)$ for general domains G with rectifiable boundary) we used special isometric transformation T_p of $K_p(G)$ onto $K_p(D)$ (Tumarkin [16]). To define the transformation T_p for a general domain G with rectifiable boundary, we use the conformal mapping $z = \phi^-(w)$ of the domain $|w| > 1$ onto G^- and the inner–outer factorization of the function $\phi^{-\prime}(w) \in H_1\{|w| > 1\}$, $\phi^{-\prime}(w) = I^-(w)Q^-(w)$.

Definition of T_p. *With any $f(z) \in K_p(G)$ we associate $T_p[f] \in K_p(D)$. The correspondence will be one-to-one:*

$$f(z) = \int_\gamma \frac{\omega(\zeta)d\zeta}{\zeta - z} \leftrightarrow F(w) = \int_{|t|=1} \frac{\Omega(t)dt}{t - w}, \tag{12}$$

where

$$\Omega(t) = \tau_p(\omega) = \omega[\phi^-(t)]\,[\phi^{-\prime}(t)]^{\frac{1}{p}}\,[I^-(t)]^{\frac{1}{q}}, \quad \frac{1}{p} + \frac{1}{q} = 1. \tag{13}$$

In [16] we proved that $T_p(f)$, as defined by (12) and (13), does not depend on the choice of the density $\omega(z)$ in the representation (1) of $f(z)$. We established (see [16]) the following important **properties** of T_p :

i) For $p \ge 1$, T_p is an isometric transformation of $K_p(G)$ onto $K_p(D)$.

ii) $T_p(G)$ transforms the set of all rational function $\{r(z)\}$ with poles in G^- onto the set of all rational functions $\{R(w)\}$ with poles in $|w| > 1$.

These facts and our previous results (see [10], [11], [12]) on rational approximation in the L_p metric on the unit circle allow us to investigate rational approximation of **CTI** in general domains. In particular, we consider the **Problem** of approximation of $f(z)$ in $K_p(G)$ ($p \ge 1$) by sequences of rational functions $\{r_k(z)\}$ with poles given by the kth row of a table $\{\alpha_{kj}\}$ (the points α_{kj} belong to G^-).

In [16], necessary and sufficient conditions were given for a table $\{\alpha_{kj}\}$ to provide approximation to any Cauchy type integral in $K_p(G)$ by rational functions $\{r_k(z)\}$ with given poles. These conditions of closeness describe a sort of "saturation" by $\{\alpha_{kj}\}$ in G^-. In the non-closeness case, the class of all functions $f(z)$ approximable by sequences $\{r_k(z)\}$) was fully described in terms of the distribution of the possible poles and the properties of γ. Similar results are valid for multiply connected domains G.

The problem of obtaining a complete geometric description of the boundaries of the domains belonging to the classes introduced above, remains open. Such characterization is known only for the class $\mathcal{K}_D$.

References

[1] G. David, *Opérateurs intégraux singuliers sur certaines courbes du plan complexe*, Ann. Scient. Éc. Norm. Sup., 4 Ser., **17** (1984), 157–189.

[2] P. L. Duren, H. S. Shapiro, and A. L. Shields, *Singular measures and domains not of Smirnov type*, Duke Math. J., **33** (1966), 247–254.

[3] P. L. Duren, *Theory of H^p Spaces*, Academic Press, New-York, **1970**.

[4] D. Gaier, *Lectures on Complex Approximation*, Birkhäuser Verlag, Basel, **1985**.

[5] M. V. Keldysh and M. A. Lavrentiev, *Sur la représentation conforme des domaines limitées par des coubes rectifiables*, Ann. Sci. École Norm. Sup., **54** (1937), 1–38.

[6] N. K. Nikolski and V. P. Khavin, *The results of V. I. Smirnov in complex analysis and their further development*, In: Smirnov V. I., *Selected Papers. Complex Analysis. Mathematical Theory of Diffraction*, Leningrad Univ. Press, Leningrad, **1988**, 111–145.

[7] I. I. Privalov, *Boundary Properties of Analytic Functions*, Gosud. Izdat. Tech. Teor. Lit., Moscow and Leningrad, **1950** (Russian); German translation: VEB Deutscher Verlag der Wissenschaften, Berlin, **1956**.

[8] V. I. Smirnov, *Sur les formules de Cauchy et Green et quelques problèmes qui s'y rattachent*, Izv. Akad. Nauk SSSR, Ser. Mat., **3** (1932), 338–372. (French)

[9] G. Ts. Tumarkin, *The properties of analytical functions representable by integrals of Cauchy-Stieltjes type and Cauchy-Lebesgue type*, Izv. Akad. Nauk Arm. SSR, **16** (1963), no. 5, 23–45. (Russian)

[10] G. Ts. Tumarkin, *Approximation with respect to various metrics of functions defined on the circumference by sequences of rational functions with fixed poles*, Izv. Akad. Nauk SSSR, Ser. Mat, **30** (1966), 721–766; English transl., Amer. Math. Soc. Transl. (2), **77** (1968), 183–233.

[11] G. Ts. Tumarkin, *Description of a class of functions admitting approximation by fractions with preassigned poles*, Izv. Akad. Nauk Arm. SSR, Ser. Mat. 1, **2** (1966), 85–109. (Russian)

[12] G. Ts. Tumarkin, *Conditions for the convergence of boundary values of analytic functions and approximation on rectifiable curves*, In: *Contemporary Problems in Theory Analytic Functions*, (International Conf., Erevan, 1965), Moscow, (1966), 283–295. (Russian)

[13] G. Ts. Tumarkin, *Conditions of convergence of boundary values of Cauchy-Type integrals*, Matem. Zametki, **5** (1969), no. 4, 441–448. (Russian)

[14] G. Ts. Tumarkin, *The boundary properties of analytical functions representable by Cauchy-Type integrals*, Matem. Sbornik, **84** (1971), no. 3, 425–439. (Russian)

[15] G. Ts. Tumarkin, *Some Problems Concerning Classes of Domains Determined by Properties of Cauchy Type Integrals*, In: *Linear and Complex Analysis. Problem Book 3*, Lecture Notes in Math., no. 1573, Springer Verlag, Berlin, **1994**, 411–413.

[16] G. Ts. Tumarkin, *On approximation of Cauchy-Type Integrals by sequences of rational functions with preassigned poles*, In: *Approximation Theory* IX, Vol. I, C. K. Chui and L. L. Schumaker (eds.), V. 1, Vanderbilt Univ. Press, Nashville, TN, **1999**.

Genrikh Tumarkin,
630 Rossmore Rd,
Goleta, CA 93117
E-mail address: tumarkin@gte.net

Operator Theory:
Advances and Applications, Vol. 113

A Dimension-free Carleson Measure Inequality

I. E. Verbitsky

Dedicated to the memory of S. A. Vinogradov

Abstract. We consider the following Carleson measure inequality for the Poisson integrals $P[f]$ in $\mathbb{R}^{n+1}_+$:

$$||P[f]||_{L^p(\mathbb{R}^{n+1}_+,\, d\mu)} \leq C\, \varkappa[\mu]^{\frac{1}{p}}\, ||f||_{L^p(\mathbb{R}^n)}, \quad f \in L^p(\mathbb{R}^n),$$

where $1 < p < \infty$, μ is a Borel measure on $\mathbb{R}^{n+1}_+$, and $\varkappa[\mu] = \sup_B \frac{\mu(\hat{B})}{|B|}$ is the Carleson norm of μ. Here $\hat{B} = \{(x,t) : x \in B,\, 0 < t < r\}$ is a cylinder whose base is the n-ball B of radius r. S. A. Vinogradov asked whether this inequality is fulfilled with a constant $C = C_p$ independent of the dimension n. We prove this inequality for $p > 2$ and discuss some related results.

1. Introduction

Let $P[f]$ be the Poisson integral (the harmonic extension of f) for the upper half-space $\mathbb{R}^{n+1}_+$:

$$P[f](x,t) = P_t \star f(x), \quad (x,t) \in \mathbb{R}^{n+1}_+,$$

where $P_t(x) = \gamma_n \frac{t}{(|x|^2+t^2)^{\frac{n+1}{2}}}$ is the Poisson kernel;

$$\gamma_n = \frac{\Gamma(\frac{n+1}{2})}{\pi^{\frac{n+1}{2}}}. \tag{1}$$

Let $1 < p < \infty$, and let μ be a Borel measure on $\mathbb{R}^{n+1}_+$. The following inequality is usually referred to as the Carleson measure theorem:

$$||P[f]||_{L^p(\mathbb{R}^{n+1}_+,\, d\mu)} \leq C\, \varkappa[\mu]^{\frac{1}{p}}\, ||f||_{L^p(\mathbb{R}^n)}, \quad f \in L^p(\mathbb{R}^n). \tag{2}$$

Here C is a constant depending on p and n, and $\varkappa[\mu]$, the "Carleson norm" of μ, is defined by the formula

$$\varkappa[\mu] = \sup_B \frac{\mu(\hat{B})}{|B|}, \tag{3}$$

where $B = B_r(x_0) = \{x \in \mathbb{R}^n : |x - x_0| < r\}$, $x_0 \in \mathbb{R}^n$, and $\hat{B} = \{(x,t) \in \mathbb{R}^{n+1}_+ : \max(|x - x_0|,\, t) < r\}$ is a cylinder with the base B. Some other equivalent

Carleson norms can be used in place of $\varkappa[\mu]$, e. g.,

$$\varkappa_1[\mu] = \sup_B \frac{\mu(\check{B})}{|B|}, \tag{4}$$

where $\check{B} = \{(x,t) \in \mathbb{R}^{n+1}_+ : |x - x_0|^2 + t^2 < r^2\}$, or

$$\varkappa_2[\mu] = \sup_B \frac{\mu(T(B))}{|B|}, \tag{5}$$

where $T(B) = \{(x,t) \in \mathbb{R}^{n+1}_+ : |x - x_0| + t < r\}$ is a "tent" over $B = B_r(x_0)$, $x_0 \in \mathbb{R}^n$. Obviously,

$$\varkappa_2[\mu] \le \varkappa_1[\mu] \le \varkappa[\mu] \le 2^n \varkappa_2[\mu].$$

See, e. g., [1], [2], [6], [8], [9], where different proofs, applications and generalizations of the Carleson measure theorem are discussed.

This paper is devoted to a question, asked by S. A. Vinogradov more than 10 years ago, whether there is a "dimension-free" version of the Carleson measure theorem; i.e., whether (2) is true with a constant $C = C_p$ depending on p only. This problem is related to the celebrated dimension-free inequality of E. M. Stein for the centered maximal function defined by

$$Mf(x) = \sup_{r>0} \frac{1}{|B_r(x)|} \int_{B_r(x)} |f(y)|\, dy, \quad x \in \mathbb{R}^n.$$

The following inequality was proved in [10], [11]:

$$||Mf||_{L^p(\mathbb{R}^n)} \le C_p\, ||f||_{L^p(\mathbb{R}^n)}, \quad f \in L^p(\mathbb{R}^n). \tag{6}$$

It is easily seen that, for $f \ge 0$,

$$\frac{c}{n}\, Mf(x) \le f^*(x) \le Mf(x), \quad x \in \mathbb{R}^n, \tag{7}$$

where $f^*(x) = \sup_{t>0} P[f](x,t)$.

We note that (2) with a constant $C = C_p$ does not seem to follow from (6). The usual proofs of the Carleson measure theorem via maximal function inequalities are based on certain covering lemmas and involve constants depending on n. It can be shown that a maximal inequality similar to (6),

$$||M_\nu f||_{L^p(\mathbb{R}^n,\nu)} \le C\, ||f||_{L^p(\mathbb{R}^n,\nu)}, \quad f \in L^p(\mathbb{R}^n, \nu),$$

for the maximal function M_ν defined by

$$M_\nu f(x) = \sup_{r>0} \frac{1}{|B_r(x)|_\nu} \int_{B_r(x)} |f|\, d\nu, \quad x \in \mathbb{R}^n,$$

where ν is a Borel measure on $\mathbb{R}^n$, would yield (2). The above inequality is known to be true with a constant C independent of ν [5]; however, it is still an open problem whether there is a dimension-free version of this inequality with $C = C_p$. (See the paper [3], where some estimates of C depending on p and n were obtained along with the best constant for the "uncentered" maximal function $\widetilde{M_\nu}$ in the one-dimensional case.)

It is not difficult to show that, conversely, maximal inequalities for f^* and Mf can be derived from the Carleson measure theorem (see, e. g., [1], [14]), apparently without sharp constants.

A simple "Hilbert space" proof of the Carleson measure theorem for $p = 2$ which does not use maximal functions was given by S. A. Vinogradov (see [8], [14]). There is a different argument that works for all $1 < p < \infty$; it uses a simple "integration by parts" lemma and is applicable to more general integral operators with "quasi-metric" kernels. (See [7], [13], where similar inequalities were treated, along with applications to nonlinear equations.) Another proof of the Carleson measure theorem that employs Green's theorem can be found in [1]. However, none of these proofs gives a constant independent of the dimension in (2).

In §2 we prove a weak-type version of (2) for $p = 2$,

$$||P[f]||_{L^{2,\infty}(\mathbb{R}^{n+1}_+,\, d\mu)} \leq C\, \varkappa_1[\mu]^{\frac{1}{2}}\, ||f||_{L^2(\mathbb{R}^n)}, \tag{8}$$

where C is a universal constant. This yields (2) with $C = C_p$ for $p > 2$. Our argument resembles Vinogradov's proof of the strong-type L^2-inequality, but we do not use the so-called Vinogradov–Senichkin test [8], which makes it possible to obtain a dimension-free estimate.

A similar proof for $1 < p \leq 2$ would require a multilinear version of the semigroup formula for the Poisson integrals; e.g., for $p = \frac{3}{2}$ one needs a "triple convolution" formula for the Poisson kernels. (See [4], where similar properties were established for the Riesz kernels.) There are different versions of these formulas, which are more cumbersome and more difficult to use than in the case where $p = 2$. Currently, this approach is investigated jointly by L. Grafakos and E. Laeng.

2. A weak-type dimension-free estimate for $p = 2$

Theorem 2.1. *Let μ be a Borel measure on $\mathbb{R}^{n+1}_+$. Then*

$$||P[f]||_{L^{2,\infty}(\mathbb{R}^{n+1}_+,\, d\mu)} \leq C\, \varkappa_1[\mu]^{\frac{1}{2}}\, ||f||_{L^2(\mathbb{R}^n)}, \quad f \in L^2(\mathbb{R}^n), \tag{9}$$

where C is a universal constant, and $\varkappa_1[\mu]$ is the Carleson norm of μ defined by (4).

Proof. We define the balayage operator P^* by

$$P^*[g\, d\mu](x) = \int_{\mathbb{R}^{n+1}_+} P_t(x-y)\, g(y,t)\, d\mu(y,t), \quad x \in \mathbb{R}^n.$$

Note that $[L^{2,1}(\mathbb{R}^{n+1}_+, \mu)]^* = L^{2,\infty}(\mathbb{R}^{n+1}_+,\, d\mu)$, where $L^{2,1}(\mathbb{R}^{n+1}_+, \mu)$ is the corresponding Lorentz space. Hence, by duality, (9) is equivalent to the inequality

$$||P^*[g\, d\mu]||_{L^2(\mathbb{R}^n)} \leq C\, \varkappa_1[\mu]^{\frac{1}{2}}\, ||g||_{L^{2,1}(\mathbb{R}^{n+1}_+, \mu)}, \quad g \in L^{2,1}(\mathbb{R}^{n+1}_+, \mu), \tag{10}$$

with the same constant C. As is well known, it suffices to prove (10) for $g = \chi_E$, where E is a measurable subset of $\mathbb{R}^{n+1}_+$ (see [12]); this is equivalent to the inequality

$$||P^*[\chi_E\, d\mu]||^2_{L^2(\mathbb{R}^n)} \le C\, \varkappa_1[\mu]\, \mu(E), \tag{11}$$

where C is a universal constant. By Fubini's theorem and the semigroup property for the Poisson kernels, the left-hand side of the preceding inequality can be rewritten as

$$||P^*[\chi_E\, d\mu]||^2_{L^2(\mathbb{R}^n)} = \iint_{E\times E} \int_{\mathbb{R}^n} P_t(x-y)\, P_s(x-z)\, dx\, d\mu(z,s)\, d\mu(y,t) =$$

$$\iint_{E\times E} P_{t+s}(y-z)\, d\mu(z,s)\, d\mu(y,t).$$

It remains to prove the inequality

$$\iint_{E\times E} P_{t+s}(y-z)\, d\mu(z,s)\, d\mu(y,t) \le C\, \varkappa_1[\mu]\, \mu(E) \tag{12}$$

for every measurable set $E \subset \mathbb{R}^{n+1}_+$. Obviously,

$$\iint_{E\times E} P_{t+s}(y-z)\, d\mu(z,s)\, d\mu(y,t) \le$$

$$\gamma_n \int_E \left[\int_{\mathbb{R}^{n+1}_+} \frac{t}{[|y-z|^2+(t+s)^2]^{\frac{n+1}{2}}}\, d\mu(z,s)\right] d\mu(y,t) +$$

$$+\,\gamma_n \int_E \left[\int_{\mathbb{R}^{n+1}_+} \frac{s}{[|y-z|^2+(t+s)^2]^{\frac{n+1}{2}}}\, d\mu(y,t)\right] d\mu(z,s).$$

Thus, it suffices to obtain uniform estimates for the expressions in square brackets on the right-hand side of the preceding inequality. In other words, (12) follows from the estimate

$$\sup_{(y,t)\in\mathbb{R}^{n+1}_+} \gamma_n \int_{\mathbb{R}^{n+1}_+} \frac{t}{[|y-z|^2+(t+s)^2]^{\frac{n+1}{2}}}\, d\mu(z,s) \le C\, \varkappa_1[\mu], \tag{13}$$

where C is a universal constant. The Fubini theorem yields

$$\gamma_n \int_{\mathbb{R}^{n+1}_+} \frac{t}{[|y-z|^2+(t+s)^2]^{\frac{n+1}{2}}}\, d\mu(z,s) =$$

$$\gamma_n\,(n+1)\,t \int_t^\infty \left[\int_{\{(y,s):\, |y-z|^2+(t+s)^2\le r^2\}} d\mu(z,s)\right] \frac{dr}{r^{n+2}}.$$

Clearly,

$$\int_{\{(y,s):\, |y-z|^2+(t+s)^2\le r^2\}} d\mu(z,s) \le$$

$$\int_{\{|y-z|^2+s^2\le r^2-t^2\}} d\mu(z,s) \le \varkappa_1[\mu]\, \Omega_n\, (r^2-t^2)^{\frac{n}{2}},$$

where

$$\Omega_n = \frac{2\pi^{\frac{n}{2}}}{n\Gamma(\frac{n}{2})}$$

is the volume of the unit ball. Hence,

$$\gamma_n \int_{\mathbb{R}^{n+1}_+} \frac{t}{[|y-z|^2+(t+s)^2]^{\frac{n+1}{2}}}\, d\mu(z,s) \le$$

$$\gamma_n\, \Omega_n\, (n+1)\, t\, \varkappa_1[\mu] \int_t^\infty (r^2-t^2)^{\frac{n}{2}} \frac{dr}{r^{n+2}} =$$

$$\gamma_n\, \Omega_n\, (n+1)\, \varkappa_1[\mu] \int_0^1 (1-\lambda)^{\frac{n}{2}} \lambda^{-\frac{1}{2}}\, d\lambda.$$

Since

$$\int_0^1 (1-\lambda)^{\frac{n}{2}} \lambda^{-\frac{1}{2}}\, d\lambda \asymp n^{-\frac{1}{2}},$$

and

$$\gamma_n\, \Omega_n = \frac{2\,\Gamma(\frac{n+1}{2})}{\pi^{\frac{1}{2}}\, n\, \Gamma(\frac{n}{2})} \asymp n^{-\frac{1}{2}},$$

it follows that the left-hand side of (12) is bounded by $C\, \varkappa_1[\mu]$, where C is a universal constant. This completes the proof of (13), which yields (12). □

The following corollary is immediate by the Marcinkiewicz interpolation theorem.

Corollary 2.2. *Let μ be a Borel measure on $\mathbb{R}^{n+1}_+$, and let $2 < p \le \infty$. Then*

$$||P[f]||_{L^p(\mathbb{R}^{n+1}_+,\, d\mu)} \le C\, \varkappa_1[\mu]^{\frac{1}{p}}\, ||f||_{L^p(\mathbb{R}^n)}, \quad f \in L^p(\mathbb{R}^n),$$

where $C = C_p$ is a constant that depends on p only.

References

[1] M. Andersson, *Topics in Complex Analysis,* Springer–Verlag, Berlin–Heidelberg–New York, 1997.

[2] L. Carleson, *Interpolation by bounded analytic functions and the corona problem,* Ann. Math., **76** (1962), 547–559.

[3] L. Grafakos and J. Kinnunen, *Sharp inequalities for maximal functions associated with general measures,* Proc. Roy. Soc. Edinburgh, Sect. A, **128** (1998), 717–723.

[4] L. Grafakos and C. Morpurgo, *A Selberg integral formula and applications,* Preprint, (1998).

[5] R. Fefferman, *Strong differentiation with respect to measures,* Amer. J. Math., **103** (1981), 33–40.

[6] L. Hörmander, *L^p estimates for (pluri) subharmonic functions,* Math. Scand., **20** (1967), 65–78.

[7] N. J. Kalton and I. E. Verbitsky, *Nonlinear equations and weighted norm inequalities,* to appear in Trans. A.M.S.

[8] N. K. Nikolskii, *Treatise on the Shift Operator,* Springer–Verlag, Berlin–Heidelberg–New York, 1986.

[9] E. M. Stein, *Harmonic Analysis: Real-Variable Methods, Orthogonality, and Oscillatory Integrals,* Princeton University Press, Princeton, New Jersey, 1983.

[10] E. M. Stein, *The development of square functions in the work of A. Zygmund,* Bull. A.M.S., **7** (1982), 359–376.

[11] E. M. Stein and J.-O. Strömberg, *Behavior of maximal functions in* $\mathbf{R}^n$ *for large* n, Ark. för Mat., **21** (1983), 259–269.

[12] E.M. Stein and G. Weiss, *Introduction to Fourier Analysis on Euclidean Spaces,* Princeton University Press, Princeton, New Jersey, 1971.

[13] I. E. Verbitsky and R. L. Wheeden, *Weighted norm inequalities for integral operators,* Trans. A.M.S., **350** (1998), 3371–3391.

[14] S. A. Vinogradov, *Free Interpolation in Spaces of Analytic Functions,* Doct. Dissertation, Leningrad (1982).

Department of Mathematics,
University of Missouri-Columbia,
Columbia, Missouri 65211,
U.S.A.
E-mail address: igor@math.missouri.edu

Operator Theory:
Advances and Applications, Vol. 113

Carleman Formula for Some Spaces of Functions Analytic in the Disk and Smooth in its Closure

Il'ya Videnskii

To the memory of my teacher
Stanislav Aleksandrovich Vinogradov

Abstract. Let X be a space of functions that are analytic in the unit disk and have boundary values satisfying some additional smoothness conditions (X may be the space of functions with derivative in the Hardy space H^1, or analytic Besov space $B_{p,1}^{1/p}$), and let E be a non-Carleson subset, i.e., $L = -\int \log \operatorname{dist}(t, E)dm < \infty$. For $f \in X$, we estimate $|f(z)|$ by a quantity depending on $\|f\|_X$, L under the assumption that f vanishes on E. Simultaneously we obtain a formula that reconstructs $f(z)$ at an arbitrary point z starting with the restriction of f to a Carleson subset E. This is an analogue of the Carleman-Golusin-Krylov formula for the functions in the Hardy space H^1 and the sets E of positive Lebesgue measure on the circle.

1. Introduction

Let X be some space of functions analytic in the unit disk $\mathbb{D} = \{z : |z| < 1\}$ and continuous in its closure $\overline{\mathbb{D}}$; we assume that the functions in X have boundary values satisfying some additional smoothness conditions. Next, let E be a closed subset of $\overline{\mathbb{D}}$. We consider the following two closely related problems.

(i) How can the value of f at an arbitrary point z be estimated under the assumption that f belongs to X and vanishes on E? We say that such an estimate is fairly good if it implies a description of the sets of uniqueness for X (by definition, E is a set of uniqueness for X if any $f \in X$ vanishing on E is identically zero).

(ii) How can the value of f at an arbitrary point z be reconstructed starting with the restriction of f to a set of uniqueness?

One of the first methods for reconstructing $f(z)$ was found by Carleman [2]. Afterwards, Golusin and Krylov [6] extended this method to the Hardy spaces $H^p (1 \le p \le \infty)$ and to the sets of positive Lebesgue measure on the unit circle

Partly supported by the RFFI grant no. 96-01-00693.

$\mathbb{T}$. They constructed a family of functions $f_\lambda(z)$ depending only on the restriction of f to E and tending to $f(z)$ uniformly on the compact subsets as λ tends to 0. Patil [10] proved that convergence holds in the H^p topology for $1 < p < \infty$.

We investigate these problems for some spaces of analytic functions smooth in $\overline{\mathbb{D}}$. A closed subset E of $\overline{\mathbb{D}}$ is called a Carleson set if

$$\int_{\mathbb{T}} \log \operatorname{dist}(t, E)\, dm = -\infty, \tag{C}$$

where $\operatorname{dist}(t, E) = \min_{z \in E} |t - z|$, $\mathbb{T}$ is the unit circle, and m is normalized Lebesgue measure on $\mathbb{T}$. The following spaces of functions analytic in $\mathbb{D}$ will be involved:

A is the disk algebra, i.e., the space of functions analytic in $\mathbb{D}$ and continuous in $\overline{\mathbb{D}}$;

$A^{(n)} = \{f : f^{(n)} \in A\}$, $n \geq 1$, are the spaces of functions with derivatives in the disk algebra, and $A^\infty = \bigcap_{n \geq 1} A^{(n)}$;

$H_p^1 = \{f; f' \in H^p\}$ are the spaces of functions with derivatives in the Hardy spaces H^p, $1 \leq p \leq \infty$;

for $1 \leq p, q \leq \infty$, $s \in \mathbb{R}$, $n \in \mathbb{Z}$, $s < n$, $B_{p,q}^s$ is the analytic Besov space of functions such that

$$\|f\|_{B_{p,q}^s} = \|(1-r)^{n-s-1/q} \|f^{(n)}(rt)\|_{L^p(\mathbb{T})}^q \|_{L^q(0,1)} < \infty.$$

Condition (C) is known to characterize the sets of uniqueness for the space $A^{(n)}$ and $E \subset \mathbb{T}$ (Carleson [3]), for the space A^∞ (Korenblum [9]), for the space H_1^1 (Vinogradov, Shirokov [15]) and for the space $B_{p,1}^{1/p}$ (Shirokov [11]) (we always assume that the set $E \cap \mathbb{D}$ satisfies the Blaschke conditions $\sum_{z \in E \cap \mathbb{D}} (1 - |z|) < +\infty$). Note (see [11]) that in the scale of the Besov spaces the spaces $B_{p,1}^{1/p}$ are most interesting from the viewpoint of the description of the sets of uniqueness, because they are contained in the disk algebra A and are not contained in any Lipschitz class.

In this paper we continue investigations started in the paper [14] where for the space H_1^1 and $E \in (C)$ (i.e., E satisfies condition (C)) an analog of the Carleman formula was obtained. In §2 we propose a general approach to the formulas of this type and we prove two abstract theorems answering questions (i) and (ii). The proof is based on an embedding theorem of Carleson type for the space of derivatives of the functions in X. In §3 we apply these abstract results to the spaces H_1^1 and $B_{p,1}^{1/p}$. As a consequence, we get a refinement of Jensen's inequality and a new proof of the fact that condition (C) characterizes the sets of uniqueness for the spaces H_1^1 and $B_{p,1}^{1/p}$. The results of this paper were partly announced in [13].

Applications of the Carleman type formulas to approximation theory were considered in [14], [7]. The monograph [1] is devoted to the Carleman formula, its generalizations, and various applications.

During all my mathematical life I was inspired by friendly, well-disposed, and helpful discussions with S. A. Vinogradov.

2. Abstract theorems

For a Banach space X of functions analytic in the unit disk, we define the following space of Borel measures on $\mathbb{D}$:

$$M_X = \left\{ \mu : \exists C_\mu > 0 \int_{\mathbb{D}} |f'| d\mu \le C_\mu \|f\|_X,\ f \in X \right\},$$

where C_μ is a constant independent on f; the infimum of C_μ for μ fixed will be denoted by $\|\mu\|$. We assume that $X \subset A$.

Let E be a closed subset of $\overline{\mathbb{D}}$ having no limit points in $\mathbb{D}$. For $t \in \mathbb{T}$, let $h(t)$ be the nearest point in E, $\operatorname{dist}(t, E) = |t - h(t)|$. The mapping $h(t)$ is uniquely determined on $\mathbb{T}$, except for at most countable set. Let $I = [0, 1]$, and let ψ be a measurable mapping such that

$$\psi : \mathbb{T} \times I \to \mathbb{D}; \quad \psi(t, 0) = t; \quad \psi(t, 1) = h(t); \quad \psi(t, s) \in \mathbb{D} \text{ for } 0 < s < 1.$$

Let φ be a function on $\mathbb{T}$, $\varphi \in L^1(m)$, $\varphi(t) \ge 0$. With the triple (E, ψ, φ) we associate a Borel measure $\mu_{(E,\psi,\varphi)}$ on $\mathbb{D}$; namely, for any Borel subset B of $\mathbb{D}$ we put

$$\mu_{(E,\psi,\varphi)}(B) = \int_{\mathbb{T}} \frac{|\psi(t, I) \cap B|}{\varphi(t)} dm,$$

where $|\cdot|$ denotes length.

Definition 2.1. *The triple (E, ψ, φ) is said to be admissible for the space X if $\mu_{(E,\psi,\varphi)} \in M_X$.*

Theorem 2.2. *If the triple (E, ψ, φ) is admissible for X, $f \in X$, f vanishes on E, and $\int_{\mathbb{T}} \log \varphi dm > -\infty$, then*

$$\exp\left(\int_{\mathbb{T}} \log |f| dm \right) \le \|f\|_X \|\mu_{(E,\psi,\varphi)}\| \exp\left(\int_{\mathbb{T}} \log \varphi dm \right).$$

Proof. The arithmetic-geometric mean inequality implies that

$$\int \log \frac{|f|}{\varphi} dm \le \log \int_{\mathbb{T}} \frac{|f|}{\varphi} dm.$$

Since $h(t) \in E$, we see that $f(h(t)) = 0$, whence

$$\int_{\mathbb{T}} \frac{|f|}{\varphi} dm = \int_{\mathbb{T}} \frac{|f(t) - f(h(t))|}{\varphi(t)} dm$$

$$\le \int_{\mathbb{T}} \frac{\int_{\psi(t,I)} |f'(s)| ds}{\varphi(t)} dm = \int_{\mathbb{D}} |f'| d\mu_{(E,\psi,\varphi)} \le \|f\|_X \|\mu_{(E,\psi,\varphi)}\|$$

(the final inequality follows from the definition of M_X). Thus,

$$\int_{\mathbb{T}} \log |f| dm = \int_{\mathbb{T}} \log \frac{|f|}{\varphi} dm + \int_{\mathbb{T}} \log \varphi dm \le \log(\|f\|_X \|\mu_{(e,\psi,\varphi)}\|) + \int_{\mathbb{T}} \log \varphi dm.$$

□

Corollary 2.3. *Assume that $E \cap \mathbb{D}$ satisfies the Blaschke condition. Under the assumption of Theorem 2.2, we have*

(i) $\log|f(0)| \le \log \prod_{\zeta \in E\cap\mathbb{D}} |\zeta| + \log(\|f\|_X \|\mu_{(E,\psi,\varphi)}\|) + \int_{\mathbb{T}} \log \varphi dm,$

(ii) $\log|f(z)| \le \left(\frac{1+|z|}{1-|z|}\right)(\log(\|f\|_X\|\mu_{(E,\psi,\varphi)}\|) + \int_{\mathbb{T}} \log\varphi dm).$

Proof. (i) follows from Jensen's inequality.

Let $P_z(t) = (1-|z|^2)|1-z\bar{t}|^{-2}$ be the Poisson kernel; then $|P_z(t) \le (1+|z|)(1-|z|)^{-1}$, and

$$\log|f(z)| \le \int_{\mathbb{T}} P_z(t)\log|f(t)|dm \le \left(\frac{1+|z|}{1-|z|}\right)\int_{\mathbb{T}}\log|f|dm$$
$$\le \left(\frac{1+|z|}{1-|z|}\right)(\log(\|f\|_X\|\mu_{(E,\psi,\varphi)}\| + \int_{\mathbb{T}}\log\varphi dm). \qquad \square$$

Corollary 2.4. *If the triple (E,ψ,φ) is admissible for X, and $\int_{\mathbb{T}}\log\varphi dm = -\infty$, then E is a set of uniqueness for X.*

Proof. For $\lambda > 0$ we put $\varphi_\lambda(t) = \max(\varphi(t),\lambda)$; then $\int_{\mathbb{T}}\log\varphi_\lambda dm > -\infty$. Since $\varphi_\lambda \ge \varphi$, the triple (E,ψ,φ_λ) is admissible for X, and $\|\mu_{(E,\psi,\varphi_\lambda)}\| \le \|\mu_{(E,\psi,\varphi)}\|$. Let $f \in X, f|_E = 0$. We apply Theorem 2.2 to the function f and the triple (E,ψ,φ_λ), obtaining

$$\exp\left(\int_{\mathbb{T}}\log|f|dm\right) \le \|f\|_X\|\mu_{(E,\psi,\varphi_\lambda)}\|\exp\left(\int_{\mathbb{T}}\log\varphi_\lambda dm\right)$$
$$\le \|f\|_X\|\mu_{(E,\psi,\varphi)}\|\exp\left(\int_{\mathbb{T}}\log\varphi_\lambda dm\right).$$

If $\lambda \to 0$, then $\int_{\mathbb{T}}\log\varphi_\lambda dm \to -\infty$; hence, $\int_{\mathbb{T}}\log|f|dm = -\infty$, which means that $f \equiv 0$. $\square$

For such a set of uniqueness we can get a reconstruction formula. For $\lambda > 0$, let $\Phi_\lambda(z)$ be an outer function in $\mathbb{D}$ such that $|\Phi_\lambda(t)| = \varphi_\lambda(t)$ a.e. on $\mathbb{T}$, i.e.,

$$\Phi_\lambda(z) = \exp\left(\int_{\mathbb{T}}\frac{t+z}{t-z}\log|\varphi_\lambda(t)|dm\right). \tag{1}$$

For $f \in X$ we define

$$f_\lambda(z) = \Phi_\lambda(z)\int_{\mathbb{T}}\frac{f(h(t))}{\Phi_\lambda(t)(1-z\bar{t})}dm, \quad z \in \mathbb{D}. \tag{2}$$

Observe that $f_\lambda(z)$ depends on $f|_E$ only.

Theorem 2.5. *If the triple (E,ψ,φ) is admissible for X, $\int_{\mathbb{T}}\log\varphi dm = -\infty$, and $f \in X$, then*

$$f(z) = \lim_{\lambda\to 0} f_\lambda(z), \tag{3}$$

the f_λ converge to f uniformly on the compact sets, and

$$\lim_{\lambda\to 0}\sup_{z\in\mathbb{D}}(1-|z|)|\log|f(z)-f_\lambda(z)|=-\infty. \tag{4}$$

Proof. For every λ, the function $\Phi_\lambda(z)$ is bounded away from zero in $\mathbb{D}$, so that we can write the Cauchy formula for f/Φ_λ:

$$\frac{f(z)}{\Phi_\lambda(z)}=\int_{\mathbb{T}}\frac{f(z)}{\Phi_\lambda(t)(1-z\bar{t})}dm=\int_{\mathbb{T}}\frac{f(h(t))}{\Phi_\lambda(t)(1-z\bar{t})}dm+\int_{\mathbb{T}}\frac{f(t)-f(h(t))}{\Phi_\lambda(t)(1-z\bar{t})}dm.$$

Hence, $f(z)=f_\lambda(z)+\Phi_\lambda(z)r_\lambda(z)$, where

$$r_\lambda(z)=\int_{\mathbb{T}}\frac{f(t)-f(h(t))}{\Phi_\lambda(t)(1-z\bar{t})}dm. \tag{5}$$

By the definition of an outer function, we have

$$\log|\Phi_\lambda(z)|=\int_{\mathbb{T}}P_z(t)\log|\varphi_\lambda(t)|dm\le\frac{2}{1-|z|}\int_{\mathbb{T}}\log\varphi_\lambda(t)dm, \tag{6}$$

whence $\lim_{\lambda\to 0}(1-|z|)\log|\Phi_\lambda(z)|=-\infty$. Now we check the boundedness of $r_\lambda(z)$ as $\lambda\to 0$. Since $|\Phi_\lambda(t)|=\varphi_\lambda(t)$ a.e., it follows that

$$\begin{aligned}|r_\lambda(z)|&\le\int_{\mathbb{T}}\frac{|f(t)-f(h(t))|}{\varphi_\lambda(t)|1-z\bar{t}|}\le\frac{1}{(1-|z|)}\int_{\mathbb{T}}\frac{\int_{\psi(t,I)}|f'(s)|ds}{\varphi(t)}dm\\&=\frac{1}{(1-|z|)}\int_{\mathbb{D}}|f'|d\mu_{(E,\psi,\varphi)}\le\frac{1}{(1-|z|)}\|f\|_X\|\mu_{(E,\psi,\varphi)}\|.\end{aligned} \tag{7}$$

From (5), (6), (7) we get (4) and, consequently, (3). □

3. Applications

In this section we construct admissible triples for H_1^1 and $B_{p,1}^{1/p}$ and apply to them the abstract theorems proved above. Let E be a closed subset of $\overline{\mathbb{D}}$, let $E\cap\mathbb{D}$ satisfy the Blaschke condition, and let t be a fixed point on $\mathbb{T}$. Denote $h(t)=r\zeta$, $0<r\le 1$, $\zeta\in\mathbb{T}$, $\rho=(\rho(t))=|t-h(t)|$. We define a path $\psi(t,I)$ as the union of two intervals $[t,(1-\rho)t]\cup[r\zeta,(1-\rho)\zeta]$ and the arc $[(1-\rho)t,(1-\rho)\zeta]$ (without loss of generality we may assume that $\rho<1$) and put $\varphi(t)=\rho^\alpha(t), 0\le\alpha\le 1$, and $\mu_\alpha=\mu_{(E,\psi,\rho^\alpha)}$.

3.1. The space H_1^1

We equip the space H_1^1 with the seminorm $\|f\|_{H_1^1}=\|f'\|_{H^1}$. Carleson [4] described the measures in $\mathbb{D}$ for which $H^1\subset L^1(\mu)$. Such measures are called Carleson measures. Thus, the space $M_{H_1^1}$ consists of Carleson measures. In [14] it was proved that μ_α is a Carleson measure for $0\le\alpha<1$. The proof in [14] shows that the following lemma is true.

Lemma 3.1. *For $0\le\alpha<1$, the measure μ_α is a Carleson measure, and $\|\mu_\alpha\|\le C/(1-\alpha)$, where C is an absolute constant.*

Observe that in Section 3 we checked condition (10) for μ_α, which implies that μ_α is Carleson. From Lemma 3.1 it follows that the triple (E, ψ, ρ^α) is admissible for H_1^1 if $0 \leq \alpha < 1$, and we can apply Theorems 2.2 and 2.5.

Theorem 3.2. *If $f \in H_1^1$, $f|_E = 0$, $L = -\int_{\mathbb{T}} \log \rho dm$, then*

$$\exp\left(\int_{\mathbb{T}} \log |f| dm\right) \leq C\|f'\|_{H^1}(1+L)exp(-L), \tag{8}$$

where C is an absolute constant.

Proof. Applying Theorem 2.2 to the triple (E, ψ, ρ^α) and using Lemma 3.1, for $0 \leq \alpha < 1$ we get

$$\exp\left(\int_{\mathbb{T}} \log |f| dm\right) \leq \|f'\|_{H^1} \frac{C}{(1-\alpha)} \exp(-\alpha L). \tag{9}$$

If $L \leq 0$, then we put $\alpha = 0$. If $L > 0$, then we put $\alpha = 1 - 1/L$ (for this α the right-hand side of (9) attains its minimum). □

Inequality (8) improves the result of [15]. From Corollary 2.4 we deduce the following statement.

Corollary 3.3 ([15]). *If $E \in (C)$, then E is a set of uniqueness for H_1^1.*

The next theorem was obtained in [14].

Theorem 3.4. *If $E \in (C), 0 < \alpha < 0, \varphi(t) = \rho^\alpha(t)$, then (4) is true for every $f \in H_1^1$.*

3.2. The spaces $B_{p,1}^{1/p}$.

First, we describe the space $M_{B_{p,1}^{1/p}}$. By the definition of a Besov space, condition $f \in B_{p,1}^{1/p}$ is equivalent to the condition $f' \in B_{p,1}^{1/p-1}$. Hence, $\mu \in M_{B_{p,1}^{1/p}}$ means that $B_{p,1}^{1/p-1} \subset L^1(\mu)$. Now we can use the following embedding theorem proved by Verbitsky in [12]:

let $1 < p < \infty$, $1/p + 1/q = 1$, and let $\Delta_{j,k}$ be the dyadic partition of the disk, i.e.,

$$\Delta_{j,k} = \{z = re^{i\theta} : \pi j 2^{-k} < \theta \leq \pi(j+1)2^{-k}; 2^{-k} < 1 - r \leq 2^{-k+1}\},$$

$k = 0, 1, \dots ; j = -2^k, -2^k + 1, \dots, 2^k - 1$. Then

$$B_{p,1}^{1/q} \subset L^1(\mu) \iff V(\mu) = \sup_{k>0} 2^k \left(\sum_{j=-2^k}^{2^k-1} \mu(\Delta_{j,k})^q\right)^{1/q} < \infty, \tag{10}$$

and the norm of the embedding operator is equivalent to $V(\mu)$ ($\|\mu\| \asymp V(\mu)$, i.e., there exist absolute constants c_1, c_2 such that $c_1 V(\mu) \leq \|\mu\| \leq c_2 V(\mu)$).

Lemma 3.5. *Let $0 \leq \alpha \leq 1/p$, $1 < p < \infty$. Then μ_α satisfies (10), and $\|\mu_\alpha\| \leq c/(1-\alpha)$, where c is an absolute constant.*

Proof. Without loss of generality, in the proof we may replace the disk $\overline{\mathbb{D}}$ by the square $Q = [0,1] \times [0,1]$. Then E is a closed subset of Q having limit points only on the real axis. If $t \in I, I = [0,1]$, then $h(t) = (x(t), y(t))$ is one of the points of E nearest to t, $\rho(t) = |t - h(t)|$, $\varphi(t) = \rho(t)^\alpha$, $\psi(t, I)$ is the broken line with the vertices $(t, 0)$, $(t, \rho(t))$, $(x(t), \rho(t))$, and $h(t)$. In the definition of μ_α we replace integration along $\mathbb{T}$ by integration along the interval [0,1]. Let O be the set of points in $I \backslash E$ for which there exists only one nearest point in E. It is clear that O is open; hence, $O = \bigcup_{j-1}^{\infty} l_j$, where the l_j are open intervals, $h(t)$ is constant on every l_j, and $h(l_j) \neq h(l_k)$ if $i \neq k$. It is convenient to replace E by the set $F = \{h(l_j)\}_{j=1}^{\infty}$; clearly, the measure μ_α associated with F is the same as the measure μ_α associated with E. We put $\mu = \mu_\alpha$,

$$\begin{aligned} I_{j,k} &= \{x : j2^{-k} < x \leq (j+1)2^{-k}\}, \\ \Delta_{j,k} &= \{(x,y) : x \in I_{j,k}, 2^{-k} < y \leq 2^{-k+1}\}, \\ S_{j,k} &= \{(x,y) : x \in I_{j,k}, 0 \leq y \leq 2^{-k+1}\}; \\ k &= 0,1,\ldots; \quad j = 0,1,\ldots, 2^k - 1. \end{aligned}$$

Now, we verify condition (10). For k fixed, we consider the sets

$$N_1 = \{j : F \cap S_{j,k} \neq \varnothing\} N_2 = \{j : F \cap S_{j,k} = \varnothing\}$$

and put $\Delta_{j,k}(t) = \Delta_{j,k} \cap \psi(t, I)$, $f_{j,k}(t) = |\Delta_{j,k}(t)| \rho^{-\alpha}(t)$.

Let $j \in N_1$. Then

$$\begin{aligned} \mu(\Delta_{j,k}) = \int_0^1 f_{j,k}(t)dt &= \left(\sum_{m=0}^{j-3} + \sum_{m=j-2}^{j+2} + \sum_{m=j+3}^{2^k-1} \right) \int_{I_{m,k}} f_{j,k}(t)dt \\ &= M_1 + M_2 + M_3. \end{aligned} \tag{11}$$

Obviously, $|\Delta_{j,k}(t)| \leq c_1 2^{-k}$. Let $E_j = \{t : \Delta_{j,k}(t) \neq \varnothing\}$, $t_1 = \inf E_j$, $t_2 = \sup E_j$. If $t \in I_{j+n,k} \cap E_j$, then $\rho(t) \asymp n2^{-k}$ for $n \in \mathbb{Z}$, $n \neq 0$; if $t \in I_{j,k} \cap E_j$, then $\rho(t) \asymp 2^{-k}$. Thus,

$$\int_{I_{j+n,k}} f_{j,k}(t)dt \leq c_2 2^{-k(2-\alpha)} n^{-\alpha}, \quad n \in \mathbb{Z}, \quad n \neq 0, \tag{12}$$

and for M_2 we get $M_2 \leq c_3 2^{-k(2-\alpha)}$. If $t_2 \in I_{j+n_j,k}$ and $n_j > 2$, then

$$M_3 \leq c_2 \sum_{m=3}^{n_j} 2^{-k(2-\alpha)} m^{-\alpha} \leq \frac{c_4}{1-\alpha} n_j^{1-\alpha} 2^{-k(2-\alpha)}$$

Similarly, if $t_1 \in I_{j-m_j,k}$ and $m_j > 2$, then

$$M_1 \leq \frac{c_4}{1-\alpha} m_j^{1-\alpha} 2^{-k(2-\alpha)},$$

whence

$$\mu(\Delta_{j,k}) \leq \frac{c_5}{(1-\alpha)} (m_j + n_j + 5)^{1-\alpha} 2^{-k(2-\alpha)}. \tag{13}$$

We denote $r_j = m_j + n_j + 5$. Observe that $(t_1, (j-2)2^{-k}) \cup ((j+3)2^{-k}, t_2) \subset E_j$. Moreover, for any fixed interval $I_{l,k}$ the number of squares $\Delta_{s,k}$ such that $\Delta_{s,k}(t) \neq \varnothing$ for some $t \in I_{l,k}$ is at most six. Hence, $\sum_{j\in N_1} r_j \leq c_6 2^k$, and

$$\sum_{j\in N_1} \mu(\Delta_{j,k})^q \leq 2^{-k(2-\alpha)q} c_5^q \frac{1}{(1-\alpha)^q} \sum_{j\in N_1} r_j^{q(1-\alpha)}$$

$$\leq c_7^q 2^{-k(2-\alpha)q} \frac{1}{(1-\alpha)^q} 2^{kq(1-\alpha)} \leq c_7^q \frac{2^{-kq}}{(1-\alpha)^q}, \tag{14}$$

because $(1-\alpha)q \geq 1$. If $t_1 \in I_{j+m_j,k}$ and $m_j > 2$, then $M_1 = 0$, and in this case we have $M_3 \leq c_8(1-\alpha)^{-1}(n_j - m_j)^{1-\alpha} 2^{-k(2-\alpha)}$, $r_j = n_j - m_j + 5$, arriving at (14) again.

Let $j \in N_2$. In this case, if $t < (j-2)2^{-k}$, or $t > (j+3)2^{-k}$, or $\rho(t) < 2^{-k}$, then $\psi(t, I)$ does not intersect $\Delta_{j,k}$, whence

$$\mu(\Delta_{j,k}) = \int_{(j-2)2^{-k}}^{(j+3)2^{-k}} \frac{|\Delta_{j,k}(t)|}{\rho^\alpha(t)} dt \leq 5 \cdot 2^{-k} \frac{2^{-k}}{(2^{-k})^\alpha} = 5 \cdot 2^{-k(2-\alpha)}$$

and

$$\sum_{j\in N_2} \mu(\Delta_{j,k})^q \leq 2^k \cdot 5^q \cdot 2^{-kq(2-\alpha)} \leq 5^q \cdot 2^{-kq}, \tag{15}$$

because $\alpha \leq 1/p$. Combining (14) and (15), we get the required estimate for $V(\mu)$. □

Lemma 3.5 means that the triple (E, ψ, ρ^α) is admissible for $B_{p,1}^{1/p}$ if $0 \leq \alpha \leq 1/p$, which allows us to apply the abstract theorems proved in Section 2.

Theorem 3.6. *If $f \in B_{p,1}^{1/p}$, $f|_E = 0$, $L = -\int_{\mathbb{T}} \log \rho dm < \infty$, $1 < p < \infty$, $1/p + 1/q = 1$, then*

$$\exp\left(\int_{\mathbb{T}} \log |f| dm\right) \leq c \|f\|_{B_{p,1}^{1/p}} g(L),$$

where $g(L) = q\exp(-L/p)$ if $q < L$, and $g(L) = (1+L)\exp(-L)$ if $q \geq L$; c is an absolute constant.

Proof. Applying Theorem 2.2 to the triple (E, ψ, ρ^α) and using Lemma 3.5, for $0 \leq \alpha \leq 1/p$ we get

$$\exp\left(\int_{\mathbb{T}} \log |f| dm\right) \leq \|f\|_{B_{p,1}^{1/p}} \frac{c}{1-\alpha} \exp(-\alpha L).$$

It remains to put $\alpha = 1/p$ if $q < L$ and $\alpha = 1 - 1/L$ if $q \geq L$ (in this case $\alpha \leq 1/p$). □

From Corollary 2.4 we get

Corollary 3.7 (see [11]). *If $E \in (C)$, then E is a set of uniqueness for $B_{p,1}^{1/p}$, $1 < p < \infty$.*

Theorem 3.8. *If $E \in (C)$, $1 < p < \infty$, $0 < \alpha \leq 1/p$, $\varphi(t) = \rho(t)^\alpha$, then* (4) *is true for every $f \in B^{1/p}_{p,1}$.*

3.3. Generalization

We formulate results more general then those of Subsections 3.1, 3.2. Let $\mathcal{E}$ be the set of all closed subsets E of $\overline{\mathbb{D}}$ such that

$$\sum_{z \in E \cap \mathbb{D}} (1 - |z|) < \infty, m(E \cap \mathbb{T}) = 0.$$

Consider the following spaces of analytic functions: for $0 < \alpha \leq 1$,

$$X_\alpha = A \cap (\bigcap_{E \in \mathcal{E}} L^1(\mu_{(E,\varphi,\rho^\alpha)}); \; X_0 = \bigcup_{0<\alpha<1} X_\alpha;$$

$$\|f\|_{X_\alpha} = \sup_{E \in \mathcal{E}} \|f\|_{L^1(\mu_{(E,\varphi,\rho^\alpha)})}; \; \|f\|_{X_0} = \inf_{0<\alpha<1} \|f\|_{X_\alpha}.$$

From Lemma 3.1 it follows that $H^1_1 \subset X_\alpha$ for every $0 < \alpha < 1$; from Lemma 3.5 it follows that $B^{1/p}_{1,p} \subset X_{1/p}$, $1 < p < \infty$; and from the Hardy–Littlewood theorem (see [5], Ch. 5) it follows that $\Lambda_\alpha \subset X_\alpha$ (Λ_α is the Lipschitz class). By definition, the triple $(E, \varphi, \rho^\alpha)$ is admissible for X_α. Applying the abstract theorems, we get following results.

Corollary 3.9. *If $f \in X_\beta$, $0 < \beta < 1$, $f|_E = 0$, $L = -\int_{\mathbb{T}} \log \rho dm < \infty$, then*

$$\exp\left(\int_{\mathbb{T}} \log|f| dm\right) \leq c\|f\|_{X_\beta} g(L),$$

where $g(L) = (1-\beta)^{-1}\exp(-L\beta)$ if $\beta > 1 - 1/L$ and $g(L) = (1+L)\exp(-L)$ if $\beta \leq 1 - 1/L$; c is an absolute constant.

Corollary 3.10. *Let $E \in (C)$. Then*
(i) *E is a set of uniqueness for X_0;*
(ii) *if $f \in X_\beta$, then relations* (4) *and* (3) *are fulfilled for $\varphi = \rho^\alpha$, $\alpha \leq \beta$.*

4. Problems

The limit space for the scale $B^{1,p}_{p,1}$ is $B^0_{\infty,1} = \{f : \int_0^1 \max_{\mathbb{T}} |f'(rt)| dr < \infty\}$. It consists of continuous functions and contains the space l^1_a of absolutely convergent Taylor series. Condition (C) does not describe the sets of uniqueness for $B^0_{\infty,1}$, because condition (C) is conformally invariant, but there exists [8] an example of a set of uniqueness for l^1_a that is not conformally invariant. A problem is to describe the sets of uniqueness for $B^0_{\infty,1}$, or to find nontrivial sufficient or necessary conditions under which E is a set of uniqueness for $B^0_{\infty,1}$. This may shed light on the structure of the sets of zeros and the sets of uniqueness for the algebra l^1_a.

It would be of interest to prove that in relation (3) or its appropriate modification convergence will take place in the topology of the Besov space. The obstruction is that the mapping h is not continuous. However, from the proof of

Theorem 2.5 we can see that in the definition of f_λ we can replace $f(h(t))$ by many other extensions of f from E to $\mathbb{T}$. Possibly, such a replacement can improve convergence.

References

[1] L. Aisenberg, *Carleman's Formulas in Complex Analysis,* Mathematics and Its Applications, **244** (1993), Kluwer Academic Press, Dordrecht-Boston-London.

[2] T. Carleman, *Les Fonctions Quasi-analytiques,* (1926), Gauthier–Villars, Paris.

[3] L. Carleson, *Sets of uniqueness for functions regular in the unit circle,* Acta Math., **87** (1952), 325–345.

[4] L. Carleson, *An interpolation problem for bounded analytic functions,* Amer. J. Math., **80** (1958), 921–930.

[5] P. L Duren, *Theory of H^p spaces,* Pure and Applied Mathematics, **38** (1970), Academic Press, New York.

[6] G. M. Golusin and V. I. Krylov, *Generalization of Carleman formula and its application to the analytic continuation of functions,* Mat. Sb., **40** (1933), 144–149.

[7] V. P. Havin and V. A. Bart, *Szegö-Kolmogorov-Krein's theorems on weight trigonometric approximation and Carleman type formulas,* Ukr. Math. J., **46** (1994), 100–127.

[8] R. Kaufman, *Zero sets of absolutely convergent Taylor series,* in: G. Weiss and S. Wainger, Eds., Harmonic Analysis in Euclidean Spaces, Part 1, Proceedings of Simposia in Pure Math. (AMS, Providence), **35** (1979), 439–443.

[9] B. I. Korenblum, *Functions holomorphic in a disc and smooth in its closure,* Dokl. Akad. Nauk SSSR, **200** (1971), 24–27.

[10] D. J. Patil, *Representation of H^p-functions,* Bull. Amer. Math. Soc., **78** (1972), 617–620.

[11] N. A. Shirokov, *Sets of zeroes of analytic functions from the space $B^{1/p}_{p,1}$ are Carlesonian,* Zap. Nauchn. Semin. LOMI, **113** (1981), 253–257.

[12] I. E. Verbitsky, *Embedding theorems for spaces of analytic functions with mixed norms,* Preprint, (1987).

[13] I. V. Videnskii, *Carleman formula for Carleson subsets of the disk,* in: Proceedings of Israel Math. Union Conference (Beer Sheva University Press), (1994), 95–99.

[14] I. V. Videnskii, E. M. Gavurina and V. P. Havin, *Analogues of Carleman-Golusin-Krylov interpolation formula,* Operator theory and function theory, Leningrad Univ., Leningrad, **1** (1983), 21–32.

[15] S. A. Vinogradov and N. A. Shirokov, *Zeroes of analytic functions with derivative in H^1*, Zap. Nauchn. Semin. LOMI, **30** (1972), 154–157.

Department of Mathematics,
SPb Electrotechnical University,
Prof. Popova st. 5
St. Petersburg, 197376, Russia
E-mail address: `ilya@viden.pdmi.ras.ru`

MIX
Papier aus verantwortungsvollen Quellen
Paper from responsible sources
FSC® C105338

If you have any concerns about our products,
you can contact us on
ProductSafety@springernature.com

In case Publisher is established outside the EU,
the EU authorized representative is:
Springer Nature Customer Service Center GmbH
Europaplatz 3, 69115 Heidelberg, Germany

Printed by Libri Plureos GmbH
in Hamburg, Germany